The Origin of Fermions and Bosons, And Their Unification

Stephen Blaha
Blaha Research

Pingree Hill Publishing

ISBN: 978-0-9984575-5-0

Cover Credits

Rev. 00/00/01 February 21, 2017

Some Other Books by Stephen Blaha

All the Megaverse! Starships Exploring the Endless Universes of the Cosmos using the Baryonic Force (Blaha Research, Auburn, NH, 2014)

SuperCivilizations: Civilizations as Superorganisms (McMann-Fisher Publishing, Auburn, NH, 2010)

Universes and Megaverses: From a New Standard Model to a Physical Megaverse; The Big Bang; Our Sister Universe's Wormhole; Origin of the Cosmological Constant, Spatial Asymmetry of the Universe, and its Web of Galaxies; A Baryonic Field between Universes and Particles; Flatverse Extended Wheeler-DeWitt Equation (Blaha Research, Auburn, NH, 2014)

PHYSICS IS LOGIC PAINTED ON THE VOID: Origin of Bare Masses and The Standard Model in Logic, U(4) Origin of the Generations, Normal and Dark Baryonic Forces, Dark Matter, Dark Energy, The Big Bang, Complex General Relativity, A Megaverse of Universe Particles (Blaha Research, Auburn, NH, 2015).

PHYSICS IS LOGIC Part II: The Theory of Everything, The Megaverse Theory of Everything, U(4)⊗U(4) Grand Unified Theory (GUT), Inertial Mass = Gravitational Mass, Unified Extended Standard Model and a New Complex General Relativity with Higgs Particles, Generation Group Higgs Particles (Blaha Research, Auburn, NH, 2015).

The Origin of Higgs ("God") Particles and the Higgs Mechanism: Physics is Logic III, Beyond Higgs – A Revamped Theory With a Local Arrow of Time, The Theory of Everything Enhanced, Why Inertial Frames are Special, Universes of the Mind (Blaha Research, Auburn, NH, 2015).

The Origin of the Eight Coupling Constants of The Theory of Everything: U(8) Grand Unified Theory of Everything (GUTE), S^8 Coupling Constant Symmetry, Space-Time Dependent Coupling Constants, Big Bang Vacuum Coupling Constants, Physics is Logic IV (Blaha Research, Auburn, NH, 2015).

New Types of Dark Matter, Big Bang Equipartition, and A New U(4) Symmetry in the Theory of Everything: Equipartition Principle for Fermions, Matter is 83.33% Dark, Penetrating the Veil of the Big Bang, Explicit QFT Quark Confinement and Charmonium, Physics is Logic V (Blaha Research, Auburn, NH, 2015).

The Periodic Table of the 192 Quarks and Leptons in The Theory of Everything: The U(4) Layer Group, Physics is Logic VI (Blaha Research, Auburn, NH, 2015).

New Boson Quantum Field Theory, Dark Matter Dynamics, Dark Matter Fermion Layer Mixing, Genesis of Higgs Particles, New Layer Higgs Masses, Higgs Coupling Constants, Non-Abelian Higgs Gauge Fields, Physics is Logic VII (Blaha Research, Auburn, NH, 2015)

Unification of the Strong Interactions and Gravitation: Quark Confinement Linked to Modified Short-Distance Gravity; Physics is Logic VIII (Blaha Research, Auburn, NH, 2016).

Unification of the Seven Boson Interactions based on the Riemann-Christoffel Curvature Tensor (Pingree Hill Publishing, Auburn, NH, 2016).

CQMechanics: A Unification of Quantum & Classical Mechanics, Quantum/Semi-Classical Entanglement, Quantum/Classical Path Integrals, Quantum/Classical Chaos (Blaha Research, Auburn, NH, 2016).

Available on Amazon.com, Amazon.co.uk, bn.com, and other international web sites as well as at better bookstores (through Ingram Distributors).

Preface

This book takes the Complex Lorentz group as our starting point and derives the four types (species) of fermions directly. It then proceeds to show the interactions of the Standard Model SU(3)⊗SU(2)⊗U(1) follow directly from Complex Lorentz group geometry. The addition of Dark matter leads to an SU(3)⊗SU(2)⊗U(1)⊗SU(2)⊗U(1) symmetry if Dark matter is included in an Extended Standard Model.

Then noting that there are particle number quanta: Baryon number, Lepton number, and their Dark equivalents it develops the U(4) Generation group that generates four generations of each species of fermion (although only three generations are known at present). The existence of four generations of each species then leads in turn to another U(4) group – the Layer group – that gives us a Periodic Table of Fermions with three additional replicas (with changes) of the layer of four generations, with which we are familiar. In total it reveals 192 different fermions – 128 quarks and 64 leptons.

Having discovered the riches of the Complex Lorentz group, and realizing that it could only exist in our slightly curved space-time, it see that Complex General Relativity and its coordinate systems must follow. It would be illogical if General Relativity were real-valued and flat space-time complex-valued. So it proceeds to consider Complex General Relativity and discover that each Complex General Coordinate Transformations can be factored into a real General Coordinate transformation and a complex coordinate transformation. This leads to another U(4) group it calls the General Relativistic Reality group. The interactions associated with this group are part of the set of interactions experienced by fermions.

Taken altogether it creates a unified theory of an Extended Standard Model and General Relativity with the symmetry SU(3)⊗SU(2)⊗U(1)⊗SU(2)⊗U(1)⊗U(4)⊗U(4)⊗U(4) plus real-valued General Coordinate transformations. Given this menagerie of interactions it develops a formalism for the 'rotation of interactions' called Ω-symmetry, which introduces a new gauge field interaction between all 192 fermions, and their interactions. This symmetry completes the unification process. For if one can rotate things into each other, they are in a sense the same (or related). It concludes by showing that the gravity potential found at solar system distances, at galactic distances, and at inter-galactic distances follow from our theory. It also shows that the theory yields a linear quark potential and the Charmonium potential. Lastly it shows the theory may explain the missing proton spin puzzle, and also the difference in the proton radius of the hydrogen and muonic hydrogen atoms.

CONTENTS

1. Introduction

1.1 A Unified Theory of Matter and Interactions

We appear to live in an almost flat space-time with real-valued coordinates. Our measuring instruments guarantee that time and spatial coorinates are real-valued. Rulers measure real values. Clocks measure real-valued times. However, it is not entirely clear that space-time is as simple as our crude measuring devices tell. It is possible that we, sitting in some reference frame, can construct a real-valued coordinate system in a more complicated space-time universe. One of the deepest studies of the mathematics underlying quantum field theory, which remains the most accurately studied theory due to its enormously accurate predictions of electromagnetic quantities in Quantum Electrodynamics, requires the use of the Complex Lorentz Group to prove theorems in axiomatic quantum field theory.[1]

Physicists have taken this need for the Complex Lorentz group rather lightly and not seen it as a hint of a deeper reality – a universe with complex-valued coordinates governed by the Complex Lorentz group. There is good reason for this reluctance to follow up on the Complex Lorentz group and its attendant coordinate systems. For if physicists accept it, then the 11th Commandment given by Einstein: "Thou shalt not exceed the speed of Light" falls, and the Gorgon of Faster-Than-Light travel appears.

We shall see that there is remarkable evidence that particles (neutrinos in particular) appear to be traveling faster than the speed of light. We will also note that this topic, of great importance for those who believe Humanity has a destiny in the stars, is definitely not being pursued by experimentalists except for a few small efforts a few years ago that had equivocal results.[2] Prior experiments had shown neutrinos traveled faster-than-light. The word 'cover up' is too often used nowadays but Einstein would have applauded the discovery of faster-than-light travel were he alive.

Be that as it may, we take the Complex Lorentz group as our starting point and derive the four types (species) of fermions directly. We then proceed to show the interactions of the Standard Model SU(3)⊗SU(2)⊗U(1) follow directly from Complex Lorentz group geometry. The addition of Dark matter leads to an SU(3)⊗SU(2)⊗U(1)⊗SU(2)⊗U(1) symmetry if Dark matter is included in an Extended Standard Model.

Then noting that there are particle number quanta: Baryon number, Lepton number, and their Dark equivalents we develop the U(4) Generation group that generates four generations of

---

[1] See Streater (2000) for this careful mathematical analysis. References appear in the back of this book.

[2] See section 1.2 below.

each species of fermion (although only three generations are known at present). The existence of four generations of each species then leads in turn to another U(4) group – the Layer group – that gives us a Periodic Table of Fermions with three additional replicas (with changes) of our layer of four generations, with which we are familiar. In total we find 192 different fermions – 128 quarks and 64 leptons – a propect that should be elating to the experimental community. So much more to discover if we only had the machines!

All these features emerge directly or indirectly from the Complex Lorentz Group!

Having discovered the riches of the Complex Lorentz group, and realizing that it could only exist in our slightly curved space-time, we see that Complex General Relativity and its coordinate systems must follow. It would be illogical if General Relativity were real-valued and flat space-time complex-valued. So we proceed to consider Complex General Relativity and discover that we can factor each Complex General Coordinate Transformation into a real General Coordinate transformation and a complex coordinate transformation, which is an element of a U(4) group in the flat space-time limit that we call the General Relativistic Reality group. The interactions associated with this group are part of the set of interactions experienced by fermions.

Taken altogether we have a unified theory of an Extended Standard Model and General Relativistic parts with the symmetry $SU(3)\otimes SU(2)\otimes U(1)\otimes SU(2)\otimes U(1)\otimes U(4)\otimes U(4)\otimes U(4)$ plus real-valued General Coordinate transformations. Given this menagerie of interactions we develop a formalism for the 'rotation of interactions' called Ω-symmetry, which introduces a further gauge field interaction between all 192 fermions, and their interactions. This symmetry completes the unification process. For if one can rotate things into each other, they are in a sense the same (or related).

The book also shows a superior formalism (Pseudoquantum field theory) for the Higgs Mechanism, and a 'new' method to eliminate infinities in perturbation theory computations – Two-Tier Quantum Field Theory. It also suggests solutions for the anomalous behavior of gravity locally, at intra-galactic distances, and at extra-galactic distances. It provides explicit quark confinement with a linear potential, suggests a solution of the Missing Proton Spin Puzzle, and may resolve the differences in measurements of the proton's radius, as well as possibly solving other problems.

1.2 Experimental Evidence for Faster-Than-Light Particles & Physics

In this section[3] we describe convincing evidence for faster than light particle physics. Until 1907 physicists thought that there was no limit on the speed of a particle or lump of matter. In 1907 Einstein and Poincaré showed that there was an inherent limit on the speed of a massive object – the speed of light – if one attempts to accelerate an object at a real-valued rate from below light speed. For the past 100 odd years physicists have generally accepted the speed

[3] Section 1.2 is extracted from Blaha (2015a) and earlier books.

of light as the limiting speed for particles with mass. Several theoretical physicists in the 1960's (E. C. Sudarshan and Gerald Feinberg) investigated the possibility of faster than light particles. They found that faster than light particles were theoretically possible but their theories – particularly their quantum field theories – had numerous discrepancies from canonical quantum field theory. These differences were taken by many to indicate that faster than light particles (called tachyons) were not present in nature. This belief was further supported by the happenings at particle accelerators where it was impossible to accelerate normal charged particles such as protons faster than the speed of light.

In the past ten years this author[4] developed a satisfactory quantum field theory of faster than light particles and found that if neutrinos and down-type quarks were faster than light particles, then one could derive the form of The Standard Model of Elementary Particles in detail. This theoretical development seems to have stimulated experimental groups at the new Linear Hadron Collider (LHC) at the CERN laboratory in Switzerland and the Gran Sasso Laboratory in Italy to measure the speed of neutrinos emitted in LHC particle collisions. The results, described below, were mixed and one can fairly say they neither proved nor disproved that neutrinos were tachyons.[5]

However there is other experimental data that strongly indicate that neutrinos are tachyons, and that quantum mechanics requires – not just faster than light behavior – but in some circumstances instantaneous effects at a distance – infinite speed of transmission!

In this chapter we will look at experimentally proven instantaneous Quantum Mechanical effects, at tritium decay experiments over the past 20 years that imply faster than light neutrinos, at neutrino speed measurements at the CERN LHC and Gran Sasso, at tachyonic particle behavior inside of Black Holes, and at the tachyonic behavior of Higgs particles, the "so-called God particle." *The cumulative result of these considerations is that faster than light particles, and physics, are a part of nature.*

1.2.1 Instantaneous Quantum Mechanical Effects

Quantum entanglement is a quantum phenomenon wherein parts of a physical system are in a certain quantum state but are separated by a space-like distance. If a change is made in part of a quantum entangled system then it is known theoretically, and experimentally, that other parts of the system change instantaneously.[6] Many experiments have shown that the change in other parts of a system is instantaneous and thus can be viewed as taking place at

---

[4] See Blaha (2012b) and earlier books extending back nine years.

[5] It is somewhat amazing that a large experiment with significant statistics to determine the whether the speed of neutrinos exceeded c has not been undertaken, and does not appear to be likely. Considering the importance of this issue it should be a priority at the LHC with a measurement distance much greater than the small experiments such as Gran Sasso so that a conclusive result could be obtained.

[6] Matson, John, "Quantum Teleportation Achieved Over Record Distances" *Nature* **13**, August 2012.

infinite speed – obviously beyond the speed of light.[7] The most recent experiment by Juan Yin et al[8] has shown directly that quantum mechanical effects travel faster than 10,000 times the speed of light. These experimental results are consistent with the instantaneous speed predicted by quantum mechanics. Thus faster than light behavior is implicit in quantum theory and is experimentally verified.

1.2.2 Tritium Decay Experiments Yielding Neutrinos

Fact: Particles with negative values for the square of their mass are tachyons – particles moving faster than light.

A series of experiments by various groups over recent years imply that electron neutrinos produced in tritium decay have negative mass squared despite the best efforts of experimenters to obtain positive values for the neutrino mass squared.

Experiment	**measured mass squared**	**Year**
Mainz	-1.6 ± 2.5 ± 2.1	2000
Troitsk	-1.0 ± 3.0 ± 2.1	2000
Zürich	-24 ± 48 ± 61	1992
Tokyo INS	- 65 ± 85 ± 65	1991
Los Alamos	- 147 ± 68 ± 41	1991
Livermore	- 130 ± 20 ± 15	1995
China	- 31 ± 75 ± 48	1995
1998 Average	-27 ± 20	1998

Table 1.1. Electron neutrino mass squared values found in various tritium decay experiments. (Masses are in units of eV.) The average mass squared is negative suggesting electron neutrinos are tachyons.

Table 1.1 summarizes the measured electron mass squared in these experiments. These experiments imply that neutrinos have negative mass squared and are thus faster-than-light particles - tachyons. However their small masses indicate that they only exceed the speed of light by a small amount.

---

[7] Francis, Matthew, "Quantum Entanglement Shows that Reality Can't be Local", *Ars Technica*, 30 October 2012.
[8] Juan Yin et al, arXiv[quant-ph]: 1303.0614V1 (March 4, 2013).

1.2.3 LHC/Gran Sasso Direct Measurements of Neutrino Speeds

Two groups performed experiments at Gran Sasso Laboratory in Italy. They detected neutrinos emitted in interactions at the CERN LHC in Switzerland. The LVD collaboration, in a study of neutrino velocities, found that the question was still open according to their data. Their refereed Physical Review Letter Abstract stated:

> We report the measurement of the time of flight of ν_μ on the CNGS baseline (732 km) with the Large Volume Detector (LVD) at the Gran Sasso Laboratory. The CERN-SPS accelerator has been operated from May 10th to May 24th 2012, with a tightly bunched-beam structure to allow the velocity of neutrinos to be accurately measured on an event-by-event basis. LVD has detected 48 neutrino events, associated with the beam, with a high absolute time accuracy. These events allow us to establish the following limit on the difference between the neutrino speed and the light velocity: $-3.8\times10^{-6} < (v_\nu - c)/c < 3.1\times10^{-6}$ (at 99% C.L.). This value is an order of magnitude lower than previous direct measurements.[9]

These results (involving at least 35 neutrino detections) slightly favor, and do not rule out, faster-than-light neutrinos. Their finding of some possible errors does not rule out the possibility.

Another experiment at the same location by the ATLAS group stated that they found several neutrino velocities above c.[10] *We conclude that the cumulative data in sections 1.2.2 and 1.2.3 appears to support faster than light neutrinos – consistent with our theory of The Standard Model.*

1.2.4 Tachyonic Behavior Within Black Holes

Inside a black hole (such as the Schwarzschild solution of General Relativity) the time coordinate effectively becomes a spatial coordinate and the radius coordinate effectively becomes a time coordinate. An in-falling particle has a constantly decreasing radial distance from the center of the black hole just as time always increases outside a black hole.

As a result of the interchange of the roles of time and radius, the velocity of a particle descending radially inside a Black Hole has a speed faster than light and is tachyonic.

---

[9] N. Yu. Agafonova et al. (LVD Collaboration), "Measurement of the Velocity of Neutrinos from the CNGS Beam with the Large Volume Detector" Phys. Rev. Lett. **109**, 070801 (15 August 2012).

[10] ICARUS Collaboration, ArXiv:1208.2629 (2012). This paper stated on p. 17: "**This measurement excludes neutrino velocities exceeding the speed of light by more than 1.35×10^{-6}c at 90% C. L.**" The issue thus remains unresolved.

1.2.5 Higgs Fields are Tachyons

Recently groups at the LHC CERN laboratory have announced the discovery of Higgs particles. The dynamic equations for Higgs bosons in The Standard Model have a negative mass squared. The mass squared must be negative or the Higgs Mechanism could not generate particle masses. Having negative mass terms implies that Higgs fields are tachyonic – faster than light particles. Their tachyonic nature is masked by a quartic self-interaction that generates a condensate and thereby the masses of other particles.

1.2.6 Conclusion: Faster-Than-Light Particles – Tachyons Exist in Nature

The bulk of the experimental and theoretical evidence presented in previous sections strongly favors the existence of faster-than-light particles such as neutrinos. Tachyonic neutrinos are an important part of our extension of The Standard Model of Elementary Particles. This form of the theory also strongly suggests that down-type quarks are also tachyonic in order to obtain the symmetries of The Standard Model.

1.2.7 Neutrinos as Tachyons?

Recently the three species of neutrinos have been found to have masses[11] in neutrino oscillation experiments although the masses are very small. If we denote the three neutrinos as m_1, m_2, and m_3; and let $\Delta m_{ij}^2 = m_j^2 - m_i^2$, then[12]

$$\Delta m_{12}^2 \cong 8 \times 10^{-5}\ \text{eV}^2$$

$$\Delta m_{23}^2 \cong 2.8 \times 10^{-3}\ \text{eV}^2$$

It is claimed that the sign of each Δm^2 is unambiguously determined.[13] However the mass difference values so obtained can be interpreted as differences in tachyon masses as well as differences in "normal" particle masses. The dependence of neutrino oscillations on masses squared results from the time evolution of mixed neutrino states. We consider mixtures of two neutrino states in the vacuum (for the sake of simplicity):

$$|\nu_a\rangle = \cos\theta\, |\nu_e\rangle - \sin\theta\, |\nu_\mu\rangle$$

$$|\nu_b\rangle = \sin\theta\, |\nu_e\rangle + \cos\theta\, |\nu_\mu\rangle$$

---

[11] Note the neutrino mass differences are approximately of the order of the mass squared estimate of an iota: $m_0^2 \approx 4 \times 10^{-4}\ \text{eV}^2$ (section 1.1.3).

[12] S. M. Bilenky, "Neutrino Masses, Mixing and Oscillations", arXiv:hep-ph/050175 (October 13, 2005).

[13] A. B. McDonald, "Evidence for Neutrino Oscillations I", p. 8 arXiv:nucl-ex/0412005 (December, 2005).

with differing masses m_a and m_b.[14] The phase factors determining the time dependence of $|v_a>$ and $|v_b>$ generate the neutrino oscillations from which the mass relations were obtained. The phase of the kth state is

$$|v_k(t)> \sim e^{-im_k^2 t/(2p)}$$

for normal neutrinos. In the case of tachyons the factor in the exponential changes sign:

Tachyonic Neutrinos

$$E = (\mathbf{p}^2 - m^2)^{1/2} \cong p - m^2/(2p)$$

"Normal" Neutrinos

$$E = (\mathbf{p}^2 + m^2)^{1/2} \cong p + m^2/(2p)$$

Thus tachyonic neutrinos would exhibit the time dependence

$$|v_k(t)> \sim e^{+im_{Tk}^2 t/(2p)}$$

where we use the mass subscript "T" to indicate a tachyon.

As a result, if neutrinos are "normal" the experimental results above would suggest the neutrino masses satisfy

$$m_3^2 > m_2^2 > m_1^2$$

On the other hand, if neutrinos are tachyons the experimental results above would suggest the value of the tachyon neutrino masses satisfies

$$m_1^2 > m_2^2 > m_3^2$$

an inverted spectrum compared to hypothetical "normal" neutrinos. When one considers the fact that tachyon masses squared are negative we see that an ordering of the tachyon neutrino mass spectrum, consistent with experimental neutrino results, is:

A Negative Neutrino m^2 Spectrum

$m^2 = 0$ ----------

m_1^2----------

[14] L. Wolfenstein, Phys. Rev. **D17**, 2369 (1978).

m_2^2----------

m_3^2-----------

1.3 Phenomena Beyond the Light Barrier

In this section[15] we will briefly survey some of the very different features of faster-than light physical phenomena.

1.3.1 Superluminal (Faster-than-Light) Transformations

We will frame our discussion in terms of the two simple reference frames depicted in Fig. 1.2. The prime frame is moving at a speed v > c (the speed of light) in the positive x direction with respect to the unprimed reference frame.

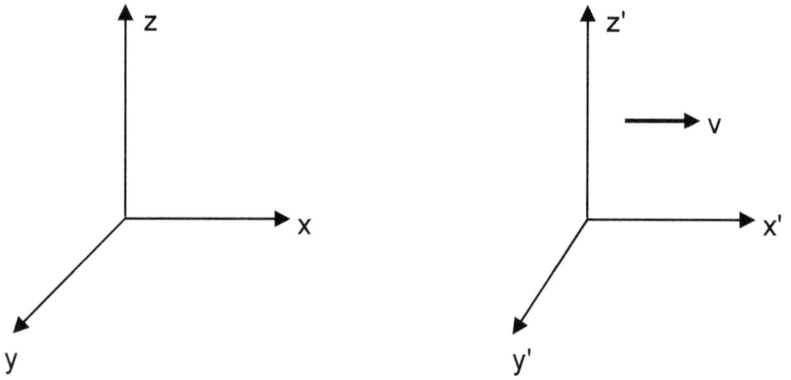

Figure 1.2. Two coordinate systems having a relative speed v in the x direction.

As shown later in the text we define a superluminal (faster-than-light) transformation between coordinates in these reference frames with

$$\begin{aligned} t' &= \gamma_s(t - \beta x/c) \\ x' &= \gamma_s(x - \beta ct) \\ y' &= iy \end{aligned} \quad (1.1)$$

[15] Extracted from Blaha (2007b).

$$z' = iz$$

where

$$\gamma_s = (\beta^2 - 1)^{-\frac{1}{2}} \tag{1.2}$$

and $\beta = v/c > 1$. The appearance of imaginary values for y' and z' is not a cause for alarm. An observer resident in the prime coordinate system will measure real y and z distances with a ruler. The only purpose of the factors of i is to relate the y and z coordinates to y' and z'. An observer in either coordinate system will view his/her coordinates as real.

The energy and momentum of a tachyon (faster-than-light) particle of mass m traveling at a speed $v > c$ is

$$E = \gamma_s mc^2 \tag{1.3}$$

and

$$\mathbf{p} = m\gamma_s\mathbf{v} \tag{1.4}$$

Note that the tachyon defining condition is satisfied:

$$E^2 - c^2p^2 = -m^2c^4 \tag{1.5}$$

Also note that in the limit $\beta \to \infty$ that

$$E = 0 \tag{1.6}$$

and

$$p = mc \tag{1.7}$$

where $p = |\mathbf{p}|$. Tachyons are always in motion. The minimal momentum of a tachyon is given by eq. 1.7. It corresponds to zero energy. It is the tachyon equivalent of Einstein's famous $E = mc^2$.

1.3.2 Length Dilations and Time Contractions

In ordinary Lorentz transformations a moving ruler will appear to be shorter in the direction of its motion when measured in another reference frame. This phenomenon is called *Lorentz contraction.*

1.3.2.1 Superluminal Length Dilation/Contraction

In the case of a superluminal transformation we find precisely the opposite effect, *superluminal length dilation,* is a possibility. Consider the case of the transformation of eq. 1.1 above (coresponding to Fig.1.2), which relates the prime reference frame traveling at speed v in the positive x direction to the unprimed reference frame. A ruler perpendicular to the x-axis will have the same length in both reference frames if its endpoints are simultaneously measured –

perhaps by photographing it. The y and z equations in eqs. 1.1 specify this fact up to an extraneous factor of i.

If the ruler is at rest in the prime reference frame and parallel to the x' axis, then a simultaneous measurement of its endpoints at the same time t_0 by an observer in the unprimed reference frame (perhaps by photographing it) will reveal both *length contraction and dilation* depending on the value of β. If the length is $L' = x'_2 - x'_1$ in the prime frame and $L = x_2 - x_1$ in the unprimed frame, then the equations:

$$x'_1 = \gamma_s(x_1 - \beta c t_0) \qquad (1.8)$$
$$x'_2 = \gamma_s(x_2 - \beta c t_0) \qquad (1.9)$$

imply

$$L' = \gamma_s L = (\beta^2 - 1)^{-1/2} L \qquad (1.10)$$

Thus we have three cases:

Case 1: $\beta \in <1, \sqrt{2}>$:	$L < L'$	Contraction	(1.11)
Case 2: $\beta = \sqrt{2}$:	$L = L'$	Equality	(1.12)
Case 3: $\beta \in <\sqrt{2}, \infty>$:	$L > L'$	Dilation	(1.13)

Superluminal Time Contraction/Dilation

In the case of a superluminal transformation we find *superluminal time contraction* is a possibility. Consider again the case of the transformation of eq. 1.1 above coresponding to Fig. 1.1 relating the prime reference frame traveling at speed v in the positive x direction to the unprimed reference frame. Consider the time interval between two events occurring at the same point x'_0 in the prime reference frame. From the viewpoint of an observer in the unprimed frame the events take place at different points x_1 and x_2. If the time interval is $T' = t'_2 - t'_1$ in the prime frame and $T = t_2 - t_1$ in the unprimed frame, then the inverse of eqs. 1.1 give:

$$t_1 = \gamma_s(t'_1 + \beta x'_0/c) \qquad (1.14)$$
$$t_2 = \gamma_s(t'_2 + \beta x'_0/c) \qquad (1.15)$$

and imply

$$T = \gamma_s T' = (\beta^2 - 1)^{-1/2} T' \qquad (1.16)$$

Again we have three cases:

Case 1: $\beta \in <1, \sqrt{2}>$:	$T > T'$	Dilation	(1.17)
Case 2: $\beta = \sqrt{2}$:	$T = T'$	Equality	(1.18)

Case 3: $\beta \in <\sqrt{2}, \infty>$: $T < T'$ Contraction (1.19)

The time interval in the unprimed frame can be less than, equal to, or greater than the time interval in the frame where the events take place at the same spatial point.

Thus superluminal transformations are more complex than Lorentz transformations with respect to space and time, dilation and contraction.

1.3.3 Tachyon Fission to More Massive Particles – Reverse Fission

Another way in which faster-than-light phenomena differ from sublight phenomena is particle fission. Normally when a particle or nucleus decays or fissions the masses of the particles produced by the decay are smaller than the mass of the original particle or nucleus. And energy is released. We are familiar with fission as the source of nuclear energy.

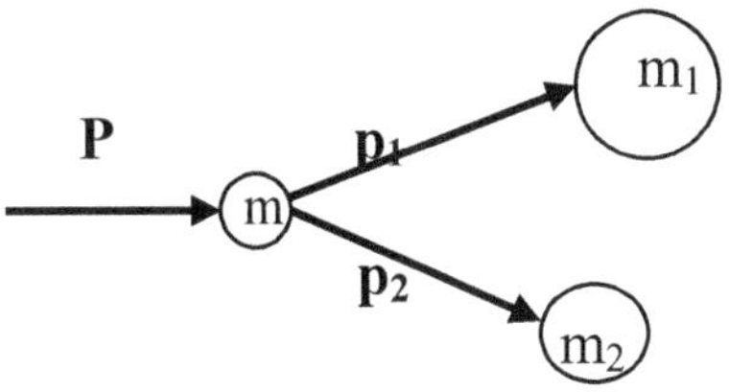

Figure 1.3. Two particle decay of a tachyon.

In the case of faster-than-light particles, tachyons, a much different possibility is present: a tachyon can decay into heavier tachyons: *a particle's spatial 3-momentum can be transformed into mass.* We will consider the specific case of a tachyon decaying into two particles to illustrate this possibility. (See Fig. 1.3.)

We will assume the initial tachyon has zero energy[16] and thus the tachyons emerging from the decay also have zero energy. The analysis is based on conservation of total energy and momentum.

Momentum conservation implies

$$\mathbf{P} = \mathbf{p}_1 + \mathbf{p}_2 \tag{1.20}$$

Since all energies are zero

$$(cP)^2 = (c\mathbf{P})^2 = m^2$$
$$(cp_1)^2 = (c\mathbf{p}_1)^2 = m_1^2 \tag{1.21}$$

[16] If a particle has zero energy its velocity is infinite so the case considered is somewhat artificial. However the results would still be approximately true for very large velocities. The simplicity of the kinematics led us to consider this case.

$$(cp_2)^2 = (\mathbf{cp_2})^2 = m_2^2$$

where $P = |\mathbf{P}|$, $p_1 = |\mathbf{p_1}|$, and $p_2 = |\mathbf{p_2}|$. If we now square eq. 1.20 and use eqs. 1.21 we obtain

$$m^2 = m_1^2 + m_2^2 + 2m_1m_2 \cos\theta \qquad (1.22)$$

where θ is the angle between the emerging particles momenta $\mathbf{p_1}$ and $\mathbf{p_2}$.

Eq. 1.22 has a number of interesting cases:

Case $\theta = 0$:

$$m = m_1 + m_2 \qquad (1.23)$$

The masses of the outgoing tachyons sum to the mass of the original tachyon.

Case $\theta = \pi/2$:

$$m^2 = m_1^2 + m_2^2 \qquad (1.24)$$

The masses of each outgoing tachyon is less than the mass of the original tachyon.

Case $\theta = \pi$:

$$m^2 = (m_1 - m_2)^2 \qquad (1.25)$$

In this case either $m_1 > m$ or $m_2 > m$. Thus one of the outgoing tachyons has a greater mass than the original tachyon. Mass is effectively created from the spatial momentum of the particle. This process is the 'inverse' of normal particle decay or fission where the sum of the outgoing masses is always less than the original particle's mass and the difference is mass converted into energy in the form of additional photons via $E = mc^2$.

This last case, where one of the outgoing particles is more massive than the original particle, is not just for $\theta = \pi$. Since

$$\cos\theta = (m^2 - m_1^2 - m_2^2)/(2m_1m_2) \qquad (1.26)$$

we see that *the sum of the outgoing tachyon masses is always greater than the original tachyon mass (except when $\theta = 0$)* since

$$\cos\theta = 1 + [m^2 - (m_1 + m_2)^2]/(2m_1m_2) \le 1 \qquad (1.27)$$

and thus

$$[m^2 - (m_1 + m_2)^2]/(2m_1m_2) \le 0 \qquad (1.28)$$

Note $m = m_1 + m_2$ only if $\theta = 0$.

Since we can transform the above discussion to the case of tachyons with a non-zero energy using an ordinary Lorentz transformation the above discussion in this subsection is general.

We therefore conclude that when a tachyon decays into two tachyons the sum of the masses of the produced tachyons is greater than the mass of the original tachyon except if the angle between the momenta of the produced tachyons is zero. In that case the sum of the masses of the produced tachyon equals the mass of the original tachyon.

Thus tachyons can engage in ***'reverse' fission*** *in which* ***momentum is converted into mass so the outgoing particles have a total mass greater than the incoming particle****.* In the case of "normal" fission part of the mass of a particle can be converted to energy and the sum of the masses of the decay product particles is less than the mass of the original particle.

1.3.4 Light Chasing Faster-than-Light Particles?

Einstein told a story that he imagined positioning himself in a (Galilean) reference frame moving at the speed of light and seeing electromagnetic waves "frozen" in time so that they were no longer vibrating. This vision inspired him to reconsider the transformation laws between coordinate systems and to derive the theory of Special Relativity. In Special Relativity the speed of light is the same in all reference frames.

In this subsection we will consider a light pulse from the points of view of two reference frames whose relative speed v is greater than the speed of light. We will use the example considered earlier and add a pulse of light traveling in the positive x direction. (See Fig. 1.4.)

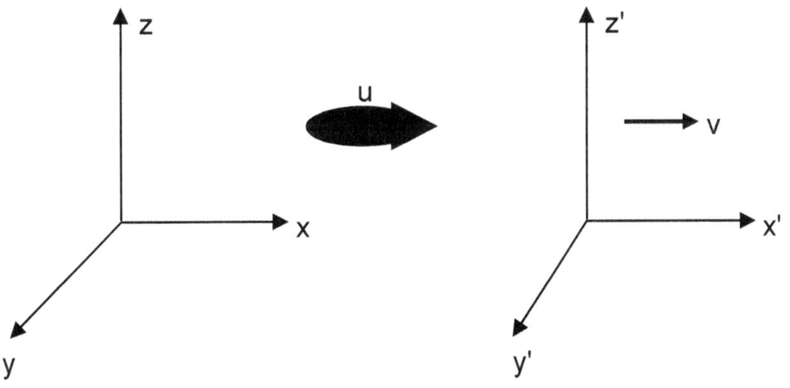

Figure 1.4. Two coordinate systems having a relative speed v in the x direction. A pulse of light is displayed as a thick arrow.

The general law for the addition of velocities in a situation such as depicted in Fig. 1.4 is well known. If we adapt it to the present example and let u be the speed of the pulse in the unprimed frame (temporarily forgetting it is a light pulse) we find it implies

$$u' = (u - \beta c)/(1 - \beta u/c) \qquad (1.29)$$

where $\beta = v/c > 1$, and u' is the speed of the pulse in the prime frame. Then if we set u = c we see that u' = c as well. *Thus our superluminal transformations preserve the constancy of the speed of light just like Lorentz transformations.*

As a result the pulse of light will intersect the z' axis eventually. However if superluminal transformations did not preserve the speed of light in all frames the pulse might never reach the z' axis. For example under a Galilean transformation the speed of the pulse would be u' = u – v = c – v and the pulse would actually be falling further and further behind the z' axis.

1.3.5 Electromagnetic Field of a Charged Tachyon – A Pancake Effect?

The electric field of a charge q at rest in a reference frame is:

$$\mathbf{E} = (q/(4\pi\varepsilon_0)\check{\mathbf{r}}/r^2 \qquad (1.30)$$

in spherical coordinates where $\check{\mathbf{r}}$ is a unit vector in the radial direction.

Sublight Charged Particle

The electric and magnetic fields of a charge q moving in the positive x direction with speed v < c are

$$\mathbf{E} = (q/(4\pi\varepsilon_0)\check{\mathbf{r}}(1 - \beta^2)/[r^2(1 - \beta^2\sin^2\theta)^{\frac{3}{2}}] \qquad (1.31)$$
$$\mathbf{B} = (q/(4\pi\varepsilon_0)\check{\mathbf{r}}\beta(1 - \beta^2)\sin\theta/[r^2(1 - \beta^2\sin^2\theta)^{\frac{3}{2}}] \qquad (1.32)$$

where $\check{\mathbf{r}}$ is the radial unit vector, $\beta = v/c$, and θ is measured with respect to the polar axis which is taken to be the x axis. As $\beta \rightarrow 1$ the electric and magnetic fields develop a "pancake" form with large field strengths in the directions perpendicular to the direction of motion similar to the transverse fields of electromagnetic quanta. This feature is the basis of the Weizsäcker-Williams method of virtual quanta.

Charged Tachyon

The electric and magnetic fields of a tachyon of charge q moving in the positive x direction with speed v > c are

$$\mathbf{E} = (q/(4\pi\varepsilon_0)\check{\mathbf{r}}(\beta^2 - 1)/[r^2(\beta^2\sin^2\theta - 1)^{\frac{3}{2}}] \quad (1.33)$$
$$\mathbf{B} = (q/(4\pi\varepsilon_0)\check{\mathbf{r}}\beta(\beta^2 - 1)\sin\theta/[r^2(\beta^2\sin^2\theta - 1)^{\frac{3}{2}}] \quad (1.34)$$

where $\beta = v/c > 1$, and θ is again measured with respect to the polar axis which is taken to be the x axis. In the case of tachyons there are three cases of interest.

Case $\beta^2\sin^2\theta - 1 < 0$:
The electric and magnetic fields are pure imaginary and are excluded from the forward and backward cones surrounding the x axis defined by $|\sin\theta| < \beta^{-1}$.

Case $\beta^2\sin^2\theta - 1 = 0$:
The electric and magnetic fields are infinite. Thus the field strengths are infinite on a cone at the angle θ with respect to the x-axis. By comparison, a magnetic monopole only has a one-dimensional, singularity line extending from the monopole to infinity.

Case $\beta^2\sin^2\theta - 1 > 0$:
The electric and magnetic fields decrease in strength as $\sin^2\theta$ increases. Thus the region of maximum field strength are the forward and backward cones where $|\sin\theta|$ is greater than but near β^{-1} in value. The pancake picture of the sublight charged particle does not hold for charged tachyons.

1.3.5.1 Are Tachyonic Cones in the Au-Au Scattering Quark-Gluon Plasma?

Cones have been observed in high energy Au-Au scattering in which a quark-gluon plasma is created. These cones have been attributed to a variety of causes such as hydrodynamically generated Mach cones, and Cherenkov radiation. The possibility exists that tachyonic excitations may transiently exist in the quark-gluon plasma and may, in part, explain the observed cones and dips. The above described cones in the case of a moving charged tachyon are remarkably similar in character. See the CERES collaboration paper arXiv:nucl-ex/0701023, and references therein, for experimental findings.

1.3.6 Superluminal (Tachyon) Physics is Different

The simple classical examples presented in this appendix demonstrate that superluminal physics has many interesting new features that are worthy of interest. Since tachyons exist in Black Holes, and, perhaps, in other contexts, their study is a worthwhile endeavor.

1.4 Superluminal (Faster Than Light) Kinetic Theory and Thermodynamics

This section[17] changes the flow of topics of this volume from gravitation and particle theory to superluminal many particle dynamics. We will progress from *superluminal* Kinetic theory to Thermodynamics. We will see that there are strong similarities with non-relativistic Thermodynamics.

1.4.1 Superluminal Kinetic Theory

Assemblages of large numbers of particles embody the Maxwell-Boltzmann distribution. The Boltzmann H theorem is the beginning point for derivations of the non-relativistic Maxwell-Boltzmann distribution. The non-relativistic Maxwell-Boltzmann distribution has the form

$$f(\mathbf{v}, \mathbf{r}) = n(m/(2\pi kT))^{3/2}\exp\{-[m(\mathbf{v} - \mathbf{v}_0)^2/2 + V(r)]/(kT)\} \qquad (1.35)$$

where n is the particle density, T is the temperature, $\mathbf{v}_0$ is the average velocity, m is the particle mass, V(r) is an external conservative force, and k is Boltzmann's constant. In terms of a hamiltonian

$$H(\mathbf{v}, \mathbf{r}) = m\mathbf{v}^2/2 + V(r) \qquad (1.36)$$

we can express the Maxwell-Boltzmann distribution as

$$f(\mathbf{v}, \mathbf{r}) = n(m/(2\pi kT))^{3/2}\exp\{-H(\mathbf{v} - \mathbf{v}_0, \mathbf{r})/(kT)\} \qquad (1.37)$$

1.4.2 Relativistic Form of the Maxwell-Boltzmann Distribution

If we assume that we have a container containing a distribution of relativistic (sublight) particles with an average velocity $\mathbf{v}_0 = 0$, and no external force, then the form of eq. 1.37 generalizes to the relativistic Maxwell-Boltzmann distribution

$$f_R(\mathbf{v}) = C_R \exp\{-H/(kT)\} \qquad (1.38)$$

where C_R is a normalization constant and H is the relativistic hamiltonian for a free particle:

$$H = c(m^2c^2 + \mathbf{p}^2)^{1/2} \qquad (1.39)$$

---

[17] This section is extracted from Blaha (2012a).

with $\mathbf{p} = \gamma m\mathbf{v}$ and $\gamma = (1 - v^2/c^2)^{-½}$. C_R is determined by the condition

$$\int d^3v f_R(\mathbf{v}) = 1 \tag{1.40}$$

1.4.3 Superluminal Form of the Maxwell-Boltzmann Distribution

The superluminal form of Maxwell-Boltzman distribution is based on the form of the mass shell condition for superluminal particles:

$$E^2 - c^2\mathbf{p}^2 = m^2c^4 \tag{1.41}$$

which implies a free hamiltonian

$$H_S = c(\mathbf{p}^2 - m^2c^2)^{½} \tag{1.42}$$

where

$$\mathbf{p} = \gamma_s m\mathbf{v} \tag{1.43}$$

and

$$\gamma_s = (v^2/c^2 - 1)^{-½}$$

The seemingly slight difference between the above equations causes major differences between superluminal and relativistic kinetic theory and thermodynamics. On the other hand relativistic kinetic theory and thermodynamics are qualitatively similar in many ways with their non-relativistic counterparts.

One major difference is the behavior of kinematic variables near the speed of light:

<u>As $v \to c$ Below the Speed of Light</u>

$p \to \infty$
$H \to \infty$

<u>As $v \to c$ From Above the Speed of Light</u>

$p \to \infty$
$H_S \to \infty$

<u>As $v \to \infty$</u>

$p \to mc$
$H_S \to 0$

Thus as v ranges from c to ∞, H_S decreases monotonically from ∞ to zero and p decreases from ∞ to mc. This behavior contrasts with H in eq. 1.39, which increases monotonically with p as v increases from 0 to c. Thus the sublight Maxwell-Boltzmann distribution decreases with v as v increases from 0 to c.

The superluminal Maxwell-Boltzmann distribution *increases* with v as v increases from c to ∞ as we see below. The superluminal Maxwell-Boltzmann distribution decreases with p as p increases from mc to ∞. *As a result the natural physical parametrization of the Maxwell-Boltzmann distribution should be in terms of the momentum rather than the velocity.* Thus Boltzmann's H function which normally is

$$H_B(t) = \int d^3v\, f(\mathbf{v}, t) \log f(\mathbf{v}, t)$$

must be replaced with[18]

$$H_{BS}(t) = \int d^3p\, f_S(\mathbf{p}, t) \log f_S(\mathbf{p}, t)$$

The equilibrium superluminal Maxwell-Boltzman distribution can be derived from $H_S(t)$. It has the same general form as the relativistic distribution

$$f_S(\mathbf{p}) = C_S \exp\{-H_S/(kT)\} \tag{1.44}$$

where C_S is a normalization constant and H_S is the superluminal hamiltonian for a free particle.

We now apply the normalization condition[19]

$$n = N/V = \int d^3p f_S(\mathbf{p}) = C_S\int d^3p \exp\{-H_S/(kT)\} \tag{1.45}$$

where n is the particle density, N is the number of particles in the system, and V is the volume of the system. We calculate C_S by evaluating the integral:

$$n = 4\pi C_S \int_m^\infty dp\, p^2 \exp\{-H_S/(kT)\} \tag{1.46}$$

Letting $x = p/(mc)$ and $\alpha = mc^2/(kT)$ we see eq. 5-D.12 becomes

$$n = 4\pi m^3c^3C_S \int_1^\infty dx\, x^2 \exp\{-\alpha(x^2 - 1)^{½}\} \tag{1.47}$$

---

[18] Note the additional factor of m^3 in $\int d^3p$ will be absorbed in the normalization (eq. 1.45).

[19] We note that using $\int d^3v$ rather than $\int d^3p$ in eq. 1.45 would result in a divergence – another reason for our choice of integration parameter.

Then letting $y^2 = x^2 - 1$ we find

-

$$n = 4\pi m^3c^3C_S \int_0^\infty dy\ y(y^2 + 1)^{½} \exp(-\alpha y)$$

$$= -m^3c^3C_S G^{31}{}_{13}(\alpha^2/4 \mid {}^{0}{}_{-3/2,0,\,½}) \qquad (1.48)$$

where $G^{31}{}_{13}(\ldots)$ is Meijer's G-Function.[20] Therefore

$$C_S = -[m^3c^3G^{31}{}_{13}((mc^2/(2kT))^2 \mid {}^{0}{}_{-3/2,0,\,½})/n]^{-1} \qquad (1.49)$$

The most probable momentum of a particle p_p is the maximum of

$$p_p = \mathrm{Max}\{p^2\exp[-H_S/(kT)]\}$$

$$= \{(2(kT)^2/c^2)[1 + (1 - m^2c^4/(kT)^2)^{½}]\}^{½} \qquad (1.50)$$

For large T or small T the maximum is

$$p_p \approx 2kT/c\ > mc$$

The velocity v_p corresponding to the maximum in the momentum is

$$v_p = cp_p/(p_p{}^2 - m^2c^2)^{½}$$

For large T or small T, the velocity v_p corresponding to the maximum in the momentum is approximately

$$v_p \approx c + ½\, m^2c^5/(2kT)^2$$

1.4.4 Superluminal Thermodynamics

Turning now to the thermodynamics of a dilute superluminal gas implied by the superluminal Maxwell-Boltzman distribution we begin by calculating the average energy per particle

$$\varepsilon = C_S \int d^3p\ H_S \exp[-H_S/(2kT)] \Big/ \int d^3p\ C_S \exp[-H_S/(kT)] \qquad (1.51)$$

---

[20] See Gradshteyn (1965) integral 3.389.2 and p. 1068 for the properties of Meijer's G-Function.

$$= (C_S/n) \int d^3p\, H_S \exp[-H_S/(kT)]$$

$$= (C_S/n)2kT\alpha\, 4\pi m^2c^3 \int_0^\infty dy\, y^2(y^2 + 1)^{½} \exp(-\alpha y)$$

$$= -(C_S/n)m^3c^5G^{31}{}_{13}(\alpha^2/4\, |^{-½}{}_{-2,0,\,½})$$

$$= mc^2\, G^{31}{}_{13}((mc^2/(2kT))^2|^{-½}{}_{-2,0,\,½})/G^{31}{}_{13}((mc^2/(2kT))^2|^{0}{}_{-3/2,0,\,½}) \quad (1.52)$$

The Maxwell-Boltzman normalization factor is related to the energy per particle by

$$C_S = -n\varepsilon/(m^3c^5G^{31}{}_{13}(\alpha^2/4\, |^{-½}{}_{-2,0,\,½})) \quad (1.53)$$

Note that C_S is proportional to the energy in contrast to the non-relativistic case where the Maxwell-Boltzman normalization factor $C = (3m/(4\pi\varepsilon))^{3/2}$.

We now calculate the superluminal pressure for the case of a distribution of superluminal particles bouncing on a wall perpendicular to the z-axis. The wall is assumed to be a perfectly reflecting plane. The pressure is the average force per unit area due to the gas of superluminal particles. The number of particles bombarding the wall per second is with $v_z > 0$ is $v_zf_S(\mathbf{p})d^3p$. Thus the pressure is

$$P = \int d^3p 2p_zv_zf_S(\mathbf{p}) \quad (1.54)$$

where the particle momentum changes by $2p_z$ due to reflection. Due to spherical symmetry one expects the average values for the various components of **v** to be equal. Consequently we can re-express eq. 1.54 as

$$P = 1/3 \int d^3p 2m\gamma_s v^2 f_S(\mathbf{p}) \quad (1.55)$$
$$= 1/3 \int d^3p 2pvf_S(\mathbf{p})$$

Since

$$v = cp/(p^2 - m^2c^2)^{½} \quad (1.56)$$

we see

$$P = 8\pi c/3 \int_m^\infty dpp^4f_S(\mathbf{p})/(p^2 - m^2c^2)^{½}$$

Following steps similar to the preceding leads to

$$P = m^4c^4\, C_S G^{31}{}_{13}((mc^2/(2kT))^2|\,{}^{1/2}{}_{-2,0,\,1/2}) \qquad (1.57)$$

The *equation of state* relating the pressure and energy is

$$P = -(m/c)\{G^{31}{}_{13}((mc^2/(4kT))^2|\,{}^{1/2}{}_{-2,0,\,1/2})/G^{31}{}_{13}(\alpha^2/4\,|^{-1/2}{}_{-2,0,\,1/2}))\}n\varepsilon \qquad (1.58)$$

Substituting for ε we find

$$P = -(nm^2c)\{G^{31}{}_{13}(\rho\,|\,{}^{1/2}{}_{-2,0,\,1/2})/\,G^{31}{}_{13}(\rho\,|^{0}{}_{-3/2,0,\,1/2})\} \qquad (1.59)$$

where

$$\rho = (mc^2/(2kT))^2 \qquad (1.60)$$

Turning now to the consideration of a dilute gas the internal energy of the gas can be defined to be[21]

$$U(t) = N\varepsilon \qquad (1.61)$$

We note that the work done by the superluminal gas if its volume increases by dV is PdV. Then the superluminal (and usual) form of the first law of thermodynamics is

$$dQ = dU + PdV \qquad (1.62)$$

where Q is the heat absorbed. The heat capacity of the system for constant volume is

$$C_V = (\partial U/\partial T)_V \qquad (1.63)$$

The second law of thermodynamics, Boltzmann's H theorem, is based on

$$H = -S/kV \qquad (1.64)$$

where H is the negative of the entropy divided k times the volume V. In systems where there are no superluminal particles, the H theorem states that the entropy never decreases for an isolated gas of fixed volume.

We can calculate H for a superluminal system under equilibrium conditions, H_e, from[22]

[21] The internal energy of a gas of non-interacting non-relativistic particles is U(t) = 3NkT/2. In the superluminal case it appears that it is eq.1.52.

$$H_e = \int d^3p f_S(\mathbf{p}) \ln(f_S(\mathbf{p})) \quad (1.65)$$

$$= \int d^3p f_S(\mathbf{p})[\ln C_S - H_S/(kT)] \quad (1.66)$$

$$= n \ln C_S - \int d^3p f_S(\mathbf{p}) H_S/(kT)$$

$$= n \ln C_S - n\varepsilon/(kT) \quad (1.67)$$

by eqs. 5-D.11 and 5-D.17. Therefore

$$S = -kVH_{Se} = -kN \ln C_S + N\varepsilon/T \quad (1.68)$$

Consequently we obtain the superluminal *and* standard non-relativistic result

$$1/T = (\partial S/\partial U)_x \quad (1.69)$$

where x represents all other extensive variables.

1.4.5 Approximate Calculation of Kinetic and Thermodynamic Quantities

We can obtain more tractable expressions for kinetic and thermodynamic quantities by assuming $\mathbf{p}^2 \gg m^2c^2$ and approximating the hamiltonian (eq. 1.42) with

$$H_{Sa} = cp \quad (1.70)$$

The approximate normalization condition is

$$n = N/V = \int d^3p f_{Sa}(\mathbf{p}) = C_{Sa} \int d^3p \exp\{-H_{Sa}/(kT)\} \quad (1.71)$$

where n is the particle density, N is the number of particles in the system, and V is the volume of the system. C_S is determined by

$$n = 4\pi C_{Sa} \int_{mc}^{\infty} dp\, p^2 \exp\{-cp/(kT)\} \quad (1.72)$$

Letting $\alpha = c/(kT)$ we see eq. 1.72 becomes

[22] We consistently assume that integrals over the momentum $\int d^3p$ are the proper integration (rather than integrations over velocity $\int d^3v$) because, for example, the calculation of the normalization constant eq. 1.45 would diverge if the integration were over $\int d^3v$.

$$n = 4\pi C_{Sa}\, d^2/d\alpha^2 \int_{mc}^{\infty} dp\, \exp(-\alpha p)$$

$$= 4\pi C_{Sa}\, d^2/d\alpha^2\, [(1/\alpha)\exp(-\alpha mc)] \qquad (1.73)$$

Therefore the normalization factor is

$$C_{Sa} = n/\{4\pi\, d^2/d\alpha^2\, [(1/\alpha)\exp(-\alpha mc)]\} \qquad (1.74)$$

The most probable momentum of a particle p_p is the maximum of

$$p_{pa} = \mathrm{Max}\{p^2 \exp[-H_{Sa}/(kT)]\}$$
$$= 2kT/c \qquad (1.75)$$

The velocity v_{pa} corresponding to the maximum in the momentum is

$$v_{pa} = c p_{pa}/(p_{pa}^{\ 2} - m^2c^2)^{1/2}$$

For large T or small T, the velocity v_{pa} corresponding to the maximum in the momentum is approximately

$$v_{pa} \approx c + \tfrac{1}{2}\, m^2c^5/(2kT)^2$$

Turning now to the thermodynamics implied by the superluminal Maxwell-Boltzman distribution we begin by calculating the average energy per particle

$$\varepsilon_a = \int d^3p\, H_{Sa} \exp[-H_{Sa}/(kT)] \Big/ \int d^3p\, \exp[-H_{Sa}/(kT)] \qquad (1.76)$$
$$= (C_{Sa}/n) \int d^3p\, H_{Sa} \exp[-H_{Sa}/(kT)]$$

$$= -(4\pi c C_{Sa}/n)\, d^3/d\alpha^3[(1/\alpha)\exp(-\alpha mc)] \rightarrow 3kT \text{ for } T \gg mc$$

where $\alpha = c/(kT)$.[23]

The Maxwell-Boltzman normalization factor is related to the energy per particle by

$$C_{Sa} = -n\varepsilon_a/\{4\pi c\, d^3/d\alpha^3[(1/\alpha)\exp(-\alpha mc)]\} \qquad (1.77)$$

[23] The Superluminal case differs from the non-relativistic case: $\varepsilon_a = 3kT/2$. An example of $\varepsilon_a = 3kT$ is a crystal with a potential energy of compression. See p. 192 Morse (1964).

Note that C_{Sa} is proportional to the energy ε_a in contrast to the non-relativistic case where the Maxwell-Boltzman normalization factor $C = (3m/(4\pi\varepsilon))^{3/2}$.

We now calculate the superluminal pressure for the case of a distribution of superluminal particles bouncing on a wall perpendicular to the z-axis. The wall is assumed to be a perfectly reflecting plane. The pressure is the average force per unit area due to the gas of superluminal particles. The number of particles bombarding the wall per second is with $v_z > 0$ is $v_z f_{Sa}(\mathbf{p})d^3p$. Thus the pressure is

$$P_a = \int d^3p 2p_z v_z f_{Sa}(\mathbf{p}) \tag{1.78}$$

where the particle momentum changes by $2p_z$ due to reflection. Due to spherical symmetry one expects the average values for the various components of **v** to be equal. Consequently we can re-express eq. 1.78 as

$$P_a = 1/3 \int d^3p 2m\gamma_s v^2 f_{Sa}(\mathbf{p}) \tag{1.79}$$
$$= 1/3 \int d^3p 2pv f_{Sa}(\mathbf{p})$$

Since

$$v = cp/(p^2 - m^2c^2)^{1/2} \tag{1.80}$$

we see

$$P_a = 8\pi c/3 \int_{mc}^{\infty} dp p^4 f_{Sa}(\mathbf{p})/(p^2 - m^2c^2)^{1/2} \tag{1.81}$$

$$\cong 8\pi c/3 \int_{mc}^{\infty} dp p^3 f_{Sa}(\mathbf{p})$$

Evaluating eq. 1.81 yields

$$P_a = -(8\pi c/3)\, C_{Sa}\, d^3/d\alpha^3[(1/\alpha)\exp(-\alpha mc)] \tag{1.82}$$

The *equation of state* relating the pressure and energy is[24]

$$P_a = 2/3\, n\varepsilon_a \tag{1.83}$$

For $T \gg mc$ we found[25]

$$\varepsilon_a \rightarrow 3kT \tag{1.84}$$

---

[24] The same equation of state as non-relativistic kinetic theory. See p. 72 Huang (1965).

[25] Later we will define temperature in terms of the entropy S as $1/T = (\partial S/\partial U)_x$ where x is all other extensive variables.

then, contrary to non-relativistic kinetic theory, we find ($T \gg mc$)

$$P_a = 2nkT \tag{1.85}$$

Turning now to the consideration of a dilute gas the internal energy of the gas for $T \gg mc$ is

$$U(t) = N\varepsilon \rightarrow 3NkT \tag{1.86}$$

We note again that the work done by the superluminal gas if its volume increases by dV is PdV. Then the superluminal (and usual) form of the first law of thermodynamics is

$$dQ = dU + PdV \tag{1.87}$$

where Q is the heat absorbed. The heat capacity of the system for constant volume is ($T \gg mc$)

$$C_V \rightarrow 3Nk \tag{1.88}$$

The second law of thermodynamics, Boltzmann's H theorem, is based on

$$H = -S/kV \tag{1.89}$$

where H is the negative of the entropy divided k times the volume V. In systems where there are no superluminal particles, the H theorem states that the entropy never decreases for an isolated gas of fixed volume.

We can calculate H_{BS} for a superluminal system under equilibrium conditions, H_{BSea}, from[26]

$$H_{BSea} = \int d^3p f_{Sa}(\mathbf{p}) \ln(f_{Sa}(\mathbf{p})) \tag{1.90}$$

$$= \int d^3p f_{Sa}(\mathbf{p})[\ln C_{Sa} - H_{Sa}/(kT)] \tag{1.91}$$

$$= n \ln C_{Sa} - \int d^3p f_{Sa}(\mathbf{p}) H_{Sa}/(kT)$$

$$= n \ln C_{Sa} - n\varepsilon_a/(kT) \tag{1.92}$$

by eqs. 1.45 and 1.51. Therefore

$$S_a = -kVH_{BSea} = -kN \ln C_{Sa} + N\varepsilon_a/T \tag{1.93}$$

[26] We consistently assume that integrals over the momentum $\int d^3p$ are the proper integration (rather than integrations over velocity $\int d^3v$) because, for example, the calculation of the normalization constant eq. 1.45 would diverge if the integration were over $\int d^3v$.

The superluminal *and* standard non-relativistic result still holds

$$1/T = (\partial S/\partial U)_x \qquad (1.94)$$

where x represents all other extensive variables.

1.4.6 Superluminal Kinetics and Themodynamics Are Similar to the Non-Relativistic Case

In this section we have shown that kinetic theory and the laws of themodynamics are usually similar in the superluminal and non-relativistic cases modulo detail differences in the values of the various quantities due to differences between superluminal kinematics and non-relativistic kinematics.

2. Basis of Four Fermion Species Derived From the Complex Lorentz Group

This chapter describes the origin of the four known fermion species (charged lepton, neutral lepton, up-type quark and down-type quark) in the Complex Lorentz group. Hitherto no natural reason has been presented for these four species other than by this author. The material in this chapter appears in Blaha (2015a) and earlier books by the author.

2.1 Basic Lorentz Group Features

2.1.1 Transformations Between Coordinate Systems

The measurement of time and space is simple in practice but raises weighty questions when their underlying basis is examined. We shall begin by measuring spatial distances with a ruler, and by measuring time with a clock. Earlier we determined that four dimensions: one space dimension and three space dimensions were required. We now define rectangular coordinate systems with x, y, and z axes as pictured in Fig. 2.1 below. We then postulate:

Postulate 2.1. Any observer can define a set of time and space coordinates called a coordinate system in which the observer is at rest. One can define a transformation that relates the coordinate systems of two observers traveling at a any constant velocity with respect to each other.

One can always relate the coordinates of two coordinate systems by having an observer in each coordinate system specify the coordinates of objects located at each spatial point, and then creating a map between the coordinates of corresponding spatial locations.

If space is flat the relation between the respective coordinates is linear. (One could reverse the logic of that statement by defining a flat space to be one in which the coordinates of a point in any coordinate system are linearly related to the coordinates of any other coordinate system moving at a constant velocity with respect to it.) Thus we can express the relation between the coordinates in the "unprimed" system to the coordinates in the "primed" system as a transformation between coordinate systems:

$$\mathbf{a}' = \mathrm{A}\mathbf{a} + \mathbf{B}\mathrm{t} + \mathbf{C} \qquad (2.1)$$
$$\mathrm{t}' = \mathrm{Dt} + \mathbf{E}\cdot\mathbf{a}$$

where A is a 3×3 matrix, **B**, **C** and **E** are 3-vectors, and D is a number (scalar value).

Having restricted the set of transformations between coordinate systems to the form of eq. 2.1 we now assert postulates that restrict the form of the transformation to Lorentz transformations and transformations similar to Lorentz transformations.

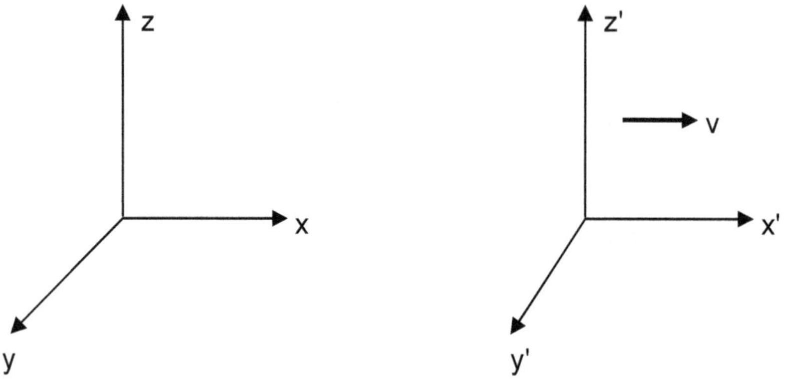

Figure 2.1. Depiction of two coordinate systems. The "primed" coordinate system is moving with velocity **v** in the positive x direction with respect to the "unprimed" coordinate system. We choose parallel axes for convenience.

Postulate 2.2. The speed of light, c, is the same in all coordinate systems.

Postulate 2.3. The invariant interval or distance $d\tau$ is defined by

$$d\tau^2 = g_{\mu\nu}dx^\mu dx^\nu \qquad (2.2)$$

It is invariant under a change of coordinate systems. The 16 quantities $g_{\mu\nu}$ are known as the metric tensor.[27] *The four quantities dx^μ are infinitesimal displacements in space and time.*

If we expand eq. 2.2 in rectangular coordinates it is equivalent to

$$d\tau^2 = g_{00}dx^0dx^0 + g_{11}dx^1dx^1 + g_{22}dx^2dx^2 + g_{33}dx^3dx^3 \qquad (2.3)$$

which equals

$$d\tau^2 = c^2dt^2 - dx^2 - dy^2 - dz^2 \qquad (2.4)$$

[27] The repeated indices indicate a summation. In this case from 0 to 3 as shown in eq. 2.3.

using the familiar form of the time and rectangular space coordinates.

2.1.2 The Lorentz Group

The metric tensor $g_{\mu\nu}$ for rectangular coordinates has the matrix form:

$$G = \begin{bmatrix} 1 & 0 & 0 & 0 \\ 0 & -1 & 0 & 0 \\ 0 & 0 & -1 & 0 \\ 0 & 0 & 0 & -1 \end{bmatrix} \qquad (2.5)$$

The invariant interval under a transformation between rectangular coordinate systems (with "primed" and "unprimed" coordinates) has the form of eq. 2.4 for the unprimed coordinates and the same form for the primed coordinates:

$$d\tau^2 = c^2dt'^2 - dx'^2 - dy'^2 - dz'^2 \qquad (2.6)$$

In matrix form we can define an "unprimed" coordinate column vector with

$$a = \begin{bmatrix} t \\ x \\ y \\ z \end{bmatrix} \qquad (2.7a)$$

and its corresponding "primed" coordinate with

$$a' = \begin{bmatrix} t' \\ x' \\ y' \\ z' \end{bmatrix} \qquad (2.7b)$$

If a and a' are the coordinates of the same point in the respective coordinate systems then, by postulates 2.2 and 2.3, they are related by a boost Lorentz transformation $\Lambda(\mathbf{v})$ with the form

$$a' = \Lambda(\mathbf{v})a \tag{2.8}$$

(and possibly a spatial rotation matrix factor), where **v** is the relative velocity of the coordinate systems. The form of the transformation eq. 2.8, which is called a Lorentz *boost,* is constrained by postulates 2.2 and 2.3 to be[28]

$$\Lambda(\mathbf{v}) = \begin{bmatrix} \gamma & -\gamma v_x & -\gamma v_y & -\gamma v_z \\ -\gamma v_x & 1 + (\gamma - 1)v_x^2/v^2 & (\gamma - 1)v_xv_y/v^2 & (\gamma - 1)v_xv_z/v^2 \\ -\gamma v_y & (\gamma - 1)v_xv_y/v^2 & 1 + (\gamma - 1)v_y^2/v^2 & (\gamma - 1)v_yv_z/v^2 \\ -\gamma v_z & (\gamma - 1)v_xv_z/v^2 & (\gamma - 1)v_yv_z/v^2 & 1 + (\gamma - 1)v_z^2/v^2 \end{bmatrix} \tag{2.9}$$

where $\gamma = (1 - v^2)^{-½}$, $\mathbf{v} = (v_x, v_y, v_z)$, $v = |\mathbf{v}|$ and we set $c = 1$ for convenience.[29] The set of all matrices of the form of $\Lambda(\mathbf{v})$, or $\Lambda(\mathbf{v})\mathcal{R}(\boldsymbol{\theta})$ or $\mathcal{R}(\boldsymbol{\theta})\Lambda(\mathbf{v})$ where $\mathcal{R}(\boldsymbol{\theta})$ is a spatial rotation with angle vector $\boldsymbol{\theta}$, for $v < c$ form a matrix representation of the Lorentz group. Elements, $\Lambda(\mathbf{v}, \boldsymbol{\theta})$, of the Lorentz group satisfy the defining relation of the Lorentz group:

$$\Lambda(\mathbf{v}, \boldsymbol{\theta})^T G \Lambda(\mathbf{v}, \boldsymbol{\theta}) = G \tag{2.10}$$

where the superscript ^{T} specifies the transpose of the matrix.

The Lorentz group, with which we are familiar, relates the coordinates of an event in two coordinate systems that differ by a a spatial rotation, and a relative velocity whose magnitude is less than the speed of light. The inhomogenous Lorentz group includes coordinate displacements.[30]

The group elements of the homogeneous Lorentz group can be expressed in terms of the generators **K** of boosts to coordinate systems moving at a constant velocity **v** and the generators **J** of purely spatial rotations by

$$\Lambda(\mathbf{v}, \boldsymbol{\theta}) = \exp[i\omega\hat{\mathbf{u}}\cdot\mathbf{K} + i\boldsymbol{\theta}\cdot\mathbf{J}] \tag{2.11}$$

---

[28] We shall consider only the proper, orthochronous Lorentz group at this point. We assume that the primed and unprimed coordinate systems have parallel axes. So there is no rotation of axes embodied in eq. 2.9.

[29] One can set c = 1 by an appropriate choice of time and spatial distance scales. The demonstration that $\Lambda(\mathbf{v})$ has the form given by eq. 2.9 can be found in many textbooks.

[30] See Weinberg (1995) for a discussion of the inhomogeneous Lorentz group.

where the vector $\boldsymbol{\theta}$ is a 3-vector specifying the rotation angles, and where $\mathbf{v} = \hat{\mathbf{u}}\tanh\omega$, $\hat{\mathbf{u}}\bullet\hat{\mathbf{u}} = 1$.

The boost transformation $\Lambda(\mathbf{v}) = \Lambda(\mathbf{v}, \mathbf{0})$ has the form

$$\Lambda(\mathbf{v}) = \exp[i\omega\hat{\mathbf{u}}\cdot\mathbf{K}] \tag{2.12}$$

Its matrix form is eq. 2.9. The matrix form can be expressed in terms of the unit normalized velocity vector $\mathbf{u} = (u_x, u_y, u_z)$ and ω as

$$\Lambda(\omega, \mathbf{u}) = \Lambda(\mathbf{v}) \tag{2.13}$$

$$= \begin{bmatrix} \cosh(\omega) & -\sinh(\omega)u_x & -\sinh(\omega)u_y & -\sinh(\omega)u_z \\ -\sinh(\omega)u_x & 1 + (\cosh(\omega) - 1)u_x^2 & (\cosh(\omega) - 1)u_xu_y & (\cosh(\omega) - 1)u_xu_z \\ -\sinh(\omega)u_y & (\cosh(\omega) - 1)u_xu_y & 1 + (\cosh(\omega) - 1)u_y^2 & (\cosh(\omega) - 1)u_yu_z \\ -\sinh(\omega)u_z & (\cosh(\omega) - 1)u_xu_z & (\cosh(\omega) - 1)u_yu_z & 1 + (\cosh(\omega) - 1)u_z^2 \end{bmatrix}$$

where $\Lambda(\omega, \mathbf{u}) = \Lambda(\omega, \mathbf{u}, \boldsymbol{\theta} = \mathbf{0})$ in the previous notation. This definition of the general form of proper, orthochronous, Lorentz boost matrices $\Lambda(\omega, \mathbf{u})$ will be used in subsequent sections to define faster-than-light boost transformations.

The vector form of a Lorentz boost transformation is

$$\begin{aligned} \mathbf{x}' &= \mathbf{x} + (\gamma - 1)\mathbf{x}\cdot\mathbf{v}\,\mathbf{v}/v^2 - \gamma\mathbf{v}t \\ t' &= \gamma(t - \mathbf{v}\cdot\mathbf{x}/c^2) \end{aligned} \tag{2.14}$$

where $\gamma = (1 - \beta^2)^{-1/2}$ with $\beta = v/c = v$ (since we set $c = 1$).

2.2 Complex Lorentz Group Features

2.2.1 The Nature of $\Lambda(\omega, u)$ for Complex ω

We now turn to the case of complex ω which includes superluminal (faster-than-light) Lorentz transformations as well as conventional Lorentz transformations. Since, for any complex value z

$$\cosh^2(z) - \sinh^2(z) = 1 \tag{2.15}$$

it follows that for any complex value of ω, Λ(ω, **u**) is a member of the Lorentz group, and/or of the complex Lorentz group[31] for complex ω:

$$\Lambda(\omega, \mathbf{u})^T G \Lambda(\omega, \mathbf{u}) = G \qquad (2.16)$$

For certain values of the imaginary part of ω the matrix Λ(ω, **u**) has a particularly simple form, similar to that of Λ(ω, **u**) for real ω, but which generates boosts to relative velocities greater than the speed of light. Among these values are:

$$\omega = \omega_{\pm} = \omega \pm i\pi/2 \qquad (2.17)$$

Later we will see that these alternate choices $\omega_{\pm}$ correspond to specific choices of parity.

2.2.2 Complex Lorentz Group

In the preceding section we saw that the parameter ω can be complex and the boost transformation will still satisfy the Lorentz condition eq. 2.10. More generally we can consider complex homogeneous Lorentz transformations Λ(**v**, **θ**) which can be represented by eq. 2.11 with complex parameters ω, **û**, and **θ** where **û** and **θ** are complex 3-vectors. **θ** specifies a rotation angle.

In general Λ(**v**, **θ**) is then a transformation between coordinate systems that have complex coordinates. One coordinate system is moving at a constant complex velocity with respect to the other. Coordinate systems do not necessarily have parallel spatial axes in general.

Within the complex Lorentz group, denoted L(C),[32] *there are subsets of boosts that play important physical roles in the derivation of the form of The Standard Model. In particular we will see that certain classes of boosts generate faster-than-light transformations. These transformations can be further divided into subclasses of "left-handed" and "right-handed" transformations based on the quantum field theories to which they lead. Further within each subclass there are subclasses of transformations that naturally lead to Dirac-like free field equations that can be described as lepton-like and quark-like.*

Thus these boosts are a key ingredient to understanding the form of The Standard Model.

2.2.3 Faster-than-Light Transformations

In this section we will substitute $\omega_{\pm}$ for ω in Λ(ω, **u**) and then show that we obtain two sets of possible transformations from sublight reference frames to faster-than-light reference frames. One set of transformations, where $\omega_L = \omega + i\pi/2$, will be called *left-handed*

[31] The complex Lorentz group is defined as the group of all complex transformations that satisfy eq. 2.16.
[32] Streater (2000) points out that the complex Lorentz group is essential to the proof of the CPT theorem.

superluminal boosts. They eventually lead to the "left-handed" part of The Standard Model. We denote members of this set, $\Lambda_L(\omega, \mathbf{u})$, with the subscript "L" for left-handed.

The other set of boosts where $\omega_R = \omega - i\pi/2$ will be called *right-handed superluminal boosts*. They eventually lead to a right-handed, unphysical,[33] version of The Standard Model. We denote members of this set of boosts, $\Lambda_R(\omega, \mathbf{u})$, with the subscript "R" for right-handed.

Before considering faster-than-light boosts we note the relation between a real-valued ω in a *conventional* Lorentz boost $\Lambda(\omega, \mathbf{u})$, and the magnitude of the relative velocity v for v < 1, is

$$\mathbf{v} = \hat{\mathbf{u}} \tanh\omega \qquad \text{with} \qquad \hat{\mathbf{u}}\cdot\hat{\mathbf{u}} = 1$$

$$\begin{aligned} \cosh(\omega) &= \gamma = (1 - v^2)^{-1/2} \\ \sinh(\omega) &= v\gamma = \beta\gamma \end{aligned} \tag{2.18}$$

where $\beta = v = |\mathbf{v}|$.

2.2.4 Left-Handed Superluminal Transformations

Left-handed (proper orthochronous) superluminal boost transformations $\Lambda_L(\mathbf{v})$ have the same form as eq. 2.9 for ordinary (proper orthochronous) Lorentz boost transformations. However the magnitude of the relative velocity **v** is greater than the speed of light. Thus $\gamma = (1 - v^2)^{-1/2}$ is pure imaginary and $\Lambda_L(\mathbf{v})$ is complex. Thus

$$\Lambda_L(\mathbf{v}) = \begin{bmatrix} \gamma & -\gamma v_x & -\gamma v_y & -\gamma v_z \\ -\gamma v_x & 1 + (\gamma - 1)v_x^2/v^2 & (\gamma - 1)v_x v_y/v^2 & (\gamma - 1)v_x v_z/v^2 \\ -\gamma v_y & (\gamma - 1)v_x v_y/v^2 & 1 + (\gamma - 1)v_y^2/v^2 & (\gamma - 1)v_y v_z/v^2 \\ -\gamma v_z & (\gamma - 1)v_x v_z/v^2 & (\gamma - 1)v_y v_z/v^2 & 1 + (\gamma - 1)v_z^2/v^2 \end{bmatrix} \tag{2.19}$$

This transformation raises several issues – the most prominent of which is the interpretation of the imaginary coordinates generated by the transformation. Imaginary coordinates would appear at first glance to be unphysical. However we view the measurement of these quantities operationally: an observer measures distances with "rulers", and time with clocks, which both give real numeric values. Thus an observer in any coordinate system will always measure real numbers for time and space distances. However an observer in another

[33] Currently the case. If a right-handed counterpart to the current Standard Model surfaces at higher energies then the features emerging from right-handed superluminal boosts then become physically important.

coordinate system that is related to the first coordinate system by a superluminal transformation will view the coordinates in the first system as complex as eq. 2.19 indicates.

The reconciliation of these points of view requires the introduction of a new transformation, called a *Reality group transformation*, in addition to a superluminal Lorentz transformation for the case of faster than light transformations. Reality group transformations maps the complex coordinates generated by a Lorentz transformation to the real coordinates seen by the observer[34] in the "faster than light" reference frame. We describe the Reality group transformations in detail later. We will show how they imply the Reality group is SU(3)⊗SU(2)⊗U(1)⊗SU(2)⊗U(1).

2.2.5 Cosh-Sinh Representation of Left-Handed Superluminal Boosts

We will now develop the representation of left-handed superluminal boost transformations in terms of cosh(ω) and sinh(ω) for later use in our discussion of tachyons. We find that we must use a complex $\omega_L \equiv \omega + i\pi/2$ to properly describe left-handed superluminal boosts. The relation between ω_L and v is different from eq. 2.18 for the case of left-handed superluminal boosts:

$$\cosh(\omega_L) = i \sinh(\omega) = -\gamma = i\gamma_s \qquad (2.20)$$
$$\sinh(\omega_L) = i \cosh(\omega) = -\beta\gamma = i\beta\gamma_s$$

where $\beta = v > 1$, $\boldsymbol{\omega \geq 0}$, and

$$\gamma_s = (\beta^2 - 1)^{-1/2} \qquad (2.21)$$

Eq. 2.20 implies

$$\sinh(\omega) = \gamma_s \qquad (2.22)$$
$$\cosh(\omega) = \beta\gamma_s$$

Upon substituting $\boldsymbol{\omega_L}$ for ω in eq. 2.13 we obtain another form for a left-handed superluminal transformation (equivalent to that of eq. 2.19):

$\Lambda_L(\omega, \mathbf{u}) = \Lambda(\omega + i\pi/2, \mathbf{u})$

[34] The linearity of the superluminal transformation makes this secondary transformation physically possible.

$$= \begin{bmatrix} \cosh(\omega_L) & -\sinh(\omega_L)u_x & -\sinh(\omega_L)u_y & -\sinh(\omega_L)u_z \\ -\sinh(\omega_L)u_x & 1+(\cosh(\omega_L)-1)u_x^2 & (\cosh(\omega_L)-1)u_xu_y & (\cosh(\omega_L)-1)u_xu \\ -\sinh(\omega_L)u_y & (\cosh(\omega_L)-1)u_xu_y & 1+(\cosh(\omega_L)-1)u_y^2 & (\cosh(\omega_L)-1)u_yu_z \\ -\sinh(\omega_L)u_z & (\cosh(\omega_L)-1)u_xu_z & (\cosh(\omega_L)-1)u_yu_z & 1+(\cosh(\omega_L)-1)u_z^2 \end{bmatrix}$$

$$= \begin{bmatrix} i\gamma_s & -i\beta\gamma_s u_x & -i\beta\gamma_s u_y & -i\beta\gamma_s u_z \\ -i\beta\gamma_s u_x & 1+(i\gamma_s-1)u_x^2 & (i\gamma_s-1)u_xu_y & (i\gamma_s-1)u_xu_z \\ -i\beta\gamma_s u_y & (i\gamma_s-1)u_xu_y & 1+(i\gamma_s-1)u_y^2 & (i\gamma_s-1)u_yu_z \\ -i\beta\gamma_s u_z & (i\gamma_s-1)u_xu_z & (i\gamma_s-1)u_yu_z & 1+(i\gamma_s-1)u_z^2 \end{bmatrix} = \Lambda_L(\mathbf{v}) \qquad (2.23)$$

A simple case that illustrates a left-handed superluminal boost is to assume the relative velocity is in the x direction. Then eq. 2.23 becomes

$$\Lambda_L(\omega, \mathbf{u} = (1,0,0)) = \begin{bmatrix} i\gamma_s & -i\beta\gamma_s & 0 & 0 \\ -i\beta\gamma_s & i\gamma_s & 0 & 0 \\ 0 & 0 & 1 & 0 \\ 0 & 0 & 0 & 1 \end{bmatrix} \qquad (2.24)$$

implementing the coordinate transformation:

$$X' = \Lambda_L(\omega, \mathbf{u} = (1,0,0))X$$

or

$$\begin{aligned} t' &= i\gamma_s(t - \beta x) \\ x' &= i\gamma_s(x - \beta t) \\ y' &= y \\ z' &= z \end{aligned} \qquad (2.25)$$

The addition rule for the x-component of velocity can be computed for infinitesimal displacements in space and time:

$$v_x' = \Delta x'/\Delta t' = (\Delta x\, \gamma_s - \Delta t\, \beta\gamma_s)/(\Delta t\, \gamma_s - \Delta x\, \beta\gamma_s)$$

$$= (v_x - \beta)/(1 - \beta v_x) \qquad (2.26)$$

in the limit $\Delta t \rightarrow 0$ where the x component of a particle's velocity in the unprimed frame is $v_x = \Delta x/\Delta t$. $\Delta t'$ is determined by

$$\Delta t' = i\Delta t\, \gamma_s(1 - \beta v_x) \tag{2.27}$$

Note the velocity of light is the same in the primed and unprimed reference frames. (If $v_x = 1$ then $v_x' = 1$.) *Thus left-handed superluminal transformations preserve the constancy of the speed of light in all reference frames.* (Postulate 2.2)

Further note that increasing the value of ω in $\Lambda_L(\omega, \mathbf{u})$ corresponds to decreasing the magnitude of the relative velocity v since

$$v = \mathrm{cotanh}(\omega) \tag{2.28}$$

by eq. 2.22. Thus when $\omega = 0$ then $v = \infty$, and when $\omega = \infty$ then $v = 1$. This is the reverse of the sublight case: by eq. 2.18 $v = \tanh(\omega)$. Thus when $\omega = 0$ then $v = 0$, and when $\omega = \infty$ then $v = \infty$.

2.2.6 General Velocity Transformation Law – Left-Handed Superluminal Boosts

The general velocity transformation law for a particle moving with velocity **v** in the unprimed reference frame and velocity **v'** in the primed reference frame is

$$\mathbf{v}' = [\mathbf{v} + (\gamma - 1)\mathbf{w}\cdot\mathbf{v}\ \mathbf{w}/w^2 - \gamma\mathbf{w}]/[\ \gamma(1 - \mathbf{w}\cdot\mathbf{v})] \tag{2.29}$$

where **w** is the relative velocity of the primed reference frame with respect to the unprimed reference frame, and $\gamma = (1 - w^2)^{-1/2}$. Eq. 2.29 is obtained by calculating the derivative $d\mathbf{x}'/dt'$ using eqs. 2.14. The relative velocity **w** can be greater or less than the speed of light. Eq. 2.29 implies

$$v'^2 = 1 + (v^2 - 1)(1 - w^2)/(1 - \mathbf{w}\cdot\mathbf{v})^2 \tag{2.30}$$

The relation of the velocities (eq. 2.30) will be used to determine the multiplication rules for subluminal and superluminal Lorentz transformations (next subsection).

2.2.7 Left-Handed Transformations Multiplication Rules

In this subsection we will determine the multiplication rules of left-handed subluminal and superluminal Lorentz boosts. To do this we will consider three reference frames: an "unprimed" frame, a "primed" frame moving with velocity **w** with respect to the unprimed frame, and a "double-primed" frame moving with velocity **v** with respect to the unprimed frame and velocity **v'** with respect to the primed frame. See Fig.2.2.

The velocity **v'** is related to **v** by eqs. 2.29 and 2.30. Think of the double-primed coordinate system as attached to a particle. In addition note that the transformation law from the unprimed to the double-primed reference frame can be viewed as the product of consecutive transformations (boosts) from the unprimed to the primed reference frames and then from the primed to the double-primed reference frames.

Thus the transformations have the general form:

$$\Lambda_?(\mathbf{v}) = \Lambda_?(\mathbf{v'})\Lambda_?(\mathbf{w}) \tag{2.31}$$

where the "?" subscripts indicate subluminal or superluminal transformations (boosts) depending on the magnitude of the relative velocity in the transformation's parentheses.

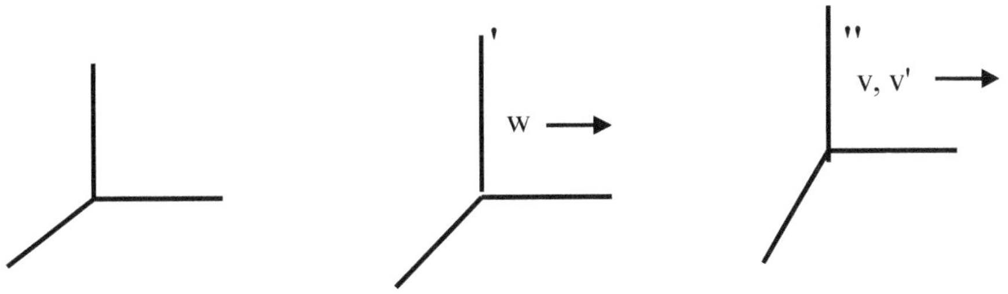

Figure 2.2. Three reference frames used to establish transformation multiplication rules.

We now consider the various cases using eq. 2.30:

1) If $w > 1$ and $v' > 1$

   then eq. 2.30 implies $v < 1$ and thus the left $\Lambda_?(\mathbf{v})$ is a subluminal transformation

$$\Lambda(\mathbf{v}) = \Lambda_L(\mathbf{v'})\Lambda_L(\mathbf{w}) \tag{2.32}$$

2) If $w > 1$, $v' < 1$

   then eq. 2.30 implies $v > 1$ and thus the left $\Lambda_?(\mathbf{v})$ is a superluminal transformation

$$\Lambda_L(\mathbf{v}) = \Lambda(\mathbf{v'})\Lambda_L(\mathbf{w}) \tag{2.33}$$

3) If $w < 1$, $v' > 1$

   then eq. 2.30 implies $v > 1$ and thus the left $\Lambda_?(\mathbf{v})$ is a superluminal transformation

$$\Lambda_L(\mathbf{v}) = \Lambda_L(\mathbf{v'})\Lambda(\mathbf{w}) \tag{2.34}$$

4) If w < 1, v' < 1

then eq. 2.30 implies v < 1 and thus the left $\Lambda_?(\mathbf{v})$ is a Lorentz transformation

$$\Lambda(\mathbf{v}) = \Lambda(\mathbf{v'})\Lambda(\mathbf{w}) \tag{2.35}$$

where, in each of the above cases, the transformation on the left side of the equation may be a boost or a combination of a boost and a spatial rotation. Thus we have obtained the multiplication rules for left-handed subluminal and superluminal Lorentz transformations.

2.2.8 Inverse of Left-Handed Transformations

The inverse of a Lorentz boost is

$$\Lambda^{-1}(\omega, \hat{\mathbf{u}}) = \exp[-i\omega\hat{\mathbf{u}}\cdot\mathbf{K}] \tag{2.36}$$

where $\boldsymbol{\omega} \geq \mathbf{0}$. Thus the inverse is generated by letting $\omega \rightarrow -\omega$. Note that since $v = \tanh\omega$, the effect of $\omega \rightarrow -\omega$ is to let $v \rightarrow -v$. In the case of superluminal left-handed boosts, since

$$\Lambda_L(\omega, \mathbf{u}) = \Lambda(\omega + i\pi/2, \mathbf{u}) = \exp[i(\omega + i\pi/2)\hat{\mathbf{u}}\cdot\mathbf{K}] \tag{2.37}$$

we find the inverse is

$$\Lambda_L^{-1}(\omega, \mathbf{u}) = \Lambda(-(\omega + i\pi/2), \mathbf{u}) = \exp[-i(\omega + i\pi/2)\hat{\mathbf{u}}\cdot\mathbf{K}] \tag{2.38}$$

where $\boldsymbol{\omega} \geq \mathbf{0}$. Since $\Lambda_L^{-1}(\omega, \mathbf{u})$ is not the hermitean conjugate of $\Lambda_L(\omega, \mathbf{u})$, superluminal boosts are not unitary. However unitarity is not required since complex Lorentz group elements satisfy the defining relation of the Lorentz group (eq. 2.10).

2.2.9 Right-Handed Superluminal Transformations

When we transform between reference frames using a *right-handed*[35] superluminal boost the relation between ω and v is different. The variable ω becomes $\omega_R = \omega - i\pi/2$ and

$$\cosh(\omega_R) = -i\sinh(\omega) = \gamma = -i\gamma_s \tag{2.39}$$
$$\sinh(\omega_R) = -i\cosh(\omega) = \beta\gamma = -i\beta\gamma_s \tag{2.40}$$

---

[35] We call these transformations right-handed because they lead eventually to an alternate right-handed Standard Model This alternate right-handed Standard Model does not appear to correspond to current experimental reality.

where $\beta = v > 1$ and $\boldsymbol{\omega} \geq \mathbf{0}$. Note that $\omega = \mathrm{Re}\, \omega_R$ with

$$\sinh(\omega) = \gamma_s \tag{2.41}$$
$$\cosh(\omega) = \beta\gamma_s \tag{2.42}$$

and with

$$\gamma_s = (\beta^2 - 1)^{-1/2} \tag{2.43}$$

Upon substituting ω_R for ω in eq. 2.13 we obtain the form of the right-handed superluminal boost:[36]

$$\Lambda_R(\omega, \mathbf{u}) = \Lambda(\omega - i\pi/2, \mathbf{u}) \tag{2.44}$$

$$= \begin{bmatrix} -i\gamma_s & i\beta\gamma_s u_x & i\beta\gamma_s u_y & i\beta\gamma_s u_z \\ i\beta\gamma_s u_x & 1 + (-i\gamma_s - 1)u_x^2 & (-i\gamma_s - 1)u_x u_y & (-i\gamma_s - 1)u_x u_z \\ i\beta\gamma_s u_y & (-i\gamma_s - 1)u_x u_y & 1 + (-i\gamma_s - 1)u_y^2 & (-i\gamma_s - 1)u_y u_z \\ i\beta\gamma_s u_z & (-i\gamma_s - 1)u_x u_z & (-i\gamma_s - 1)u_y u_z & 1 + (-i\gamma_s - 1)u_z^2 \end{bmatrix}$$

A simple case that illustrates right-handed superluminal transformations is, assuming a relative velocity in the x direction, we find eq. 2.44 becomes

$$\Lambda_R(\omega, \mathbf{u} = (1,0,0)) = \begin{bmatrix} -i\gamma_s & i\beta\gamma_s & 0 & 0 \\ i\beta\gamma_s & -i\gamma_s & 0 & 0 \\ 0 & 0 & 1 & 0 \\ 0 & 0 & 0 & 1 \end{bmatrix} \tag{2.45}$$

implementing the coordinate transformation:

$$X' = \Lambda_R(\omega, \mathbf{u})X$$

or

$$\begin{aligned} t' &= -i\gamma_s(t - \beta x) \\ x' &= -i\gamma_s(x - \beta t) \\ y' &= y \\ z' &= z \end{aligned} \tag{2.46}$$

---

[36] We note the singularities at $\beta = \pm 1$ or $\omega = \pm\infty$. **As a result we have a branch cut in the complex ω-plane consisting of the entire real ω axis. Therefore three left-handed boosts are not equivalent to a right-handed boost but rather appear on a different Riemann sheet.**

Comparing eq. 2.45 with eq. 2.24 for a left-handed superluminal boost we see that

$$PT\Lambda_L(\omega, \mathbf{u} = (1,0,0)) = \begin{bmatrix} -i\gamma_s & i\beta\gamma_s & 0 & 0 \\ i\beta\gamma_s & -i\gamma_s & 0 & 0 \\ 0 & 0 & -1 & 0 \\ 0 & 0 & 0 & -1 \end{bmatrix}$$

where P is the parity operator and T is the time reversal operator. If we now apply a spatial rotation $\mathfrak{R}$ of π radians around the x axis then we obtain

$$\mathfrak{R}PT\Lambda_L(\omega, \mathbf{u} = (1,0,0))\mathfrak{R}^{-1} = \begin{bmatrix} -i\gamma_s & i\beta\gamma_s & 0 & 0 \\ i\beta\gamma_s & -i\gamma_s & 0 & 0 \\ 0 & 0 & 1 & 0 \\ 0 & 0 & 0 & 1 \end{bmatrix} \tag{2.47}$$

$$= \Lambda_R(\omega, \mathbf{u} = (1,0,0))$$

Since P and T commute with spatial rotations we find

$$\Lambda_R(\omega, \mathbf{u} = (1,0,0)) = PT\mathfrak{R}\Lambda_L(\omega, \mathbf{u} = (1,0,0))\mathfrak{R}^{-1} \tag{2.48}$$

or, more generally, performing additional spatial rotations:

$$\Lambda_R(\omega, \mathbf{u}) = PT\mathfrak{R}_u\mathfrak{R}\mathfrak{R}_w\Lambda_L(\omega, \mathbf{w})\mathfrak{R}_w^{-1}\mathfrak{R}^{-1}\mathfrak{R}_u^{-1} \tag{2.49}$$

or,

$$\Lambda_R(\omega, \mathbf{u}) = PT\mathfrak{R}_{tot}\Lambda_L(\omega, \mathbf{w})\mathfrak{R}_{tot}^{-1} \tag{2.50}$$

where **u** and **w** are unit vectors, and $\mathfrak{R}_{tot} = \mathfrak{R}_u\mathfrak{R}\mathfrak{R}_w$. Alternately,

$$\Lambda_L(\omega, \mathbf{w}) = PT\mathfrak{R}_{tot}^{-1}\Lambda_R(\omega, \mathbf{u})\mathfrak{R}_{tot} \tag{2.51}$$

or

$$\Lambda_L(\omega, \mathbf{w}) = PT\Lambda_R(\omega, \mathbf{u}') \tag{2.52}$$

for some unit vector **u'**.

Thus we have shown that PT can be used to relate left-handed and right-handed boosts in a one-to-one fashion. *The appearance of the* parity *operator P takes on great significance when we derive features of the Standard Model. The appearance of left-handed form of The*

Standard Model stems directly from the implicit parity dependence of the left-handed sector of the superluminal part of the complex Lorentz group.

For a right-handed boost the addition rule for the x-component of velocity can be computed for infinitesimal displacements in space and time:

$$\begin{aligned} v_x' = \Delta x'/\Delta t' &= (\Delta x\, \gamma_s - \Delta t\, \beta\gamma_s)/(\Delta t\, \gamma_s - \Delta x\, \beta\gamma_s) \\ &= (v_x - \beta)/(1 - \beta v_x) \end{aligned} \tag{2.53}$$

in the limit $\Delta t \rightarrow 0$ where the x component of a particle's velocity in the unprimed frame is $v_x = \Delta x/\Delta t$. Note if $v_x = 1$ then $v_x' = 1$. *Thus right-handed superluminal transformations also preserve the constancy of the speed of light in all reference frames.*

2.2.10 Inhomogeneous Left-Handed Lorentz Group Transformations

The *Left-Handed transformations of the complex Lorentz group* consist of the elements of the real Lorentz group plus left-handed superluminal boost transformations, and combinations of boosts and spatial rotations. Thus the homogeneous left-handed superluminal transformations have the general form:

$$\Lambda_L(\mathbf{v}, \boldsymbol{\theta}) = \exp[i\omega_L \hat{\mathbf{u}}\cdot\mathbf{K} + i\boldsymbol{\theta}\cdot\mathbf{J}] \tag{2.54}$$

where $\omega_L' = \omega + i\pi/2$, $\boldsymbol{\theta}$ is the angular vector, and $\mathbf{J}$ is the angular momentum operator vector. Inhomogeneous left-handed superluminal transformations, which include displacements, can be expressed as

$$\Lambda_L(\mathbf{v}, \boldsymbol{\theta}, \mathbf{d}) = \exp[i\omega_L \hat{\mathbf{u}}\cdot\mathbf{K} + i\boldsymbol{\theta}\cdot\mathbf{J} - i\mathbf{d}\cdot\mathbf{P}] \tag{2.55}$$

where $\mathbf{P}$ is the momentum operator vector and $\mathbf{d}$ is a displacement vector.

We note

$$\det \Lambda_L(\omega, \mathbf{u}) = \pm 1 \tag{2.56}$$

The ordinary Lorentz group is divided into four disjoint subgroups that are often denoted:

$$\begin{aligned} &L^{\uparrow}_{+}: \quad \det \Lambda(\omega, \mathbf{u}) = +1; \;\; \mathrm{sgn}\, \Lambda(\omega, \mathbf{u})^0{}_0 = +1 \\ &L^{\uparrow}_{-}: \quad \det \Lambda(\omega, \mathbf{u}) = -1; \;\; \mathrm{sgn}\, \Lambda(\omega, \mathbf{u})^0{}_0 = +1 \\ &L^{\downarrow}_{+}: \quad \det \Lambda(\omega, \mathbf{u}) = +1; \;\; \mathrm{sgn}\, \Lambda(\omega, \mathbf{u})^0{}_0 = -1 \end{aligned} \tag{2.57}$$

$L_-^\downarrow$: det $\Lambda(\omega, \mathbf{u}) = -1$; sgn $\Lambda(\omega, \mathbf{u})^0{}_0 = -1$

where sgn $\Lambda(\omega, \mathbf{u})^0{}_0$ is the sign of the 00 component of the $\Lambda(\omega, \mathbf{u})$ matrix. The various subgroups are related by the discrete transformations of parity P and time reversal T:

$$L_+^\uparrow \xrightarrow{P} L_-^\uparrow$$
$$L_+^\uparrow \xrightarrow{PT} L_+^\downarrow$$
$$L_+^\uparrow \xrightarrow{T} L_-^\downarrow$$

The left-handed superluminal transformations are disjoint in a somewhat different way. By eq. 2.56 the determinants are ±1. However the 0-0 matrix element of eq. 2.16 gives

$$\Lambda_L{}^0{}_0{}^2 - \Sigma_i (\Lambda_L{}^i{}_0)^2 = 1 \qquad (2.58)$$

The representation of superluminal boosts shows that each factor in eq. 2.58 is imaginary. Thus eq. 2.58 implies

$$\Sigma_i |\Lambda_L{}^i{}_0|^2 \geq 1 \qquad (2.59)$$

$$|\Lambda_L{}^0{}_0| \geq 0 \qquad \text{(not } \geq 1) \qquad (2.60)$$

where || indicates absolute value since the quantities in eq. 2.58 are squares – not in absolute value. Thus the magnitude of $\Lambda_L{}^0{}_0$ does not have a gap. Therefore left-handed superluminal transformations can be divided into two categories:

$$_L L_+\text{: det } \Lambda_L(\omega, \mathbf{u}) = +1 \qquad (2.61)$$
$$_L L_-\text{: det } \Lambda_L(\omega, \mathbf{u}) = -1$$

as one expects for complex Lorentz group transformations.[37]

Earlier we saw that under a PT transformation a left-handed superluminal transformation becomes a right handed superluminal transformation. Again, as in the left-handed case, the various disjoint pieces are related by the discrete transformations of parity P and time reversal T:

$$_L L_+ \xrightarrow{P} {}_L L_-$$
$$_L L_+ \xrightarrow{T} {}_L L_- \qquad (2.62)$$

[37] Streater (2000) p. 13.

2.2.11 Inhomogeneous Right-Handed Extended Lorentz Group

The inhomogeneous right-handed part of the complex Lorentz group[38] consists of the real Lorentz group plus right-handed superluminal transformations plus rotations and displacements that have the form:

$$\Lambda_R(\mathbf{v}, \boldsymbol{\theta}, \mathbf{d}) = \exp[i\omega_R\hat{\mathbf{u}}\cdot\mathbf{K} + i\boldsymbol{\theta}\cdot\mathbf{J} - i\mathbf{d}\cdot\mathbf{P}] \tag{2.63}$$

in general where $\omega_R = \omega - i\pi/2$.

2.2.12 General Forms of Superluminal Boosts

The group elements of the homogeneous complex Lorentz group L(C) can be expressed in terms of the group generators as

$$\Lambda_C = \exp[i(\omega_r\hat{\mathbf{u}}_r + i\omega_i\hat{\mathbf{u}}_i)\cdot\mathbf{K} + i\boldsymbol{\theta}_c\cdot\mathbf{J}] \tag{2.64}$$

where the vector $\boldsymbol{\theta}_c$ is a complex 3-vector, $\omega_r \geq 0$ and $\omega_i \geq 0$ are real numbers, and $\hat{\mathbf{u}}_r$ and $\hat{\mathbf{u}}_i$ are real normalized 3-vectors such that $\hat{\mathbf{u}}_r\cdot\hat{\mathbf{u}}_r = 1 = \hat{\mathbf{u}}_i\cdot\hat{\mathbf{u}}_i$. The generators of the homogeneous complex Lorentz group are **K**, and **J** just as for the homogeneous real Lorentz group.

We now focus on boosts because they will be crucial in the determination of the equations of motion of various types of spin ½ particles. A boost has the form

$$\Lambda_C(\mathbf{v}_c) = \exp[i\omega\hat{\mathbf{w}}\cdot\mathbf{K}] \tag{2.65}$$

where

$$\omega = (\omega_r^2 - \omega_i^2 + 2i\omega_r\omega_i\,\hat{\mathbf{u}}_r\cdot\hat{\mathbf{u}}_i)^{1/2} \tag{2.66}$$

and

$$\hat{\mathbf{w}} = (\omega_r\hat{\mathbf{u}}_r + i\omega_i\hat{\mathbf{u}}_i)/\omega \tag{2.67}$$

Since $\hat{\mathbf{u}}_r\cdot\hat{\mathbf{u}}_r = 1 = \hat{\mathbf{u}}_i\cdot\hat{\mathbf{u}}_i$ we see

$$\hat{\mathbf{w}}\cdot\hat{\mathbf{w}} = 1 \tag{2.68}$$

The complex relative velocity is

$$\mathbf{v}_c = \hat{\mathbf{w}}\tanh(\omega) \tag{2.69}$$

---

[38] **Since γ_s has branch points at $v = \pm 1$ (which corresponds to $\omega = \pm\infty$ for both the left-handed and right-handed groups) there is a cut along the real ω axis between $-\infty$ and $+\infty$ in the ω complex plane. Therefore, we note, the product of three left-handed Lorentz transformations does not yield a right-handed transformation (as might be supposed from eqs. 2.43 and 2.51) but rather a left-handed transformation on the second sheet. A transformation with $\omega + 3i\pi/2$ is not equivalent to a transformation with $\omega - i\pi/2$.**

Having placed boost transformations in the form of eq. 2.12 we can take advantage of the form of real proper orthchronous Lorentz boost transformations, eq. 2.13, and analytically continue to complex ω and complex unit vectors $\hat{\mathbf{w}}$ provided eq. 2.69 is satisfied. The resulting complex generalization will be the matrix form of proper boosts:

$$\Lambda_C(\mathbf{v}_c) = \exp[i\omega\hat{\mathbf{w}}\cdot\mathbf{K}] \equiv \Lambda_C(\omega, \hat{\mathbf{w}})$$

$$= \begin{bmatrix} \cosh(\omega) & -\sinh(\omega)\hat{w}_x & -\sinh(\omega)\hat{w}_y & -\sinh(\omega)\hat{w}_z \\ -\sinh(\omega)\hat{w}_x & 1+(\cosh(\omega)-1)\hat{w}_x^2 & (\cosh(\omega)-1)\hat{w}_x\hat{w}_y & (\cosh(\omega)-1)\hat{w}_x\hat{w}_z \\ -\sinh(\omega)\hat{w}_y & (\cosh(\omega)-1)\hat{w}_x\hat{w}_y & 1+(\cosh(\omega)-1)\hat{w}_y^2 & (\cosh(\omega)-1)\hat{w}_y\hat{w}_z \\ -\sinh(\omega)\hat{w}_z & (\cosh(\omega)-1)\hat{w}_x\hat{w}_z & (\cosh(\omega)-1)\hat{w}_y\hat{w}_z & 1+(\cosh(\omega)-1)\hat{w}_z^2 \end{bmatrix} \quad (2.70)$$

Since analytic continuations are unique, the above form for $\Lambda_C(\mathbf{v}_c)$ is well-defined and unique. It spans the complete set of proper complex Lorentz boosts.

2.3 The Four Species of Fermions

We now will study six classes of boosts that have the property that they boost from a coordinate system with real time and space coordinates to a coordinate system with either a purely real or purely imaginary time, and real, imaginary or complex spatial coordinates. These boosts produce left-handed lepton-like and "quark-like" free Dirac-like equations. They also produce right-handed lepton-like and "quark-like" free Dirac-like equations. We will discuss these Dirac-like equations in detail later. First we describe the four categories of boosts that have the property that they transform the reference frame of a particle at rest to a reference frame where the energy is either purely real or purely imaginary – the distinguishing feature of these four sets of transformations.

2.3.1 "Lepton-like" Left-Handed Boosts

If we let

$$\hat{\mathbf{u}}_i = \hat{\mathbf{u}}_r \equiv \hat{\mathbf{u}} \quad (2.71)$$

so that the vector $\hat{\mathbf{u}}_i$ is parallel to $\hat{\mathbf{u}}_r$, and let

$$\omega_i = \pi/2 \quad (2.72)$$

then $\Lambda_C(\mathbf{v}_c)$ becomes a lepton-like left-handed boost:[39]

$$\Lambda_C = \exp[i(\omega_r + i\,\pi/2)\hat{\mathbf{u}}_r\cdot\mathbf{K}] \qquad (2.73)$$

2.3.2 "Lepton-like" Right-Handed Boosts

If we let

$$\hat{\mathbf{u}}_i = -\hat{\mathbf{u}}_r \equiv -\hat{\mathbf{u}} \qquad (2.74)$$

so that the vector $\hat{\mathbf{u}}_i$ is anti-parallel to $\hat{\mathbf{u}}_r$, and

$$\omega_i = -\pi/2 \qquad (2.75)$$

then $\Lambda_C(\mathbf{v}_c)$ becomes a right-handed boost:

$$\Lambda_C = \exp[i(\omega_r - i\,\pi/2)\hat{\mathbf{u}}_r\cdot\mathbf{K}] \qquad (2.76)$$

2.3.3 "Quark-like" Left-Handed Boosts

If the real and imaginary relative vectors parts of $\hat{\mathbf{w}}$, namely $\hat{\mathbf{u}}_r$ and $\hat{\mathbf{u}}_i$, are perpendicular, $\hat{\mathbf{u}}_r\cdot\hat{\mathbf{u}}_i = 0$, then by eq. 2.66

$$\omega = (\omega_r^2 - \omega_i^2)^{1/2} \qquad (2.77)$$

Thus ω is either pure real ($\omega_r \geq \omega_i$) or pure imaginary ($\omega_r < \omega_i$). We choose ω real, and then reset

$$\omega = (\omega_r^2 - \omega_i^2)^{1/2} \rightarrow \omega' = (\omega_r^2 - \omega_i^2)^{1/2} + i\pi/2 = \omega + i\pi/2 \qquad (2.78)$$

by adding $i\pi/2$ to the ω factor in eq. 2.65 since ω is a free parameter. Then the resulting Lorentz transformation then becomes a "quark-like" left-handed boost:[40]

$$\Lambda_C = \exp[i((\omega_r^2 - \omega_i^2)^{1/2} + i\pi/2)(\omega_r\hat{\mathbf{u}}_r + i\omega_i\hat{\mathbf{u}}_i)\cdot\mathbf{K}/\omega] \qquad (2.79)$$

[39] We say "lepton-like" because we obtain a lepton-like Dirac-like equation using these boosts later. Similarly for "quark-like.'

[40] We say "quark-like" because we will later obtain a quark-like left-handed Dirac-like equation with complex spatial momentum terms using these boosts.

2.3.4 "Quark-like" Right-Handed Boosts

If the real and imaginary relative vectors parts of $\hat{\mathbf{w}}$, namely $\hat{\mathbf{u}}_r$ and $\hat{\mathbf{u}}_i$, are perpendicular, $\hat{\mathbf{u}}_r \cdot \hat{\mathbf{u}}_i = 0$, then by eq. 2.66

$$\omega = (\omega_r^2 - \omega_i^2)^{1/2} \tag{2.80}$$

Thus ω again starts out either pure real ($\omega_r \geq \omega_i$) or pure imaginary ($\omega_r < \omega_i$). In this case we also choose ω real, and then reset

$$\omega = (\omega_r^2 - \omega_i^2)^{1/2} \rightarrow \omega' = (\omega_r^2 - \omega_i^2)^{1/2} - i\pi/2 \tag{2.81}$$

by subtracting $i\pi/2$ from ω in eq. 2.65 since ω is a free parameter. The resulting Lorentz boost

$$\Lambda_C = \exp[i((\omega_r^2 - \omega_i^2)^{1/2} - i\pi/2)(\omega_r\hat{\mathbf{u}}_r + i\omega_i\hat{\mathbf{u}}_i)\cdot\mathbf{K}/\omega] \tag{2.82}$$

becomes a quark-like right-handed boost.[41]

2.3.5 "Quark-like" Boosts

If the real and imaginary relative vectors parts of $\hat{\mathbf{w}}$, namely $\hat{\mathbf{u}}_r$ and $\hat{\mathbf{u}}_i$, are perpendicular, $\hat{\mathbf{u}}_r \cdot \hat{\mathbf{u}}_i = 0$, then by eq. 2.66

$$\omega = (\omega_r^2 - \omega_i^2)^{1/2} \tag{2.83}$$

Thus ω again starts out either pure real ($\omega_r \geq \omega_i$) or pure imaginary ($\omega_r < \omega_i$). In this case choose ω_r real and use ω as defined by eq. 2.83. Then the resulting Lorentz boost is

$$\Lambda_C = \exp[i(\omega_r^2 - \omega_i^2)^{1/2}(\omega_r\hat{\mathbf{u}}_r + i\omega_i\hat{\mathbf{u}}_i)\cdot\mathbf{K}/\omega] \tag{2.84}$$

becomes a quark-like boost without handedness.[42]

2.3.6 Conventional "Dirac" Boosts

If we let

$$\hat{\mathbf{u}}_i = \hat{\mathbf{u}}_r \equiv \hat{\mathbf{u}} \tag{2.85}$$

---

[41] We say "quark-like" because we obtain a quark-like right-handed Dirac-like equation with complex spatial momentum terms using these boosts later.

[42] We again say "quark-like" because we obtain a quark-like Dirac equation with complex spatial momentum terms using these boosts later.

so that the vector $\hat{\mathbf{u}}_i$ is parallel to $\hat{\mathbf{u}}_r$, and let

$$\omega_i = 0 \tag{2.86}$$

then $\Lambda_C(\mathbf{v_c})$ becomes a Dirac boost:[43]

$$\Lambda = \exp[i\omega_r\hat{\mathbf{u}}_r\cdot\mathbf{K}] \tag{2.87}$$

This boost can be used to generate the free Dirac equation.

[43] We say "Dirac" because we obtain a Dirac equation using this boost later.

3. Dirac-like Equations and Second Quantization for the Four Species Particles and Antiparticles

Having defined coordinates, gauge fields, and covariant driuvatives, we now derive the Dirac-like equations and quantization for the four species (types) of fermions found in nature[44] by a consideration of *Lorentz group boosts. These boosts have the important feature that they transform a real-valued energy in one reference frame to a real-valued energy in the transformed reference frame.* The energies of *free* fundamental elementary particles (Ineractions are introduced later.) must be real-valued or they would not be fundamental but would be subject to decay to yet more fundamental particles – contrary to our assumption. The four species of fermions will be identified as charged leptons, neutral leptons – neutrinos, up-type quarks, and down-type quarks.[45] (Each lepton species will be seen later to have four generations. Each quark species consists of four generations – each generation consisting of three variants.[46] (We will discuss the four generations of each species, and the three variants in each generation of each quark species in a later chapter.[47])

In developing our theory of fermions (and the previously defined gauge vector bosons associated with the Reality group) we will assume all particles are quantum fields and conform to the rules of canonical quantum field theory.

We begin by defining energy and momentum as Fourier transform variables for functions of coordinates. For any function g(x) where x is a real or complex 4-vector we define its Fourier transform h(p) using an inner product of the coordinates with a 4-vector p that we call the momentum 4-vector consisting of an energy component and a spatial 3-momentum:

$$h(p) = \int d^4x \exp(ip\cdot x)\, g(x) \qquad (3.1)$$

where $p\cdot x = g_{\mu\nu}p^\mu x^\nu$.

---

[44] This chapter was extracted from Blaha (2007b).

[45] And their anti-particles.

[46] We will also discuss our proposed Dark fermion spectrum later. It is similar in overall form to the normal fermion spectrum except that each quark species generation has only *one* 'variant.'

[47] A lepton species has only one variant (color) in each generation.

3.1 Matrix Representation of Complex Lorentz Group L_C Boosts

We shall now turn our attention to L_C boosts[48] because they will be crucial in the determination of the equations of motion of various types of spin ½ particles. An L_C boost can be expressed in the form

$$\Lambda_C(\mathbf{v}_c) = \exp[i\omega\hat{\mathbf{w}}\cdot\mathbf{K}] \tag{3.2}$$

where

$$\omega = (\omega_r^2 - \omega_i^2 + 2i\omega_r\omega_i\, \hat{\mathbf{u}}_r\cdot\hat{\mathbf{u}}_i)^{1/2} \tag{3.3}$$

and

$$\hat{\mathbf{w}} = (\omega_r\hat{\mathbf{u}}_r + i\omega_i\hat{\mathbf{u}}_i)/\omega \tag{3.4}$$

Since $\hat{\mathbf{u}}_r\cdot\hat{\mathbf{u}}_r = 1 = \hat{\mathbf{u}}_i\cdot\hat{\mathbf{u}}_i$

$$\hat{\mathbf{w}}\cdot\hat{\mathbf{w}} = 1 \tag{3.5}$$

and the complex relative velocity is

$$\mathbf{v}_c = \hat{\mathbf{w}}\tanh(\omega) \tag{3.6}$$

We now analytically continue to complex ω and complex unit vectors $\hat{\mathbf{w}}$. The resulting complex generalization will be the matrix form of proper L_C boosts:

$$\Lambda_C(\mathbf{v}_c) = \exp[i\omega\hat{\mathbf{w}}\cdot\mathbf{K}] \equiv \Lambda_C(\omega, \hat{\mathbf{w}})$$

$$= \begin{bmatrix} \cosh(\omega) & -\sinh(\omega)\hat{w}_x & -\sinh(\omega)\hat{w}_y & -\sinh(\omega)\hat{w}_z \\ -\sinh(\omega)\hat{w}_x & 1 + (\cosh(\omega) - 1)\hat{w}_x^2 & (\cosh(\omega) - 1)\hat{w}_x\hat{w}_y & (\cosh(\omega) - 1)\hat{w}_x\hat{w}_z \\ -\sinh(\omega)\hat{w}_y & (\cosh(\omega) - 1)\hat{w}_x\hat{w}_y & 1 + (\cosh(\omega) - 1)\hat{w}_y^2 & (\cosh(\omega) - 1)\hat{w}_y\hat{w}_z \\ -\sinh(\omega)\hat{w}_z & (\cosh(\omega) - 1)\hat{w}_x\hat{w}_z & (\cosh(\omega) - 1)\hat{w}_y\hat{w}_z & 1 + (\cosh(\omega) - 1)\hat{w}_z^2 \end{bmatrix} \tag{3.7}$$

Since analytic continuations are unique, the above form for $\Lambda_C(\mathbf{v}_c)$ is well-defined and unique. It spans the complete set of proper L_C boosts.

---

[48] This section, and the following section, summarizes parts of 2.2.12 and 2.3 to for the reader's convenience.

3.2 Left-handed and Right-handed Parts of L_C

We now describe the Left-handed and Right-handed parts[49] of L_C boosts.

Left-handed Part of L_C

If we let

$$\hat{\mathbf{u}}_i = \hat{\mathbf{u}}_r \equiv \hat{\mathbf{u}} \tag{3.8}$$

so that the vector $\hat{\mathbf{u}}_i$ is parallel to $\hat{\mathbf{u}}_r$, and

$$\omega_i = \pi/2 \tag{3.9}$$

then $\Lambda_C(\mathbf{v}_c)$ becomes a Left-handed L_C boost:

$$\Lambda_C(\mathbf{v}_c) = \Lambda_L(\omega_r, \mathbf{u}) \tag{3.10}$$

Right-handed part of L_C

If we let

$$\hat{\mathbf{u}}_i = -\hat{\mathbf{u}}_r \equiv -\hat{\mathbf{u}} \tag{3.11}$$

so that the vector $\hat{\mathbf{u}}_i$ is anti-parallel to $\hat{\mathbf{u}}_r$, and

$$\omega_i = -\pi/2 \tag{3.12}$$

then $\Lambda_C(\mathbf{v}_c)$ becomes a Right-handed L_C boost:

$$\Lambda_C(\mathbf{v}_c) = \Lambda_R(\omega_r, \mathbf{u}) \tag{3.13}$$

as described in Blaha (2007b).

3.3 Difference between the Parts of L_C Reduced to Parallelism of $\hat{\mathbf{u}}_r$ and $\hat{\mathbf{u}}_i$

Since the Left-handed L_C part leads to the Standard Model's left-handed features, it seems that the parallel case $\hat{\mathbf{u}}_i = \hat{\mathbf{u}}_r \equiv \hat{\mathbf{u}}$ is more favored by Nature.[50] To some extent this concept

[49] The designations Left-handed and Right-handed are chosen to reflect the Left-handed and Right-handed fermion fields that will be used to construct The Standard Model later. See Blaha (2007b) for more detail.

[50] It is possible that parity violation might disappear at ultra-high energies. Then we would view the parity symmetric theory as broken to the left-handed Standard Model currently established by experiment with possible right-handed parts at higher energy.

of parallel vectors $\hat{\mathbf{u}}_i$ and $\hat{\mathbf{u}}_r$, which leads to the Left-handed L_C, is more intuitively satisfying then the anti-parallel case that leads to the Right-handed L_C part. However, a deeper reason for Nature's choice remains to be found.

3.4 Free Spin ½ Particles – Leptons & Quarks

In this section we begin by developing dynamical equations for spin ½ particles based on the L_C parts. These spin ½ particles are conventional Dirac particles (Majorana particles are also allowed but not discussed here), spin ½ tachyons, and "color" versions of both types totalling four species. We will identify leptons and quarks with these fields.[51]

3.4.1 Introduction

Tachyons are particles that move faster than the speed of light. As we saw in earlier books tachyons exist inside Black Holes, and within many current theories – particularly SuperString theories. There are also experimental indications that neutrinos are tachyons.

Attempts to create canonical tachyon quantum field theories began in the 1960's. These attempts were made within the framework of the Lorentz group and, consequently, were limited to spin 0 theories since there are no finite dimensional representations of the Lorentz group for negative m^2 except for the one-dimensional representation. None of these attempts, or attempts since then, succeeded in creating a canonically quantized spin 0 tachyon quantum field theory.[52]

In this section we will formulate a free spin ½ tachyon Quantum Field Theory. We choose to develop a normal spin ½ theory first. Then we develop a free spin ½ tachyon theory because, as we will see, spin ½ tachyon particles (quarks and leptons) play an extraordinary role in the Standard Model.

We will develop our spin ½ tachyon theory from the "ground up" by applying a Left-Handed L_C boost to the Dirac equation, and its Dirac spinor wave function, for a particle at rest. This procedure will give a tachyon spinor wave function, and the momentum space tachyon equation equivalent of the Dirac equation. Then we will obtain the coordinate space tachyon Dirac equation, define a lagrangian, and proceed to create a canonical quantum field theory for spin ½ tachyons.

---

[51] We use the notation of Bjorken (1965) because of its clarity. It is similar to that of more recent books such as Kaku (1993).

[52] Except Blaha (2006).

3.4.2 First Step - Deriving the Conventional Dirac Equation

In this section we will review a method of obtaining the equation of motion of a particle using a free Dirac equation that is obtained by a Lorentz boost of a spinor wave function[53] of a particle at rest.

In the case of a Lorentz transformation the 4×4 matrix form of a Lorentz transformation of Dirac matrices is

$$S^{-1}(\Lambda(v))\gamma^{\nu}S(\Lambda(v)) = \Lambda^{\nu}{}_{\mu}(v)\gamma^{\mu} \qquad (3.14)$$

where $S(\Lambda(v))$ is

$$S(\Lambda(v)) = \exp(-i\omega\sigma_{0i}v_i/(2|\mathbf{v}|)) = \exp(-\omega\gamma^0\boldsymbol{\gamma}\cdot\mathbf{v}/(2|\mathbf{v}|))$$
$$= \cosh(\omega/2)I + \sinh(\omega/2)\gamma^0\boldsymbol{\gamma}\cdot\mathbf{p}/|\mathbf{p}| \qquad (3.15)$$

with $\omega = \text{arctanh}(|\mathbf{v}|)$, $\cosh(\omega/2) = [(E+m)/(2m)]^{1/2}$ and $\sinh(\omega/2) = |\mathbf{p}|[2m(E+m)]^{-1/2}$. Also

$$S^{-1}(\Lambda(v)) = \gamma^0 S^{\dagger}(\Lambda(v))\gamma^0 = \exp(\omega\gamma^0\boldsymbol{\gamma}\cdot\mathbf{v}/(2|\mathbf{v}|))$$
$$= \cosh(\omega/2)I - \sinh(\omega/2)\gamma^0\boldsymbol{\gamma}\cdot\mathbf{p}/|\mathbf{p}| \qquad (3.16)$$

In constructing fermion dynamical equations *we shall assume that they are linear in derivatives* (although a quadratic form is possible.) We will use the sixteen 4×4 Dirac matrices. Since by theorem[54] all 4 × 4 γ matrices are equivalent up to a unitary transformation we can rotate any constant matrix into a multiple of γ^0 without loss of generality.

We begin by defining a generic positive energy plane wave solution of the Dirac equation for a normal fermion particle at rest with rest energy m as

$$\psi(x) = e^{-imt}w(0) \qquad (3.17)$$

with w(0) a four component spinor column vector. *For a free particle at rest, we set the rest mass-energy $m = m_0$.* The wave function satisfies the momentum space Dirac equation for a fermion at rest:

$$(m\gamma^0 - m)e^{-imt}w(0) = 0 \qquad (3.18)$$

Subsequently we will use a similar procedure to construct the free tachyonic Dirac equation. If we now apply $S(\Lambda(v))$ we find

$$0 = S(\Lambda(v))(m\gamma^0 - m)e^{-imt}w(0) = [mS(\Lambda(v))\gamma^0 S^{-1}(\Lambda(v)) - m]S(\Lambda(v))w(0)$$

---

[53] The spinor wave function of a particle at rest is a 4-vector of the 4×4 matrix representation of 4-valued Asynchronous Logic.

[54] R. H. Good, Rev. Mod. Phys., **27**, 187 (1955).

A straightforward evaluation shows

$$mS(\Lambda(v))\gamma^0 S^{-1}(\Lambda(v)) = g_{\mu\nu}p^\mu\gamma^\nu = \not{p} \tag{3.19}$$

where $p^0 = (p^2 + m^2)^{1/2}$, $\mathbf{p} = \gamma m\mathbf{v}$, and $p = |\mathbf{p}|$. In addition

$$S(\Lambda(v))w(0) = w(p) \tag{3.20}$$

is a positive energy Dirac spinor. Therefore the Dirac equation for a fermion in motion in momentum space has the form:

$$(\not{p} - m)e^{-ip\cdot x}w(p) = 0 \tag{3.21}$$

where the exponential factor, mt, is also boosted to p·x. Eq. 3.21 implies the well-known free, coordinate space Dirac equation:

$$(i\gamma^\mu\partial/\partial x^\mu - m)\psi(x) = 0 \tag{3.22}$$

3.4.3 Derivation of the Left-Handed Tachyon Dirac Equation

The Left-handed boost has the form:

$$\Lambda_L(\omega, \mathbf{u}) = \Lambda(\omega + i\pi/2, \mathbf{u}) = \exp[i\omega_L\hat{\mathbf{u}}\cdot\mathbf{K}] \tag{3.23}$$

where $\omega_L = \omega + i\pi/2$ and

$$\begin{aligned}\cosh(\omega_L) &= i\sinh(\omega) = -\gamma = i\gamma_s \\ \sinh(\omega_L) &= i\cosh(\omega) = -\beta\gamma = i\beta\gamma_s\end{aligned} \tag{3.24}$$

with, $\beta = v > 1$, $\gamma_s = (\beta^2 - 1)^{-1/2}$, and $\boldsymbol{\omega} \geq \mathbf{0}$. Thus

$$\begin{aligned}\sinh(\omega) &= \gamma_s \\ \cosh(\omega) &= \beta\gamma_s\end{aligned} \tag{3.25}$$

The corresponding spinor transformation is:

$$\begin{aligned}S_L(\Lambda_L(\omega, \mathbf{u})) &= \exp(-i\omega_L\sigma_{0i}v_i/(2|\mathbf{v}|)) = \exp(-\omega_L\gamma^0\boldsymbol{\gamma}\cdot\mathbf{v}/(2|\mathbf{v}|)) \\ &= \cosh(\omega_L/2)I + \sinh(\omega_L/2)\gamma^0\boldsymbol{\gamma}\cdot\mathbf{p}/|\mathbf{p}|\end{aligned} \tag{3.26}$$

The inverse transformation is

$$S_L^{-1}(\Lambda_L(\omega, \mathbf{u})) = \gamma^2\gamma^0 K^{-1} S_L^\dagger K \gamma^0\gamma^2 = \gamma^2\gamma^0 S_L{}^T \gamma^0\gamma^2 = \exp(\omega_L\gamma^0\boldsymbol{\gamma}\cdot\mathbf{v}/(2|\mathbf{v}|))$$
$$= \cosh(\omega_L/2)I - \sinh(\omega_L/2)\gamma^0\boldsymbol{\gamma}\cdot\mathbf{p}/|\mathbf{p}| \qquad (3.27)$$

where the superscript T denotes the transpose and K is the complex conjugation operator (that also appears in the time-reversal operator). Note that S_L is not unitary just as the equivalent spinor Lorentz transformation $S(\Lambda(v))$ is not unitary.

We can now apply a left-handed superluminal transformation to the generic positive energy plane wave solution of the Dirac equation for a particle of mass m at rest. The result is

$$0 = S_L(\Lambda_L(\omega, \mathbf{u}))(m\gamma^0 - m)e^{-imt}w(0)$$
$$= [mS_L\gamma^0 S_L^{-1} - m]e^{-imt}S_L w(0)$$

where $S_L = S_L(\Lambda_L(\omega, \mathbf{u}))$. After some algebra

$$mS_L\gamma^0 S_L^{-1} = m[\cosh(\omega_L)\gamma^0 - \sinh(\omega_L)\boldsymbol{\gamma}\cdot\mathbf{p}/|\mathbf{p}|]$$
$$= i\gamma^0 E - i\boldsymbol{\gamma}\cdot\mathbf{p} = i\not{p} \qquad (3.28)$$

using the tachyon energy and momentum expressions

$$\mathbf{p} = m\mathbf{v}\gamma_s \qquad\qquad E = m\gamma_s \qquad (3.29)$$

Also

$$S_L w(0) = w_T(p) \qquad (3.30)$$

is a tachyon spinor. See Appendix 3-A (at the end of this section) for a discussion of tachyon spinors.

The momentum space tachyonic Dirac equation is

$$(i\not{p} - m)e^{ip\cdot x}w_T(p) = 0 \qquad (3.31)$$

where $p\cdot x = Et - \mathbf{p}\cdot\mathbf{x}$ after performing a corresponding left-handed superluminal coordinate transformation in the exponential factor. Thus a positive energy wave is transformed into a negative energy wave by the superluminal transformation.

If we apply $i\not{p}$ to we find the tachyon mass condition is satisfied

$$-E^2 + \mathbf{p}^2 = m^2 \qquad (3.32)$$

Transforming back to coordinate space we obtain the *tachyon Dirac equation*:

$$(\gamma^\mu\partial/\partial x^\mu - m)\psi_T(x) = 0 \tag{3.33}$$

The "missing" factor of i in the first term of eq. 3.33 requires the lagrangian to be different from the conventional Dirac lagrangian in order for the lagrangian to be real. The simplest, physically acceptable, free spin ½ tachyon lagrangian density is:

$$\mathcal{L}_T = \psi_T^{\ S}(\gamma^\mu\partial/\partial x^\mu - m)\psi_T(x) \tag{3.34}$$

where

$$\psi_T^{\ S} = \psi_T^{\ \dagger}\, i\gamma^0\gamma^5 \tag{3.35}$$

The corresponding action is

$$I = \int d^4x \mathcal{L}_T \tag{3.36}$$

Appendix 3-B of Blaha (2007b) proves I is real. The Hamiltonian density is

$$\mathcal{H} = \pi_T\dot{\psi}_T - \mathcal{L} = i\psi_T^{\ \dagger}\gamma^5(\boldsymbol{\alpha}\cdot\nabla + \beta m)\psi_T = -i\psi_T^{\ \dagger}\gamma^5\dot{\psi}_T \tag{3.37}$$

using the tachyon Dirac equation to obtain the last equality. The reader will note that the tachyon hamiltonian is hermitean by explicit calculation up to an irrelevant total spatial divergence.

Probability Conservation Law

The tachyon Dirac equation implies a probability conservation law:

$$\partial\rho_5/\partial t = \nabla\cdot\mathbf{j}_5 \tag{3.38}$$

where

$$\rho_5 = \psi_T^{\ \dagger}\gamma^5\psi_T \qquad\qquad \mathbf{j}_5 = \psi_T^{\ \dagger}\gamma^5\boldsymbol{\alpha}\psi_T \tag{3.39}$$

We are thus led to define the conserved axial charge Q_5

$$Q_5 = \int d^3x\, \psi_T^{\ \dagger}\gamma^5\psi_T \tag{3.40}$$

Energy-Momentum Tensor

The tachyon energy-momentum tensor is

$$\mathfrak{T}_{T\mu\nu} = - g_{\mu\nu} \mathfrak{L}_T + \partial\mathfrak{L}_T/\partial(\partial\psi_T/\partial x_\mu)\, \partial\psi_T/\partial x^\nu \qquad (3.41)$$
$$= i\psi_T^{\dagger}\gamma^0\gamma^5\gamma_\mu\partial\psi_T/\partial x^\nu \qquad (3.42)$$

and thus the conserved energy and momentum are

$$P^0 = H = \int d^3x\, \mathfrak{T}_T^{00} = i\int d^3x \psi_T^{\dagger}\gamma^5(\boldsymbol{\alpha}\cdot\nabla + \beta m)\psi_T \qquad (3.43)$$

and

$$P^i = \int d^3x\, \mathfrak{T}_T^{0i} = - i\int d^3x\, \psi_T^{\dagger}\gamma^5\partial\psi_T/\partial x_i \qquad (3.44)$$

Both the energy and momentum differ significantly from the corresponding quantities for conventional Dirac fields.

3.4.4 Tachyon Canonical Quantization

Having defined a suitable tachyon lagrangian we can now proceed to its canonical quantization. The conjugate momentum can be calculated from the above lagrangian density:

$$\pi_{Ta} = \partial\mathfrak{L}_T/\partial\dot{\psi}_{Ta} \equiv \partial\mathfrak{L}_T/\partial(\partial\psi_{Ta}/\partial t) = -i(\psi_T^{\dagger}\gamma^5)_a \qquad (3.45)$$

The resulting non-zero, canonical anti-commutation relations are

$$\{\pi_{Ta}(x), \psi_{Tb}(x')\} = i\,\delta_{ab}\,\delta^3(x - x')$$

or

$$\{\psi_{T}{}^{\dagger}{}_a(x), \psi_{Tb}(x')\} = - [\gamma^5]_{ab}\,\delta^3(x - x') \qquad (3.46)$$

At this point we might attempt to complete the canonical quantization procedure in the conventional manner by fourier expanding the quantum field and specifying anti-commutation relations for the fourier component amplitudes. However the incompleteness of the set of plane waves, which are limited by the restriction $|p| \geq m$*, causes the anti-commutator of the fields not to yield a* $\delta^3(x - x')$*. Thus the conventional approach fails to yield the required anti-commutation relations.55*

Other approaches: 1) decompose the tachyon field into left-handed and right-handed parts and then second quantize each part; and 2) second quantize in light-front coordinates ($x^{\pm} = (x^0 \pm x^3)/\sqrt{2}$). These approaches also both fail.[56]

---

[55] See G. Feinberg, Phys. Rev. **159**, 1089 (1967) for example.
[56] See the first edition Blaha (2006) where these possibilities were considered and found to fail.

The only approach that does succeed[57] *is to decompose the tachyon field into left-handed and right-handed parts and then second quantize in light-front coordinates. We follow that procedure in the following subsections.*

Separation into Left-Handed and Right-Handed Fields

We will use a transformed set of Dirac matrices to develop our left-handed and right-handed tachyon formulations:

$$\gamma^0 = \begin{bmatrix} 0 & -I \\ -I & 0 \end{bmatrix} \qquad \gamma^i = \begin{bmatrix} 0 & \sigma_i \\ -\sigma_i & 0 \end{bmatrix} \qquad \gamma^5 = \begin{bmatrix} I & 0 \\ 0 & -I \end{bmatrix} \tag{3.47}$$

which are obtained from the Dirac matrices by applying the unitary transformation $U = 2^{-1/2}(I + \gamma^5\gamma^0)$. *I is the 4×4 identity matrix in eq. 3.47.* The γ^5 chirality operator's eigenvalues define handedness: +1 corresponds to right-handed; and –1 corresponds to left-handed:

$$\gamma^5\psi_L = -\psi_L \qquad\qquad \gamma^5\psi_R = \psi_R \tag{3.48}$$

Consequently, we can define left-handed and right-handed tachyon fields with the projection operators:

$$\begin{aligned} C^{\pm} &= ½(I \pm \gamma^5) \\ C^{+} + C^{-} &= I \\ C^{\pm\,2} &= C^{\pm} \\ C^{+}C^{-} &= 0 \end{aligned} \tag{3.49}$$

with the result

$$\begin{aligned} \psi_{TL} &= C^{-}\psi_T \\ \psi_{TR} &= C^{+}\psi_T \end{aligned} \tag{3.50}$$

We can calculate the commutation relations of the left-handed and right-handed tachyon fields from eq. 3.46 by pre-multiplying and post-multiplying by $½(1 - \gamma^5)$ and $½(1 + \gamma^5)$. The results are:

$$\{\psi_{TLa}^{\dagger}(x), \psi_{TLb}(x')\} = ½(1 - \gamma^5)_{ab}\,\delta^3(x - x') \tag{3.51}$$

---

[57] Blaha (2006) discusses this case in detail.

$$\{\psi_{TRa}{}^{\dagger}(x), \psi_{TRb}(x')\} = -\tfrac{1}{2}(1+\gamma^5)_{ab}\,\delta^3(x - x') \qquad (3.52)$$

$$\{\psi_{TLa}{}^{\dagger}(x), \psi_{TRb}(x')\} = \{\psi_{TRa}{}^{\dagger}(x), \psi_{TLb}(x')\} = 0 \qquad (3.53)$$

The lagrangian density above decomposes into left-handed and right-handed parts:

$$\mathcal{L}_T = \psi_{TL}{}^{\dagger}\gamma^0 i\gamma^\mu\partial_\mu\psi_{TL} - \psi_{TR}{}^{\dagger}\gamma^0 i\gamma^\mu\partial_\mu\psi_{TR} - im[\psi_{TR}{}^{\dagger}\gamma^0\psi_{TL} - \psi_{TL}{}^{\dagger}\gamma^0\psi_{TR}] \qquad (3.54)$$

Further Separation into + and – Light-Front Fields

There have been many studies of light-front (infinite momentum frame) physics in the past forty years.[58] Light-front coordinates *cannot* be obtained by a Lorentz transformation, or by a superluminal transformation, from a standard set of coordinate system variables even in a limiting sense. Instead they are a defined set of variables that have been used to develop quantum field theories that have been shown to be equivalent to quantum field theories based on conventional coordinates. In particular, light-front quantum field theories have been shown to yield fully Lorentz covariant S matrix elements that are the same as S matrix elements calculated in the conventional way.

Light-front variables can be defined by:

$$x^{\pm} = (x^0 \pm x^3)/\surd 2 \qquad (3.55)$$
$$\partial/\partial x^{\pm} \equiv \partial^{\mp} \equiv (\partial/\partial x^0 \pm \partial/\partial x^3)/\surd 2$$

with the "transverse" coordinate variables, x^1 and x^2, unchanged.

The inner product of two 4-vectors has the form

$$x{\cdot}y = x^+y^- + y^+x^- - x^1y^1 - x^2y^2 \qquad (3.56)$$

and the light-front definition of Dirac matrices is:

$$\gamma^{\pm} = (\gamma^0 \pm \gamma^3)/\surd 2 \qquad (3.57)$$

with transverse matrices γ^1 and γ^2 defined as usual. Note the useful identity:

---

[58] L. Susskind, Phys. Rev. **165**, 1535 (1968); K. Bardakci and M. B. Halpern Phys. Rev. **176**, 1686 (1968), S. Weinberg, Phys. Rev. **150**, 1313 (1966); J. Kogut and D. Soper, Phys. Rev. **D1**, 2901 (1970); J. D. Bjorken, J. Kogut, and D. Soper, Phys. Rev. **D3**, 1382 (1971); R. A. Neville and F. Rohrlich, Nuov. Cim. **A1**, 625 (1971); F. Rohrlich, Acta Phys Austr. Suppl. **8**, 277 (1971); S-J Chang, R. Root, and T-M Yan, Phys. Rev. **D7**, 1133 (1973); S-J Chang, and T-M Yan, Phys. Rev. **D7**, 1147 (1973); T-M Yan, Phys. Rev. **D7**, 1761 (1973); T-M Yan, Phys. Rev. **D7**, 1780 (1973); C. Thorn, Phys. Rev. **D19**, 639 (1979); and references therein.

$$\gamma^{\pm 2} = 0$$

We define "+" and "–" tachyon fields with the projection operators:

$$R^{\pm} = ½(I \pm \gamma^0\gamma^3) \tag{3.58}$$

They are:

Left-handed, ± light-front fields: $\psi_{TL}{}^{\pm} = R^{\pm}C^{-}\psi_T$

Right-handed, ± light-front fields: $\psi_{TR}{}^{\pm} = R^{\pm}C^{+}\psi_T$ (3.59)

Now if we transform to light-front variables and fields as above we obtain the light-front free tachyon lagrangian:

$$\begin{aligned}\mathcal{L}_T = {} & 2^{½}\psi_{TL}{}^{+\dagger}i\partial^{-}\psi_{TL}{}^{+} + 2^{½}\psi_{TL}{}^{-\dagger}i\partial^{+}\psi_{TL}{}^{-} - \psi_{TL}{}^{+\dagger}\gamma^0 i\gamma^j\partial^j\psi_{TL}{}^{-} - \psi_{TL}{}^{-\dagger}\gamma^0 i\gamma^j\partial^j\psi_{TL}{}^{+} - \\ & - 2^{½}\psi_{TR}{}^{+\dagger}i\partial^{-}\psi_{TR}{}^{+} - 2^{½}\psi_{TR}{}^{-\dagger}i\partial^{+}\psi_{TR}{}^{-} + \psi_{TR}{}^{+\dagger}\gamma^0 i\gamma^j\partial^j\psi_{TR}{}^{-} + \psi_{TR}{}^{-\dagger}\gamma^0 i\gamma^j\partial^j\psi_{TR}{}^{+} - \\ & - im[\psi_{TR}{}^{+\dagger}\gamma^0\psi_{TL}{}^{-} - \psi_{TL}{}^{+\dagger}\gamma^0\psi_{TR}{}^{-} + \psi_{TR}{}^{-\dagger}\gamma^0\psi_{TL}{}^{+} - \psi_{TL}{}^{-\dagger}\gamma^0\psi_{TR}{}^{+}]\end{aligned} \tag{3.60}$$

with an implied sum over j = 1, 2. In contrast to the light-front tachyon lagrangian we note the corresponding light-front "normal" Dirac fermion lagrangian is

$$\begin{aligned}\mathcal{L}_{Dirac} = {} & 2^{½}\psi_{L}{}^{+\dagger}i\partial^{-}\psi_{L}{}^{+} + 2^{½}\psi_{L}{}^{-\dagger}i\partial^{+}\psi_{L}{}^{-} - \psi_{L}{}^{+\dagger}\gamma^0 i\gamma^j\partial^j\psi_{L}{}^{-} - \psi_{L}{}^{-\dagger}\gamma^0 i\gamma^j\partial^j\psi_{L}{}^{+} - \\ & - 2^{½}\psi_{R}{}^{+\dagger}i\partial^{-}\psi_{R}{}^{+} + 2^{½}\psi_{R}{}^{-\dagger}i\partial^{+}\psi_{R}{}^{-} - \psi_{R}{}^{+\dagger}\gamma^0 i\gamma^j\partial^j\psi_{R}{}^{-} - \psi_{R}{}^{-\dagger}\gamma^0 i\gamma^j\partial^j\psi_{R}{}^{+} - \\ & - im[\psi_{R}{}^{+\dagger}\gamma^0\psi_{L}{}^{-} + \psi_{L}{}^{+\dagger}\gamma^0\psi_{R}{}^{-} + \psi_{R}{}^{-\dagger}\gamma^0\psi_{L}{}^{+} + \psi_{L}{}^{-\dagger}\gamma^0\psi_{R}{}^{+}]\end{aligned} \tag{3.61}$$

The difference in signs between these lagrangians will turn out to be a crucial factor in the derivation of features of the Standard Model later.

Returning to the tachyon lagrangian eq. 3.60 we obtain equations of motion through the standard variational techniques:

$$\begin{aligned} 2^{½}i\partial^{-}\psi_{TL}{}^{+} - \gamma^0 i\gamma^j\partial^j\psi_{TL}{}^{-} + im\gamma^0\psi_{TR}{}^{-} &= 0 \\ 2^{½}i\partial^{-}\psi_{TR}{}^{+} - \gamma^0 i\gamma^j\partial^j\psi_{TR}{}^{-} + im\gamma^0\psi_{TL}{}^{-} &= 0 \\ 2^{½}i\partial^{+}\psi_{TL}{}^{-} - \gamma^0 i\gamma^j\partial^j\psi_{TL}{}^{+} + im\gamma^0\psi_{TR}{}^{+} &= 0 \\ 2^{½}i\partial^{+}\psi_{TR}{}^{-} - \gamma^0 i\gamma^j\partial^j\psi_{TR}{}^{+} + im\gamma^0\psi_{TL}{}^{+} &= 0 \end{aligned} \tag{3.62}$$

Eqs. 3.62 show that ψ_{TL}^{-} and ψ_{TR}^{-} are dependent fields that are functions of ψ_{TL}^{+} and ψ_{TR}^{+} on the light-front where x^{+} equals a constant. They can be expressed in an integral form as well. (The independent fields ψ_{TL}^{+} and ψ_{TR}^{+} play a fundamental role in tachyon theory and are used to define "in" and "out" tachyon states in perturbation theory.)

The conjugate momenta are

$$\pi_{TL}^{+} = \partial\mathcal{L}/\partial(\partial^{-}\psi_{TL}^{+}) = 2^{\frac{1}{2}}i\psi_{TL}^{+\dagger} \tag{3.63}$$
$$\pi_{TL}^{-} = \partial\mathcal{L}/\partial(\partial^{-}\psi_{TL}^{-}) = 0$$
$$\pi_{TR}^{+} = \partial\mathcal{L}/\partial(\partial^{-}\psi_{TR}^{+}) = -2^{\frac{1}{2}}i\psi_{TR}^{+\dagger} \tag{3.64}$$
$$\pi_{TR}^{-} = \partial\mathcal{L}/\partial(\partial^{-}\psi_{TR}^{-}) = 0$$

Quantization on surfaces of constant x^{+} (light-front surfaces) has been shown to support satisfactory formulations of Quantum Electrodynamics and other quantum field theories. Thus x^{+} plays the role of the "time" variable in light-front quantized theories. So we will define canonical equal x^{+} anti-commutation relations for spin ½ tachyons.

The resulting canonical equal-light-front ($x^{+} = y^{+}$) anti-commutation relations of the independent fields are:

$$\{\psi_{TL}^{+\dagger}{}_{a}(x), \psi_{TL}^{+}{}_{b}(y)\} = 2^{-1}[C^{-}R^{+}]_{ab}\,\delta(x^{-} - y^{-})\delta^{2}(x - y) \tag{3.65}$$
$$\{\psi_{TR}^{+\dagger}{}_{a}(x), \psi_{TR}^{+}{}_{b}(y)\} = -2^{-1}[C^{+}R^{+}]_{ab}\,\delta(x^{-} - y^{-})\delta^{2}(x - y) \tag{3.66}$$
$$\{\psi_{TL}^{+\dagger}{}_{a}(x), \psi_{TR}^{+}{}_{b}(y)\} = \{\psi_{TR}^{+\dagger}{}_{a}(x), \psi_{TL}^{+}{}_{b}(y)\} = 0 \tag{3.67}$$
$$\{\psi_{TL}^{+}{}_{a}(x), \psi_{TR}^{+}{}_{b}(y)\} = \{\psi_{TR}^{+\dagger}{}_{a}(x), \psi_{TL}^{+\dagger}{}_{b}(y)\} = 0 \tag{3.68}$$

where the factors of 2^{-1} are the result of the $2^{\frac{1}{2}}$ factor in eqs. 3.63 and 3.64, and the factor of $2^{-\frac{1}{2}}$ in the definition of x^{-} above.

If we compare eqs. 3.65 and 3.66 with the corresponding anti-commutation relations of *conventional* <u>*Dirac*</u> quantum fields:

$$\{\psi_{L}^{+\dagger}{}_{a}(x), \psi_{L}^{+}{}_{b}(y)\} = 2^{-1}[C^{-}R^{+}]_{ab}\,\delta(x^{-} - y^{-})\delta^{2}(x - y) \tag{3.69}$$
$$\{\psi_{R}^{+\dagger}{}_{a}(x), \psi_{R}^{+}{}_{b}(y)\} = 2^{-1}[C^{+}R^{+}]_{ab}\,\delta(x^{-} - y^{-})\delta^{2}(x - y) \tag{3.70}$$

we see that the right-handed tachyon anti-commutation relation has a minus sign relative to the corresponding right-handed conventional anti-commutation relation. The right-handed tachyon anti-commutation relation with its minus sign will require compensating minus signs in its creation and annihilation Fourier component operators' anti-commutation relations.

The sign differences between the lagrangian terms in eqs. 3.63 and 3.64 ultimately lead to parity violating features in the Standard Model lagrangian and thus resolve the long-standing question:

Why parity violation? Answer: Nature preferentially chooses the Left-handed part of the complex Lorentz group..

Left-Handed Tachyons

The free, "+" light-front, left-handed tachyon wave function Fourier expansion is:

$$\psi_{TL}{}^{+}(x) = \sum_{\pm s}\int d^2p dp^+ N_{TL}{}^{+}(p)\theta(p^+)[b_{TL}{}^{+}(p, s)u_{TL}{}^{+}(p, s)e^{-ip\cdot x} + d_{TL}{}^{+\dagger}(p, s)v_{TL}{}^{+}(p, s)e^{+ip\cdot x}] \quad (3.71)$$

and its hermitean conjugate is

$$\psi_{TL}{}^{+\dagger}(x) = \sum_{\pm s}\int d^2p dp^+ N_{TL}{}^{+}(p)\theta(p^+) [b_{TL}{}^{+\dagger}(p, s)u_{TL}{}^{+\dagger}(p,s)e^{+ip\cdot x} + d_{TL}{}^{+}(p, s)v_{TL}{}^{+\dagger}(p, s)e^{-ip\cdot x}] \quad (3.72)$$

where † indicates hermitean conjugate, where

$$N_{TL}{}^{+}(p) = [2m|\mathbf{p}|/((2\pi)^3(p^+(p^+ - p^-) + p_\perp{}^2))]^{1/2} \quad (3.73)$$

where the anti-commutation relations of the Fourier coefficient operators are

$$\begin{aligned}
&\{b_{TL}{}^{+}(q,s), b_{TL}{}^{+\dagger}(p,s')\} = \delta_{ss'}\delta^2(\mathbf{q} - \mathbf{p})\delta(q^+ - p^+) \\
&\{d_{TL}{}^{+}(q,s), d_{TL}{}^{+\dagger}(p,s')\} = \delta_{ss'}\delta^2(\mathbf{q} - \mathbf{p})\delta(q^+ - p^+) \\
&\{b_{TL}{}^{+}(q,s), b_{TL}{}^{+}(p,s')\} = \{d_{TL}{}^{+}(q,s), d_{TL}{}^{+}(p,s')\} = 0 \\
&\{b_{TL}{}^{+\dagger}(q,s), b_{TL}{}^{+\dagger}(p,s')\} = \{d_{TL}{}^{+\dagger}(q,s), d_{TL}{}^{+\dagger}(p,s')\} = 0 \\
&\{b_{TL}{}^{+}(q,s), d_{TL}{}^{+\dagger}(p,s')\} = \{d_{TL}{}^{+}(q,s), b_{TL}{}^{+\dagger}(p,s')\} = 0 \\
&\{b_{TL}{}^{+\dagger}(q,s), d_{TL}{}^{+\dagger}(p,s')\} = \{d_{TL}{}^{+}(q,s), b_{TL}{}^{+}(p,s')\} = 0
\end{aligned} \quad (3.74)$$

and where the spinors are

$$\begin{aligned}
&u_{TL}{}^{+}(p, s) = C^- R^+ S_L(\Lambda_L(\mathbf{p}))w^1(0) \\
&u_{TL}{}^{+}(p, -s) = C^- R^+ S_L(\Lambda_L(\mathbf{p}))w^2(0) \\
&v_{TL}{}^{+}(p, s) = C^- R^+ S_L(\Lambda_L(\mathbf{p}))w^3(0) \\
&v_{TL}{}^{+}(p, -s) = C^- R^+ S_L(\Lambda_L(\mathbf{p}))w^4(0) \\
&u_{TL}{}^{+\dagger}(p, s) = w^{1T}(0)S_L{}^{\dagger}(\Lambda_L(\mathbf{p}))R^+C^-
\end{aligned} \quad (3.75)$$

$$u_{TL}{}^{+\dagger}(p, -s) = w^{2T}(0)S_L{}^{\dagger}(\Lambda_L(\mathbf{p}))R^+C^-$$
$$v_{TL}{}^{+\dagger}(p, s) = w^{3T}(0)S_L{}^{\dagger}(\Lambda_L(\mathbf{p}))R^+C^-$$
$$v_{TL}{}^{+\dagger}(p, -s) = w^{4T}(0)S_L{}^{\dagger}(\Lambda_L(\mathbf{p}))R^+C^-$$

where the superscript "T" indicates the transpose. (These spinors are described in Appendix 3-A.)

The canonical left-handed, light-front anti-commutation relation results in:

$$\{\psi_{TL}{}^{+}{}_{a}(x), \psi_{TL}{}^{+\dagger}{}_{b}(y)\} = \sum_{\pm s,s'} \int d^2pdp^+\int d^2p'dp'^+ N_{TL}{}^{+}(p)N_{TL}{}^{+}(p')\theta(p^+)\theta(p'^+)\cdot$$

$$\cdot[\{b_{TL}{}^{+\dagger}(p',s'),b_{TL}{}^{+}(p,s)\}u_{TL}{}^{+}{}_{a}(p,s)u_{TL}{}^{+\dagger}{}_{b}(p',s')e^{+ip'\cdot y - ip\cdot x} +$$

$$+ \{d_{TL}{}^{+}(p',s'),d_{TL}{}^{+\dagger}(p,s)\}v_{TL}{}^{+}{}_{a}(p,s)v_{TL}{}^{+\dagger}{}_{b}(p',s')e^{-ip'\cdot y + ip\cdot x}]$$

$$= \sum_{\pm s}\int d^2pdp^+ N_{TL}{}^{+2}(p)\theta(p^+)[u_{TL}{}^{+}{}_{a}(p,s)u_{TL}{}^{+\dagger}{}_{b}(p,s)e^{+ip\cdot(y-x)} + $$
$$+ v_{TL}{}^{+}{}_{a}(p,s)v_{TL}{}^{+\dagger}{}_{b}(p,s)e^{-ip\cdot(y-x)}]$$

$$= -i\int d^2pdp^+\theta(p^+)N_{TL}{}^{+2}(p)(2m|\mathbf{p}|)^{-1}\{[C^-R^+(i\not{p} - m)\boldsymbol{\gamma}\cdot\mathbf{p}R^+C^-]_{ab}e^{+ip\cdot(y-x)} +$$
$$+ [C^-R^+(i\not{p} + m)\boldsymbol{\gamma}\cdot\mathbf{p}R^+C^-]_{ab}e^{-ip\cdot(y-x)}\}$$

$$= -i\int d^2p_\perp\int_0^\infty dp^+N_{TL}{}^{+2}(p)\{[C^-R^+(ip^+(p^+ - p^-) + ip_\perp{}^2 - mp_\perp\cdot\gamma_\perp)C^-]_{ab}e^{+ip^+(y^- - x^-) - ip_\perp\cdot(y_\perp - x_\perp)} -$$
$$- [C^-R^+(-ip^+(p^+ - p^-) - ip_\perp{}^2 - mp_\perp\cdot\gamma_\perp)C^-]_{ab}e^{-ip^+(y^- - x^-) + ip_\perp\cdot(y_\perp - x_\perp)}\}/(2m|\mathbf{p}|)$$

$$= \int d^2p_\perp\int_{-\infty}^{\infty} dp^+N_{TL}{}^{+2}(p)[C^-R^+(p^+(p^+ - p^-) + p_\perp{}^2)]_{ab}\, e^{+ip^+(y^- - x^-) - ip_\perp\cdot(y_\perp - x_\perp)}/(2m|\mathbf{p}|)$$

upon letting $p^+ \rightarrow -p^+$ and $\mathbf{p}_\perp \rightarrow -\mathbf{p}_\perp$ in the second term after using $N_{TL}{}^{+2}(p)(p^+(p^+ - p^-) + p_\perp{}^2) = 1$. The result

$$= \tfrac{1}{2}\int d^2p_\perp\int_{-\infty}^{\infty} dp^+(2\pi)^{-3}[C^-R^+]_{ab}e^{+ip^+(y^- - x^-) - ip_\perp\cdot(y_\perp - x_\perp)}$$

$$= 2^{-1}[C^-R^+]_{ab}\,\delta(y^- - x^-)\delta^2(\mathbf{y} - \mathbf{x}) \qquad (3.76)$$

Therefore we have left-handed, light-front quantized tachyons with canonical commutation relations and localized tachyons. As a result we have a canonical Tachyon Quantum Field Theory unlike previous efforts.

Right-Handed Tachyons

The case of right-handed tachyons is similar to the left-handed case with only two differences: a minus sign in the creation and annihilation operator anti-commutation relations, and the use of right-handed projection operators. The right-handed tachyon wave function light-front Fourier expansion is:

$$\psi_{TR}^{+}(x) = \sum_{\pm s} \int d^2p dp^+ N_{TR}^{+}(p)\theta(p^+)[b_{TR}^{+}(p, s)u_{TR}^{+}(p, s)e^{-ip\cdot x} + d_{TR}^{+\dagger}(p, s)v_{TR}^{+}(p, s)e^{+ip\cdot x}] \quad (3.77)$$

and its hermitean conjugate is

$$\psi_{TR}^{+\dagger}(x) = \sum_{\pm s} \int d^2p dp^+ N_{TR}^{+}(p)\theta(p^+) [b_{TR}^{+\dagger}(p, s)u_{TR}^{+\dagger}(p, s)e^{+ip\cdot x} + d_{TR}^{+}(p, s)v_{TR}^{+\dagger}(p, s)e^{-ip\cdot x}] \quad (3.78)$$

where $N_{TR}^{+}(p) = N_{TL}^{+}(p)$, where the anti-commutation relations of the Fourier coefficient operators are

$$\begin{aligned}
\{b_{TR}^{+}(q,s), b_{TR}^{+\dagger}(p,s')\} &= -\delta_{ss'}\delta^2(\mathbf{q} - \mathbf{p})\delta(q^+ - p^+) \quad (3.79)\\
\{d_{TR}^{+}(q,s), d_{TR}^{+\dagger}(p,s')\} &= -\delta_{ss'}\delta^2(\mathbf{q} - \mathbf{p})\delta(q^+ - p^+)\\
\{b_{TR}^{+}(q,s), b_{TR}^{+}(p,s')\} &= \{d_{TR}^{+}(q,s), d_{TR}^{+}(p,s')\} = 0\\
\{b_{TR}^{+\dagger}(q,s), b_{TR}^{+\dagger}(p,s')\} &= \{d_{TR}^{+\dagger}(q,s), d_{TR}^{+\dagger}(p,s')\} = 0\\
\{b_{TR}^{+}(q,s), d_{TR}^{+\dagger}(p,s')\} &= \{d_{TR}^{+}(q,s), b_{TR}^{+\dagger}(p,s')\} = 0\\
\{b_{TR}^{+\dagger}(q,s), d_{TR}^{+\dagger}(p,s')\} &= \{d_{TR}^{+}(q,s), b_{TR}^{+}(p,s')\} = 0
\end{aligned}$$

and where the spinors are

$$u_{TR}^{+}(p, s) = C^+R^+u_T(p,s) \quad (3.80)$$
$$v_{TR}^{+}(p, s) = C^+R^+v_T(p,s) \quad (3.81)$$

by Appendix 3-A (eq. 3-A.7).

The right-handed anti-commutation relation with the minus sign follows in particular because of the minus signs found earlier.

3.4.5 Interpretation of Tachyon Creation and Annihilation Operators

To properly discuss the physical interpretation of tachyon creation and annihilation operators we must first determine the Hamiltonian and momentum operators in terms of creation and annihilation operators.

The energy-momentum tensor density is the symmetrized version of

$$\mathfrak{T}^{\mu\nu} = \sum_i \partial\mathfrak{L}/\partial(\partial\chi_i/\partial x_\mu)\, \partial\chi_i/\partial x_\nu - g^{\mu\nu}\mathfrak{L} \tag{3.82}$$

where the sum over i is over the fields. The light-front hamiltonian is

$$H \equiv P^- = T^{+-} = \int dx^- d^2x\, \mathfrak{T}^{+-} \tag{3.83}$$

and the "momenta" are

$$P^+ = T^{++} = \int dx^- d^2x\, \mathfrak{T}^{++} \tag{3.84}$$

$$P^i = T^{+i} = \int dx^- d^2x\, \mathfrak{T}^{+i} \tag{3.85}$$

for i = 1, 2.

The light-front, left-handed and right-handed tachyon lagrangian $\mathfrak{L}_T$ and its equations of motion imply

$$\begin{aligned} H = i2^{-1/2}\int dx^- d^2x\, [&\psi_{TL}{}^{+\dagger}\partial^-\psi_{TL}{}^{+} - \partial^-\psi_{TL}{}^{+\dagger}\psi_{TL}{}^{+} + \psi_{TL}{}^{-\dagger}\partial^+\psi_{TL}{}^{-} - \partial^+\psi_{TL}{}^{-\dagger}\psi_{TL}{}^{-} - \\ &- \psi_{TR}{}^{+\dagger}\partial^-\psi_{TR}{}^{+} + \partial^-\psi_{TR}{}^{+\dagger}\psi_{TR}{}^{+} - \psi_{TR}{}^{-\dagger}\partial^+\psi_{TR}{}^{-} + \partial^+\psi_{TR}{}^{-\dagger}\psi_{TR}{}^{-} + \text{mass terms}] \end{aligned} \tag{3.86}$$

After substituting for the various fields we find the *independent fields* (which create the in and out particle states) have the hamiltonian terms:

$$\begin{aligned} H = \sum_{\pm s}\int d^2p\,dp^+\, p^-[&b_{TL}{}^{+\dagger}(p,s)b_{TL}{}^{+}(p,s) - d_{TL}{}^{+}(p,s)d_{TL}{}^{+\dagger}(p,s) - b_{TR}{}^{+\dagger}(p,s)b_{TR}{}^{+}(p,s) + \\ &+ d_{TR}{}^{+}(p,s)d_{TR}{}^{+\dagger}(p,s)] \end{aligned} \tag{3.87}$$

$$\begin{aligned} = \sum_{\pm s}\int d^2p\,dp^+\, p^-[&b_{TL}{}^{+\dagger}(p,s)b_{TL}{}^{+}(p,s) + d_{TL}{}^{+\dagger}(p,s)d_{TL}{}^{+}(p,s) - b_{TR}{}^{+\dagger}(p,s)b_{TR}{}^{+}(p,s) - \\ &- d_{TR}{}^{+\dagger}(p,s)d_{TR}{}^{+}(p,s)] \end{aligned} \tag{3.88}$$

up to the usual infinite constants due to left-handed operator rearrangement and right-handed operator rearrangement that are discarded. Eq. 3.88 is the basis for our particle interpretation of tachyon creation and annihilation operators based on Dirac's hole theory. Dirac hole theory as applied in light-front coordinates assumes all negative p^- ("energy") states are filled.

Left-Handed Tachyon Creation and Annihilation Operators

1. We identify $b_{TL}^{+\dagger}(p,s)$ and $d_{TL}^{+}(p,s)$ as creation operators for left-handed tachyons. $b_{TL}^{+\dagger}(p,s)$ creates a positive p^- ("energy") state and $d_{TL}^{+}(p,s)$ creates a negative p^- ("energy") state.

2. $b_{TL}^{+}(p,s)$ and $d_{TL}^{+\dagger}(p,s)$ are the corresponding annihilation operators for left-handed tachyons. $b_{TL}^{+}(p,s)$ annihilates a positive p^- ("energy") state and $d_{TL}^{+\dagger}(p,s)$ annihilates a negative p^- ("energy") state.

3. We assume Dirac hole theory holds for the left-handed tachyon vacuum with all negative energy states filled. There is no tachyon energy gap as there is for Dirac fermions. There is also the problem that the left-handed tachyon vacuum is not invariant under ordinary Lorentz transformations or Superluminal transformations. *However if we confine ourselves to light-front coordinates for computations no ambiguity can result and the Lorentz covariant quantities that we calculate, such as the S matrix, are well-defined.*

4. Using tachyon hole theory we identify $b_{TL}^{+}(p,s)$ and $d_{TL}^{+\dagger}(p,s)$ as annihilation operators for left-handed tachyons. $b_{TL}^{+}(p,s)$ annihilates a positive p^- ("energy") state and $d_{TL}^{+\dagger}(p,s)$ annihilates a negative p^- ("energy") state – thus creating a hole in the tachyon sea that we view as the creation of a positive p^- ("energy"), left-handed antitachyon. $d_{TL}^{+}(p,s)$ annihilates a positive p^- ("energy"), left-handed antitachyon.

Right-Handed Tachyon Creation and Annihilation Operators

The anti-commutation relations of right-handed tachyon creation and annhilation operators and the right-handed Hamiltonian terms have the "wrong" sign compared to corresponding Dirac operators and left-handed tachyon operators. This situation is completely analogous to the situation of time-like photons in the covariant formulation of quantum Electrodynamics.[59] In the case of time-like photons it was possible to introduce an indefinite metric (Gupta-Bleuler formulation), and then to use the subsidiary condition $\partial A^\nu/\partial x^\nu = 0$ to reduce the dynamics of QED to the transverse components. Thus the time-like photons were

[59] Bogoliubov (1959) pp. 130-136.

intermediate artifacts needed to have a manifestly covariant formulation while QED observables depended solely on the transverse components of the electromagnetic field.

In the present case of free tachyons, and in leptonic ElectroWeak Theory there is no evident "subsidiary condition" to eliminate the right-handed tachyon fields. But since the only manner in which the right-handed leptonic tachyon fields[60] interact is through mass terms, which can be easily 'integrated out", right-handed leptonic tachyon fields are removed from the observable part of the leptonic ElectroWeak Theory by their "lack of interaction" with left-handed fields.

In the case of quark ElectroWeak Theory right-handed tachyon quark fields have charge (–1/3) and thus experience an electromagnetic interaction as well as a Z interaction. However, *since quarks are totally confined, right-handed tachyon quarks will not be able to continuously emit photons or Z's due to energy conservation and their confinement to bound states of fixed positive energy.* Earlier, when we consider complex Lorentz group boosts, we will suggest that quarks may not consist of Dirac particles or tachyons of the type considered up to this point in this chapter. Rather they may be variants on Dirac particles and tachyons satisfying different dynamical equations. However, the preceding comments on quarks would still apply.

Thus right-handed tachyons are analogous to time-like photons – necessary theoretically but prevented from causing a negative energy disaster by the forms of their interactions. We discuss this subject in more detail in the following chapters.

3.4.6 Tachyon Feynman Propagator

In this section we develop the light-front propagator for tachyons. We begin with a subsection describing the light-front propagators of Dirac fields.

Dirac Field Light-Front Propagators

The light-front Feynman propagator for the ψ^+ field of a Dirac fermion is

$$iS^+_F(x,y)\gamma^0 = \theta(x^+ - y^+)<0|\psi^+(x)\psi^{+\dagger}(y)|0> - \theta(y^+ - x^+)<0|\psi^{+\dagger}(y)\psi^+(x)|0> \quad (3.89)$$

and does not contain a non-covariant piece due to the projection operators:

$$\begin{aligned} iS^+_F(x,y) &= \int d^2p dp^+ \theta(p^+)[1/(2(2\pi)^3 p^+)]\{\theta(x^+ - y^+)[R^+(\not{p} + m)R^-]e^{-ip\cdot(x-y)} + \\ &\quad + \theta(y^+ - x^+)[R^+(-\not{p} + m)R^-]e^{+ip\cdot(x-y)}\} \\ &= R^+ iS_F(x,y)R^- \end{aligned} \quad (3.90)$$

where $S_F(x,y)$ is the usual Feynman propagator.

---

[60] The tachyon fields are provisionally assumed to be neutrino fields in the leptonic sector, and d, s and b quarks in the quark sector.

The light-front Feynman propagator for a *left-handed* Dirac field ψ^+ is

$$iS^+_{LF}(x,y) = \int d^2p dp^+ \theta(p^+)[1/(2(2\pi)^3 p^+)]\{\theta(x^+ - y^+)[C^-R^+(\not{p}+m)R^-C^-]e^{-ip\cdot(x-y)} + \\ + \theta(y^+ - x^+)[C^-R^+(-\not{p}+m)R^-C^-]e^{+ip\cdot(x-y)}\}$$

$$= C^-R^+ iS_F(x,y)R^-C^- \qquad (3.91)$$

Tachyon Field Light Front Propagators

Turning now to tachyons, the light-front Feynman propagator for the left-handed ψ_{TL}^+ *tachyon* field is (using the previous Fourier expansion of the left-handed tachyon field):

$$iS^+_{TLF}(x,y) = \theta(x^+ - y^+)<0|\psi_{TL}{}^+(x)\psi_{TL}{}^{+\dagger}(y)\gamma^0|0> - \theta(y^+ - x^+)<0|\psi_{TL}{}^{+\dagger}(y)\gamma^0\psi_{TL}{}^+(x)|0>$$

$$= -i\int d^2p dp^+ \theta(p^+) N_{TL}^{+2}(2m|\mathbf{p}|)^{-1} C^-R^+\{\theta(x^+ - y^+)[(i\not{p} - m)\boldsymbol{\gamma}\cdot\mathbf{p}]e^{-ip\cdot(x-y)} + \\ + \theta(y^+ - x^+)[(i\not{p} + m)\boldsymbol{\gamma}\cdot\mathbf{p}]e^{+ip\cdot(x-y)}\}R^+C^-\gamma^0$$

If we define the on-shell momentum variable

$$p_0{}^- = (p_0{}^1 p_0{}^1 + p_0{}^2 p_0{}^2 - m^2)/(2p_0{}^+),\ p_0{}^+ = p^+,\ p_0{}^j = p^j \text{ (for } j = 1, 2),\ p_{\perp 0}{}^2 = p_0{}^j p_0{}^j \text{ and } \not{p}_0 = p_0\cdot\gamma$$

then the above equation can be rewritten as

$$iS^+_{TLF}(x,y) = -iC^-R^+\int d^4p[32\pi^4(p_0{}^+(p_0{}^+ - p_0{}^-) + p_{0\perp}{}^2)]^{-1}e^{-ip\cdot(x-y)}\{\theta(p^+)(i\not{p}_0 - m)\boldsymbol{\gamma}\cdot\mathbf{p}_0]/[p^- - p_0{}^- + i\varepsilon] + \\ + \theta(-p^+)(i\not{p}_0 + m)\boldsymbol{\gamma}\cdot\mathbf{p}_0]/[p^- + p_0{}^- - i\varepsilon]\}R^+C^-\gamma^0$$

$$= -\tfrac{1}{2}\, i\int d^4p(2\pi)^{-4}[C^-R^+(i\not{p} - m)\boldsymbol{\gamma}\cdot\mathbf{p}R^+C^-\gamma^0]e^{-ip\cdot(x-y)}[(p^2 + m^2 + i\varepsilon)(p^+(p^+ - p^-) + p_\perp{}^2))]^{-1}$$

and using $C^-R^+(i\not{p} - m)\boldsymbol{\gamma}\cdot\mathbf{p}R^+C^- = i\,C^-R^+(p^+(p^+ - p^-) + p_\perp{}^2)$ we find

$$iS^+_{TLF}(x,y) = \tfrac{1}{2}C^-R^+\gamma^0\int d^4p(2\pi)^{-4}\, p^+ e^{-ip\cdot(x-y)}/(p^2 + m^2 + i\varepsilon) \qquad (3.92)$$

Similarly the light-front Feynman propagator for the right-handed ψ_{TR}^+ tachyon field is

$$iS^+_{TRF}(x,y) = \theta(x^+ - y^+)<0|\psi_{TR}{}^+(x)\psi_{TR}{}^{+\dagger}(y)\gamma^0|0> - \theta(y^+ - x^+)<0|\psi_{TR}{}^{+\dagger}(y)\gamma^0\psi_{TR}{}^+(x)|0>$$

$$= -\tfrac{1}{2}C^+R^+\gamma^0\int d^4p(2\pi)^{-4}\, p^+ e^{-ip\cdot(x-y)}/(p^2 + m^2 + i\varepsilon) \qquad (3.93)$$

where the relative minus sign between eqs. 3.92 and 3.93 is due to the relative minus signs of the Fouier component operator anti-commutation relations.

Thus we find *tachyon* pole terms in the tachyon propagators as one would expect.

3.5 Complex Space and 3-Momentum & Real-Valued Energy Fermions (Quarks)

In this section we will use L_C boosts to develop a wider set of dynamical equations for free spin ½ fermions with real-valued energy and complex-valued 3-momentum.[61] Earlier we defined L_C boosts with

$$\Lambda_C(\mathbf{v_c}) = \exp[i\omega\hat{\mathbf{w}}\cdot\mathbf{K}] \tag{3.94}$$

$$\omega = (\omega_r^{\,2} - \omega_i^{\,2} + 2i\omega_r\omega_i\,\hat{\mathbf{u}}_r\cdot\hat{\mathbf{u}}_i)^{1/2} \tag{3.95}$$

$$\hat{\mathbf{w}} = (\omega_r\hat{\mathbf{u}}_r + i\omega_i\hat{\mathbf{u}}_i)/\omega \tag{3.96}$$

$$\hat{\mathbf{w}}\cdot\hat{\mathbf{w}} = \hat{\mathbf{u}}_r\cdot\hat{\mathbf{u}}_r = \hat{\mathbf{u}}_i\cdot\hat{\mathbf{u}}_i = 1 \tag{3.97}$$

$$\mathbf{v_c} = \hat{\mathbf{w}}\tanh(\omega) \tag{3.98}$$

3.5.1 L_C Spinor "Normal" Lorentz Boosts & More Spin ½ Particle Types

Spinor boost transformations were used in previous sections to develop the dynamical equations for Dirac fields and tachyon fields.

The form of the L_C spinor boost transformation corresponding to the coordinate transformation is:

$$\begin{aligned} S_C(\omega, \mathbf{v_c}) &= \exp(-i\omega\sigma_{0k}\hat{w}_k/2) = \exp(-\omega\gamma^0\boldsymbol{\gamma}\cdot\hat{\mathbf{w}}/2) \\ &= \cosh(\omega/2)I + \sinh(\omega/2)\gamma^0\boldsymbol{\gamma}\cdot\hat{\mathbf{w}} \end{aligned} \tag{3.99}$$

The inverse transformation is

$$\begin{aligned} S_C^{-1}(\omega, \mathbf{v_c}) &= \gamma^2\gamma^0 K^{-1} S_C^{\dagger} K \gamma^0\gamma^2 = \gamma^2\gamma^0 S_C^{\,T}\gamma^0\gamma^2 = \exp(\omega\gamma^0\boldsymbol{\gamma}\cdot\hat{\mathbf{w}}/2) \\ &= \cosh(\omega/2)I - \sinh(\omega/2)\gamma^0\boldsymbol{\gamma}\cdot\hat{\mathbf{w}} \end{aligned} \tag{3.100}$$

where the superscript T denotes the transpose and *K* is the complex conjugation operator (that also appears in the time-reversal operator). Note that S_C is not unitary just as in previous cases considered in this chapter.

We now redo the development of spin ½ dynamical equations of motion of earlier sections for this more general case of complex ω and $\hat{\mathbf{w}}$. Again we apply a boost to a Dirac equation for a positive energy plane wave particle of mass m at rest:

---

[61] The complexon theory that we develop and use for quark dynamics in the Standard Model is not required. Our Standard Model could use Dirac fermion dynamics for the up-type quarks and tachyon dynamics for down-type quarks. We choose to use complexon dynamics for all quark types because they have an internal SU(3)-like structure suggestive of color SU(3). More importantly, their spin dynamics is different and thus may resolve the differences between theory and experiment – particularly for the deep inelastic parton spin-dependent structure functions.

$$0 = S_C(\omega, \mathbf{v_c}))(m\gamma^0 - m)e^{-imt}w(0)$$
$$= [mS_C\gamma^0 S_C^{-1} - m]e^{-imt}S_C w(0) \qquad (3.101)$$

where $S_C = S_C(\omega, \hat{\mathbf{w}})$. After some algebra

$$mS_C\gamma^0 S_C^{-1} = m[\cosh(\omega)\gamma^0 - \sinh(\omega)\boldsymbol{\gamma}\cdot\hat{\mathbf{w}}] \qquad (3.102)$$

Case 1: Parallel Real and Imaginary Relative Vectors

If the real and imaginary relative vectors parts of $\hat{\mathbf{w}}$, namely $\hat{\mathbf{u}}_r$ and $\hat{\mathbf{u}}_i$, are parallel, then $\hat{\mathbf{u}}_r\cdot\hat{\mathbf{u}}_i = 1$ and

$$\omega = \omega_r + i\omega_i \qquad (3.103)$$

Eq. 3.102 can be re-expressed as

$$mS_C\gamma^0 S_C^{-1} = m[\cosh(\omega_r)\cos(\omega_i) + i\sinh(\omega_r)\sin(\omega_i)]\gamma^0 - m[\sinh(\omega_r)\cos(\omega_i) + i\cosh(\omega_r)\sin(\omega_i)]\boldsymbol{\gamma}\cdot\hat{\mathbf{u}}_r \qquad (3.104)$$

or equivalently

$$mS_C\gamma^0 S_C^{-1} = \cos(\omega_i)\gamma\cdot p_r + i\sin(\omega_i)\gamma\cdot p_i \qquad (3.105)$$

where

$$p_r^0 = m\cosh(\omega_r) \qquad p_i^0 = m\sinh(\omega_r) \qquad (3.106)$$

and

$$\mathbf{p}_r = m\hat{\mathbf{u}}_r\sinh(\omega_r) \qquad \mathbf{p}_i = m\hat{\mathbf{u}}_r\cosh(\omega_r) \qquad (3.107)$$

If $\omega_i = 0$, then we recover the momentum space Dirac equation. If $\omega_i = \pi/2$, then we obtain the left-handed momentum space tachyon equation. Since the range of ω_i is $[0, \infty>$ (due to the cut along the real ω-plane axis) eq. 3.105 corresponds to the results of the Left-Handed Lorentz boost part discussed earlier.

Case 2: Anti-Parallel Real and Imaginary Relative Vectors

If the real and imaginary relative vectors parts of $\hat{\mathbf{w}}$, $\hat{\mathbf{u}}_r$ and $\hat{\mathbf{u}}_i$, are anti-parallel $\hat{\mathbf{u}}_r = -\hat{\mathbf{u}}_i$, then $\hat{\mathbf{u}}_r\cdot\hat{\mathbf{u}}_i = -1$ and

$$\omega = \omega_r - i\omega_i \qquad (3.108)$$

We can then express eq. 3.105 as

$$mS_C\gamma^0 S_C^{-1} = m[\cosh(\omega_r)\cos(\omega_i) - i\sinh(\omega_r)\sin(\omega_i)]\gamma^0 - m[\sinh(\omega_r)\cos(\omega_i) - i\cosh(\omega_r)\sin(\omega_i)]\boldsymbol{\gamma}\cdot\hat{\mathbf{u}}_r \qquad (3.109)$$

or

$$mS_C\gamma^0 S_C^{-1} = \cos(\omega_i)\gamma\cdot p_r - i\sin(\omega_i)\gamma\cdot p_i \qquad (3.110)$$

where

$$p_r^0 = m\cosh(\omega_r) \qquad p_i^0 = m\sinh(\omega_r) \qquad (3.111)$$

and

$$\mathbf{p}_r = m\hat{\mathbf{u}}_r \sinh(\omega_r) \qquad \mathbf{p}_i = m\hat{\mathbf{u}}_r \cosh(\omega_r) \tag{3.112}$$

If $\omega_i = 0$, then we again recover the momentum space Dirac equation, If $\omega_i = \pi/2$, then we obtain the right-handed momentum space tachyon equation. (The range of ω_i is again $[0, \infty>$.)

Note: Since the matrix elements in the boost depend on $\gamma = (1 - \beta^2)^{-1/2}$ with a singularity at $\beta = \pm 1$, which in turn corresponds to $\omega = \pm\infty$, there is a branch cut along the ω axis in the complex ω-plane. Therefore we point out again the product of three Left-handed transformations is not equivalent to a Right-handed transformation.

Case 3: Complexons: A New Type of Particle with Perpendicular Real and Imaginary 3-Momenta

If the real and imaginary relative vectors parts of $\hat{\mathbf{w}}$, namely $\hat{\mathbf{u}}_r$ and $\hat{\mathbf{u}}_i$, are perpendicular, $\hat{\mathbf{u}}_r \cdot \hat{\mathbf{u}}_i = 0$, then

$$\omega = (\omega_r^2 - \omega_i^2)^{1/2} \tag{3.113}$$

Thus ω is either purely real ($\omega_r \geq \omega_i$) or purely imaginary ($\omega_r < \omega_i$).

The momentum space equation generated by the corresponding L_C spinor boost is

$$\{m \cosh(\omega)\gamma^0 - m \sinh(\omega)\gamma\cdot(\omega_r\hat{\mathbf{u}}_r + i\omega_i\hat{\mathbf{u}}_i)/\omega - m\}e^{-ip\cdot x}w_c(p) = 0 \tag{3.114}$$

Defining the momentum 4-vector

$$p = (p^0, \mathbf{p}) \tag{3.115}$$

where

$$p^0 = m \cosh(\omega) \qquad \mathbf{p} = \mathbf{p}_r + i\mathbf{p}_i \tag{3.116}$$

$$\mathbf{p}_r = m\omega_r\hat{\mathbf{u}}_r \sinh(\omega)/\omega \qquad \mathbf{p}_i = m\omega_i\hat{\mathbf{u}}_i \sinh(\omega)/\omega \tag{3.117}$$

$$\mathbf{p}_r\cdot\mathbf{p}_i = 0 \tag{3.118}$$

then we obtain a positive energy Dirac-like equation with complex 3-momentum

$$[p\cdot\gamma - m]e^{-ip\cdot x}w_c(p) = 0$$

or, explicitly, (3.119)

$$[p^0\gamma^0 - (\mathbf{p}_r + i\mathbf{p}_i)\cdot\gamma - m]e^{-ip\cdot x}w_c(p) = 0$$

with a complex 3-momentum $\mathbf{p}$ and the 4-momentum mass shell condition:

$$p^2 = p^{0\,2} - \mathbf{p}_r\cdot\mathbf{p}_r + \mathbf{p}_i\cdot\mathbf{p}_i = m^2 \tag{3.120}$$

Note

$$|\mathbf{v}| = |\mathbf{p}|/p^0 = [(\mathbf{p}_r + i\mathbf{p}_i)\cdot(\mathbf{p}_r + i\mathbf{p}_i)]^{1/2}/p^0 = \tanh(\omega) \quad (3.121)$$

and thus the Lorentz factor

$$\gamma = \cosh(\omega) \quad (3.122)$$

Eq. 3.119 is the momentum space equivalent of the wave equation

$$[i\gamma^0\partial/\partial t + i\boldsymbol{\gamma}\cdot(\nabla_r + i\nabla_i) - m]\psi_C(t, \mathbf{x}_r, \mathbf{x}_i) = 0 \quad (3.123)$$

where

$$x_c = (t, \mathbf{x}_r - i\mathbf{x}_i) \quad (3.123a)$$

and where the grad operators ∇_r and ∇_i are with respect to $\mathbf{x}_r$ and $\mathbf{x}_i$ respectively. Since $\hat{\mathbf{u}}_r\cdot\hat{\mathbf{u}}_i = 0$, we see that there is a subsidiary condition on the wave function

$$\nabla_r\cdot\nabla_i\,\psi_C(t, \mathbf{x}_r, \mathbf{x}_i) = 0 \quad (3.124)$$

We will call the particles satisfying eqs.3.123 and 3.124 *complexons*. In addition eq. 3.118 implies the anti-commutation relation

$$\{\boldsymbol{\gamma}\cdot\mathbf{p}_r, \boldsymbol{\gamma}\cdot\mathbf{p}_i\} = 0 \quad (3.125)$$

which in turn implies

$$\boldsymbol{\gamma}\cdot\nabla_r\boldsymbol{\gamma}\cdot\nabla_i\psi_C(t, \mathbf{x}_r, \mathbf{x}_i) = \boldsymbol{\gamma}\cdot\nabla_i\boldsymbol{\gamma}\cdot\nabla_r\psi_C(t, \mathbf{x}_r, \mathbf{x}_i) = 0 \quad (3.126)$$

We note that eq. 3.125 is covariant under the real Lorentz group and eq. 3.126 can be easily put into covariant form since the difference of these 4-vectors squared is a real Lorentz group invariant: $[\gamma^0\partial/\partial t + \boldsymbol{\gamma}\cdot(\nabla_r + i\nabla_i)]^2 - [\gamma^0\partial/\partial t + i\boldsymbol{\gamma}\cdot(\nabla_r - i\nabla_i)]^2 = 4\nabla_r\cdot\nabla_i$.

Before considering a lagrangian formulation and the Fourier operator representation of $\psi_C(t, \mathbf{x}_r, \mathbf{x}_i)$ we will define the spinors and associated real and imaginary spin operators. The spinor generated from a spin up Dirac spinor at rest by a complex boost is

$$w_c(p) = S_C(p)w(0) = [\cosh(\omega/2)I + \sinh(\omega/2)\gamma^0\boldsymbol{\gamma}\cdot\hat{\mathbf{w}}]w(0) \quad (3.127)$$

Following a procedure similar to Appendix 3-A (which the reader may wish to examine first) we define four spinors for Dirac particles at rest:

$$w^k(0) = \begin{bmatrix} \delta_{1k} \\ \delta_{2k} \\ \delta_{3k} \\ \delta_{4k} \end{bmatrix} \quad (3\text{-A}.2)$$

where Kronecker deltas appear in the brackets. Then by applying eq. 3.127 to the spinors defined by eq. 5-A.2 we find the L_C spinors

$$S_C w^k(0) = w_{Cr}{}^k(p) + i w_{Ci}{}^k(p) \qquad (3.128)$$

where

$$S_{Cr} = \cosh(\omega/2)I + (\omega_r/\omega)\sinh(\omega/2)\gamma^0\boldsymbol{\gamma}\cdot\hat{\mathbf{u}}_r$$
$$= [(m + E)/(2m)]^{½}I + [m(m + E)]^{-½}\gamma^0\boldsymbol{\gamma}\cdot\mathbf{p}_r = aI + b\gamma^0\boldsymbol{\gamma}\cdot\mathbf{p}_r \qquad (3.129)$$

Thus the "real" spinors $w_{Cr}{}^k(p)$ are the columns of

$$S_{Cr} = \begin{bmatrix} \underline{w_{Cr}{}^1(p)} & \underline{w_{Cr}{}^2(p)} & \underline{w_{Cr}{}^3(p)} & \underline{w_{Cr}{}^4(p)} \\ a & 0 & bp_{r\,z} & bp_{r-} \\ 0 & a & bp_{r\,+} & -bp_{r\,z} \\ bp_{r\,z} & bp_{r-} & a & 0 \\ bp_{r\,+} & -bp_{r\,z} & 0 & a \end{bmatrix} \qquad (3.130)$$

where $p_{r\,\pm} = p_{r\,x} \pm ip_{r\,y}$. The "imaginary" spinors are the columns of

$$S_{Ci} = (\omega_i/\omega)\sinh(\omega/2)\gamma^0\boldsymbol{\gamma}\cdot\hat{\mathbf{u}}_i = [m(m + E)]^{-½}\gamma^0\boldsymbol{\gamma}\cdot\mathbf{p}_i = b\gamma^0\boldsymbol{\gamma}\cdot\mathbf{p}_i \qquad (3.131)$$

$$S_{Ci} = \begin{bmatrix} \underline{w_{Ci}{}^1(p)} & \underline{w_{Ci}{}^2(p)} & \underline{w_{Ci}{}^3(p)} & \underline{w_{Ci}{}^4(p)} \\ 0 & 0 & bp_{i\,z} & bp_{i-} \\ 0 & 0 & bp_{i\,+} & -bp_{i\,z} \\ bp_{i\,z} & bp_{i-} & 0 & 0 \\ bp_{i\,+} & -bp_{i\,z} & 0 & 0 \end{bmatrix} \qquad (3.132)$$

where $p_{i\,\pm} = p_{i\,x} \pm ip_{i\,y}$.

Eqs. 3.127 through 3.132 imply that the wave function solution of eq. 3.123, subject to the subsidiary condition eq. 3.124, is[62, 63]

---

[62] Note that when $|\mathbf{p}_i| \geq |\mathbf{p}_r|$ (for imaginary $\omega = (\omega_r{}^2 - \omega_i{}^2)^{½}$) the 3-momentum becomes imaginary $\mathbf{p}\cdot\mathbf{p} < 0$. However, since we will be identifying confined quarks with this type of particle – much modified by a confining color quark interaction – the issue of an imaginary 3-momentum in the hypothetical free quark case becomes moot. We note the energy gap between positive and negative energy states disappears so $E = 0$ is possible. Thus real Lorentz transformations can mix positive and negative energy states. The solution is to do all calculations in the light-front

$$\psi_C(x_r, x_i) = \sum_{\pm s} \int d^3p_r d^3p_i\, N_C(p)\delta(\mathbf{p_r}\cdot\mathbf{p_i}/m^2)[b_C(p,s)u_C(p, s)e^{-i(p\cdot x + p^*\cdot x^*)/2} + \\ + d_C^\dagger(p,s)v_C(p, s)e^{+i(p\cdot x + p^*\cdot x^*)/2}] \qquad (3.133)$$

where $\mathbf{p} = \mathbf{p_r} + i\mathbf{p_i}$ (eq. 3.95), $\mathbf{x} = \mathbf{x_r} - i\mathbf{x_i}$, $p\cdot x = p^0x^0 - \mathbf{p}\cdot\mathbf{x}$, and where we use

$$(p\cdot x + p^*\cdot x^*)/2 = p^0x^0 - \mathbf{p_r}\cdot\mathbf{x_r} - \mathbf{p_i}\cdot\mathbf{x_i} \qquad (3.134)$$

in the exponentials in order to avoid divergences that would appear in the calculation of the equal-time commutator, the Feynman propagator, and other quantities of interest, after second quantization. Note that

$$(\nabla_r + i\nabla_i)e^{-i(p\cdot x + p^*\cdot x^*)/2} = i(\mathbf{p_r} + i\mathbf{p_i})e^{-i(p\cdot x + p^*\cdot x^*)/2} \qquad (3.135)$$

and

$$(\nabla_r + i\nabla_i)e^{-ip^*\cdot x^*} = 0 \qquad (3.136)$$

for all p.

The wave function's conjugate (the hermitean conjugate modified by letting $\mathbf{x_i} \to -\mathbf{x_i}$ in addition to hermitean conjugation) is

$$\psi_C^\dagger(x) = \psi_C^\dagger(x_r, -x_i) = \sum_{\pm s} \int d^3p_r d^3p_i\, \delta(\mathbf{p_r}\cdot\mathbf{p_i}/m^2)N_C(p^*)\cdot \\ \cdot[b_C^\dagger(p^*,s)u_C^\dagger(p^*,s)e^{+i(p\cdot x^* + p^*\cdot x)/2} + d_C(p^*,s)v_C^\dagger(p^*,s)e^{-i(p\cdot x^* + p^*\cdot x)/2}] \qquad (3.137)$$

where $\mathbf{p} = \mathbf{p_r} + i\mathbf{p_i}$, $\mathbf{x} = \mathbf{x_r} - i\mathbf{x_i}$, $p\cdot x = p^0x^0 - \mathbf{p}\cdot\mathbf{x}$, and † indicates hermitean hermitean conjugation.

The spinors are

$$\begin{aligned} u_C(p, s) &= S_C(p)w^1(0) \\ u_C(p, -s) &= S_C(p)w^2(0) \\ v_C(p, s) &= S_C(p)w^3(0) \\ v_C(p, -s) &= S_C(p)w^4(0) \\ u_C^\dagger(p^*, s) &= w^{1T}(0)S_C^\dagger(p^*) = w^{1T}(0)S_C(p) \\ u_C^\dagger(p^*, -s) &= w^{2T}(0)S_C^\dagger(p^*) = w^{2T}(0)S_C(p) \end{aligned} \qquad (3.138)$$

---

frame as we do for tachyons. Then the mixing issue is resolved. In the present case we second quantize on the "time-front" for illustrative purposes.

[63] We scale $\mathbf{p_r}\cdot\mathbf{p_i}$ with m^2 in the delta function for convenience. All fermions have at least a minimal mass – the mass of the iota.

$$v_C^{\dagger}(p^*, s) = w^{3T}(0)S_C^{\dagger}(p^*) = w^{3T}(0)S_C(p)$$
$$v_C^{\dagger}(p^*, -s) = w^{4T}(0)S_C^{\dagger}(p^*) = w^{4T}(0)S_C(p)$$

with the superscript "T" indicating the transpose. Note that

$$S_C^{\dagger}(p^*) = [S_C(p^*)]^{\dagger} = S_C(p) \tag{3.139}$$

The normalization factor $N_C(p)$ is

$$N_C(p) = [2m/((2\pi)^6 p^0)]^{1/2} \tag{3.140}$$

Since $\mathbf{p}_r = \mathbf{p}_i = 0$ in the particle rest frame prior to the complex group boost, the boosted particle spin 4-vector s^{μ} satisfies

$$s^{\mu}p_r^{\mu} = s^{\mu}p_i^{\mu} = 0 \tag{3.141}$$

Thus s^{μ}, p_r^{μ} and p_i^{μ} form an 'orthogonal 4-triplet. Note that s^{μ} is itself complex[64] and, if the spin points in the z-direction prior to the complex boost, then the boosted s^{μ} has the form

$$s^{\mu} = (-\sinh(\omega)\hat{w}_z, (0,0,1) + (\cosh(\omega) - 1)\hat{w}_z\hat{\mathbf{w}}) \tag{3.142}$$

with $\hat{\mathbf{w}}$ defined earlier: $\hat{\mathbf{w}} = (\omega_r\hat{\mathbf{u}}_r + i\omega_i\hat{\mathbf{u}}_i)/\omega = \mathbf{p}/(m\sinh(\omega))$.

A Global SU(3) Symmetry Revealed

Before proceding to consider the second quantization of this case, we will consider a global SU(3) symmetry implicit in the previous equations. The defining property of the group SU(3) is that it preserves the invariance of inner products of complex 3-vectors of the form:

$$u^*\cdot v = u^{1*}v^1 + u^{2*}v^2 + u^{3*}v^3 \tag{3.143}$$

If we examine the dynamical equation eq. 3.123 we see that the differential operator is invariant under an SU(3) transformation U (using $\nabla_c = (\nabla_c^*)^* = \mathbf{D}_c^*$)

$$[i\gamma^0\partial/\partial t + i\mathbf{D}_c^*\cdot\gamma - m] = [i\gamma^0\partial/\partial t + i\mathbf{D}_c'^*\cdot\gamma' - m] \tag{3.144}$$

where

$$\mathbf{D}_c^* = \nabla_c = \nabla_r + i\nabla_i$$

and

$$\gamma'^a = U^{ab}\gamma'^b$$

---

[64] This feature of partons, which is not present in ordinary Dirac particles, might be the source of the discrepancies between theory and experiment in deep inelstic parton spin physics which is based on conventional real parton spins.

$$D_c'^{*a} = D_c'^{*b}U^{\dagger ab}$$

where U is a global SU(3) transformation and $U^\dagger = U^{-1}$. By theorem[65] all 4×4 γ matrices such as **γ'** are equivalent up to a unitary transformation V. Thus $V^\dagger\boldsymbol{\gamma}'V = \boldsymbol{\gamma}$ and eq. 3.144 is equivalent to

$$[i\gamma^0\partial/\partial t + i\mathbf{D}_c^*\cdot\boldsymbol{\gamma} - m] = [i\gamma^0\partial/\partial t + i\mathbf{D}_c'^*\cdot\boldsymbol{\gamma} - m] \qquad (5.145)$$

$$= [i\gamma^0\partial/\partial t + i\nabla_c'\cdot\boldsymbol{\gamma} - m]$$

where $\nabla_c'{}_a = U^{ab}\nabla_{cb}$, This demonstrates that eq. 3.123 is invariant under an SU(3) transformation if

$$\psi_C(t, \mathbf{x}_c) = \psi_C(t, U\mathbf{x}_c) = \psi_C'(t, \mathbf{x}_c') \qquad (3.146)$$

where $\psi_C(t, \mathbf{x}_c) \equiv \psi_C(t, \mathbf{x}_r, \mathbf{x}_i)$.

The subsidiary condition eq. 3.124 can be seen to transform as

$$\nabla_r\cdot\nabla_i\, \psi_C(t, \mathbf{x}_c) = \nabla_r^*\cdot\nabla_i\, \psi_C(t, \mathbf{x}_c) = \nabla_r'^*\cdot\nabla_i'\psi_C'(t, \mathbf{x}_c') = 0 \qquad (3.147)$$

under an SU(3) rotation. The invariance of the orthogonality condition is preserved.

The wave function (eq. 3.123) transforms in the following way under the SU(3) transformation U. If we define

$$q^{*\mu} = (q^0, \mathbf{q}^*) = (p^0, \mathbf{p}_r + i\mathbf{p}_i) = (p^0, \mathbf{p}) = p^\mu \qquad (3.148)$$

then eq. 3.133 can be rewritten in an invariant form under a global SU(3) transformation:

$$\psi_C(x) = \sum_{\pm s}\int d^3q_r d^3q_i N_C(p^0)\delta(\mathbf{q}_r^*\cdot\mathbf{q}_i/m^2)[b_C(q^*,s)u_C(q^*,s)e^{-i(q^*\cdot x+q\cdot x^*)/2} + d_C^\dagger(q^*,s)v_C(q^*,s)e^{+i(q^*\cdot x+q\cdot x^*)/2}] \qquad (3.149)$$

where x = x_c (subject to an examination of the transformation properties of the fourier coefficients and spinors done below.) Note both terms in each exponential are separately invariant under global SU(3). (Note also $\mathbf{q}_r^* = \mathbf{q}_r$ since $\mathbf{q}_r$ is real.)

From the form of S_C above it is clear that an argument similar to that for the dynamical equations shows S_C is invariant under an SU(3) transformation and thus their spinors are also invariant under SU(3) transformations. The fourier coefficients, if second quantized in a direct

[65] R. H. Good, Rev. Mod. Phys., **27**, 187 (1955).

generalization of the usual manner, have covariant anti-commutation relations under an SU(3) transformation. For example

$$\{b_C(q,s),\ b_C^{\dagger}(q'^*,s')\} = \delta_{ss'}\delta^3(q_r - q'_{r'})\delta^3(q_i - q'_{i'}) \tag{3.150}$$

Under an SU(3) transformation, z = Uq and z' = Uq', the right side of eq. 3.150 transforms to

$$\delta^3(q_r - q'_{r'})\delta^3(q_i - q'_{i'}) \rightarrow \delta^3(z_r - z'_{r'})\delta^3(z_i - z'_{i'})/|\partial(q)/\partial(z)| = \delta^3(z_r - z'_{r'})\delta^3(z_i - z'_{i'}) \tag{3.151}$$

where

$$|\partial(q)/\partial(z)| = |\partial(q_r^{\,1},q_r^{\,2},q_r^{\,3},q_i^{\,1},\ q_i^{\,2},\ q_i^{\,3})/\partial(z_r^{\,1},z_r^{\,2},z_r^{\,3},z_i^{\,1},\ z_i^{\,2},\ z_i^{\,3})| = 1 \tag{3.152}$$

is the Jacobian of the transformation U. Thus the fourier coefficients transform trivially under SU(3). For example,

$$b_C(q^*,s) \rightarrow b_C(z^*,s) \tag{3.153}$$

Since the integrand transforms as

$$\int d^3q_r d^3q_i \ \rightarrow \int d^3z_r d^3z_i\, |\partial(q)/\partial(z)| = \int d^3z_r d^3z_i \tag{3.154}$$

the wave function $\psi_C(t, \mathbf{x})$ transforms as an SU(3) scalar up to an inessential unitary transformation V of γ matrices: $\psi_C(t, \mathbf{x}) \rightarrow V\psi_C(t, \mathbf{x})$.[66]

Global SU(3) Spin ½ Complexon Fields

Having uncovered an SU(3) symmetry in the scalar field equations of Case 3 above the generalization of the scalar field equations to the <u>3</u> representation of SU(3) is direct:

$$\psi_C^{\,a}(x) = \sum_{\pm s}\int d^3p_r d^3p_i N_C(p)\delta(\mathbf{p}_r\cdot\mathbf{p}_i/m^2)[b_C(p,a,s)u_C^{\,a}(p,\ s)e^{-i(p\cdot x+p^*\cdot x^*)/2} + d_C^{\dagger}(p,a,s)v_C^{\,a}(p,\ s)e^{+i(p\cdot x+p^*\cdot x^*)/2}] \tag{3.155}$$

where x = x_c for a = 1,2, 3 with $u_C^{\,a}(p, s)$ and $v_C^{\,a}(p, s)$ being the product a spinor of type eq. 3.138 and a 3 element column vector c^a with b^{th} element

$$c^a(b) = \delta^{ab} \tag{3.156}$$

Under a global SU(3) transformation U the <u>3</u> complexon wave functions transform as

---

[66] The spinors $u_C(q^*,s)$ and $v_C(q^*,s)$ are unchanged up to a unitary transformation of the γ matrices ($V^{\dagger}\gamma' V = \gamma$). Thus the term $(U\mathbf{w})^*\cdot\gamma' = \mathbf{w}^*\cdot V\gamma V^{\dagger} \equiv \mathbf{w}^*\cdot\gamma$ in the expressions for the $u_C(q^*,s)$ and $v_C(q^*,s)$ spinors.

$$\psi_C{}'^a(x) = U^{ab}\psi_C{}^b(x) \tag{3.157}$$

In a subsequent discussion we will extend the global SU(3) symmetry described in these subsections to be color *local* SU(3) upon the introduction of the Yang-Mills color gluon interaction.

Lagrangian Formulation and Second Quantization of Complexons

In this subsection we will outline the canonical quantization of SU(3) singlet complexons with the quantum field equation

$$[i\gamma^0\partial/\partial t + i\gamma\cdot(\nabla_r + i\nabla_i) - m]\psi_C(t, \mathbf{x_r}, \mathbf{x_i}) = 0 \tag{3.158}$$

and subsidiary condition

$$\nabla_r\cdot\nabla_i\, \psi_C(t, \mathbf{x_r}, \mathbf{x_i}) = 0 \tag{3.159}$$

We begin with the Lagrangian density

$$\mathcal{L} = \bar{\psi}_C(i\gamma^\mu D_\mu - m)\psi_C(x) \tag{3.160}$$

where $\bar{\psi}_C = \psi_C{}^\dagger\gamma^0$:

$$\psi_C{}^\dagger = [\psi_C(\mathbf{x_r}, \mathbf{x_i})]^\dagger\,|_{\mathbf{x_i} = -\mathbf{x_i}} \tag{3.161}$$

$$D_0 = \partial/\partial x^0$$
$$D_k = \partial/\partial x^k + i\,\partial/\partial x_i{}^k \tag{3.162}$$

with $x^k = x_r{}^k$ for k = 1, 2, 3. The invariant action (under real Lorentz transformations) is

$$I = \int d^7x\mathcal{L} \tag{3.163}$$

It is easy to show that the action is real

$$I^* = I \tag{3.164}$$

in a manner similar to the case considered in Appendix 3-A due to the form of $\psi_C{}^\dagger$ in eq. 3.161. (One has to change the integration over $\mathbf{x_i}$ to $-\mathbf{x_i}$ after taking the complex conjugate of I and performing manipulations similar to those in Appendix 3-A.)

The conjugate momentum is

$$\pi_{Ca} = \partial\mathcal{L}/\partial\dot{\psi}_{Ca} \equiv \partial\mathcal{L}/\partial(\partial\psi_{Ca}/\partial x^0) = i\psi_{C\,a}^\dagger \tag{3.165}$$

where a is a spinor index. It yields the non-zero anti-commutation relation

$$\{\psi_{C\,a}^{\dagger}(x), \psi_{Cb}(y)\} = \delta_{ab}\,\delta^3(x_r - y_r)\delta^3(x_i - y_i) \tag{3.166}$$

where x and y are complex. However we will see that the constraint eq. 3.159 is required. So the correct anti-commutator turns out to be

$$\{\psi_{C\,a}^{\dagger}(x), \psi_{Cb}(y)\} = -\delta_{ab}\delta'(\nabla_r\cdot\nabla_i/m^2)[\delta^3(x_r - y_r)\delta^3(x_i - y_i)] \tag{3.167}$$

where all ∇_r and ∇_i are ∇ derivatives with respect to x, and where $\delta'(\nabla_r\cdot\nabla_i)$ is the derivative of a delta function with the argument being differential operators such as those in eq. 3.159. The minus sign is due to the presence of a *derivative* of a delta-function and is not an issue.

The hamiltonian density is

$$\mathcal{H} = \pi_C\dot{\psi}_C - \mathcal{L} = \psi_C^{\dagger}(-i\boldsymbol{\alpha}\cdot\mathbf{D} + \beta m)\psi_C \tag{3.168}$$

and the (unsymmetrized) energy-momentum tensor is

$$\mathfrak{T}_{\mu\nu} = -\,g_{\mu\nu}\,\mathcal{L} + \partial\mathcal{L}/\partial(D^{\mu}\psi_C)D_{\nu}\psi_C \tag{3.169}$$

The conserved energy and momentum are

$$P^0 = H = \int d^3x_r d^3x_i\ \mathfrak{T}^{00} = \int d^3x_r d^3x_i\ \mathcal{H} \tag{3.170}$$

and

$$P^i = \int d^3x_r d^3x_i\ \mathfrak{T}^{0i} \tag{3.171}$$

We now proceed to establish the canonical anti-commutation relations. First, the second quantization of the complexon field uses the above fourier coefficient anti-commutation relations (suitably rewritten):

$$\begin{aligned}
&\{b_C(p,s), b_C^{\dagger}(p'^*,s')\} = \delta_{ss'}\delta^3(\mathbf{p}_r - \mathbf{p}'_{r'})\delta^3(\mathbf{p}_i + \mathbf{p}'_{i'})\\
&\{d_C(p,s), d_C^{\dagger}(p'^*,s')\} = \delta_{ss'}\,\delta^3(\mathbf{p}_r - \mathbf{p}'_{r'})\delta^3(\mathbf{p}_i + \mathbf{p}'_{i'})\\
&\{b_C(p,s), b_C(p'^*,s')\} = \{d_C(p,s), d_C(p'^*,s')\} = 0\\
&\{b_C^{\dagger}(p,s), b_C^{\dagger}(p'^*,s')\} = \{d_C^{\dagger}(p,s), d_C^{\dagger}(p'^*,s')\} = 0\\
&\{b_C(p,s), d_C^{\dagger}(p'^*,s')\} = \{d_C(p,s), b_C^{\dagger}(p'^*,s')\} = 0\\
&\{b_C^{\dagger}(p,s), d_C^{\dagger}(p'^*,s')\} = \{d_C(p,s), b_C(p'^*,s')\} = 0
\end{aligned} \tag{3.172}$$

The delta-function arguments $\delta^3(\mathbf{p}_i + \mathbf{p}'_{i'})$ above have a positive sign in order to obtain $\delta^3(\mathbf{x}_i - \mathbf{y}_i)$ in the field anti-commutator eq. 3.167.

The spinors, eq. 3.138, satisfy

$$\sum_{\pm s} u_\alpha(p, s)\bar{u}_\beta(p^*, s) = (2m)^{-1}(\not{p} + m)_{\alpha\beta} \tag{3.173}$$

$$\sum_{\pm s} v_\alpha(p, s)\bar{v}_\beta(p^*, s) = (2m)^{-1}(\not{p} - m)_{\alpha\beta}$$

remembering

$$\bar{u}_C(p^*,s) = w^{1T}(0)S_C(p)\gamma^0 = w^{1T}(0)[\cosh(\omega/2)I + \sinh(\omega/2)\gamma^0\boldsymbol{\gamma}\cdot\hat{\mathbf{w}}]\gamma^0 \tag{3.174}$$

by eqs. 3.137 since $\hat{\mathbf{w}}^{**} = \hat{\mathbf{w}}$.

We will now evaluate the equal-time anti-commutation relation using eqs. 3.136 and 3.137:

$$\{\psi_{C\,a}^{\dagger}(x), \psi_{Cb}(y)\} = \sum_{\pm s, s'} \int d^3p_r d^3p_i\, d^3p'_r d^3p'_i\, \delta(\mathbf{p}_r\cdot\mathbf{p}_i/m^2)\delta(\mathbf{p}'_r\cdot\mathbf{p}'_i/m^2)\, N_C(p')N_C(p)\cdot$$
$$\cdot[\{b_C^{\dagger}(p^*,s)u_{Ca}^{\dagger}(p^*,s)e^{+i(p\cdot x^* + p^*\cdot x)/2}, b_C(p',s')u_{Cb}(p', s')e^{-i(p'\cdot y + p'^*\cdot y^*)/2}\}+$$
$$+ \{d_C(p^*,s)v_{Ca}^{\dagger}(p^*,s)e^{-i(p\cdot x^* + p^*\cdot x)/2}, d_C^{\dagger}(p',s')v_{Cb}(p', s')e^{+i(p'\cdot y + p'^*\cdot y^*)/2}\}]$$

$$= \int d^3p_r d^3p_i\, N_C^2(p)[\delta(\mathbf{p}_r\cdot\mathbf{p}_i/m^2)]^2[((\not{p} + m)\gamma^0)_{ba}e^{+i(p\cdot x^* + p^*\cdot x)/2 - i(p^*\cdot y + p\cdot y^*)/2} +$$
$$+((\not{p} - m)\gamma^0)_{ba}e^{-i(p\cdot x^* + p^*\cdot x)/2 + i(p^*\cdot y + p\cdot y^*)/2}]/(2m)$$

Next we use eq. 3.140 and the identity

$$[\delta(x - y)]^2 = -\tfrac{1}{2}\,\delta'(x - y) \equiv -\tfrac{1}{2}\, d\delta(x - y)/dx \tag{3.175}$$

which can be derived from the step function identity $\theta(x - y) = [\theta(x - y)]^2$ to obtain

$$\{\psi_{C\,a}^{\dagger}(x),\psi_{Cb}(y)\} = -\tfrac{1}{2}\int d^3p_r d^3p_i N_C^2(p)\delta'(\mathbf{p}_r\cdot\mathbf{p}_i/m^2)[((\not{p}+m)\gamma^0)_{ba}e^{-i\mathbf{pr}\cdot(\mathbf{xr} - \mathbf{yr}) + i\mathbf{pi}\cdot(\mathbf{xi} - \mathbf{yi})} +$$
$$+ ((\not{p} - m)\gamma^0)_{ba}e^{+i\mathbf{pr}\cdot(\mathbf{xr} - \mathbf{yr}) - i\mathbf{pi}\cdot(\mathbf{xi} - \mathbf{yi})}]/(2m)$$

$$= -\tfrac{1}{2}\delta_{ba}\int d^3p_r d^3p_i N_C^2(p)\delta'(\mathbf{p}_r\cdot\mathbf{p}_i/m^2)p^0 e^{-i\mathbf{pr}\cdot(\mathbf{xr} - \mathbf{yr}) + i\mathbf{pi}\cdot(\mathbf{xi} - \mathbf{yi})}/m$$

$$= -\,\delta_{ab}\,\delta'(\nabla_r\cdot\nabla_i/m^2)[\delta^3(x_r - y_r)\delta^3(x_i - y_i)] \tag{3.176}$$

The grad operators, ∇_r and ∇_i, are derivatives with respect to x in the Dirac delta functions. The factor[67] $\delta'(\nabla_r\cdot\nabla_i)$ expresses the orthogonality constraint in coordinate space on the momenta. It is analogous to the transversality constraint on the electromagnetic vector potential commutator:

$$[\pi_A{}^j(x), A_k(y)] = -i\,\delta^{tr}{}_{jk}(x - y) \qquad (3.177)$$

with

$$\delta^{tr}{}_{jk}(x - y) = (\delta_{jk} - \partial_j\partial_k/\nabla^2)\,\delta^3(x - y) \qquad (3.178)$$

where $\partial_k = \partial/\partial x_k$.

Complexon Feynman Propagator

The complexon Feynman propagator for ψ_C is[68]

$$iS_C(x, y) = \theta(x^0 - y^0)\langle 0|\psi_C(x)\psi_C{}^\dagger(y)\gamma^0|0\rangle - \theta(y^0 - x^0)\langle 0|\psi_C{}^\dagger(y)\gamma^0\psi_C(x)|0\rangle \qquad (3.179)$$

$$= \int d^3p_r d^3p_i N_C{}^2(p)[\delta(\mathbf{p_r}\cdot\mathbf{p_i}/m^2)]^2\{\theta(x^0 - y^0)(\not{p} + m)e^{-i(p^*\cdot(x-y) + p\cdot(x^*-y^*))/2} - \\ - \theta(y^0 - x^0)(\not{p} - m)e^{+i(p^*\cdot(x-y) + p\cdot(x^*-y^*))/2}\}/(2m)$$

$$= -(4\pi)^{-1}\int dp^0 d^3p_r d^3p_i (2\pi)^{-6}\delta'(\mathbf{p_r}\cdot\mathbf{p_i}/m^2)(\not{p}+m)e^{-i(p^*\cdot(x-y) + p\cdot(x^*-y^*))/2}/(p^2-m^2+i\varepsilon)$$

$$= -\tfrac{1}{2}\int dp^0 d^3p_r d^3p_i\,\delta'(\mathbf{p_r}\cdot\mathbf{p_i}/m^2)(\not{p} + m)(2\pi)^{-7}\exp[-ip^0(x^0-y^0)+i\mathbf{p_r}\cdot(\mathbf{x_r}-\mathbf{y_r})-i\mathbf{p_i}\cdot(\mathbf{x_i}-\mathbf{y_i})]/(p^2 - m^2 + i\varepsilon) \qquad (3.180)$$

The integral can be written in the form:

$$I = \int dp^0 d^3p_r d^3p_i \delta'(\mathbf{p_r}\cdot\mathbf{p_i}/m^2)(\not{p}+m)\exp[-ip^0(x^0-y^0)+i\mathbf{p_r}\cdot(\mathbf{x_r}-\mathbf{y_r})-i\mathbf{p_i}\cdot(\mathbf{x_i}-\mathbf{y_i})]/(p^2 - m^2 + i\varepsilon)$$
$$= \int d^4p_r dM^2\delta'(\nabla_r\cdot\nabla_i/m^2)(p^0\gamma^0-(\mathbf{p_r}-\nabla_i)\cdot\gamma+m)\exp[-ip^0(x^0-y^0)+i\mathbf{p_r}\cdot(\mathbf{x_r}-\mathbf{y_r})]J(\mathbf{x_i} - \mathbf{y_i},M^2)/(p_r{}^2-M^2 + i\varepsilon) \qquad (3.181)$$

where $p_r{}^2 = p^{0\,2} - \mathbf{p_r}\cdot\mathbf{p_r}$ and

---

[67] A derivative of a delta function containing grad operators.

[68] The reader, upon seeing the additional integrations $\int d^3p_i$ might suspect that they would ultimately lead to divergence issues in perturbation theory calculations. However the $\delta'(\mathbf{p_r}\cdot\mathbf{p_i}/m^2)$ term compensates in part for the additional integrations by four powers of momentum since $\delta'(\mathbf{p_r}\cdot\mathbf{p_i}/m^2) = (|\mathbf{p_r}||\mathbf{p_i}|/m^2)^{-2}\delta'(\cos\theta_{ri})$ where θ_{ri} is the angle between the momenta. As a result only 2 fermion and 3 fermion loop integrations would potentially have difficulties if one uses the conventional approach to perturbation theory. If one uses the approach of Blaha (2003) and (2005a) then there are no divergences.

$$J(\mathbf{x_i} - \mathbf{y_i}, M^2) = (2\pi)^{-3}\int d^3p_i\,\delta(M^2 + \mathbf{p_i}^2 - m^2)\exp[-i\mathbf{p_i}\cdot(\mathbf{x_i} - \mathbf{y_i})] \qquad (3.182)$$
$$= (2\pi)^{-2}|\mathbf{x_i} - \mathbf{y_i}|^{-1}\theta(m^2 - M^2)\sin((m^2 - M^2)^{1/2}|\mathbf{x_i} - \mathbf{y_i}|)$$

The complexon Feynman propagator can be rearranged into the form of a spectral integral:

$$iS_C(x, y) = -\int dM\,(i\gamma^0\partial/\partial x^0 - i(\nabla_r - i\nabla_i)\cdot\gamma + m)\delta'(\nabla_r\cdot\nabla_i/m^2)J(\mathbf{x_i} - \mathbf{y_i}, M^2)\Delta_F(x - y, M) \quad (3.183)$$

where

$$\Delta_F(x - y, M) = (2\pi)^{-4}\int d^4p_r\,\exp[-ip^0(x^0 - y^0) + i\mathbf{p_r}\cdot(\mathbf{x_r} - \mathbf{y_r})]/(p_r^2 - M^2 + i\varepsilon) \qquad (3.184)$$

Case 4: Left-handed Tachyon Complexons

In this case $\hat{\mathbf{u}}_r\cdot\hat{\mathbf{u}}_i = 0$ again. However we add an imaginary term to ω to obtain a manifest Left-handed L_C boost[69]

$$\Lambda_{CL}(\mathbf{v_c}) = \exp[i(\omega + i\pi/2)\hat{\mathbf{w}}\cdot\mathbf{K}] \qquad (3.185)$$

where ω remains

$$\omega = (\omega_r^2 - \omega_i^2)^{1/2} \qquad (3.186)$$

and

$$\hat{\mathbf{w}} = (\omega_r\hat{\mathbf{u}}_r + i\omega_i\hat{\mathbf{u}}_i)/\omega \qquad (3.187)$$
$$\hat{\mathbf{w}}\cdot\hat{\mathbf{w}} = \hat{\mathbf{u}}_r\cdot\hat{\mathbf{u}}_r = \hat{\mathbf{u}}_i\cdot\hat{\mathbf{u}}_i = 1 \qquad (3.188)$$
$$\mathbf{v_c} = \hat{\mathbf{w}}\tanh(\omega + i\pi/2) = \hat{\mathbf{w}}\,\mathrm{cotanh}(\omega) \qquad (3.189)$$

Letting $\omega_L = \omega + i\pi/2$ we find, as before,

$$\cosh(\omega_L) = i\sinh(\omega) = -\gamma = i\gamma_s$$
$$\sinh(\omega_L) = i\cosh(\omega) = -\beta\gamma = i\beta\gamma_s \qquad (3.190)$$

with, $\beta = v_c = |\mathbf{v_c}| > 1$, $\gamma_s = (\beta^2 - 1)^{-1/2}$, and

$$\sinh(\omega) = \gamma_s$$
$$\cosh(\omega) = \beta\gamma_s \qquad (3.191)$$

Thus we denote $\Lambda_{CL}(\mathbf{v_c})$ by

$$\Lambda_{CL}(\mathbf{v_c}) \equiv \Lambda_{CL}(\omega, \hat{\mathbf{w}}) \qquad (3.192)$$

---

[69] The reader can readily verify the form is consistent that generated by an L_C boost transformation.

The corresponding spinor boost transformation is:

$$S_{CL}(\Lambda_{CL}(\omega, \hat{\mathbf{w}})) = \exp(-i\omega_L\sigma_{0i}\hat{w}_i/2) = \exp(-\omega_L\gamma^0\boldsymbol{\gamma}\cdot\hat{\mathbf{w}}/2)$$
$$= \cosh(\omega_L/2)I + \sinh(\omega_L/2)\gamma^0\boldsymbol{\gamma}\cdot\hat{\mathbf{w}} \qquad (3.193)$$

The momentum space equation generated by $S_{CL}(\Lambda_{CL}(\omega, \hat{\mathbf{w}}))$ is

$$\{m\cosh(\omega_L)\gamma^0 - m\sinh(\omega_L)\boldsymbol{\gamma}\cdot(\omega_r\hat{\mathbf{u}}_r + i\omega_i\hat{\mathbf{u}}_i)/\omega - m\}e^{+ip\cdot x}w_{cL}(p) = 0 \qquad (3.194)$$

or

$$\{im\sinh(\omega)\gamma^0 - im\cosh(\omega)\boldsymbol{\gamma}\cdot(\omega_r\hat{\mathbf{u}}_r + i\omega_i\hat{\mathbf{u}}_i)/\omega - m\}e^{+ip\cdot x}w_{cL}(p) = 0 \qquad (3.195)$$

where $p\cdot x = Et - \mathbf{p}\cdot\mathbf{x}$ after performing a corresponding left-handed superluminal coordinate transformation in the exponential factor. Thus the positive energy wave is transformed into a negative energy wave by the transformation.

The momentum 4-vector is defined by

$$p = (p^0, \mathbf{p}) \qquad (3.196)$$

where

$$p^0 = m\sinh(\omega) \qquad \mathbf{p} = \mathbf{p}_r + i\mathbf{p}_i \qquad (3.197)$$

with

$$\mathbf{p}_r = m\omega_r\hat{\mathbf{u}}_r\cosh(\omega)/\omega \qquad \mathbf{p}_i = m\omega_i\hat{\mathbf{u}}_i\cosh(\omega)/\omega \qquad (3.198)$$

and

$$\mathbf{p}_r\cdot\mathbf{p}_i = 0 \qquad (3.199)$$

then eq. 3.195 becomes the complexon tachyon equation

$$[ip\cdot\gamma - m]e^{+ip\cdot x}w_{cL}(p) = 0 \qquad (3.200)$$

with a complex 3-momentum **p** and the tachyon 4-momentum mass shell condition:[70]

$$p^2 = p^{0\,2} - \mathbf{p}_r^{\,2} + \mathbf{p}_i^{\,2} = -m^2 \qquad (3.201)$$

Eq. 3.200 is the momentum space equivalent of the wave equation

$$[\gamma^0\partial/\partial t + \boldsymbol{\gamma}\cdot(\nabla_r + i\nabla_i) - m]\psi_{CL}(t, \mathbf{x}_r, \mathbf{x}_i) = 0 \qquad (3.202)$$

or

[70] Note that the presence of the $\mathbf{p}_i^2$ term does not change the tachyon requirement that $\mathbf{p}_r^2 \geq m^2$ as seen in the previous cases.

$$[\gamma\cdot\nabla - m]\psi_{CL}(t, \mathbf{x}_r, \mathbf{x}_i) = 0 \qquad (3.203)$$

with the subsidiary condition on the wave function

$$\nabla_r\cdot\nabla_i\, \psi_{CL}(t, \mathbf{x}_r, \mathbf{x}_i) = 0 \qquad (3.204)$$

also holds. We note that eq. 3.202 is covariant under the real Lorentz group and eq. 3.204 can be easily put into (real Lorentz group) covariant form.

Before considering a lagrangian formulation and the Fourier operator representation of $\psi_{CL}(t, \mathbf{x}_r, \mathbf{x}_i)$ we will define the tachyon spinors, and its associated real and imaginary spin operators.

The spinor generated from a spin up Dirac spinor at rest by the L_C spinor boost eq. 3.193 is

$$w_{cL}(p) = S_{CL}w(0) = [\cosh(\omega_L/2)I + \sinh(\omega_L/2)\gamma^0\boldsymbol{\gamma}\cdot\hat{\mathbf{w}}]w(0) \qquad (3.205)$$

Following a procedure similar to Appendix 3-A (which the reader may wish to examine first) we define four spinors for Dirac particles at rest with eq. 3-A.2. Then by applying a boost to these rest spinors we find the L_C tachyon spinors:

$$S_{CL}w^k(0) = w_{cL}{}^k(p) \qquad (3.206)$$

and from these tachyon spinors we generalize to tachyon spinors $u_{CL}(p, s)$ and $v_{CL}(p, s)$ in a manner similar tothat of the previous case.

Eqs. 3.200 through 3.204 imply that the wave function solution of eq. 3.200, subject to the subsidiary condition eq. 3.204, has the form

$$\psi_{CL}(x) = \sum_{\pm s} \int_{\mathbf{p}_r^2 \geq m^2} d^3p_r d^3p_i N_{CL}(p)\delta(\mathbf{p}_r\cdot\mathbf{p}_i/m^2)[b_{CL}(p,s)u_{CL}(p, s)e^{-i(p\cdot x+p^*\cdot x^*)/2} + d_{CL}{}^\dagger(p,s)v_{CL}(p, s)e^{+i(p\cdot x+p^*\cdot x^*)/2}] \qquad (3.207)$$

where $\mathbf{p} = \mathbf{p}_r + i\mathbf{p}_i$, $\mathbf{x} = \mathbf{x}_r - i\mathbf{x}_i$, $p\cdot x = p^0x^0 - \mathbf{p}\cdot\mathbf{x}$, and $b_{CL}(p, s)$ and $d_{CL}(p,s)$ are tachyon fourier coefficients.

Global SU(3) Symmetry

We can show that there is also a global SU(3) symmetry present here as shown in the previous case. The demonstration is similar to that of eqs. 3.143 – 3.156.

Light-Front Quantization of Tachyonic Complexons

Because of the momentum constraint $\mathbf{p_r}^2 \geq m^2$ the set of solutions of the form of eq. .207 is incomplete and the result of second quantization would not be an equal time anti-commutator expression consisting of derivatives of delta functions (eq. 3.176) but rather an analogue to previous unsuccessful attempts to create a second quantized tachyon theory.[71]

Therefore we will use light-front coordinates, and left and right handed field operators (as previously) to obtain a successful second quantization of this new type of tachyon.

The "missing" factor of i in the first term of eq. 3.203 requires the lagrangian to be different from the conventional Dirac lagrangian in order for the lagrangian to be real. The simplest, physically acceptable, free spin ½ tachyon lagrangian density for ψ_{CL} is:

$$\mathcal{L}_{CL} = \psi_{CL}{}^{C}(x)(\gamma\cdot\nabla - m)\psi_{CL}(x) \tag{3.208}$$

where

$$\psi_{CL}{}^{C}(x) = [\psi_{CL}(x)]^{\dagger}|_{\mathbf{x_i} = -\mathbf{x_i}}\, i\gamma^0\gamma^5 \tag{3.209}$$

is similar to eq. 3.161. In words, eq. 3.209 states: take the hermitean conjugate of $\psi_{CL}(x)$; change $\mathbf{x_i}$ to $-\mathbf{x_i}$; and then post-multiply by the indicated factors.

The free complexon invariant action (under real Lorentz transformations) is

$$I = \int d^7x\mathcal{L}_{CL} \tag{3.210}$$

The action can be shown to be real

$$I^* = I \tag{3.211}$$

in a manner similar to the case considered in Appendix 3-A. The tachyonic complexon's energy-momentum tensor is

$$\begin{aligned}\mathfrak{T}_{CL\mu\nu} &= -\, g_{\mu\nu}\,\mathcal{L}_{CL} + \partial\mathcal{L}_{CL}/\partial(D^{\mu}\psi_{CL})\, D_{\nu}\psi_{CL} \\ &= i\psi_{CL}{}^{C}\gamma^0\gamma^5\gamma_{\mu}D_{\nu}\psi_{CL}\end{aligned} \tag{3.212}$$

where

$$\begin{aligned}D_0 &= \partial/\partial x^0 \\ D_k &= \partial/\partial x_r{}^k + i\,\partial/\partial x_i{}^k\end{aligned} \tag{3.213}$$

and thus the conserved energy and momentum are

$$P^0 = H = \int d^3x_r d^3x_i\, \mathfrak{T}_{CL}{}^{00} = i\int d^3x_r d^3x_i \psi_{CL}{}^{C}\gamma^5(\boldsymbol{\alpha}\cdot\mathbf{D} + \beta m)\psi_{CL} \tag{3.214}$$

---

[71] Such as G. Feinberg, Phys. Rev. **159**, 1089 (1967).

$$P^k = \int d^3x_r d^3x_i\, \mathfrak{T}_{CL}{}^{0k} = -\, i\int d^3x_r d^3x_i\, \psi_{CL}{}^{C}\gamma^5 D^k \psi_{CL} \qquad (3.215)$$

Having defined a suitable tachyon lagrangian we can now proceed to its canonical quantization. The conjugate momentum can be calculated from the lagrangian density eq. 3.212:

$$\pi_{CLa} = \partial\mathcal{L}_{CL}/\partial\dot{\psi}_{CLa} \equiv \partial\mathcal{L}_{CL}/\partial(\partial\psi_{CLa}/\partial t) = -i([\psi_{CL}(x)]^\dagger|_{\mathbf{x_i} = -\mathbf{x_i}}\gamma^5)_a \qquad (3.216)$$

The resulting non-zero, canonical anti-commutation relations are

$$\{\pi_{CL\alpha}(x), \psi_{CLb}(y)\} = i\,\delta_{ab}\,\delta^3(x_r - y_r)\delta^3(x_i - y_i)$$

based on locality in both real and imaginary coordinates:

$$\{\psi_{CL}{}^{\dagger}{}_{\alpha}(x)|_{\mathbf{x_i} = -\mathbf{x_i}}, \psi_{Tb}(y)\} = -\,[\gamma^5]_{ab}\,\delta^3(x_r - y_r)\delta^3(x_i - y_i) \qquad (3.217)$$

At this point we might attempt to complete the canonical quantization procedure in the conventional manner by Fourier expanding the field and specifying anti-commutation relations for the fourier component amplitudes. However the incompleteness of the set of plane waves, which are limited by the restriction $\mathbf{p_r}^2 \geq m^2$, causes the equal time anti-commutator of the fields *not* to yield a δ-functions.

Therefore we turn to the previous successful approach to tachyon quantization[72] and decompose the tachyonic complexon field into left-handed and right-handed parts and then second quantize in light-front coordinates.

Separation into Left-Handed and Right-Handed Fields

As before we will use a transformed set of Dirac matrices to develop our left-handed and right-handed tachyon formulations. The γ^5 chirality operator's eigenvalues define handedness: +1 corresponds to right-handed; and −1 corresponds to left-handed:

$$\gamma^5\psi_{CLL} = -\,\psi_{CLL} \qquad\qquad \gamma^5\psi_{CLR} = \psi_{CLR} \qquad (3.218)$$

We define left-handed and right-handed tachyon fields with the projection operators:

$$C^{\pm} = \tfrac{1}{2}(I \pm \gamma^5)$$

[72] Blaha (2006) discusses this case in detail.

$$C^+ + C^- = I \qquad (3.219)$$
$$C^{\pm 2} = C^{\pm}$$
$$C^+C^- = 0$$

with the result

$$\psi_{CLL} = C^-\psi_{CL} \qquad (3.220)$$
$$\psi_{CLR} = C^+\psi_{CL}$$

We can calculate the commutation relations of the left-handed and right-handed tachyonic complexon fields from eq. 3.217 by pre-multiplying and post-multiplying by ½(1 – γ^5) and ½(1 + γ^5). The results are:

$$\{\psi_{CLLa}{}^{\dagger}(x)|_{\mathbf{x_i} = -\mathbf{x_i}}, \psi_{CLLb}(y)\} = C^-{}_{ab}\,\delta^6(x - y) \qquad (3.221)$$

$$\{\psi_{CLRa}{}^{\dagger}(x)|_{\mathbf{x_i} = -\mathbf{x_i}}, \psi_{CLRb}(y)\} = -C^+{}_{ab}\,\delta^6(x - y) \qquad (3.222)$$

$$\{\psi_{CLLa}{}^{\dagger}(x)|_{\mathbf{x_i} = -\mathbf{x_i}}, \psi_{CLRb}(y)\} = \{\psi_{CLRa}{}^{\dagger}(x)|_{\mathbf{x_i} = -\mathbf{x_i}}, \psi_{CLLb}(x')\} = 0 \qquad (3.223)$$

where

$$\delta^6(x - y) = \delta^3(x_r - y_r)\delta^3(x_i - y_i) \qquad (3.224)$$

The lagrangian density of eq. 3.208 decomposes into left-handed and right-handed parts: (The change $\mathbf{x_i}$ to $-\mathbf{x_i}$ will be understood in $\psi_{CLL}{}^{\dagger}(x)$ and $\psi_{CLR}{}^{\dagger}(x)$ in the following.)

$$\mathcal{L}_{CL} = \psi_{CLL}{}^{\dagger}\gamma^0 i\gamma^{\mu}\partial_{\mu}\psi_{CLL} - \psi_{CLR}{}^{\dagger}\gamma^0 i\gamma^{\mu}\partial_{\mu}\psi_{CLR} - im[\psi_{CLR}{}^{\dagger}\gamma^0\psi_{CLL} - \psi_{CLL}{}^{\dagger}\gamma^0\psi_{CLR}] \qquad (3.225)$$

Further Separation into + and – Light-Front Complexon Fields

As previously, we now use light-front coordinates and quantization to obtain a successful second quantization of this form of tachyon field. Light-front variables, in the present case where we have to contend with complex 3-vectors, are defined by real coordinates and derivatives:

$$x^{\pm} = (x^0 \pm x_r{}^3)/\sqrt{2}$$
$$\partial/\partial x^{\pm} \equiv \partial^{\mp} \equiv (\partial/\partial x^0 \pm \partial/\partial x_r{}^3)/\sqrt{2} \qquad (3.226)$$

with the "transverse" real coordinate variables, $x_r{}^1$ and $x_r{}^2$, and imaginary coordinate variables $x_i{}^1$, $x_i{}^2$, and $x_i{}^3$.

The inner product of two 4-vectors has the form

$$x\cdot y = x^+y^- + y^+x^- + i[y_i^3(x^+ - x^-) + x_i^3(y^+ - y^-)]/\sqrt{2} + x_i^3y_i^3 - (\mathbf{x}_{r\perp} - i\mathbf{x}_{i\perp})\cdot(\mathbf{y}_{r\perp} - i\mathbf{y}_{i\perp}) \qquad (3.227)$$

with

$$\mathbf{x}_{r\perp} = (x_r^1, x_r^2) \qquad \mathbf{x}_{i\perp} = (x_i^1, x_i^2) \qquad (3.228)$$
$$\mathbf{y}_{r\perp} = (y_r^1, y_r^2) \qquad \mathbf{y}_{i\perp} = (y_i^1, y_i^2)$$

where $x = (x^0, \mathbf{x} = \mathbf{x}_r - i\mathbf{x}_i)$ and $y = (y^0, \mathbf{y} = \mathbf{y}_r - i\mathbf{y}_i)$. Momenta are always defined as $p = (p^0, \mathbf{p} = \mathbf{p}_r + i\mathbf{p}_i)$.

The light-front definition of Dirac matrices is:

$$\gamma^\pm = (\gamma^0 \pm \gamma^3)/\sqrt{2} \qquad (3.229)$$

with transverse matrices γ^1 and γ^2 defined as usual. Note:

$$\gamma^{\pm\,2} = 0$$

We define "+" and "–" tachyon fields with the projection operators:

$$R^\pm = \tfrac{1}{2}(I \pm \gamma^0\gamma^3) \qquad (3.230)$$

Left-handed, ± light-front fields: $\psi_{CLL}{}^\pm = R^\pm C^-\psi_{CL}$ (3.231)

Right-handed, ± light-front fields: $\psi_{CLR}{}^\pm = R^\pm C^+\psi_{CL}$

Transforming to light-front variables and fields as above we obtain the light-front free tachyon lagrangian:

$$\begin{aligned}\mathcal{L}_{CL} = {} & 2^{1/2}\psi_{CLL}{}^{+\dagger}i\partial^-\psi_{CLL}{}^+ + 2^{1/2}\psi_{CLL}{}^{-\dagger}i\partial^+\psi_{CLL}{}^- - \psi_{CLL}{}^{+\dagger}\gamma^0[i\boldsymbol{\gamma}_\perp\cdot\boldsymbol{\nabla}_{r\perp} - \boldsymbol{\gamma}\cdot\boldsymbol{\nabla}_i]\psi_{CLL}{}^- - \\ & - \psi_{CLL}{}^{-\dagger}\gamma^0[i\boldsymbol{\gamma}_\perp\cdot\boldsymbol{\nabla}_{r\perp} - \boldsymbol{\gamma}\cdot\boldsymbol{\nabla}_i]\psi_{CLL}{}^+ - 2^{1/2}\psi_{CLR}{}^{+\dagger}i\partial^-\psi_{CLR}{}^+ - 2^{1/2}\psi_{CLR}{}^{-\dagger}i\partial^+\psi_{CLR}{}^- + \\ & + \psi_{CLR}{}^{+\dagger}\gamma^0[i\boldsymbol{\gamma}_\perp\cdot\boldsymbol{\nabla}_{r\perp} - \boldsymbol{\gamma}\cdot\boldsymbol{\nabla}_i]\psi_{CLR}{}^- + \psi_{CLR}{}^{-\dagger}\gamma^0[i\boldsymbol{\gamma}_\perp\cdot\boldsymbol{\nabla}_{r\perp} - \boldsymbol{\gamma}\cdot\boldsymbol{\nabla}_i]\psi_{CLR}{}^+ - \\ & - im[\psi_{CLR}{}^{+\dagger}\gamma^0\psi_{CLL}{}^- - \psi_{CLL}{}^{+\dagger}\gamma^0\psi_{CLR}{}^- + \psi_{CLR}{}^{-\dagger}\gamma^0\psi_{CLL}{}^+ - \psi_{CLL}{}^{-\dagger}\gamma^0\psi_{CLR}{}^+]\end{aligned} \qquad (3.232)$$

(Note the similarity to the previous tachyon case.) Again the difference in signs between the left-handed and right-handed terms will be a crucial factor in the derivation of the left-handed features of the Standard Model.

Eq. 3.232 generates the equations of motion:

$$2^{1/2}i\partial^-\psi_{CLL}{}^+ - \gamma^0[i\boldsymbol{\gamma}_\perp\cdot\boldsymbol{\nabla}_{r\perp} - \boldsymbol{\gamma}\cdot\boldsymbol{\nabla}_i]\psi_{CLL}{}^- + im\gamma^0\psi_{CLR}{}^- = 0 \qquad (3.233)$$

$$2^{½}i\partial^-\psi_{CLR}{}^+ - \gamma^0[i\gamma_\perp\cdot\nabla_{r\perp} - \gamma\cdot\nabla_i]\psi_{CLR}{}^- + im\gamma^0\psi_{CLL}{}^- = 0$$
$$2^{½}i\partial^+\psi_{CLL}{}^- - \gamma^0[i\gamma_\perp\cdot\nabla_{r\perp} - \gamma\cdot\nabla_i]\psi_{CLL}{}^+ + im\gamma^0\psi_{CLR}{}^+ = 0$$
$$2^{½}i\partial^+\psi_{CLR}{}^- - \gamma^0[i\gamma_\perp\cdot\nabla_{r\perp} - \gamma\cdot\nabla_i]\psi_{CLR}{}^+ + im\gamma^0\psi_{CLL}{}^+ = 0$$

Eqs. 3.233 show that $\psi_{CLL}{}^-$ and $\psi_{CLR}{}^-$ are dependent fields that are functions of $\psi_{CLL}{}^+$ and $\psi_{CLR}{}^+$ on the light-front where x^+ equals a constant. They can be expressed in an integral form as well. (The independent fields $\psi_{CLL}{}^+$ and $\psi_{CLR}{}^+$ play a fundamental role in tachyonic complexon theory and are used to define "in" and "out" tachyon states in perturbation theory.)

The conjugate momenta implied by eq. 3.232 are

$$\pi_{CLL}{}^+ = \partial\mathcal{L}/\partial(\partial^-\psi_{CLL}{}^+) = 2^{½}i\psi_{CLL}{}^{+\dagger} \tag{3.234}$$
$$\pi_{CLL}{}^- = \partial\mathcal{L}/\partial(\partial^-\psi_{CLL}{}^-) = 0$$
$$\pi_{CLR}{}^+ = \partial\mathcal{L}/\partial(\partial^-\psi_{CLR}{}^+) = -2^{½}i\psi_{CLR}{}^{+\dagger} \tag{3.235}$$
$$\pi_{CLR}{}^- = \partial\mathcal{L}/\partial(\partial^-\psi_{CLR}{}^-) = 0$$

x^+ plays the role of the "time" variable in light-front quantized theories. So we define canonical equal x^+ anti-commutation relations for spin ½ tachyonic complexons also.

The canonical equal-light-front ($x^+ = y^+$) anti-commutation relations of the independent fields would normally be:

$$\{\psi_{CLL}{}^{+\dagger}{}_a(x), \psi_{CLL}{}^+{}_b(y)\} = 2^{-1}[C^-R^+]_{ab}\delta(x^- - y^-)\delta^2(x_r - y_r)\delta^3(x_I - y_i) \tag{3.236}$$
$$\{\psi_{CLR}{}^{+\dagger}{}_a(x), \psi_{CLR}{}^+{}_b(y)\} = -2^{-1}[C^+R^+]_{ab}\,\delta(x^- - y^-)\delta^2(x_r - y_r)\delta^3(x_I - y_i) \tag{3.237}$$
$$\{\psi_{CLL}{}^{+\dagger}{}_a(x), \psi_{CLR}{}^+{}_b(y)\} = \{\psi_{CLR}{}^{+\dagger}{}_a(x), \psi_{CLL}{}^+{}_b(y)\} = 0 \tag{3.238}$$
$$\{\psi_{CLL}{}^+{}_a(x), \psi_{CLR}{}^+{}_b(y)\} = \{\psi_{CLR}{}^{+\dagger}{}_a(x), \psi_{CLL}{}^{+\dagger}{}_b(y)\} = 0 \tag{3.239}$$

But, as in the previous case, they will be modified.

Again we see that the right-handed tachyon anti-commutation relation (eq. 3.237) has a minus sign relative to the corresponding conventional right-handed anti-commutation relation.

The sign differences between the left-handed and right-handed lagrangian terms ultimately lead to parity violating features in the Standard Model lagrangian.

Left-Handed Tachyonic Complexons

The free, "+" light-front, left-handed tachyonic complexon Fourier expansion is:

$$\psi_{CLL}{}^{+}(x_r, x_i) = \sum_{\pm s}\int d^2p_r dp^+ d^3p_i\, N_{CLL}{}^{+}(p)\theta(p^+)\delta((p_i{}^3(p^+ - p^-)/\sqrt{2} + \mathbf{p}_{r\perp}\cdot\mathbf{p}_{i\perp})/m^2)\cdot$$

$$\cdot[b_{CLL}{}^{+}(p, s)u_{CLL}{}^{+}(p, s)e^{-i(p\cdot x + p^*\cdot x^*)/2} + d_{CLL}{}^{+\dagger}(p, s)v_{CLL}{}^{+}(p, s)e^{+i(p\cdot x + p^*\cdot x^*)/2}] \tag{3.240}$$

Its hermitean conjugate is

$$\psi_{CLL}{}^{+\dagger}(x_r, x_i) = \sum_{\pm s}\int d^2p_r dp^+ d^3p_i\, N_{CLL}{}^{+}(p)\theta(p^+)\delta((p_i{}^3(p^+ - p^-)/\sqrt{2} + \mathbf{p}_{r\perp}\cdot\mathbf{p}_{i\perp})/m^2)\cdot$$
$$\cdot[b_{CLL}{}^{\dagger}(p^*,s)u_{CLL}{}^{\dagger}(p^*,s)e^{+i(p^*\cdot x + p\cdot x^*)/2} + d_{CLL}(p^*,s)v_{CLL}{}^{\dagger}(p^*,s)e^{-i(p^*\cdot x + p\cdot x^*)/2}] \tag{3.241}$$

where $\mathbf{p} = \mathbf{p_r} + i\mathbf{p_i}$, $\mathbf{x} = \mathbf{x_r} - i\mathbf{x_i}$, $p\cdot x = p^0x^0 - \mathbf{p}\cdot\mathbf{x}$, and † indicates hermitean conjugate. The spinors are

$$\begin{aligned}
u_{CLL}{}^{+}(p, s) &= C^- R^+ S_{CL}w^1(0)\\
u_{CLL}{}^{+}(p, -s) &= C^- R^+ S_{CL}w^2(0)\\
v_{CLL}{}^{+}(p, s) &= C^- R^+ S_{CL}w^3(0)\\
v_{CLL}{}^{+}(p, -s) &= C^- R^+ S_{CL}w^4(0)\\
u_{CLL}{}^{+\dagger}(p^*, s) &= w^{1T}(0)S_{CL}R^+C^-\\
u_{CLL}{}^{+\dagger}(p^*, -s) &= w^{2T}(0)S_{CL}R^+C^-\\
v_{CLL}{}^{+\dagger}(p^*, s) &= w^{3T}(0)S_{CL}R^+C^-\\
v_{CLL}{}^{+\dagger}(p^*, -s) &= w^{4T}(0)S_{CL}R^+C^-
\end{aligned} \tag{3.242}$$

where the superscript "T" indicates the transpose (These spinors are described in Appendix 3-A.) and

$$N_{CLL}{}^{+}(p) = (2\pi)^{-3}(2m/p^+)^{1/2} \tag{3.243}$$

The anti-commutation relations of the Fourier coefficient operators are

$$\begin{aligned}
\{b_{CLL}(p,s), b_{CLL}{}^{\dagger}(p'^*,s')\} &= 2^{-1/2}\delta_{ss'}\delta(p^+ - p'^+)\delta^2(\mathbf{p_r} - \mathbf{p'}_{r'})\delta^3(\mathbf{p_i} + \mathbf{p'}_{i'})\\
\{d_{CLL}(p,s), d_{CLL}{}^{\dagger}(p'^*,s')\} &= 2^{-1/2}\delta_{ss'}\delta(p^+ - p'^+)\delta^2(\mathbf{p_r} - \mathbf{p'}_{r'})\delta^3(\mathbf{p_i} + \mathbf{p'}_{i'})\\
\{b_{CLL}(p,s), b_{CLL}(p'^*,s')\} &= \{d_{CLL}(p,s), d_{CLL}(p'^*,s')\} = 0\\
\{b_{CLL}{}^{\dagger}(p,s), b_{CLL}{}^{\dagger}(p'^*,s')\} &= \{d_{CLL}{}^{\dagger}(p,s), d_{CLL}{}^{\dagger}(p'^*,s')\} = 0\\
\{b_{CLL}(p,s), d_{CLL}{}^{\dagger}(p'^*,s')\} &= \{d_{CLL}(p,s), b_{CLL}{}^{\dagger}(p'^*,s')\} = 0\\
\{b_{CLL}{}^{\dagger}(p,s), d_{CLL}{}^{\dagger}(p'^*,s')\} &= \{d_{CLL}(p,s), b_{CLL}(p'^*,s')\} = 0
\end{aligned} \tag{3.244}$$

The delta-function arguments $\delta^3(\mathbf{p}_i + \mathbf{p'}_{i'})$ above have a positive sign in order to obtain $\delta^3(\mathbf{x}_i - \mathbf{y}_i)$ in the field anti-commutators.

The spinors, eq. 3.242, satisfy

$$\sum_{\pm s} u_{CLL\ \alpha}^{\ +}(p, s)\bar{u}_{CLL\ \beta}^{\ +}(p^*, s) = (2m)^{-1}[C^-R^+(i\not{p} + m)R^-C^+]_{\alpha\beta} \tag{3.245}$$

$$\sum_{\pm s} v_{CLL\ \alpha}^{\ +}(p, s)\bar{v}_{CLL\ \beta}^{\ +}(p^*, s) = (2m)^{-1}[C^-R^+(i\not{p} - m)R^-C^+]_{\alpha\beta}$$

where $\bar{u}_{CLL}^{\ +} = u_{CLL}^{\ +\dagger}\gamma^0$ and $\bar{v}_{CLL}^{\ +} = v_{CLL}^{\ +\dagger}\gamma^0$.

We now evaluate the canonical left-handed, light-front anti-commutation relation:

$$\{\psi_{CLL\ a}^{\ +}(x), \psi_{CLL\ b}^{\ +\dagger}(y)\} = \sum_{\pm s,s'} \int d^3p_i d^2p dp^+ \int d^3p_i' d^2p' dp'^+ N_{CLL}^{\ +}(p)\, N_{CLL}^{\ +}(p')\cdot$$
$$\cdot\theta(p^+)\theta(p'^+)\delta((p_i^{\ 3}(p^+ - p^-)/\surd 2 + \mathbf{p}_{r\perp}\cdot\mathbf{p}_{i\perp})/m^2)\, \delta((p_i'^{\ 3}(p'^- - p'^-)/\surd 2 + \mathbf{p'}_{r\perp}\cdot\mathbf{p'}_{i\perp})/m^2)\cdot$$
$$\cdot[\{b_{CLL}^{\ +\dagger}(p'^*,s'), b_{CLL}^{\ +}(p,s)\}u_{CLL\ a}^{\ +}(p,s)u_{CLL\ b}^{\ +\dagger}(p'^*,s')e^{+i(p'^*\cdot y + p'\cdot y^*)2 - i(p\cdot x + p^*\cdot x^*)/2} +$$
$$+\{d_{CLL}^{\ +}(p'^*,s'), d_{CLL}^{\ +\dagger}(p,s)\}v_{CLL\ a}^{\ +}(p,s)v_{CLL\ b}^{\ +\dagger}(p'^*,s')e^{-i(p'^*\cdot y + p'\cdot y^*)/2 + i(p\cdot x + p^*\cdot x^*)/2}]$$

$$= 2^{-1/2}\sum_{\pm s} \int d^3p_i d^2p_r dp^+\, [N_{CLL}^{\ +}(p)]^2\theta(p^+)[\delta((p_i^{\ 3}(p^+ - p^-)/\surd 2 + \mathbf{p}_{r\perp}\cdot\mathbf{p}_{i\perp})/m^2)]^2\cdot$$
$$\cdot[u_{CLL\ a}^{\ +}(p,s)u_{CLL\ b}^{\ +\dagger}(p^*,s)e^{+i(p^*\cdot(y - x) + p\cdot(y^* - x^*))/2} + v_{CLL\ a}^{\ +}(p,s)v_{CLL\ b}^{\ +\dagger}(p^*,s)e^{-i(p^*\cdot(y - x) + p\cdot(y^* - x^*))/2}]$$

$$= -2^{-3/2}\int d^3p_i d^2p dp^+ \theta(p^+)[N_{CLL}^{\ +}(p)]^2\delta'((p_i^{\ 3}(p^+ - p^-)/\surd 2 + \mathbf{p}_{r\perp}\cdot\mathbf{p}_{i\perp})/m^2)(2m)^{-1}\cdot$$
$$\cdot\{[C^-R^+(i\not{p} + m)\gamma^0R^+C^-]_{ab}e^{+i(p^*\cdot(y - x) + p\cdot(y^* - x^*))/2} + [C^-R^+(i\not{p} - m)\gamma^0R^+C^-]_{ab}e^{-i(p^*\cdot(y - x) + p\cdot(y^* - x^*))/2}\}$$

$$= -(1/2)C^-R^+\delta_{ab}\int d^3p_i\, d^2p_\perp \int_0^\infty dp^+\, \delta'((p_i^{\ 3}(p^+ - p^-)/\surd 2 + \mathbf{p}_\perp\cdot\mathbf{p}_{i\perp})/m^2)(2\pi)^{-6}\cdot$$
$$\cdot\{e^{+i\{p^+(y^- - x^-) - \mathbf{p}_{r\perp}\cdot(\mathbf{y}_{r\perp} - \mathbf{x}_{r\perp}) + \mathbf{p}_i\cdot(\mathbf{y}_i - \mathbf{x}_i)\}} + e^{-i\{p^+(y^- - x^-) - \mathbf{p}_{r\perp}\cdot(\mathbf{y}_{r\perp} - \mathbf{x}_{r\perp}) + \mathbf{p}_i\cdot(\mathbf{y}_i - \mathbf{x}_i)\}}\}$$

$$= -C^-R^+\delta_{ab}(4\pi)^{-1}\int_0^\infty dp^+\delta'(\nabla_r\cdot\nabla_i/m^2)\delta^3(y_i - x_i)\,\delta^2(y_r - x_r)\{e^{+ip^+(y^- - x^-)} + e^{-ip^+(y^- - x^-)}\}$$

whereupon we revert back to the original form of the constraint: $\delta(\nabla_r\cdot\nabla_i/m^2)$

$$\{\psi_{CLL\ a}^{\ +}(x), \psi_{CLL\ b}^{\ +\dagger}(y)\} = -(1/2)C^-R^+\delta_{ab}\,\delta'(\nabla_r\cdot\nabla_i/m^2)\delta(y^- - x^-)\delta^2(y_r - x_r)\delta^3(y_i - x_i) \tag{3.246}$$

The result is the left-handed, light-front equivalent of the earlier non-tachyon result. Again the constraint is apparent in the anti-commutator. (The factor of 2 difference is due to light-front coordinate definitions.)

Therefore we have left-handed, light-front quantized tachyonic complexons with the equivalent of canonical anti-commutation relations, and with localized tachyonic complexons. As a result we have a canonical tachyonic complexon Quantum Field Theory.

Left-handed Case 4: Tachyonic Complexon Feynman Propagator

The light-front Feynman propagator for the left-handed $\psi_{CLL}{}^{+}$ *tachyonic* complexon field is

$$iS^{+}{}_{CLLF}(x,y) = \theta(x^{+} - y^{+})\langle 0|\psi_{CLL}{}^{+}(x)\psi_{CLL}{}^{++}(y)\gamma^{0}|0\rangle - \theta(y^{+} - x^{+})\langle 0|\psi_{CLL}{}^{++}(y)\gamma^{0}\psi_{CLL}{}^{+}(x)|0\rangle \quad (3.247)$$

$$= -\tfrac{1}{2}\int d^3p_i d^2p_r dp^{+}\theta(p^{+})N_{CLL}{}^{+2}\delta'((p_i{}^3(p^{+} - p^{-})/\surd 2 + \mathbf{p}_{r\perp}\cdot\mathbf{p}_{i\perp})/m^2)(2m)^{-1}C^{-}R^{+}\cdot$$
$$\cdot\{\theta(x^{+} - y^{+})[(i\not p + m)\gamma^{0}]e^{+i(p^{*}\cdot(y - x)+p\cdot(y^{*}-x^{*}))/2} +$$
$$+ \theta(y^{+} - x^{+})[(i\not p - m)\gamma^{0}]e^{-i(p^{*}\cdot(y - x)+p\cdot(y^{*}-x^{*}))/2}\}R^{+}C^{-}\gamma^{0}$$

If we define the on-shell momentum variables

$$p_0{}^{-} = (p_{r0}{}^1 p_{r0}{}^1 + p_{r0}{}^2 p_{r0}{}^2 - \mathbf{p}_{i0}\cdot\mathbf{p}_{i0} - m^2)/(2p_0{}^{+})$$
$$p_0{}^{+} = p^{+},\ p_{r0}{}^{j} = p_r{}^{j} \qquad (\text{for } j = 1, 2),$$
$$\mathbf{p}_{i0} = \mathbf{p}_i,\ p_{r\perp 0}{}^2 = p_{r0}{}^{j} p_{r0}{}^{j}$$
$$\not p_0 = p_0\cdot\gamma$$

with $p_0 = (p^0, \mathbf{p}_{r0} + \mathbf{i}\mathbf{p}_{r0})$ then the above equation can be rewritten as

$$= -\tfrac{1}{2}C^{-}R^{+}\int d^4pd^3p_iN_{CLL}{}^{+2}\delta'((p_{i0}{}^3(p_0{}^{+} - p_0{}^{-})/\surd 2 + \mathbf{p}_{r\perp 0}\cdot\mathbf{p}_{i\perp 0})/m^2)(4\pi m)^{-1}e^{+i(p^{*}\cdot(y - x)+p\cdot(y^{*}-x^{*}))/2}\cdot$$
$$\cdot\{\theta(p^{+})(i\not p_0 + m)\gamma^{0}]/[p^{-} - p_0{}^{-} + i\varepsilon] + \theta(-p^{+})(i\not p_0 - m)\gamma^{0}]/[p^{-} + p_0{}^{-} - i\varepsilon]\}R^{+}C\gamma^{0}$$

$$= -\tfrac{1}{2}\int d^4p_r d^3p_i N_{CLL}{}^{+2}\delta'((p_{i0}{}^3(p^{+} - p^{-})/\surd 2 + \mathbf{p}_{r\perp}\cdot\mathbf{p}_{i\perp})/m^2)(p^{+}/4\pi m)\, e^{+i(p^{*}\cdot(y - x)+p\cdot(y^{*}-x^{*}))/2}\cdot$$
$$\cdot[C^{-}R^{+}(i\not p + m)\gamma^{0}R^{+}C^{-}\gamma^{0}][(p^2 + m^2 + i\varepsilon)]^{-1}$$

with $p_r = (p^0, \mathbf{p}_r)$ and $p = (p^0, \mathbf{p}_r + \mathbf{i}\mathbf{p}_r)$. Substituting for N_{CLL} and using $x\delta'(x) = -\delta(x)$ we obtain

$$= -\tfrac{1}{2}\int d^4p_r d^3p_i(2\pi)^{-7}\delta'(\mathbf{p}_r\cdot\mathbf{p}_i/m^2)\exp[ip^0(y^0 - x^0) - i\mathbf{p}_r\cdot(\mathbf{y}_r - \mathbf{x}_r) + i\mathbf{p}_i\cdot(\mathbf{y}_i - \mathbf{x}_i)]\cdot$$
$$\cdot[C^{-}R^{+}(i\not p + m)R^{-}C^{+}]/(p^2 + m^2 + i\varepsilon)$$

since $C^{-}R^{+}(i\not p + m)\gamma^{0}R^{+}C^{-}\gamma^{0} = C^{-}R^{+}(i\not p + m)R^{-}C^{+}$. The integral can be written:

$$= \int d^4p_r d^3p_i \delta'(\mathbf{p}_r\cdot\mathbf{p}_i/m^2) C^- R^+(i\not{p} + m) R^- C^+ \cdot$$
$$\cdot \exp[-ip^0(x^0 - y^0) + i\mathbf{p}_r\cdot(\mathbf{x}_r - \mathbf{y}_r) - i\mathbf{p}_i\cdot(\mathbf{x}_i - \mathbf{y}_i)]/(p^2 + m^2 + i\varepsilon)$$
$$= \int d^4p_r dM^2 \delta'(\nabla_r\cdot\nabla_i/m^2) C^- R^+(ip^0\gamma^0 - (\nabla_r - i\nabla_i)\cdot\gamma + m) R^- C^+ \cdot$$
$$\cdot \exp[-ip^0(x^0 - y^0) + i\mathbf{p}_r\cdot(\mathbf{x}_r - \mathbf{y}_r)] J_2(\mathbf{x}_i - \mathbf{y}_i, M^2)/(p_r^2 + M^2 + i\varepsilon)$$

where

$$J_2(\mathbf{x}_i - \mathbf{y}_i, M^2) = (2\pi)^{-3}\int d^3p_i \delta(M^2 - \mathbf{p}_i^2 - m^2) \exp[-i\mathbf{p}_i\cdot(\mathbf{x}_i - \mathbf{y}_i)] \tag{3.248}$$
$$= (2\pi)^{-2}|\mathbf{x}_i - \mathbf{y}_i|^{-1}\theta(M^2 - m^2)\sin((M^2 - m^2)^{1/2}|\mathbf{x}_i - \mathbf{y}_i|)$$

This tachyonic complexon Feynman propagator can be rearranged into the form of a spectral integral:

$$iS^+_{CLLF}(x, y) = -\int dM\, C^- R^+(\gamma^0\partial/\partial x^0 + (\nabla_r - i\nabla_i)\cdot\gamma - m) R^- C^+ \delta'(\nabla_r\cdot\nabla_i/m^2) J_2(\mathbf{x}_i - \mathbf{y}_i, M^2) \triangle_{FT}(x - y, M) \tag{3.249}$$

with ∇_r and ∇_i derivatives with respect to $\mathbf{x}_r$ and $\mathbf{x}_i$ and where

$$\triangle_{FT}(x - y, M) = (2\pi)^{-4}\int d^4p_r \exp[-ip^0(x^0 - y^0) + i\mathbf{p}_r\cdot(\mathbf{x}_r - \mathbf{y}_r)]/(p_r^2 + M^2 + i\varepsilon) \tag{3.250}$$

Case 5: Right-Handed Tachyonic Complexons

The case of right-handed tachyonic complexons is similar to left-handed complexons with only one difference: a minus sign in the canonical right-handed equal-time commutation relations resulting in a minus sign in the creation and annihilation operator anti-commutation relations. The right-handed tachyonic complexon wave function light-front Fourier expansion is:

$$\psi_{CLR}{}^+(x_r, x_i) = \sum_{\pm s}\int d^2p_r dp^+ d^3p_i\, N_{CLR}{}^+(p)\theta(p^+)\delta((p_i{}^3(p^+ - p^-)/\surd 2 + \mathbf{p}_{r\perp}\cdot\mathbf{p}_{i\perp})/m^2)\cdot$$
$$\cdot[b_{CLR}{}^+(p, s)u_{CLR}{}^+(p, s)e^{-i(p\cdot x + p^*\cdot x^*)/2} + d_{CLR}{}^{+\dagger}(p, s)v_{CLR}{}^+(p, s)e^{+i(p\cdot x + p^*\cdot x^*)/2}] \tag{3.251}$$

where

$$N_{CLR}{}^+(p) = (2\pi)^{-3}(2m/p^+)^{1/2} \tag{3.252}$$

Its hermitean conjugate is

$$\psi_{CLR}{}^{+\dagger}(x_r, x_i) = \sum_{\pm s}\int d^2p_r dp^+ d^3p_i\, N_{CLR}{}^+(p)\theta(p^+)\delta((p_i{}^3(p^+ - p^-)/\surd 2 + \mathbf{p}_{r\perp}\cdot\mathbf{p}_{i\perp})/m^2)\cdot$$
$$\cdot[b_{CLR}{}^\dagger(p^*, s)u_{CLR}{}^\dagger(p^*, s)e^{+i(p^*\cdot x + p\cdot x^*)/2} + d_{CLR}(p^*, s)v_{CLR}{}^\dagger(p^*, s)e^{-i(p^*\cdot x + p\cdot x^*)/2}] \tag{3.253}$$

where $\mathbf{p} = \mathbf{p}_r + i\mathbf{p}_i$, $\mathbf{x} = \mathbf{x}_r - i\mathbf{x}_i$, $p\cdot x = p^0x^0 - \mathbf{p}\cdot\mathbf{x}$, and † indicates hermitean conjugate. The right-handed spinors are

$$\begin{aligned}
u_{CLR}^{+}(p, s) &= C^+ R^+ S_{CR}w^1(0)\\
u_{CLR}^{+}(p, -s) &= C^+ R^+ S_{CR}w^2(0)\\
v_{CLR}^{+}(p, s) &= C^+ R^+ S_{CR}w^3(0)\\
v_{CLR}^{+}(p, -s) &= C^+ R^+ S_{CR}w^4(0)\\
u_{CLR}^{+\dagger}(p^*, s) &= w^{1T}(0)S_{CR}R^+C^+\\
u_{CLR}^{+\dagger}(p^*, -s) &= w^{2T}(0)S_{CR}R^+C^+\\
v_{CLR}^{+\dagger}(p^*, s) &= w^{3T}(0)S_{CR}R^+C^+\\
v_{CLR}^{+\dagger}(p^*, -s) &= w^{4T}(0)S_{CR}R^+C^+
\end{aligned} \qquad (3.254)$$

where the superscript "T" indicates the transpose. The anti-commutation relations of the Fourier coefficient operators are

$$\begin{aligned}
\{b_{CLR}(p,s), b_{CLR}^\dagger(p'^*,s')\} &= -2^{-1/2}\delta_{ss'}\delta(p^+ - p'^+)\delta^2(\mathbf{p}_r - \mathbf{p}'_{r'})\delta^3(\mathbf{p}_i + \mathbf{p}'_{i'})\\
\{d_{CLR}(p,s), d_{CLR}^\dagger(p'^*,s')\} &= -2^{-1/2}\delta_{ss'}\delta(p^+ - p'^+)\delta^2(\mathbf{p}_r - \mathbf{p}'_{r'})\delta^3(\mathbf{p}_i + \mathbf{p}'_{i'})\\
\{b_{CLR}(p,s), b_{CLR}(p'^*,s')\} &= \{d_{CLR}(p,s), d_{CLR}(p'^*,s')\} = 0\\
\{b_{CLR}^\dagger(p,s), b_{CLR}^\dagger(p'^*,s')\} &= \{d_{CLR}^\dagger(p,s), d_{CLR}^\dagger(p'^*,s')\} = 0\\
\{b_{CLR}(p,s), d_{CLR}^\dagger(p'^*,s')\} &= \{d_{CLR}(p,s), b_{CLR}^\dagger(p'^*,s')\} = 0\\
\{b_{CLR}^\dagger(p,s), d_{CLR}^\dagger(p'^*,s')\} &= \{d_{CLR}(p,s), b_{CRR}(p'^*,s')\} = 0
\end{aligned} \qquad (3.255)$$

The spinors satisfy

$$\sum_{\pm s} u_{CLR\,\alpha}^{+}(p, s)\bar{u}_{CLR\,\beta}^{+}(p^*, s) = (2m)^{-1}[C^+R^+(-i\not{p} + m)R^-C^-]_{\alpha\beta} \qquad (3.256)$$

$$\sum_{\pm s} v_{CLR\,\alpha}^{+}(p, s)\bar{v}_{CLR\,\beta}^{+}(p^*, s) = (2m)^{-1}[C^+R^+(-i\not{p} - m)R^-C^-]_{\alpha\beta}$$

where $\bar{u}_{CLR}^{+} = u_{CLR}^{+\dagger}\gamma^0$ and $\bar{v}_{CLR}^{+} = v_{CLR}^{+\dagger}\gamma^0$.

The right-handed anti-commutation relation with a minus sign follows in particular because of the minus signs in eqs. 3.255.

Right-handed Case 5: Tachyonic Complexon Feynman Propagator

The Feynman propagator for right-handed tachyonic complexons can be obtained from eqs. 3.249 and 3.250 by changing the parity projection operator and some numerator signs in the integral (basically $p \rightarrow -p$) resulting in

$$iS^{+}_{CLRF}(x, y) = \int dM\, C^{+}R^{+}(\gamma^0\partial/\partial x^0 + (\nabla_r - i\nabla_i)\cdot\boldsymbol{\gamma} - m)R^{-}C^{-}\delta'(\nabla_r\cdot\nabla_i/m^2)J_2(\mathbf{x_i} - \mathbf{y_i}, M^2)\Delta_{FT}(x - y, M) \quad (3.257)$$

with $\nabla_r + i\nabla_i$ derivatives with respect to $\mathbf{x_r}$ and $\mathbf{x_i}$ and where

$$\Delta_{FT}(x - y, M) = (2\pi)^{-4}\int d^4p_r \exp[-ip^0(x^0 - y^0) + i\mathbf{p}_r\cdot(\mathbf{x_r} - \mathbf{y_r})]/(p_r^2 + M^2 + i\varepsilon) \quad (3.258)$$

Other Cases? No

The four cases considered above are the only cases having symmetry under the real Lorentz group L and *a single real-valued energy* (with a corresponding single real time parameter) that is independent of the direction of the boost, thus preserving (real) spatial rotation invariance. The reality of the time variable survives the breakdown to conventional Lorentz invariance.

One might think that using the other type of spinor boost operator.

$$S_{CR}(\Lambda_{CR}(\omega, \hat{\mathbf{w}})) = \exp(-i\omega_R\sigma_{0i}w_i/2) = \exp(-\omega_R\gamma^0\boldsymbol{\gamma}\cdot\hat{\mathbf{w}}/2) \quad (3.259)$$
$$= \cosh(\omega_R/2)I + \sinh(\omega_R/2)\gamma^0\boldsymbol{\gamma}\cdot\hat{\mathbf{w}}$$

where $\omega_R = \omega - i\pi/2$ might lead to more possible forms of spin ½ wave equations and particles. In fact it merely leads to the same particle types but with the role of the left-handed and right-handed fields reversed. The result would be a "right-handed" Standard Model contrary to experiment.

3.6 Spinor Boosts Generate 4 Species of Particles: Leptons and Quarks

In this chapter we have found four types of fermions using complex Lorentz boosts that correspond in a natural way with the four general species (types) of known fermions: charged leptons, neutrinos, up-type color quarks and down-type color quarks.[73]

Charged lepton fermions

The conventional Dirac equation and solutions.

Neutrinos

Simple tachyons with real energy and 3-momentum. Their free field equation is:

$$(\gamma^\mu\partial/\partial x^\mu - m)\psi_T(x) = 0 \quad (3.260)$$

[73] We call each type of fermion a *species*. Each species has three known generations.

and their left-handed ψ_{TL}^{+} Feynman propagator is:

$$iS^{+}_{TLF}(x, y) = \tfrac{1}{2}C^{-}R^{+}\gamma^{0}\int d^{4}p(2\pi)^{-4}\, p^{+}e^{-ip\cdot(x-y)}/(p^{2} + m^{2} + i\varepsilon) \qquad (3.261)$$

Similarly the light-front Feynman propagator for the right-handed ψ_{TR}^{+} tachyon field is

$$iS^{+}_{TRF}(x,y) = -\tfrac{1}{2}C^{+}R^{+}\gamma^{0}\int d^{4}p(2\pi)^{-4}\, p^{+}e^{-ip\cdot(x-y)}/(p^{2} + m^{2} + i\varepsilon) \qquad (3.262)$$

Up-type Color Quarks

Up-type quarks are assumed[74] to be fermions with complex 3-momenta - complexons, and an internal color SU(3) symmetry, that satisfy $p^2 = m^2$. Their field equation with a color SU(3) index, denoted a, inserted is

$$[i\gamma^{0}\partial/\partial t + i\gamma\cdot(\nabla_r + i\nabla_i) - m]\psi_C^{a}(t, \mathbf{x}_r, \mathbf{x}_i) = 0 \qquad (3.263)$$

with the subsidiary condition

$$\nabla_r\cdot\nabla_i\, \psi_C^{a}(t, \mathbf{x}_r, \mathbf{x}_i) = 0 \qquad (3.264)$$

The free field solution is:

$$\psi_C^{a}(x) = \sum_{\pm s}\int d^3p_r d^3p_i N_C(p)\delta(\mathbf{p}_r\cdot\mathbf{p}_i/m^2)[b_C(p,a,s)u_C^{a}(p, s)e^{-i(p\cdot x+p^*\cdot x^*)/2} + d_C^{\dagger}(p,a,s)v_C^{a}(p, s)e^{+i(p\cdot x+p^*\cdot x^*)/2}] \qquad (3.265)$$

The free Feynman propagator arranged into the form of a spectral integral is

$$iS_C^{ab}(x,y) = -\delta^{ab}\int dM\,(i\gamma^{0}\partial/\partial x^{0} - i(\nabla_r - i\nabla_i)\cdot\gamma + m)\delta'(\nabla_r\cdot\nabla_i/m^2)J(\mathbf{x}_i - \mathbf{y}_i, M^2)\Delta_F(x - y, M) \qquad (3.266)$$

where

$$\Delta_F(x - y, M) = (2\pi)^{-4}\int d^4p_r \exp[-ip^0(x^0 - y^0) + i\mathbf{p}_r\cdot(\mathbf{x}_r - \mathbf{y}_r)]/(p_r^{2} - M^2 + i\varepsilon) \qquad (3.267)$$

and

$$J(\mathbf{x}_i, M^2) = (2\pi)^{-3}\int d^3p_i\,\delta(M^2 + \mathbf{p}_i^{2} - m^2)\exp[-i\mathbf{p}_i\cdot(\mathbf{x}_i - \mathbf{y}_i)] \qquad (3.268)$$
$$= (2\pi)^{-2}|\mathbf{x}_i - \mathbf{y}_i|^{-1}\theta(m^2 - M^2)\sin((m^2 - M^2)^{1/2}|\mathbf{x}_i - \mathbf{y}_i|)$$

---

[74] The complexon theory that we develop and use for quark dynamics in the Standard Model is <u>not</u> required. Our Standard Model could use Dirac fermion dynamics for the up-type quarks and tachyon dynamics for down-type quarks. Then the (broken) Left-handed complex Lorentz boosts would have the basic space-time group rather than L_C. We choose to use complexon dynamics for quarks because they have an internal SU(3)-like structure suggestive of color SU(3). More importantly, their spin dynamics is different and thus may resolve the differences between theory and experiment for the deep inelastic parton spin-dependent structure functions.

Down-type Color Quarks

Tachyonic complexons with complex 3-momenta, and an internal global SU(3) symmetry, that have mass shell condition $p^2 = -m^2$. Their field equation with a color SU(3) index, denoted a, inserted is

$$[\gamma^0\partial/\partial t + \boldsymbol{\gamma}\cdot(\nabla_r + i\nabla_i) - m]\psi_{CL}{}^a(t, \mathbf{x_r}, \mathbf{x_i}) = 0 \quad (3.269)$$

with the subsidiary condition on the wave function

$$\nabla_r\cdot\nabla_i\ \psi_{CL}{}^a(t, \mathbf{x_r}, \mathbf{x_i}) = 0 \quad (3.270)$$

Its free field left-handed solution is:

$$\psi_{CLL}{}^{+a}(x_r, x_i) = \sum_{\pm s}\int d^2p_r dp^+ d^3p_i\, N_{CLL}{}^+(p)\theta(p^+)\delta((p_i{}^3(p^+ - p^-)/\surd 2 + \mathbf{p_{r\perp}}\cdot\mathbf{p_{i\perp}})/m^2)\cdot$$
$$\cdot[b_{CLL}{}^+(p,a,s)u_{CLL}{}^a(p,a,s)e^{-i(p\cdot x + p^*\cdot x^*)/2} + d_{CLL}{}^{+\dagger}(p,a,s)v_{CLL}{}^{+a}(p,a,s)e^{+i(p\cdot x + p^*\cdot x^*)/2}] \quad (3.271)$$

and its right-handed solution is

$$\psi_{CLR}{}^{+a}(x_r, x_i) = \sum_{\pm s}\int d^2p_r dp^+ d^3p_i\, N_{CLR}{}^+(p)\theta(p^+)\delta((p_i{}^3(p^+ - p^-)/\surd 2 + \mathbf{p_{r\perp}}\cdot\mathbf{p_{i\perp}})/m^2)\cdot$$
$$\cdot[b_{CLR}{}^+(p,a,s)u_{CLR}{}^{+a}(p,a,s)e^{-i(p\cdot x+p^*\cdot x^*)/2} + d_{CLR}{}^{+\dagger}(p,a,s)v_{CLR}{}^{+a}(p,a,s)e^{+i(p\cdot x+p^*\cdot x^*)/2}] \quad (3.272)$$

The free left-handed Feynman propagator arranged into the form of a spectral integral is

$$iS^+{}_{CLLF}{}^{ab}(x,y) = -\delta^{ab}\int dM\, C^-R^+(\gamma^0\partial/\partial x^0 + (\nabla_r - i\nabla_i)\cdot\boldsymbol{\gamma} - m)R^-C^+\delta'(\nabla_r\cdot\nabla_i/m^2)J_2(\mathbf{x_i} - \mathbf{y_i}, M^2)\Delta_{FT}(x-y,M) \quad (3.273)$$

with ∇_r and ∇_i derivatives with respect to $\mathbf{x_r}$ and $\mathbf{x_i}$ and where

$$\Delta_{FT}(x - y, M) = (2\pi)^{-4}\int d^4p_r \exp[-ip^0(x^0 - y^0) + i\mathbf{p_r}\cdot(\mathbf{x_r} - \mathbf{y_r})]/(p_r{}^2 + M^2 + i\varepsilon) \quad (3.274)$$

and

$$J_2(\mathbf{x_i}, M^2) = (2\pi)^{-3}\int d^3p_i\, \delta(M^2 - \mathbf{p_i}^2 - m^2)\exp[-i\mathbf{p_i}\cdot(\mathbf{x_i} - \mathbf{y_i})] \quad (3.275)$$

$$= (2\pi)^{-2}|\mathbf{x_i} - \mathbf{y_i}|^{-1}\theta(M^2 - m^2)\sin((M^2 - m^2)^{1/2}|\mathbf{x_i} - \mathbf{y_i}|)$$

The free right-handed Feynman propagator arranged into the form of a spectral integral is

$$iS^{+}{}_{CLRF}{}^{ab}(x, y) = \delta^{ab}\int dM\, C^{+}R^{+}(\gamma^{0}\partial/\partial x^{0} + (\nabla_r - i\nabla_i)\cdot\boldsymbol{\gamma} - m)R^{-}C^{-}\,\delta'(\nabla_r\cdot\nabla_i/m^2)J_2(\mathbf{x}_i - \mathbf{y}_i, M^2)\Delta_{FT}(x-y,M) \quad (3.276)$$

with ∇_r and ∇_i derivatives with respect to $\mathbf{x}_r$ and $\mathbf{x}_i$, and where

$$\Delta_{FT}(x - y, M) = (2\pi)^{-4}\int d^4p_r \exp[-ip^0(x^0 - y^0) + i\mathbf{p}_r\cdot(\mathbf{x}_r - \mathbf{y}_r)]/(p_r^2 + M^2 + i\varepsilon) \quad (3.277)$$

3.7 First Step towards The Standard Model

Thus we have found a set of four fermion species that corresponds to the known fermions of one fermion generation. In subsequent chapters we will derive the one generation model in detail. Then we will introduce three additional generations with mixing to complete the derivation of the form of The Standard Model sector.

The overall pattern that begins to emerge from the developments in this chapter divides particles and interactions into two categories (as seen in Nature):

Particles with real 4-Momenta	Complexons (Complex 3-Momenta)
Leptons	color quarks
SU(2)⊗U(1) Vector Bosons	Color SU(3) gluons
Higgs Particles	

We will explore additional Standard Model issues in detail in the following chapters. But basically the leptons, SU(2)⊗U(1) vector bosons and a set of Higgs particles appear to be based on the Left-handed boosts. These particles have real energies and momenta although some are "normal" and some are tachyons.

Another category of particles, complexons, emerges from our study of L_C. These particles have real energies and complex 3-momenta. In perturbation theory the loop integrations of loops of these particles would consist of a 7-fold integration over energy and complex 3-momenta with corresponding 7-fold delta functions to enforce energy-momentum conservation. As pointed out earlier the complex 3-momenta of these types of fermions has an SU(3) symmetry that it is natural to generalize to local color SU(3). (The other category of fermions lacks this global SU(3) symmetry – just as leptons lack color SU(3).) Thus we see the beginnings of the structure of the Standard Model in this chapter on spin ½ particles. The following chapters lead to a detailed derivation of the form of The Standard Model.

3.8 Why Second Quantization of Fields?

One might have argued that the fermion field types that we have found could be treated as ordinary c-number fields and not be second quantized. However, particles are discrete entities that can be enumerated with integers. Second quantization implements the discrete particle concept in the most direct way and thus second quantization is the best mechanism for obtaining particle discreteness.

Appendix 3-A. Leptonic Tachyon Spinors

The general form of the solutions of the free tachyon Dirac equation eq. 3.18 can be written

$$\psi_T{}^r(x) = e^{-i\chi_r p\cdot x} w^r(p) \qquad (3\text{-A}.1)$$

where $\chi_r = +1$ for r = 1, 2 and $\chi_r = -1$ for r = 3, 4. Denoting the spinors $w^r(p) = w^r(0)$ for a particle is at rest in a frame (E = m) we see they can take the form

$$w^r(0) = \begin{bmatrix} \delta_{1r} \\ \delta_{2r} \\ \delta_{3r} \\ \delta_{4r} \end{bmatrix} \qquad (3\text{-A}.2)$$

where Kronecker deltas appear in the brackets. From eq. 3.30 we find

$$S_L(\Lambda_L(\omega, \mathbf{u}))w^r(0) = w_T{}^r(p) \qquad (3\text{-A}.3)$$

Using eq. 3.11 for $S_L(\Lambda_L(\omega, \mathbf{u}))$ and

$$\mathbf{p} = m\mathbf{v}\gamma_s \qquad E = m\gamma_s \qquad (3\text{-A}.4)$$

we see that eq. 3-A.3 implies the columns of the resulting $S_L(\Lambda_L(\omega, \mathbf{u}))$ matrix are

$$\begin{matrix} \underline{w_T{}^3(p)} & \underline{w_T{}^4(p)} & \underline{w_T{}^1(p)} & \underline{w_T{}^2(p)} \end{matrix} \qquad (3\text{-A}.5)$$

$$S_L(\Lambda_L(\omega, \mathbf{u})) = \begin{bmatrix} \cosh(\omega_L/2) & 0 & \sinh(\omega_L/2)p_z/p & \sinh(\omega_L/2)p_-/p \\ 0 & \cosh(\omega_L/2) & \sinh(\omega_L/2)p_+/p & -\sinh(\omega_L/2)p_z/p \\ \sinh(\omega_L/2)p_z/p & \sinh(\omega_L/2)p_-/p & \cosh(\omega_L/2) & 0 \\ \sinh(\omega_L/2)p_+/p & -\sinh(\omega_L/2)p_z/p & 0 & \cosh(\omega_L/2) \end{bmatrix}$$

based on the superluminal transformation of positive energy states to negative energy states (eqs. 3.30 and 3.31) with $p_\pm = p_x \pm ip_y$ and where $p = |\mathbf{p}|$. It is easy to verify

$$(i\not{p} - \chi_r m)w_T{}^r(p) = 0 \tag{3-A.6}$$

where $\chi_r = -1$ for r = 1, 2 and $\chi_r = +1$ for r = 3, 4.

The spinors that we defined earlier can be generalized in a manner similar to Dirac spinors. We will use a similar notation to the Dirac spinor notation:

$$\begin{aligned} u_T(p, s) &= w_T{}^1(p) \\ u_T(p, -s) &= w_T{}^2(p) \\ v_T(p, s) &= w_T{}^3(p) \\ v_T(p, -s) &= w_T{}^4(p) \end{aligned} \tag{3-A.7}$$

We define "double dagger" spinors:

$$\begin{aligned} u_T{}^\ddagger(p, s) &= u_T{}^\dagger(p, s)i\gamma\cdot\mathbf{p}/|\mathbf{p}| \\ u_T{}^\ddagger(p, -s) &= u_T{}^\dagger(p, -s)i\gamma\cdot\mathbf{p}/|\mathbf{p}| \\ v_T{}^\ddagger(p, s) &= v_T{}^\dagger(p, s)i\gamma\cdot\mathbf{p}/|\mathbf{p}| \\ v_T{}^\ddagger(p, -s) &= v_T{}^\dagger(p, -s)i\gamma\cdot\mathbf{p}/|\mathbf{p}| \end{aligned} \tag{3-A.8}$$

.

where † indicates hermitean conjugate in spinor "completeness" sums:

$$\sum_{\pm s} u_{T\alpha}(p, s)u_T{}^\ddagger{}_\beta(p, s) = (2m)^{-1}(i\not{p} - m)_{\alpha\beta} \tag{3-A.9}$$

$$\sum_{\pm s} v_{T\alpha}(p, s)v_T{}^\ddagger{}_\beta(p, s) = (2m)^{-1}(i\not{p} + m)_{\alpha\beta} \tag{3-A.10}$$

or

$$\sum_{\pm s} u_{T\alpha}(p, s)u_T{}^\dagger{}_\beta(p, s) = -i(2m)^{-1}[(i\not{p} - m)\gamma\cdot\mathbf{p}/|\mathbf{p}|]_{\alpha\beta} \tag{3-A.11}$$

$$\sum_{\pm s} v_{T\alpha}(p, s)v_T{}^\dagger{}_\beta(p, s) = -i(2m)^{-1}[(i\not{p} + m)\gamma\cdot\mathbf{p}/|\mathbf{p}|]_{\alpha\beta} \tag{3-A.12}$$

Lastly we define light-front, left-handed tachyon spinors by

$$\begin{aligned} u_{TL}{}^+(p, s) &= C^- R^+ S_L(\Lambda_L(\omega, \mathbf{u}))w^1(0) \\ u_{TL}{}^+(p, -s) &= C^- R^+ S_L(\Lambda_L(\omega, \mathbf{u}))w^2(0) \\ v_{TL}{}^+(p, s) &= C^- R^+ S_L(\Lambda_L(\omega, \mathbf{u}))w^3(0) \\ v_{TL}{}^+(p, -s) &= C^- R^+ S_L(\Lambda_L(\omega, \mathbf{u}))w^4(0) \end{aligned} \tag{3-A.13}$$

$$
\begin{aligned}
u_{TL}^{+\dagger}(p, s) &= w^{1T}(0)\, S_L^{\dagger}(\Lambda_L(\omega, \mathbf{u}))\, R^+C^- \\
u_{TL}^{+\dagger}(p, -s) &= w^{2T}(0)\, S_L^{\dagger}(\Lambda_L(\omega, \mathbf{u}))R^+C^- \\
v_{TL}^{+\dagger}(p, s) &= w^{3T}(0)\, S_L^{\dagger}(\Lambda_L(\omega, \mathbf{u}))R^+C^- \\
v_{TL}^{+\dagger}(p, -s) &= w^{4T}(0)\, S_L^{\dagger}(\Lambda_L(\omega, \mathbf{u}))R^+C^-
\end{aligned}
\qquad (3\text{-}A.14)
$$

where the superscript "T" indicates the transpose and † indicates hermitean conjugate.

4. ElectroWeak Doublets from Rotations between Tachyon and Normal Particles

4.0 Introduction

In the previous chapters we established a basis for four species of fermions based on Complex Lorentz group boosts. This justification for the number of fermion species had not been noted before the author's work in his previous books.

In this chapter[75] we will introduce the Weak interactions based again on Complex Lorentz group considerations. We begin by generalizing the free Dirac equation to a 2×2 matrix of Dirac-like equations that have a larger group covariance. This matrix equation is applied to a doublet consisting of a normal Dirac particle wave function and a tachyon wave function. We will identify these doublets with Weak lepton doublets initially.

Then starting in section 4.5 we consider a generalized 2×2 equation matrix (covariant under the L_C group) for doublets of complexon particles (quarks) with complex 3-momenta consisting of an up-type complexon and a down-type tachyonic complexon. Because of an inherent SU(3) symmetry we will identify these doublets with quark Weak doublets. SU(3) symmetry leads us to identify each complexon quark in a doublet as a color SU(3) triplet, and leads thence to SU(3) color quark confinement (described in a subsequent chapter).

4.1 Transformations of Dirac and Tachyon Equations

A Left-handed boost of the Dirac equation transforms the Dirac equation into the spin ½ tachyon equation, and vice versa:

$$S_L(\Lambda_L(\omega, \mathbf{u}))\psi(x) \rightarrow \psi_T'(x') \qquad (4.1a)$$
$$S_L(\Lambda_L(\omega, \mathbf{u}))\psi_T(x) \rightarrow \psi'(x')$$

Also, noting the appearance of a γ^5,

$$S_L(\Lambda_L(\omega, \mathbf{u}))(\gamma^\mu\partial/\partial x^\mu - m)S_L^{-1}(\Lambda_L(\omega, \mathbf{u})) = (i\gamma^\mu\partial/\partial x'^\mu - m) \qquad (4.1b)$$
$$S_L(\Lambda_L(\omega, \mathbf{u}))\gamma^5(i\gamma^\mu\partial/\partial x^\mu - m)\gamma^5 S_L^{-1}(\Lambda_L(\omega, \mathbf{u})) = (\gamma^\mu\partial/\partial x'^\mu - m)$$

where

[75] This chapter is extracted from Blaha (2007b).

$$x'^{\mu} = i\Lambda_L{}^{\mu}{}_{\nu}(\omega, \mathbf{u})x^{\nu} \qquad (4.1c)$$
$$\partial/\partial x'^{\mu} = -i\Lambda_L{}^{\nu}{}_{\mu}(\omega, \mathbf{u})\partial/\partial x^{\nu}$$

with

$$x' = E(\mathbf{v})x = i\Lambda_L(\mathbf{v})x$$

Eqs. 4.1a – 4.1c imply

$$S_L(\Lambda_L(\omega, \mathbf{u}))(\gamma^{\mu}\partial/\partial x^{\mu} - m)\psi_T(x) = (i\gamma^{\mu}\partial/\partial x'^{\mu} - m)S_L(\Lambda_L(\omega, \mathbf{u}))\psi_T(x)$$
$$= (i\gamma^{\mu}\partial/\partial x'^{\mu} - m)\psi'(x') \qquad (4.1d)$$

and

$$S_L(\Lambda_L(\omega, \mathbf{u}))\gamma^5(i\gamma^{\mu}\partial/\partial x^{\mu} - m)\psi(x) = (\gamma^{\mu}\partial/\partial x'^{\mu} - m)S_L(\Lambda_L(\omega, \mathbf{u}))\gamma^5\psi(x)$$
$$= (\gamma^{\mu}\partial/\partial x'^{\mu} - m)\psi_T'(x') \qquad (4.1e)$$

where

$$\psi'(x') = S_L(\Lambda_L(\omega, \mathbf{u}))\psi_T(x) \qquad (4.1f)$$

and

$$\psi_T'(x') = S_L(\Lambda_L(\omega, \mathbf{u}))\gamma^5\psi(x) \qquad (4.1g)$$

Note the Dirac equation is not left-handed complex Lorentz covariant.

4.2 Doublet Extended Dirac Equations

We will now consider the issue of generalizing the Dirac equation so that the extended equation is covariant under both Lorentz transformations and Left-handed complex Lorentz transformations.

The only obvious method to obtain an extended Dirac equation that is covariant under complex Lorentz transformations is to define an 8 × 8 matrix generalization. Let

$$đ(x) = \begin{bmatrix} (\gamma^{\mu}\partial/\partial x^{\mu} - m) & 0 \\ 0 & (i\gamma^{\mu}\partial/\partial x^{\mu} - m) \end{bmatrix} \qquad (4.2)$$

be an 8 × 8 matrix operator with the 4 × 4 matrix elements shown, and let

$$\Psi(x) = \begin{bmatrix} \psi_T(x) \\ \psi(x) \end{bmatrix} \qquad (4.3)$$

be an 8 component column vector composed of a Dirac field and a tachyon field. Then the extended Dirac equation is

$$\text{đ}(x)\Psi(x) = 0 \tag{4.4}$$

We now define the 8 × 8 Left-handed complex Lorentz transformation

$$S_{L8}(\Lambda_L(v)) = \begin{bmatrix} 0 & S_L(\Lambda_L(v))\gamma^5 \\ S_L(\Lambda_L(v)) & 0 \end{bmatrix} \tag{4.5}$$

with inverse transformation

$$S_{L8}^{-1}(\Lambda_L(v)) = \begin{bmatrix} 0 & S_L^{-1}(\Lambda_L(v)) \\ \gamma^5 S_L^{-1}(\Lambda_L(v)) & 0 \end{bmatrix} \tag{4.6}$$

Note: we use the notations $S_L(\Lambda_L(v))$ and $S_L(\Lambda_L(\omega, \mathbf{u}))$ interchangeably. Applying S_{L8} to eq. 4.4 yields

$$0 = S_{L8}(\Lambda_L(v))\text{đ}(x)\Psi(x) = \text{đ}(x')\Psi'(x') \tag{4.7}$$

where

$$\Psi'(x') = \begin{bmatrix} S_L\gamma^5\psi(x) \\ S_L\psi_T(x) \end{bmatrix} = \begin{bmatrix} \psi_T'(x') \\ \psi'(x') \end{bmatrix} \tag{4.8}$$

Thus the extended Dirac equation is covariant under generalized Left-handed complex Lorentz transformations such as eqs. 4.5-4.6. Covariance requires the tachyon and the Dirac particles must have the same absolute value for the mass which is the iotal mass in the free fermion case.

It is easy to show that the extended Dirac equation eq. 4.4 is also covariant under conventional Lorentz transformations in the 8 × 8 representation:

$$S_8(\Lambda(v)) = \begin{bmatrix} S(\Lambda(v)) & 0 \\ 0 & S(\Lambda(v)) \end{bmatrix} \tag{4.9}$$

with inverse

$$S_8^{-1}(\Lambda(v)) = \begin{bmatrix} S^{-1}(\Lambda(v)) & 0 \\ 0 & S^{-1}(\Lambda(v)) \end{bmatrix} \tag{4.10}$$

and non-diagonal Lorentz transformations:

$$S_{8A}(\Lambda(v)) = \begin{bmatrix} 0 & S(\Lambda(v)) \\ S(\Lambda(v)) & 0 \end{bmatrix} \tag{4.11}$$

with inverse transformation

$$S_{8A}^{-1}(\Lambda(v)) = \begin{bmatrix} 0 & S^{-1}(\Lambda(v)) \\ S^{-1}(\Lambda(v)) & 0 \end{bmatrix} \tag{4.12}$$

Under a conventional Lorentz transformation we find

$$0 = S_8(\Lambda(v))đ(x)\Psi(x) = đ(x')\Psi'(x') \tag{4.13}$$

$$0 = S_{8A}(\Lambda(v))đ(x)\Psi(x) = đ(x')\Psi'(x')$$

The lagrangian density that corresponds to our 8-dimensional construction is

$$\mathcal{L}_8 = \overline{\Psi}(x)đ(x)\Psi(x) \tag{4.14}$$

where

$$\overline{\Psi}(x) = \Psi^{\dagger}\Gamma^0 \tag{4.15}$$

And

$$\Gamma^0 = \begin{bmatrix} i\gamma^0\gamma^5 & 0 \\ 0 & \gamma^0 \end{bmatrix} \tag{4.16}$$

The action is

$$I = \int d^4x \mathcal{L}_8 \tag{4.17}$$

is invariant under Lorentz transformations S_8 and S_{8A}.

The Hamiltonian density for the 8-dimensional theory is

$$\mathcal{H}_8(x) = \begin{bmatrix} i\psi_T^{\dagger}\gamma^5(\boldsymbol{\alpha}\cdot\nabla + \beta m)\psi_T & 0 \\ 0 & \psi^{\dagger}(-i\boldsymbol{\alpha}\cdot\nabla + \beta m)\psi \end{bmatrix} \tag{4.18}$$

4.3 Non-Invariance of the Extended Free Action under a Left-handed Extended Lorentz Transformation

The action 4.17 is not invariant under Left-handed complex Lorentz transformations. The fundamental cause of this non-invariance is the three dimensional nature of space. In the case of Dirac particles one can define a Lorentz invariant action because time is one-dimensional. Thus one can use $\psi^{\dagger}\gamma^0 = \bar{\psi}$ to form the Dirac field lagrangian and action. A key factor in Lorentz invariance is the relation between the inverse and hermitean conjugate of the spinor boost operator

$$\gamma^0 S^{-1}\gamma^0 = S^{\dagger} \tag{4.19}$$

In the case of the tachyon lagrangian and action, Left-handed complex Lorentz invariance is not possible because the tachyonic equivalent to eq. 4.19 is[76]

$$S_L^{-1}(\Lambda(\mathbf{v}))\boldsymbol{\gamma}\cdot\mathbf{p}/|\mathbf{p}| = i\gamma^0 S_L^{\dagger}(\Lambda(\mathbf{v})) \tag{4.20}$$

where $\mathbf{p} = m\gamma_s\mathbf{v}$. The appearance of $\boldsymbol{\gamma}\cdot\mathbf{p}/|\mathbf{p}|$ in eq. 4.20 precludes the invariance of the free tachyon action.

[76] This relation is derivable from eqs. 3.11 and 3.12.

We will now show the effect of a Left-handed complex Lorentz transformation (eqs. 4.5 and 4.6) on the lagrangian density eq. 4.14. The two non-zero parts of the lagrangian density $\mathcal{L}_8$ (eq. 4.14) are

$$\mathcal{L}_1 = \psi_T^{\dagger} i\gamma^0\gamma^5(\gamma^\mu\partial/\partial x^\mu - m)\psi_T(x) \tag{4.21}$$

and

$$\mathcal{L}_2 = \psi^{\dagger}\gamma^0(i\gamma^\mu\partial/\partial x^\mu - m)\psi(x) \tag{4.22}$$

The effect of the transformation, eqs. 4.5-4.6, on these terms is

$$\begin{aligned}\mathcal{L}_1{}' &= \psi_T^{\dagger} i\gamma^0\gamma^5 S_L^{-1} S_L(\gamma^\mu\partial/\partial x^\mu - m)\, S_L^{-1} S_L\psi_T(x)\\ &= \psi_T^{\dagger} i\gamma^0\gamma^5 S_L^{-1}(i\gamma^\mu\partial/\partial x'^\mu - m)S_L\psi_T(x)\\ &= -\psi_T^{\dagger} S_L^{\dagger}\gamma^5(\boldsymbol{\gamma}\cdot\mathbf{p}/|\mathbf{p}|)(i\gamma^\mu\partial/\partial x'^\mu - m)S_L\psi_T(x)\\ &= \psi'^{\dagger}(x')(\boldsymbol{\gamma}\cdot\mathbf{p}/|\mathbf{p}|)\gamma^5(i\gamma^\mu\partial/\partial x'^\mu - m)\psi'(x')\end{aligned} \tag{4.23}$$

and

$$\begin{aligned}\mathcal{L}_2{}' &= \psi^{\dagger}\gamma^0\gamma^5 S_L^{-1} S_L\gamma^5(i\gamma^\mu\partial/\partial x^\mu - m)\gamma^5 S_L^{-1} S_L\gamma^5\psi(x)\\ &= \psi^{\dagger}\gamma^0\gamma^5 S_L^{-1}(\gamma^\mu\partial/\partial x'^\mu - m)S_L\gamma^5\psi(x)\\ &= i\psi^{\dagger}\gamma^5 S_L^{\dagger}(\boldsymbol{\gamma}\cdot\mathbf{p}/|\mathbf{p}|)(\gamma^\mu\partial/\partial x'^\mu - m)S_L\gamma^5\psi(x)\\ &= i\psi_T'^{\dagger}(x')(\boldsymbol{\gamma}\cdot\mathbf{p}/|\mathbf{p}|)(\gamma^\mu\partial/\partial x'^\mu - m)\psi_T'(x')\end{aligned} \tag{4.24}$$

using eqs. 4.20, 4.1f and 4.1g, where $\psi'(x')$ is a solution of the Dirac equation obtained by Left-handed complex Lorentz boosting (by $\mathbf{v} = \mathbf{p}/(\gamma m)$) of a tachyon field and where $\psi_T'(x')$ is a solution of the tachyon equation obtained by Left-handed complex Lorentz boosting (by $\mathbf{v} = \mathbf{p}/(\gamma m)$) of a Dirac field. Eqs. 4.23-4.24 clearly show that $\mathcal{L}_8$ is *not* invariant under Left-handed complex Lorentz transformations.

Consequently the action of eq. 4.17 is only invariant under inhomogeneous Lorentz transformations. *This state of affairs is actually an advantage when we derive features of the Standard Model because it will be seen to prevent any interplay between unbroken Weak SU(2) rotations and Left-handed complex Lorentz transformations.*

4.4 The Diracian Dilemma – To what do Left-handed Extended Lorentz Boost Particles Correspond? Answer: Leptons

The development of this 8-dimensional formalism, and in particular, the "bi-spinor" wave function consisting of a Dirac spinor and and a tachyon spinor, raises the question, "Is there a particle interpretation for the "bi-spinor" wave function?" Dirac faced a similar issue in 1928-1930 with the negative energy states of the Dirac equation. He developed "hole theory" which eventually led to the interpretation of holes in the sea of filled negative energy states as *positrons*. We now face the same problem: with what pairs of particles do we identify the doublets consisting of a Dirac particle and a tachyon?

The obvious natural interpretation of these 8-spinors is ElectroWeak isodoublets such as:

$$\Psi_\ell(x) = \begin{bmatrix} \psi_{\ell T} \\ \\ \psi_\ell \end{bmatrix} \sim \begin{bmatrix} \nu \\ \\ e \end{bmatrix} \qquad (4.25)$$

is for leptons to have "e" represent a charged lepton and ν represent a neutrino. With this interpretation we can introduce SU(2) gauge interactions and develop one-generation, leptonic Weak theory naturally.

4.5 To what do Complexons Correspond? Quarks

We have identified two of the four types of spin ½ fermions as leptons. The remaining two types of spin ½ fermions – complexons – ψ_C and ψ_{CT} seem to naturally correspond to quarks since their equations of motion and wave functions have a natural SU(3) symmetry as we pointed out earlier. We therefore associate a color SU(3) symmetry with these two types of spin ½ complexons. The Electroweak doublet of quarks then is

$$\Psi_q^{\,a}(x) = \begin{bmatrix} \psi_C^{\,a} \\ \\ \psi_{CT}^{\,a} \end{bmatrix} \sim \begin{bmatrix} u^a \\ \\ d^a \end{bmatrix} \qquad (4.26)$$

where u is an "up" type quark and d is a "down" type quark.[77]

---

[77] While the lepton situation is clear in the sense that charged leptons cannot be tachyons since their masses are known (Thus only tachyonic neutrinos are the only currently allowed possibility.), the quark situation is somewhat unclear. We have provisionally chosen the "down" type of quark (d, s, and b) as tachyonic. The association of bound states of these quarks such as the K^0 and B^0 systems which are known to have CP violation, and the CP violation engendered by tachyons, encourages this interpretation.

In addition, $W^\pm$ charge asymmetry in $p\bar{p}$ collisions indicate the d sea in a proton is greater than the u sea (K. Abe et al, PRL **74**, 850 (1995)) as does the asymmetry of Drell-Yan production in deep inelastic scattering on p and n targets (A. Baldit et al, Phys. Lett. **B332**, 244 (1994)). These results are to be expected since there is no mass gap for a d tachyon sea while there is a mass gap for a u Dirac particle sea. Complexon quarks may explicate the discrepancies between theory and experiment in the spin structure functions of the parton model for nucleons.

The rationale for constructing quark doublets is the same as in the leptonic case: We wish to define a generalization of the "Dirac-like" equations of motion that is covariant under L_C boosts.

4.6 Quark Doublets

We assume that quark doublets consist of a complexon[78] and a tachyonic complexon[79] and to this extent they mirror lepton doublets. In this section we will develop a generalized free complexon equation and describe its features.

Summary of L_C Boosts to Generate Spin ½ Equations

We begin by recapitulating L_C boost features for coordinates and spinors:[80]

$$\Lambda_C(\mathbf{v_c}) = \exp[i\omega\hat{\mathbf{w}}\cdot\mathbf{K}] \tag{2.61}$$
$$\omega = (\omega_r^{\,2} - \omega_i^{\,2} + 2i\omega_r\omega_i\ \hat{\mathbf{u}}_r\cdot\hat{\mathbf{u}}_i)^{1/2} \tag{2.62}$$
$$\hat{\mathbf{w}} = (\omega_r\hat{\mathbf{u}}_r + i\omega_i\hat{\mathbf{u}}_i)/\omega \tag{2.63}$$
$$\hat{\mathbf{w}}\cdot\hat{\mathbf{w}} = \hat{\mathbf{u}}_r\cdot\hat{\mathbf{u}}_r = \hat{\mathbf{u}}_i\cdot\hat{\mathbf{u}}_i = 1 \tag{2.64a}$$
$$\mathbf{v_c} = \hat{\mathbf{w}}\tanh(\omega) \tag{2.64b}$$

The corresponding L_C spinor boost for $m^2 > 0$ particles with complex 3-momenta is

$$\begin{aligned} S_C(\omega, \mathbf{v_c}) &= \exp(-i\omega\sigma_{0k}\hat{w}_k/2) = \exp(-\omega\gamma^0\boldsymbol{\gamma}\cdot\hat{\mathbf{w}}/2) \\ &= \cosh(\omega/2)I + \sinh(\omega/2)\gamma^0\boldsymbol{\gamma}\cdot\hat{\mathbf{w}} \end{aligned} \tag{3.78}$$

with inverse transformation

$$\begin{aligned} S_C^{-1}(\omega, \mathbf{v_c}) &= \gamma^2\gamma^0 K^{-1} S_C^{\dagger} K \gamma^0\gamma^2 = \gamma^2\gamma^0 S_C^{\,T}\gamma^0\gamma^2 = \exp(\omega\gamma^0\boldsymbol{\gamma}\cdot\hat{\mathbf{w}}/2) \\ &= \cosh(\omega/2)I - \sinh(\omega/2)\gamma^0\boldsymbol{\gamma}\cdot\hat{\mathbf{w}} \end{aligned} \tag{3.79}$$

---

[78] **An "ordinary" complexon can "exceed the speed of light" just like a tachyonic complexon because a complexon has a complex valued velocity enabling it to evade the real-valued singularity at v = c.**

[79] The global SU(3) symmetry of complexons makes their identification with quarks reasonable. However, the complexon theory that we develop and use for quark dynamics in the Standard Model is <u>not</u> required. Our Standard Model could use Dirac fermion dynamics for the up-type quarks and tachyon dynamics for down-type quarks. Then the (broken) Left-handed Extended Lorentz group would be the basic space-time group rather than L_C. We choose to use complexon dynamics for quarks because they have an internal SU(3)-like structure suggestive of color SU(3). More importantly, their spin dynamics is different and thus may resolve the differences between theory and experiment for the deep inelastic parton spin-dependent structure functions. Nevertheless, quarks could be similar to leptons in this regard and form a doublet of a Dirac fermion and an ordinary tachyon. Whether quarks are complexons or not is an experimental question!

[80] The equation numbering of this subsection follows that of Blaha (2007b).

The Dirac-like complexon equation resulting from the boost is

$$[i\gamma^0\partial/\partial t + i\boldsymbol{\gamma}\cdot(\nabla_r + i\nabla_i) - m]\psi_C(t, \mathbf{x}_r, \mathbf{x}_i) = 0 \tag{3.101}$$

where $\mathbf{x} = \mathbf{x}_r - i\mathbf{x}_i$. The subsidiary condition is

$$\nabla_r\cdot\nabla_i\ \psi_C(t, \mathbf{x}_r, \mathbf{x}_i) = 0 \tag{3.102a}$$

The L_C coordinate boost that leads to $m^2 < 0$ tachyonic complexons with complex 3-momenta is

$$\Lambda_{CL}(\mathbf{v_c}) \equiv \Lambda_{CL}(\omega, \hat{\mathbf{w}}) = \exp[i(\omega + i\pi/2)\hat{\mathbf{w}}\cdot\mathbf{K}] \tag{3.151}$$

where

$$\omega = (\omega_r^2 - \omega_i^2)^{1/2} \tag{2.62}$$
$$\hat{\mathbf{w}} = (\omega_r\hat{\mathbf{u}}_r + i\omega_i\hat{\mathbf{u}}_i)/\omega \tag{2.63}$$
$$\hat{\mathbf{w}}\cdot\hat{\mathbf{w}} = \hat{\mathbf{u}}_r\cdot\hat{\mathbf{u}}_r = \hat{\mathbf{u}}_i\cdot\hat{\mathbf{u}}_i = 1 \tag{2.64a}$$
$$\mathbf{v_c} = \hat{\mathbf{w}}\ \tanh(\omega + i\pi/2) = \hat{\mathbf{w}}\ \mathrm{cotanh}(\omega) \tag{3.152}$$
$$\omega_L = \omega + i\pi/2$$

The L_C spinor boost for tachyonic complexons is

$$S_{CL}(\Lambda_{CL}(\omega, \hat{\mathbf{w}})) = \exp(-i\omega_L\sigma_{0i}\hat{w}_i/2) = \exp(-\omega_L\gamma^0\boldsymbol{\gamma}\cdot\hat{\mathbf{w}}/2)$$
$$= \cosh(\omega_L/2)I + \sinh(\omega_L/2)\gamma^0\boldsymbol{\gamma}\cdot\hat{\mathbf{w}} \tag{3.154}$$

The resulting Dirac-like tachyonic complexon equation is

$$[\gamma^0\partial/\partial t + \boldsymbol{\gamma}\cdot(\nabla_r + i\nabla_i) - m]\psi_{CL}(t, \mathbf{x}_r, \mathbf{x}_i) = 0 \tag{3.163a}$$

with the subsidiary condition

$$\nabla_r\cdot\nabla_i\ \psi_{CL}(t, \mathbf{x}_r, \mathbf{x}_i) = 0 \tag{3.164}$$

L_C Boosts between Complexons and Tachyonic Complexons

An L_C spinor boost of a complexon can change it into a tachyonic complexon and vice versa:

$$S_{CL}(\Lambda_{CL}(\omega, \hat{\mathbf{w}}))\psi_C(x) \rightarrow \psi_{CT}'(x') \tag{4.27}$$
$$S_{CL}(\Lambda_{CL}(\omega, \hat{\mathbf{w}}))\psi_{CT}(x) \rightarrow \psi_C'(x')$$

Similarly the differential operator used in the equations of motion can also be transformed.

$$S_{CL}(\Lambda_{CL}(\omega, \hat{\mathbf{w}}))(\gamma^{\mu}D_{\mu} - m)S_{CL}^{-1}(\Lambda_{CL}(\omega, \hat{\mathbf{w}})) = (i\gamma^{\mu}D'_{\mu} - m) \tag{4.28}$$
$$S_{CL}(\Lambda_{CL}(\omega, \hat{\mathbf{w}}))\gamma^5(i\gamma^{\mu}D_{\mu} - m)\gamma^5 S_{CL}^{-1}(\Lambda_{CL}(\omega, \hat{\mathbf{w}})) = (\gamma^{\mu}D'_{\mu} - m)$$

where

$$x'^{\mu} = i\Lambda_{CL}{}^{\mu}{}_{\nu}(\omega, \mathbf{u})x^{\nu} \tag{4.29}$$
$$D'_{\mu} = -i\Lambda_{CL}{}^{\nu}{}_{\mu}(\omega, \mathbf{u})D_{\nu}$$

or in matrix form

$$X' = E_{CL}(\omega, \hat{\mathbf{w}})X \equiv i\Lambda_{CL}(\omega, \hat{\mathbf{w}})X \tag{4.30}$$

Eqs. 4.27 – 4.29 imply

$$S_{CL}(\Lambda_{CL}(\omega, \hat{\mathbf{w}}))(\gamma^{\mu}D_{\mu} - m)\psi_{CT}(x) = (i\gamma^{\mu}D'_{\mu} - m)S_{CL}(\Lambda_{CL}(\omega, \hat{\mathbf{w}}))\psi_{CT}(x)$$
$$= (i\gamma^{\mu}D'_{\mu} - m)\psi_C'(x') \tag{4.31}$$

and

$$S_{CL}(\Lambda_{CL}(\omega,\hat{\mathbf{w}}))\gamma^5(i\gamma^{\mu}D_{\mu} - m)\psi_C(x) = (\gamma^{\mu}D'_{\mu} - m)S_{CL}(\Lambda_{CL}(\omega,\hat{\mathbf{w}}))\gamma^5\psi_C(x)$$
$$= (\gamma^{\mu}D'_{\mu} - m)\psi_{CT}'(x') \tag{4.32}$$

where

$$\psi_C'(x') = S_{CL}(\Lambda_{CL}(\omega, \hat{\mathbf{w}}))\psi_{CT}(x) \tag{4.33}$$

and

$$\psi_{CT}'(x') = S_{CL}(\Lambda_{CL}(\omega, \hat{\mathbf{w}}))\gamma^5\psi_C(x) \tag{4.34}$$

Thus neither complexon dynamical equation is L_C covariant.

Doublet Dynamical Equation for Complexons

We will now consider the issue of generalizing the complexon dynamical equations so that the generalized equation is covariant under both Lorentz transformations and L_C boosts.

The only obvious method to obtain a generalized equation that is covariant under L_C boosts is to define an 8×8 matrix generalization. Let

$$đ_C(x) = \begin{bmatrix} (i\gamma^{\mu}D_{\mu} - m) & 0 \\ 0 & (\gamma^{\mu}D_{\mu} - m) \end{bmatrix} \tag{4.35}$$

be an 8×8 matrix operator with the 4×4 matrix elements shown, and let

$$\Psi_C(x) = \begin{bmatrix} \psi_C(x) \\ \psi_{CT}(x) \end{bmatrix} \tag{4.36}$$

be an 8 component column vector composed of a complexon field and a tachyonic complexon field. Then the generalized complexon equation is

$$đ_C(x)\Psi_C(x) = 0 \tag{4.37}$$

We now define the 8 × 8 Left-handed L_C boost transformation

$$S_{CL8} \equiv S_{CL8}(\Lambda_{CL}(\omega, \hat{\mathbf{w}})) = \begin{bmatrix} 0 & S_{CL}(\Lambda_{CL}(\omega, \hat{\mathbf{w}})) \\ S_{CL}(\Lambda_{CL}(\omega, \hat{\mathbf{w}}))\gamma^5 & 0 \end{bmatrix} \tag{4.38}$$

with inverse transformation

$$S_{CL8}^{-1} \equiv S_{CL8}^{-1}(\Lambda_{CL}(\omega, \hat{\mathbf{w}})) = \begin{bmatrix} 0 & \gamma^5 S_{CL}^{-1}(\Lambda_{CL}(\omega, \hat{\mathbf{w}})) \\ S_{CL}^{-1}(\Lambda_{CL}(\omega, \hat{\mathbf{w}})) & 0 \end{bmatrix} \tag{4.39}$$

Applying S_{CL8} to eq. 4.37 yields

$$0 = S_{CL8}đ_C(x)\Psi_C(x) = đ_C(x')\Psi_C'(x') \tag{4.40}$$

where

$$\Psi_C'(x') = \begin{bmatrix} S_{CL8}\psi_{CT}(x) \\ S_{CL8}\gamma^5\psi_C(x) \end{bmatrix} = \begin{bmatrix} \psi_C'(x') \\ \psi_{CT}'(x') \end{bmatrix} \tag{4.41}$$

Thus the generalized complexon equation is covariant under L_C boosts. Covariance requires the complexon, and the tachyonic complexon, must have the same absolute value for the mass.

It is easy to show that the generalized complexon equation is also covariant under conventional Lorentz transformations represented as 4 × 4 diagonal blocks in an 8 × 8 matrix representation. (The demonstration is analogous to eqs. 4.9 – 4.13.)

The lagrangian density that corresponds to our 8-dimensional construction is

$$\mathcal{L}_{C8} = \overline{\Psi}_C(x)\not{d}_C(x)\Psi_C(x) \qquad (4.42)$$

where

$$\overline{\Psi}_C(x) = \Psi_C^{\dagger}|_{\mathbf{x}_i = -\mathbf{x}_i}\Gamma_C^{0} \qquad (4.43)$$

and

$$\Gamma_C^{0} = \begin{bmatrix} \gamma^0 & 0 \\ 0 & i\gamma^0\gamma^5 \end{bmatrix} \qquad (4.44)$$

The action

$$I = \int d^4x \mathcal{L}_{C8} \qquad (4.45)$$

is invariant under Lorentz transformations S_8 and S_{8A} (eqs. 4.9 – 4.12).

The Hamiltonian density for the 8-dimensional theory is

$$\mathcal{H}_{C8}(x) = \begin{bmatrix} \psi_C^{\dagger}(-i\boldsymbol{\alpha}\cdot\nabla_C + \beta m)\psi_C & 0 \\ 0 & i\psi_{CT}^{\dagger}\gamma^5(\boldsymbol{\alpha}\cdot\nabla_C + \beta m)\psi_{CT} \end{bmatrix} \qquad (4.46)$$

where the spatial vector part of D^{μ} is

$$\nabla_C = \mathbf{D} \qquad (4.47)$$

Non-Invariance of the Generalized Free Complexon Action under an L_C Boost

The action 4.45 is not invariant under L_C boosts. The reason is similar to that of section 4.3 for the "leptonic" type of particle: there is no simple relation between the hermitean conjugate of an L_C spinor boost and its inverse (a situation similar to eq. 4.19 for the Dirac boost case).

Consequently the action of eq. 4.45 is only invariant under inhomogeneous Lorentz transformations. *This state of affairs is again an advantage when we derive features of the Standard Model because it prevents any interplay between unbroken Weak SU(2) rotations and L_C transformations in the complexon (quark) sector.*

5. ElectroWeak SU(2)⊗U(1) from Real Superluminal Velocities

In the preceding chapter we developed a fermion doublet framework for the Weak interactions. We now turn to develop the Weak and Electromagnetic interactions[81] with an SU(2) ⊗U(1) group structure, from the geometry of Complex Lorentz transformations.

In the discussions up to this point we have not considered imaginary (and more generally complex) coordinates resulting from a superluminal Lorentz transformation. In this chapter we show that the coordinates generated from real-valued coordinates are complex-valued in general and require us to introduce another transformation that maps complex coordinates to real coordinates.[82] This transformation, which we will call a *Reality group* transformation, will be of significance because it has an SU(2)⊗U(1) group symmetry. It emerges when we consider superluminal transformations but is not required for ordinary sublight Lorentz transformations. This new SU(2)⊗U(1) symmetry is identified with the SU(2)⊗U(1) symmetry of the Weak and Electromagnetic sectors of The Standard Model.

We introduce this new transformation by reconsidering the previous simple example wherein one coordinate system is traveling at a speed v in the x direction with respect to the "laboratory" system. See Fig. 5.1.

The (left-handed[83]) Lorentz transformation is given by eq. 5.1, and the coordinates in the two reference frames are related by eq. 5.2.

$$\Lambda_L(\omega, \mathbf{u} = (1,0,0)) = \begin{bmatrix} i\gamma_s & -i\beta\gamma_s & 0 & 0 \\ -i\beta\gamma_s & i\gamma_s & 0 & 0 \\ 0 & 0 & 1 & 0 \\ 0 & 0 & 0 & 1 \end{bmatrix} \qquad (5.1)$$

implementing the coordinate transformation:

---

[81] For reasons that will become evident much later we treat the Weak and Electromagnetic interactions as separate but unitable. The reason for the separation is our introduction of the new concept of 'rotating' interactions – Ω transformations.

[82] The complex coordinates resulting from a superluminal transformation are physically viewed as real-valued by an observer in the new coordinate system. The apparent complexity of the coordinates resulting from a superluminal transformation are an artifact of the transformation.

[83] The right-handed Lorentz transformation case is analogous.

$$X' = \Lambda_L(\omega, \mathbf{u} = (1,0,0))X$$

or

$$\begin{aligned} t' &= i\gamma_s(t - \beta x) \\ x' &= i\gamma_s(x - \beta t) \\ y' &= y \\ z' &= z \end{aligned} \qquad (5.2)$$

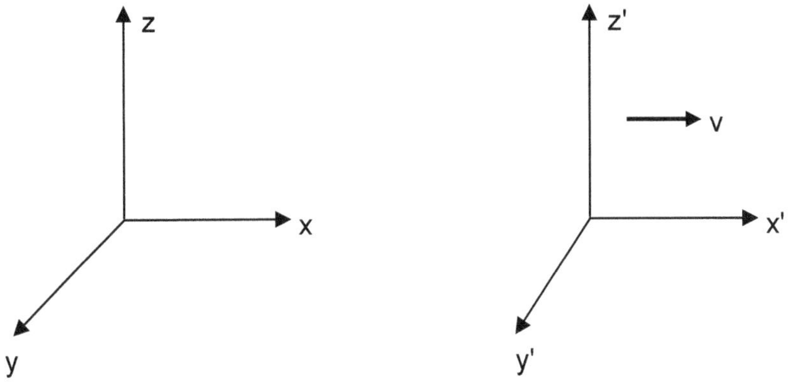

Figure 5.1. Depiction of two coordinate systems. The "primed" coordinate system is moving with velocity v in the positive x direction with respect to the "unprimed" coordinate system. We choose parallel axes for convenience.

We now define a Reality group transformation $\Pi_L(\mathbf{u})$ that maps the real coordinates of the unprimed reference frame to real coordinates in the primed reference frame.

$$\Pi_L(\mathbf{u}) = \begin{bmatrix} -i & 0 & 0 & 0 \\ 0 & -i & 0 & 0 \\ 0 & 0 & 1 & 0 \\ 0 & 0 & 0 & 1 \end{bmatrix} \qquad (5.3)$$

where $\mathbf{u}$ is the unit vector corresponding to the direction of $\mathbf{v}$ (the positive x direction in this example). Using $\Pi_L(\mathbf{u})$ we obtain an overall transformation from real coordinates to real coordinates in the case considered:

$$X'' = \Pi_L(\mathbf{u})\Lambda_L(\omega, \mathbf{u} = (1,0,0))X$$

or

$$\begin{aligned} t'' &= \gamma_s(t - \beta x) \\ x'' &= \gamma_s(x - \beta t) \\ y'' &= y \\ z'' &= z \end{aligned} \tag{5.4}$$

where $\gamma_s = (\beta^2 - 1)^{-½}$. An observer in the primed reference frame would consider his/her time to be real when measured on a clock, and distances along the x axis to be real when measured with a ruler. Thus eq. 5.4 makes good sense physically because in any reference frame, observers measure real distances and real times. For this reason we will call combined transformations of the type of eq. 5.4 – from real coordinates to real coordinates – *physical* superluminal transformations for real-valued velocities.

It is important to note that $\Pi_L(\mathbf{u})$ is position dependent in general for more complicated Λ_L transformations and so the Reality group is a local group of the Yang-Mills type. This is clear from eqs. 5.1 and 5.2. We will see the Reality group contains the local (Yang-Mills) SU(2)⊗U(1) Weak and Electromagnetic symmetry.

This simple example generalizes to arbitrary relative real velocities **v**. First we note that the Lorentz transformation for a velocity **v** that is a rotation of the velocity in the x-direction (**v** = |**v**|R**u** where R is the relevant rotation matrix) has the form

$$\Lambda_L(\omega, \mathbf{v}) = \mathcal{R}(\mathbf{v}/v, \mathbf{u})\Lambda_L(\omega, \mathbf{u} = (1,0,0))\mathcal{R}^{-1}(\mathbf{v}/v, \mathbf{u}) \tag{5.5}$$

where $\mathcal{R}(\mathbf{v}/v, \mathbf{u})$ is a rotation from the velocity direction **u** to direction **v**/v.

The original transformation (eq. 5.2) can be written as

$$\Pi_L(\mathbf{u})\Lambda_L(\omega, \mathbf{u} = (1,0,0)) = \Pi_L(\mathbf{u})\mathcal{R}^{-1}(\mathbf{v}/v, \mathbf{u})\Lambda_L(\omega, \mathbf{v})\mathcal{R}(\mathbf{v}/v, \mathbf{u}) \tag{5.6}$$

Consequently the combined transformation for velocity **v** is

$$\begin{aligned} &\mathcal{R}(\mathbf{v}/v, \mathbf{u})\Pi_L(\mathbf{u})\Lambda_L(\omega, \mathbf{u} = (1,0,0))\mathcal{R}^{-1}(\mathbf{v}/v, \mathbf{u}) \\ &\quad = \mathcal{R}(\mathbf{v}/v, \mathbf{u})\Pi_L(\mathbf{u})\mathcal{R}^{-1}(\mathbf{v}/v, \mathbf{u})\Lambda_L(\omega, \mathbf{v}) \\ &\quad = \Pi_L(\mathbf{v}/v)\Lambda_L(\omega, \mathbf{v}) \end{aligned} \tag{5.7}$$

Thus for a Lorentz transformation $\Lambda_L(\omega, \mathbf{v})$ for velocity **v** we see that we can define a subsidiary transformation $\Pi_L(\mathbf{v}/v)$ of the form

$$\Pi_L(\mathbf{v}/v) = \mathcal{R}(\mathbf{v}/v, \mathbf{u})\Pi_L(\mathbf{u})\mathcal{R}^{-1}(\mathbf{v}/v, \mathbf{u}) \tag{5.8}$$

The general form of $\mathcal{R}(\mathbf{v}/v, \mathbf{u})$, is

$$\mathfrak{R}(\mathbf{v}/\mathrm{v}, \mathbf{u}) = \begin{bmatrix} 1 & 0 & 0 & 0 \\ 0 & & & \\ 0 & & \mathfrak{R}_3(\mathbf{v}/\mathrm{v}, \mathbf{u}) & \\ 0 & & & \end{bmatrix} \tag{5.9}$$

where $\mathfrak{R}_3(\mathbf{v}/\mathrm{v}, \mathbf{u})$ is a 3 × 3 rotation matrix that can be expressed in terms of the generators of the 3-dimensional rotation group as

$$\mathfrak{R}_3(\mathbf{v}/\mathrm{v}, \mathbf{u}) = \exp(i\boldsymbol{\theta}\cdot\mathbf{J}) \tag{5.10}$$

The rotation angles $\boldsymbol{\theta}$ are real numbers since we are rotating the real vector **u** to the real vector **v**/v. Given the form of eq. 5.10 then we see that the form of $\Pi_L(\mathbf{v}//\mathrm{v})$ is

$$\Pi_L(\mathbf{v}/\mathrm{v}) = \begin{bmatrix} -i & 0 & 0 & 0 \\ 0 & & & \\ 0 & & \mathfrak{R}_3(\mathbf{v}/\mathrm{v}, \mathbf{u})\Pi_{L3}(\mathbf{u})\mathfrak{R}_3^{-1}(\mathbf{v}/\mathrm{v}, \mathbf{u}) & \\ 0 & & & \end{bmatrix} \tag{5.11}$$

where

$$\Pi_{L3}(\mathbf{u}) = \begin{bmatrix} -i & 0 & 0 \\ 0 & 1 & 0 \\ 0 & 0 & 1 \end{bmatrix} \tag{5.12}$$

If we consider the case of an infinitesimal rotation $\boldsymbol{\theta}$ to first order in $\boldsymbol{\theta}$

$$\mathfrak{R}_3(\mathbf{v}/\mathrm{v}, \mathbf{u}) \simeq I + i\boldsymbol{\theta}\cdot\mathbf{J} \tag{5.13}$$

then

$$\begin{aligned}\Pi_{L3}(\mathbf{v}/\mathrm{v}) = \mathfrak{R}_3(\mathbf{v}/\mathrm{v}, \mathbf{u})\Pi_{L3}(\mathbf{u})\mathfrak{R}_3^{-1}(\mathbf{v}/\mathrm{v}, \mathbf{u}) &\simeq \Pi_{L3}(\mathbf{u}) + i\boldsymbol{\theta}\cdot\mathbf{J}\Pi_{L3}(\mathbf{u}) - i\Pi_{L3}(\mathbf{u})\boldsymbol{\theta}\cdot\mathbf{J} \\ &\simeq \Pi_{L3}(\mathbf{u})[I + i\Pi_{L3}^{-1}(\mathbf{u})[\boldsymbol{\theta}\cdot\mathbf{J}, \Pi_{L3}(\mathbf{u})]\end{aligned} \tag{5.14}$$

where $\Pi_{L3}^{-1}(\mathbf{u})$ is the inverse of $\Pi_{L3}(\mathbf{u})$ and [...] represents the commutator. Thus for arbitrary rotations eq. 5.14 implies

$$\Pi_{L3}(\mathbf{v}/\mathrm{v}) = \mathfrak{R}_3(\mathbf{v}/\mathrm{v}, \mathbf{u})\Pi_{L3}(\mathbf{u})\mathfrak{R}_3^{-1}(\mathbf{v}/\mathrm{v}, \mathbf{u}) = \Pi_{L3}(\mathbf{u})\exp\{i\Pi_{L3}^{-1}(\mathbf{u})[\boldsymbol{\theta}\cdot\mathbf{J}, \Pi_{L3}(\mathbf{u})]\} \tag{5.15}$$

We can find the general form of $\Pi_{L3}(\mathbf{v}/\mathrm{v})$ by considering the case of eq. 5.6 in more detail. The exponentiated matrix expression in 5.15 can be written

$$\Pi_{L3}^{-1}(\mathbf{u})[\boldsymbol{\theta}\cdot\mathbf{J}, \Pi_{L3}(\mathbf{u})] = \Pi_{L3}^{-1}(\mathbf{u})\boldsymbol{\theta}\cdot\mathbf{J}\Pi_{L3}(\mathbf{u}) - \boldsymbol{\theta}\cdot\mathbf{J} = \boldsymbol{\theta}\cdot\mathbf{Q} \tag{5.16}$$

where

$$\mathbf{Q} = \Pi_{L3}^{-1}(\mathbf{u})\mathbf{J}\Pi_{L3}(\mathbf{u}) - \mathbf{J} = \mathbf{Q}' - \mathbf{J} \tag{5.17}$$

The matrices Q_i can be evaluated using eq. 5.12 and the matrix representations of rotation generators J_i: which are equivalent in form to the SU(2) generators T_i:

$$J_1 = \begin{bmatrix} 0 & 0 & 0 \\ 0 & 0 & -i \\ 0 & i & 0 \end{bmatrix} = T_1 \tag{5.18}$$

$$J_2 = \begin{bmatrix} 0 & 0 & i \\ 0 & 0 & 0 \\ -i & 0 & 0 \end{bmatrix} = T_2 \tag{5.19}$$

$$J_3 = \begin{bmatrix} 0 & -i & 0 \\ i & 0 & 0 \\ 0 & 0 & 0 \end{bmatrix} = T_3 \tag{5.20}$$

The rotation generators satisfy the commutation relations

$$[J_i, J_j] = i\epsilon_{ijk}J_k \tag{5.21}$$

as do the SU(2) generators:

$$[T_i, T_j] = i\epsilon_{ijk}T_k \tag{5.22}$$

We can calculate Q' and obtain

$$Q'_1 = \begin{bmatrix} 0 & 0 & 0 \\ 0 & 0 & -i \\ 0 & i & 0 \end{bmatrix} \tag{5.23}$$

$$Q'_2 = \begin{bmatrix} 0 & 0 & -1 \\ 0 & 0 & 0 \\ -1 & 0 & 0 \end{bmatrix} \tag{5.24}$$

$$Q'_3 = \begin{bmatrix} 0 & 1 & 0 \\ 1 & 0 & 0 \\ 0 & 0 & 0 \end{bmatrix} \tag{5.25}$$

We note that each Q'_i is hermitean and the Q'_i satisfy the commutation relations:

$$[Q'_i, Q'_j] = i\epsilon_{ijk}Q'_k \tag{5.26}$$

Consequently the set of Q'_i are also equivalent to SU(2) generators. As a result the exponential factor

$$\Pi_{L3}(\mathbf{v}/v) = \Pi_{L3}(\mathbf{u})\exp\{i\boldsymbol{\theta}\cdot(\mathbf{Q}' - \mathbf{J})\} \tag{5.27}$$

is equivalent to a combination of SU(2) rotations not only in this case but in general for superluminal transformations. The factor $\Pi_{L3}(\mathbf{u})$ is not an SU(2) matrix since its determinant is not 1. However

$$\Pi'_{L3}(\mathbf{u}) = -i\Pi_{L3}(\mathbf{u}) \tag{5.28}$$

is an SU(2) matrix since

$$\Pi'^{-1}_{L3}(\mathbf{u}) = \Pi'^{\dagger}_{L3}(\mathbf{u}) \tag{5.29}$$

$$\det \Pi'_{L3}(\mathbf{u}) = 1 \tag{5.30}$$

and

$$\Pi'_{L3}(\mathbf{v}/v) = \Pi'_{L3}(\mathbf{u})\exp\{i\boldsymbol{\theta}\cdot(\mathbf{Q}' - \mathbf{J})\} \tag{5.31}$$

is similarly an SU(2) rotation.

Thus the general form of superluminal, *real* velocity, transformation from a real set of coordinates to a real set of coordinates is[84]

$$\Pi_L(\mathbf{v}/v)\Lambda_L(\omega, \mathbf{v}) \tag{5.32}$$

where

$$\Pi_L(\mathbf{v}/v) = \begin{bmatrix} -i & 0 & 0 & 0 \\ 0 & & & \\ 0 & & \Pi_{L3}(\mathbf{u})\exp\{i\,\boldsymbol{\theta}\cdot(\mathbf{Q}' - \mathbf{J})\} & \\ 0 & & & \end{bmatrix} \tag{5.33}$$

The Lorentz condition for real to real physical transformations generalizes to

$$\Lambda(\mathbf{v})^T\Pi_L(\mathbf{v}/v)^\dagger G\, \Pi_L(\mathbf{v})\Lambda(\mathbf{v}/v) = G \tag{5.34}$$

---

84 The choice of the unit vector **u** and the angle vector **θ** must be such that applying the transformation to a real set of coordinates yields a real set of coordinates.

Since superluminal transformations $\Lambda_L(\omega, \mathbf{v})$ transform real coordinates to complex coordinates in general, we can generalize the form of a real-to-real superluminal transformation to

$$e^{i\varphi}\Pi_L(\mathbf{v'}/v')\Lambda_L(\omega, \mathbf{v}) \tag{5.35}$$

where φ is a constant phase and $\mathbf{v'}$ is an arbitrary velocity. This generalization will satisfy the generalized Lorentz condition

$$\Lambda(\mathbf{v})^T\Pi_L(\mathbf{v'}/v')^\dagger e^{-i\varphi}G\, e^{i\varphi}\Pi_L(\mathbf{v'}/v')\Lambda(\mathbf{v}) = G \tag{5.36}$$

but the transformation will, in general, yield a complex set of coordinates when applied to a set of real coordinates.

These considerations imply:

1. Any observer in a coordinate system will treat a complex 4-dimensional coordinate system as if it were a real 4-dimensional coordinate system with complex-valued straight lines along each dimension (assuming rectangular coordinates).

2. The transformation $e^{i\varphi}\Pi'_{L3}(\mathbf{v}/v)$ is a SU(2)⊗U(1) transformation that takes complex 3-dimensional spatial coordinates to complex 3-dimensional spatial coordinates. In particular straight lines map to straight lines.

3. Physical observations in the observer's coordinate system are invariant under SU(2)⊗U(1) rotations of the spatial coordinates and the multiplication of the time component by an arbitrary phase.

4. The matrix

$$\Pi'_L(\mathbf{v}/v, \chi, \varphi) = \begin{bmatrix} e^{i\chi} & 0 & 0 & 0 \\ 0 & & & \\ 0 & & e^{i\varphi}\Pi'_{L3}(\mathbf{u})\exp\{i\,\boldsymbol{\theta}\cdot(\mathbf{Q}' - \mathbf{J})\} & \\ 0 & & & \end{bmatrix} \tag{5.37}$$

(where χ and φ are real numbers and $\mathbf{u}$ is a unit vector along any convenient coordinate axis) is a SU(2)⊗U(1) transformation that transforms complex 4-dimensional coordinates to complex 4-dimensional coordinates. Note, $\Pi_L(\mathbf{v}/v) = \Pi'_L(\mathbf{v}/v, 3\pi/2, \pi/2)$ is a special case of $\Pi'_L(\mathbf{v}/v, \chi, \varphi)$. Due to the manifest form of 5.37 we see

$$\Pi'_L{}^{\mu}{}_{\alpha}{}^{*}\Pi'_L{}^{\mu}{}_{\beta} = [\Pi'_L{}^{\dagger}\Pi'_L]_{\alpha\beta} = I_{\alpha\beta} \tag{5.38}$$

(with an implied sum over μ) or, in matrix form,

$$\Pi'_L{}^{\dagger}\,\Pi'_L = I \tag{5.39}$$

and also[85]

$$\Pi'_L{}^{\dagger}G\Pi'_L = G \tag{5.40}$$

5. Complex coordinate values of the type generated by superluminal transformations with real-valued velocities are transformable to real coordinates. The complex coordinates are thus physically equivalent to corresponding real coordinate values in the sense that an observer in that frame would automatically use the real coordinates so obtained since rulers and clocks always measure real spatial coordinates and times. *Therefore physical theory is invariant under global SU(2)⊗U(1) coordinate transformations since complex coordinates, so generated, can be rotated back to real coordinates.*

6. The complex coordinates of any point obtained through a superluminal transformation can be transformed to a real set of coordinates by the above SU(2)⊗U(1) transformation. This SU(2)⊗U(1) invariance is the SU(2)⊗U(1) symmetry of the Weak and Electromagnetic interactions.

---

[85] Eq.7.40 is close to the defining condition for a Lorentz group element but the presence of complex conjugation rather than a transpose means Π'_L is outside the real and complex Lorentz groups.

6. Dark Weak and Dark Electromagnetic SU(2)⊗U(1) from *Complex* Superluminal Velocities

In this chapter[86] we will consider superluminal transformations based on complex-valued velocities. The previous chapter considered the case of real-valued velocities. This case led to Weak and Electromagnetic SU(2)⊗U(1).

We now consider Complex Lorentz transformations for complex-valued relative velocities. These transformations will require us to introduce another Reality group with transformations that map complex coordinates to real coordinates. These transformations will be of significance because they lead to a hitherto unstated SU(2)⊗U(1) symmetry. We identify this SU(2)⊗U(1) symmetry as a symmetry of the *Dark* Weak and *Dark* Electromagnetic interactions of an extended The Standard Model. Dark matter and interactions remain to be found experimentally but there may be some preliminary suggestive data from the CERN LHC.

We introduce these new Dark transformations by extending the previous simple example to Fig. 6.1 wherein one coordinate system is traveling at a complex-valued velocity u in the x direction with respect to the "laboratory" system. In these new transformations the relative velocity is complex-valued and has two components: a real-valued component in the x direction and an imaginary-valued component in the y direction. $\mathbf{u} = u_x\mathbf{i} + iu_y\mathbf{j}$. In the complex case $\beta = \tanh(\omega_L)$ is real-valued by eqs. 2.20 – 2.22 where $\omega_L = \omega + i\pi/2$ and ω is real.

The (left-handed[87]) Lorentz transformation is given by eq. 2.23:

$$\Lambda_L(\omega, \mathbf{u}) = \begin{bmatrix} i\gamma_s & -i\beta\gamma_s u_x & \beta\gamma_s u_y & 0 \\ -i\beta\gamma_s u_x & 1 + (i\gamma_s - 1)u_x^2 & i(i\gamma_s - 1)u_x u_y & 0 \\ \beta\gamma_s u_y & i(i\gamma_s - 1)u_x u_y & 1 - (i\gamma_s - 1)u_y^2 & 0 \\ 0 & 0 & 0 & 1 \end{bmatrix} \quad (6.1)$$

$$= \Lambda(\omega + i\pi/2, \mathbf{u})$$

implementing the coordinate transformation:

$$X' = \Lambda_L(\omega, \mathbf{u} = (u_x, iu_y, 0))X$$

or

[86] Most of the material in this chapter appeared in Blaha (2011c) originally.

[87] The right-handed Lorentz transformation case is analogous.

$$
\begin{aligned}
t' &= i\gamma_s(t - \beta u_x - i\beta u_y) \\
x' &= -i\gamma_s\beta u_x t + i\gamma_s x + u_x x - u_x^2 x + i(i\gamma_s - 1)u_x u_y y \\
y' &= \gamma_s\beta u_y t + i(i\gamma_s - 1)u_x u_y x + [1 - (i\gamma_s - 1)u_y^2]y \\
z' &= z
\end{aligned}
\tag{6.2}
$$

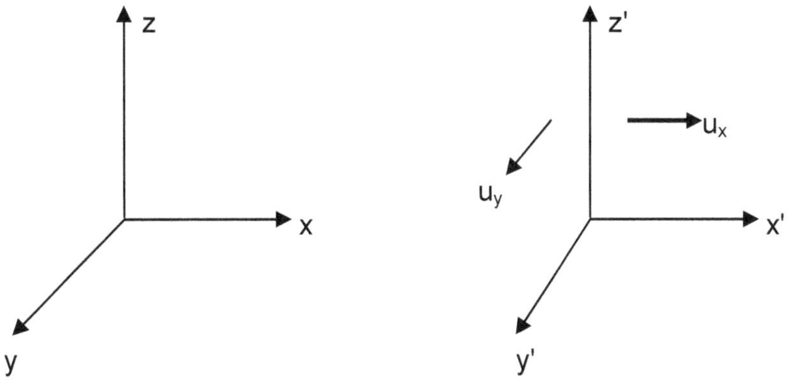

Figure 6.1. Depiction of two coordinate systems. The "primed" coordinate system is moving with velocity $u = u_x i + i u_y j$ with respect to the "unprimed" coordinate system. We choose parallel axes for convenience.

We now define a transformation that maps the real coordinates of the unprimed reference frame to real coordinates in the primed reference frame.

$$
\Pi_L(\mathbf{u}, \mathbf{X}) = \begin{bmatrix} e^{ia} & e^{ib} & e^{ic} & 0 \\ e^{id} & e^{ie} & e^{if} & 0 \\ e^{ig} & e^{ih} & e^{ij} & 0 \\ 0 & 0 & 0 & 1 \end{bmatrix} \tag{6.3}
$$

where **u** is the unit vector corresponding to the direction of the relative velocity. It is important to note that $\Pi_L(\mathbf{u}, \mathbf{X})$ is position dependent and so the Reality group is a local group of the Yang-Mills type. This is clear from eqs. 6.1 and 6.2. The Reality group here is SU(2)⊗U(1). It appears similar to the Reality group of chapter 5 except for the crucial difference that the Reality group here mixes time and spatial rows while the Reality group of chapter 5 only mixed spatial rows (See eqs. 5.35 and 5.37.).Again we have a local Yang-Mills theory. Thus the combined Reality group from this chapter and chapter 5 is SU(2)⊗U(1)⊗DSU(2)⊗DU(1) where we prepend 'D' to the Dark Reality group parts.

Earlier we defined complex boosts. We summarize the definition below:

$$\Lambda_L(\omega, \mathbf{u}) = \Lambda_L(\mathbf{v}_c) = \exp[i(\omega + i\pi/2)\mathbf{u}\cdot\mathbf{K}] \tag{6.4}$$

where ω remains

$$\omega = (\omega_r^2 - \omega_i^2)^{1/2}$$

and

$$\mathbf{u} = (\omega_r\mathbf{u}_r + i\omega_i\mathbf{u}_i)/\omega$$
$$\mathbf{u}\cdot u = \mathbf{u}_r\cdot\mathbf{u}_r = \mathbf{u}_i\cdot\mathbf{u}_i = 1$$
$$\mathbf{v}_c = \mathbf{u}\tanh(\omega + i\pi/2) = \mathbf{u}\,\mathrm{cotanh}(\omega)$$

In the example, that we are considering, we set $\mathbf{u}_x = \omega_r\mathbf{u}_r$ and $\mathbf{u}_y = \omega_i\mathbf{u}_i$.

Using $\Pi_L(\mathbf{u}, \mathbf{X})$ we obtain an overall transformation from real coordinates to real coordinates:

$$X'' = \Pi_L(\mathbf{u}, \mathbf{X})\Lambda_L(\omega, \mathbf{u} = (u_x, u_y, 0))X \tag{6.5}$$

with the coordinates of X and X" real-valued. An observer in the double primed reference frame would consider his/her time to be real when measured on a clock, and distances along the x and y axes to be real when measured with a ruler.

The velocity vectors: $u_x\mathbf{i}$ and $iu_y\mathbf{j}$ in our example define a plane in space. There are two types of rotations that are possible. 1) An angular rotation in the plane defined by the vectors. This is a U(1) transformation. 2) a spatial rotation of the plane that is an SU(2) rotation. Thus the joint rotations of **u** have an SU(2)⊗U(1) symmetry group. The R group – the Reality group for 4-dimensions – has two SU(2)⊗U(1) factors that we denote SU(2)⊗U(1)⊗DSU(2)⊗DU(1). We see that this "newly found" group can be *assumed* to be the Dark Weak and Dark Electromagnetic symmetry group. This group of transformations has associated Dark interactions. Very recent data from the CERN LHC suggests that Dark Matter has interactions. We note that a further Reality group part SU(3) will be introduced in the next chapter.

We discuss Dark Matter and its interactions in detail later.

7. Color SU(3)

7.1 Two Possible Approaches to Color SU(3)

There are two approaches to obtaining the Strong interaction and Color SU(3) symmetry:

1. Assume up-type and down-type quarks are in $\underline{3}$ representations of Color SU(3). This assumption sheds no light on a deeper origin of the Strong interaction and Color SU(3). It simply assumes the color SU(3) of the Strong interaction sector of the Standard Model. Thus our understanding is not deepened. A postulate corresponding to this assumption is:

Postulate: Quarks are in the $\underline{3}$ representation of Color SU(3). The SU(3) symmetry is gauged with local Yang-Mills SU(3) fields called gluons that constitute the Strong interaction of the quark sector. Quarks are minimally coupled to the gluons in a gauge covariant fashion.

2. In the preceding chapters the Weak and Electromagnetic interactions of the extended Standard Model (modulo generations and their mixing) was shown to follow from the Reality group associated with the Complex Lorentz group. The Reality group included SU(2)⊗U(1)⊗DSU(2)⊗DU(1). Thus we found a significant geometrical basis for the form of the Weak and Electromagnetic interactions of normal and Dark matter.

3. We now establish a similar geometrical basis for the Strong interaction and Color SU(3). *If we now extend the parameters to be real functions of the space-time coordinates (i.e. local SU(3) transformations), then we obtain a geometrical basis for color SU(3). A key factor in this interpretation is the global covariance of complexon equations of motion under global SU(3).*[88]

7.2 A Global SU(3) Symmetry of Complexon Quarks

We will now consider a global SU(3) covariance implicit in eqs. 3.123 – 3.127. The defining property of the group SU(3) is that it preserves the invariance of inner products of complex 3-vectors of the form:

$$u^{*}\cdot v = u^{1*}v^{1} + u^{2*}v^{2} + u^{3*}v^{3} \tag{7.1}$$

[88] This chapter was extracted from chapter 17 of Blaha (2011c) with some changes.

If we examine the dynamical equation eq. 3.123 we see that the differential operator is covariant under a global SU(3) transformation U of the complex spatial 3-coordinates:

$$[i\gamma^0\partial/\partial t + i\mathbf{D_c}^*\cdot\gamma - m] = [i\gamma^0\partial/\partial t + i\mathbf{D_c}'^*\cdot\gamma' - m] \tag{7.2}$$

where

$$\mathbf{D_c}^* = \nabla_c = \nabla_r + i\nabla_i$$

and

$$\gamma^a = U^{ab}\gamma'^b \tag{7.3a}$$

$$D_c{}^{*a} = D_c{}'^{*b}U^{ab*} \tag{7.3b}$$

where $U^\dagger = U^{-1}$. We now exhibit the covariance of eq. 3.123. Since we can view the three spatial γ-matrices as SU(3) 3-vectors, we can express eq. 7.3 as the result of a SU(3) rotation V of the γ-matrices (on the spinor indices)

$$V\gamma^aV^{-1} = U^{ab}\gamma'^b \tag{7.4}$$

where V is a 4×4 reducible representation of SU(3), namely, $\underline{3}\oplus\underline{1}$. Since V commutes with γ^0 in the Pauli matrix representation of the γ matrices we see that V can have the form

$$V = \begin{bmatrix} A\exp(i\alpha_i\sigma_i) & 0 \\ 0 & B\exp(i\beta_i\sigma_i) \end{bmatrix}$$

where A, B, α_i and β_i are constants, and the zeroes represent 2×2 zero matrices. The inverse of V is $V^\dagger$. Thus eq. 7.4 becomes

$$V\gamma^aV^{-1} = \begin{bmatrix} 0 & AB^*\exp(i\alpha_i\sigma_i)\sigma_a\exp(-i\beta_i\sigma_i) \\ -A^*B\exp(i\beta_i\sigma_i)\sigma_a\exp(-i\alpha_i\sigma_i) & 0(-i\beta_i\sigma_i) \end{bmatrix}$$

We now note the generators of the global SU(3) symmetry under discussion have a 4×4 matrix reducible representation ($\underline{3}\oplus\underline{1}$). The generators of this reducible representation are F_i and F_0 (a diagonal matrix diag(0,0,0,0,0,0,0,0,1).with F_i being the Gell-Mann SU(3) generators for i = 1, 2, …, 8.

Projection operators can be defined to project out the $\underline{3}$ representation piece P_3 and the $\underline{1}$ representations piece P_1 of the complexon spinor fields:

Thus the $\underline{3}$ complexon field is

$$\psi_{C3}(t, \mathbf{x_r}, \mathbf{x_i}) = P_3\psi_C(t, \mathbf{x_r}, \mathbf{x_i}) \tag{7.5}$$

while the 1 complexon field is

$$\psi_{C1}(t, \mathbf{x_r}, \mathbf{x_i}) = P_1\psi_C(t, \mathbf{x_r}, \mathbf{x_i}) \tag{7.6}$$

Since P_1 and P_3 do not commute with Lorentz transformations, a Lorentz transformation mixes ψ_{C1} and ψ_{C3}.[89] Since P_1 and P_3 do not commute with $\gamma 5$, left-handed and right-handed complexons would also be mixed by these projection operators. The matrix V has a $\underline{3}\oplus\underline{1}$ reducible representation.

In a manner similar to the covariance proof of the Dirac equation[90] we see that eq. 7.2 is covariant under SU(3) transformations:

$$V[i\gamma^0\partial/\partial t + i\gamma\cdot\mathbf{D_c}^* - m]V^{-1}V\psi_C(t, \mathbf{x_r}, \mathbf{x_i}) = 0$$

or

$$[i\gamma^{0\prime}\partial/\partial t' + i\mathbf{D_c}'^*\cdot\gamma' - m]V\psi_C(t, \mathbf{x_r}, \mathbf{x_i}) = 0 \tag{7.7}$$

(Note $\gamma^{0\prime} = V\gamma^0V^{-1}$ and $t' = t$.) The SU(3) transformed wave function $\psi_C'(t, \mathbf{x}')$ is

$$\psi_C'(t', \mathbf{x}') = V\psi_C(t, \mathbf{x}) = V\psi_C(t', U\mathbf{x}') \tag{7.8}$$

Thus the complexon Dirac equation is covariant under global SU(3).

The subsidiary condition,

$$\nabla_r\cdot\nabla_i\,\psi_{Cu}(t, \mathbf{x_r}, \mathbf{x_i}) = 0 \tag{7.9}$$

is also covariant under an SU(3) rotation:

$$\nabla_r'^*\cdot\nabla_i'\psi_C'(t, \mathbf{x}') = \nabla_r\cdot\nabla_i\,V\psi_C(t, \mathbf{x}) = V\nabla_r^*\cdot\nabla_i\,\psi_C(t, \mathbf{x}) = 0 \tag{7.10}$$

We now examine the transformation of the wave function eq. 7.8 under the SU(3) transformation U. If we define

---

[89] At this point it is worth noting that the construction of complexon fields, based on a boost from a particle rest state, guarantees that a reference frame exists in which any complexon particle has a single real time variable. Similarly a reference frame exists for a set of complexon particles (that is within a Lorentz of the center of momentum frame) with a single real time variable. The time variables of the individual complexon particles in the set are complex in general but are functions of the center of momentum real time variable. So there is only one real time variable for each complexon in the set although the time variable of an individual particle may be a complex function of the real center of momentum time variable.

[90] For example see Bjorken (1964) pp. 18 – 20.

$$q^{*\mu} = (q^0, \mathbf{q}^*) = (p^0, \mathbf{p}_r + i\mathbf{p}_i) = (p^0, \mathbf{p}) = p^\mu \tag{7.11}$$

then $\psi_C(t, \mathbf{x})$ will be seen to be covariant form under an SU(3) transformation:

$$\psi_C(t, x) = \sum_{\pm s} \int d^3q_r d^3q_i\, N_C(p^0)\delta(\mathbf{q}_r^*\cdot\mathbf{q}_i/m^2)[b_C(q^*,s)u_C(q^*,s)e^{-i(q^*\cdot x + q\cdot x^*)/2} + \\ + d_C^\dagger(q^*,s)v_C(q^*,s)e^{+i(q^*\cdot x + q\cdot x^*)/2}] \tag{7.12}$$

Note both terms in each exponential are separately invariant under global SU(3). ($\mathbf{q}_r^* = \mathbf{q}_r$ since $\mathbf{q}_r$ is real.)

Eq. 7.8 implies that the spinors appearing in eq. 7.12 are covariant under SU(3) transformations

$$u_C'(q'^*,s') = Vu_C(q^*,s) \tag{7.13}$$
$$v_C'(q'^*,s') = Vv_C(q^*,s) \tag{7.14}$$

The fourier coefficients, if second quantized in a complex spatial coordinate generalization of the usual manner, also have covariant anti-commutation relations under an SU(3) transformation:

$$\{b_C(q,s), b_C^\dagger(q'^*,s')\} = \delta_{ss'}\delta^3(q_r - q'_{r'})\delta^3(q_i - q'_{i'}) \tag{7.15}$$

Under an SU(3) transformation, z = Uq and z' = Uq', the right side of eq. 7.15 transforms to

$$\delta^3(q_r - q'_{r'})\delta^3(q_i - q'_{i'}) \rightarrow \delta^3(z_r - z'_{r'})\delta^3(z_i - z'_{i'})/|\partial(q)/\partial(z)| = \delta^3(z_r - z'_{r'})\delta^3(z_i - z'_{i'}) \tag{7.16}$$

where

$$|\partial(q)/\partial(z)| = |\partial(q_r^1, q_r^2, q_r^3, q_i^1, q_i^2, q_i^3)/\partial(z_r^1, z_r^2, z_r^3, z_i^1, z_i^2, z_i^3)| = 1 \tag{7.17}$$

is the Jacobian of the transformation U. The fourier coefficients transform trivially under SU(3):

$$b_C(q^*,s) \rightarrow b_C(z^*,s) \tag{7.18}$$

Since the integrand transforms as

$$\int d^3q_r d^3q_i \rightarrow \int d^3z_r d^3z_i\, |\partial(q)/\partial(z)| = \int d^3z_r d^3z_i \tag{7.19}$$

we see that the wave function $\psi_C(t, \mathbf{x})$ transforms covariantly.

7.3 Local Color SU(3) and the Strong Interactions

In the previous section we showed that the equations of motion of free Dirac-like, complexon, up-type quarks are covariant under global SU(3).. The free, tachyon, complexon, down-type quark equations of motion are also easily seen to be covariant under this SU(3) subgroup. In this section we will show this covariance is the "source" of local Color SU(3) symmetry of quarks, and then we will introduce the Strong interaction via minimal coupling to SU(3) Yang-Mills gluons in gauge covariant derivatives.

We now introduce a complexon field with a global SU(3) index a which takes values from 1 to 3 making the field a member of the 3 representation of global SU(3):

$$\psi_C{}^a(t, \mathbf{x}) \tag{7.20}$$

Due to the SU(3) index the transformation property of $\psi_C{}^a(t, \mathbf{x})$ changes from eq. 7.8 to

$$\psi_C{}''^a(t, \mathbf{x}') = U^{ab}V\psi_C{}^b(t, \mathbf{x}) = U^{ab}V\psi_C{}^b(t, U\mathbf{x}') \tag{7.21}$$

where U^{ab} is an SU(3) rotation of 3 representation "vectors" such as $\psi_C{}^b$ and **x.** V is the corresponding rotation of the spinor indices of $\psi_C{}^b(t, \mathbf{x})$.

Note that the SU(3) rotation of the field factorizes into an SU(3) rotation of the three fields $\psi_C{}^b$ by U^{ab} and an SU(3) rotation of the four spinor components of each individual field $\psi_C{}^b$ by V.

This factorization enables us to consider a global SU(3) rotation of the $\psi_C{}^b$ fields while holding the coordinates fixed:

$$\psi_C{}'^a(t, \mathbf{x}) = U^{ab}\psi_C{}^b(t, \mathbf{x}) \tag{7.22}$$

The equations of motion are covariant under this global transformation

$$0 = U^{ab}[i\gamma^0\partial/\partial t + i\gamma\cdot\mathbf{D_c}^* - m]\psi_C{}^b(t, \mathbf{x_r}, \mathbf{x_i})$$

$$= [i\gamma^0\partial/\partial t + i\gamma\cdot\mathbf{D_c}^* - m]\psi'_C{}^a(t, \mathbf{x_r}, \mathbf{x_i}) \tag{7.23}$$

We now note the form of eq. 7.22 is the same as that of a *local* Yang-Mills rotation:

$$\psi_C{}'^a(t, \mathbf{x}) = \Theta^{ab}(t, \mathbf{x})\psi_C{}^b(t, \mathbf{x}) \tag{7.24}$$

where $\mathbf{x} = \mathbf{x_r} + i\mathbf{x_i}$. Therefore if we introduce a local SU(3) Yang-Mills field $A_{C\nu}(t, \mathbf{x_r}, \mathbf{x_i})$ and define a covariant derivative we can generalize eq. 7.21 to the case of local, color SU(3) if we do <u>not</u> perform the spinor rotation V.[91]

$$\mathcal{D}_\nu = D_\nu - igA_{C\nu} \tag{7.25}$$

where

$$A_{C\nu} = A_C{}^a{}_\nu t^a \tag{7.26}$$

and where $D_\nu = D_{q\nu}$ is given by

$$D_0 = \partial/\partial x^0$$
$$D_k = \partial/\partial x_r{}^k + i\,\partial/\partial x_i{}^k \tag{7.27}$$

The SU(3) 3×3 matrix generators satisfy

$$[t^a, t^b] = if^{abc}t^c \tag{7.28}$$

We can represent $\Theta_{ab}(x)$ in the form:

$$\Theta_{ab}(x) = [\exp(-i\varphi_c(x)t^c)]_{ab} \tag{7.29}$$

where $\varphi_c(x)$ is a local parameter dependent on $x = (x^0, \mathbf{x} = \mathbf{x_r} + i\mathbf{x_i})$, and t^c is an SU(3) generator.

Applying a gauge transformation to the gauge covariant derivative of a complexon fermion field $\mathcal{D}_\nu\psi_C(x)$:

$$\Theta\mathcal{D}_\nu\psi_C(x) = \Theta D_\nu\psi_C(x) - ig\Theta A_{C\nu}\Theta^{-1}\Theta\psi_C(x) \tag{7.30}$$
$$= D_\nu\psi_C'(x) - igA_{C'\nu}\psi_C'(x) = (\mathcal{D}_\nu\psi_C(x))'$$

where

$$\psi_C'(x) = \Theta(x)\psi_C(x) \tag{7.31}$$

we find

$$A_{C'\nu} = (-i/g)(D_\nu\Theta(x))\Theta^{-1}(x) + \Theta(x)A_{C\nu}(x)\Theta^{-1}(x) \tag{7.32}$$

The reader will note that the form of eqs. 7.25 – 7.31 is identical to those associated with a conventional non-abelian gauge interaction with the replacement:

$$\partial/\partial x^\nu \rightarrow D_\nu \tag{7.33}$$

---

[91] This approach enables us to avoid the dilemmas associated with mixing coordinate and internal symmetries as described by Coleman, S., Phys. Rev. **138** B1262 (1965) and others in the case of SU(6) in the 1960's. Note that the spinor rotation V is expressed in terms of numerical matrices while, in the second quantized formulation, the U^{ab} rotation is expressed in terms of second quantized fields as well as numeric matrices. Thus the factorization is reflected in the form of the transformation.

with D_ν given by eq. 7.27. Note that $\varphi_c(x)$, the local parameter in eq. 7.29 is dependent in general, on time, and the real and imaginary parts of the complex spatial 3-vector.

Introducing the SU(3) gauge covariant derivative transforms eq. 7.23 to

$$0 = [i\gamma^\nu \mathcal{D}_\nu - m]\psi_C{}^a(t, \mathbf{x_r}, \mathbf{x_i}) \tag{7.34}$$

The preceding argumentation supports the following postulates:

Postulate: Quarks are in a $\underline{3}$ *representation of a global SU(3) group.. Their transformation law under SU(3) transformations is eq. 19.6 wherein the space-time coordinates are not transformed.*

Postulate: The covariance of the quark equations of motion under global SU(3) transformations becomes covariance under local color SU(3) transformations when the equations of motion are generalized to gauge covariant derivatives. We assume the equations of motion are so generalized. The interaction terms introduced constitute the Strong interaction.

We note the case of tachyon complexon quarks differs only in small details from the above discussion of Dirac-type complexon quarks.

7.4 Interactions Resulting from Complex Space-Time Projected to Real Physical Space-Time

This chapter has shown that the Complex Lorentz group and the Reality group generate the familiar interactions of The Standard Model: SU(3)⊗SU(2)⊗U(1) plus an additional set of DSU(2)⊗DU(1) interactions that we take to be the interactions of Dark Matter. The total Reality group that we have found is thus SU(3)⊗SU(2)⊗U(1)⊗SU(2)⊗U(1).

8. Five Interactions Result From Complex Lorentz Group Considerations

In the preceding three chapters we have found that the Complex Lorentz group geometry leads directly an SU(3)⊗SU(2)⊗U(1)⊗SU(2)⊗U(1) Reality group. This group is necessary to make the resultant coordinate system of each Complex Lorentz group transformation real-valued.

We have denoted the Reality group as SU(3)⊗SU(2)⊗U(1)⊗DSU(2)⊗DU(1) to distinguish between the Weak and Electromagnetic interaction parts for normal matter and the Weak and Electromagnetic interaction parts for Dark matter. We have associated the SU(3)⊗SU(2)⊗U(1) factors with normal matter since we see only an SU(2)⊗U(1) symmetry (broken) for normal matter, and there are astronomical suggestions that Dark matter does not clump into stars and planets like normal matter leading us to view Dark matter as not supporting the Strong SU(3) interaction. (If Dark matter did support the Strong interaction then it would have been experimentally 'visible' by now,)

We are thus led to five interactions for an extended Standard Model that includes Dark matter. Several questions still remain for discussion:

1. Are there more interactions implied by the Complex Lorentz group. No. The 4×4 complex-valued matrices of Complex Lorentz group transformations require transformations based on sixteen generators. The Reality group has sixteen generators and thus fulfill the requirement for transformations (matrices) of Complex Lorentz group transformations to real-valued matrices. Thus no more interactions are possible/required by Complex Lorentz group geometry.

2. Can the factors of the Reality group be combined into an overall symmetry? Since the five factors of the Reality group do not generally commute it is not possible to combine them into a single group such as U(4).

3. We thus conclude we have a complete geometrical derivation of the Reality group (the extended Standard Model interactions) from Complex Loretz group geometry. \Complex Lorentz group geometry also yields an understanding of the four species of fermions – a previously unexplained fact.

4. We will next proceed to determine the dynamical equations of the one generation Standard Model. Subsequently we will introduce groups to expand the fermion spectrum to 192 fermions. We will also consider General Relativity and develop a comprehensive theory for what appears to be a universe governed by eleven interactions with 192 fermions.

9. Derivation of One Generation Leptonic Standard Model Features

This chapter[92] uses the (one generation) spectrum of four fermion species, and the Reality group of interactions SU(3)⊗SU(2)⊗U(1)⊗SU(2)⊗U(1) developed in preceding chapters, to formulate the dynamical equations of a one generation extended Standard Model of Elementary particles. In subsequent chapters we will expand this one generation Standard Model to four generations, and then to four layers of four generations, using a 'new' U(4) Generation group and a 'new' U(4) Layer group and their interactions respectively. These groups have been described in earlier books by the author in 2015 and 2016.

9.1 The Standard Model Leptonic Sector as a Consequence of the Complex Lorentz Group Coordinates Projected to Real Values

The development of the Standard Model in the 1960's and 1970's was a major step forward in our understanding of the forces of nature. However the strange, and, in many physicists' opinions, unattractive form of the theory led particle theorists to conclude that it was a provisional theory that would eventually be replaced by a more elegant fundamental theory. Work in these directions has focussed upon 1) embedding the Standard Model within a larger ("elegant") symmetry group; 2) embedding the Standard Model within a space-time with extra dimensions that generate the Standard Model through some mechanism such as the Kaluza-Klein mechanism; and 3) viewing the Standard Model as somehow a "low energy" phenomenology that emerges from a Superstring Theory.

In this chapter we take an alternate view. We will derive the leptonic sector of the Standard Model based on the Complex Lorentz group, L_c, and Reality group covariance of the equations of motion. Thus, unlike previous efforts, we view the Standard Model as in a natural form dictated by L_c and Reality group covariance and certain other fundamental physical requirements. Beauty being in the eye of the beholder we shall endeavor to show the attractiveness of the derivation of the Standard Model based on this covariance.

The naturalness of the derivation, and its close connection to the left-handed L_c *group, strongly suggest the Standard Model, which was grown by theorists from experiment, has an undeniable quality of genuineness that will likely survive the passage of time. The basis of the*

[92] This chapter, and chapter 10 below, are extracted from chapters 7 – 11 of Blaha (2007b).

derivation, in the Complex Lorentz group,encourages the hope that we have found a new, deeper level of understanding of elementary particle dynamics.

9.2 Assumptions for the Leptonic Sector of the Standard Model

We will make certain assumptions, consistent with our construction approach, that provide a basis for the derivation of most aspects of the leptonic sector the Standard Model (The quark sector is derived in chapter 10.). These assumptions, some of which will be derived later, connect our theory with physical reality.

1. One generation of leptons is assumed in this chapter. (The multi-generation case is treated later.)
2. The form of the equations of motion of the unbroken leptonic sector of the Standard Model is determined by covariance under the L_c and Reality groups.
3. Leptonic matter is composed of spin ½ Dirac particles (electrons) with charge -1 and tachyons (neutrinos) of charge zero as well as their anti-particles.[93]
4. A neutrino is a tachyon with a non-zero bare mass that may be changed by symmetry breaking effects.
5. Gauge fields are massless "before" spontaneous symmetry breaking and are thus conventional gauge fields without a tachyon equivalent in the theory.
6. L_c and Reality group covariance of the dynamical equations of motion, are spontaneously broken through the appearance of additional fermion and vector boson mass terms generated by a mechanism such as the Higgs mechanism.
7. Spatial coordinates are rotated to real values (from complex values) using the Reality group.

9.3 Derivation of the Leptonic Sector of the Standard Model

The steps of the construction/derivation are:

A. L_c and Reality group covariance of the dynamical equations of motion requires that spin ½ particles be described by a generalization of the Dirac equation to an 8 × 8 matrix formbased on a doublet consisting of a Dirac particle and a tachyon (Chapter 4).

B. We identify the Dirac particle with a charged lepton and the tachyon with a neutrino. The bare masses of these particles have the same numeric value (before symmetry breaking).

C. The leptonic sector free field lagrangian (without gauge fields introduced as yet) is explicitly

---

[93] If neutrinos are Majorana particles then the derivation must be modified.

$$\mathcal{L}_{\text{freelep}} = \Psi^{\dagger}(x)\begin{bmatrix} \gamma^0\gamma^5 i(\gamma^\mu\partial/\partial x^\mu - m) & 0 \\ 0 & \gamma^0(i\gamma^\mu\partial/\partial x^\mu - m) \end{bmatrix}\Psi(x) \qquad (9.1)$$

Focussing on the derivative term we see that it can be put in the form

$$\Psi^{\dagger}(x)\gamma^0\begin{bmatrix} \gamma^5 & 0 \\ 0 & I_4 \end{bmatrix} i\gamma^\mu\partial/\partial x^\mu\,\Psi(x) = \Psi^{\dagger}(x)\gamma^0[C^+I - C^-\sigma_3]i\gamma^\mu\partial/\partial x^\mu\Psi(x) \qquad (9.2)$$

where I_4 is a 4×4 identity matrix, where I and σ_3 are 2 × 2 matrices, and where C^+ and C^- are defined in eq. 3.219 of chapter 3. The Pauli matrices σ_i are

$$\sigma_1 = \begin{bmatrix} 0 & 1 \\ 1 & 0 \end{bmatrix} \qquad \sigma_2 = \begin{bmatrix} 0 & -i \\ i & 0 \end{bmatrix} \qquad \sigma_3 = \begin{bmatrix} 1 & 0 \\ 0 & -1 \end{bmatrix} \qquad (9.3)$$

Expression 9.2 can be re-expressed in terms of left-handed and right-handed fields as

$$\Psi_L^{\dagger}(x)\gamma^0 i\gamma^\mu\partial/\partial x^\mu\Psi_L(x) - \Psi_R^{\dagger}(x)\gamma^0 i\gamma^\mu\partial/\partial x^\mu\sigma_3\Psi_R(x) \qquad (9.4)$$

D. At this point we are in a position to introduce couplings to gauge fields. In view of the doublet nature of the fields $\Psi_L(x)$ and $\Psi_R(x)$ it would appear, at first glance, that the symmetry group of the gauge fields would be $SU(2)_L \otimes SU(2)_R$. However the right-handed tachyon field in expression 9.4 has the wrong sign in the lagrangian, as has been noted in the previous discussion of the free tachyon lagrangian and anti-commutator. Consequently the right-handed tachyon field *cannot* have trilinear or higher order couplings. If it did have such interactions then it would rapidly degrade to lower and lower energy by the emission of particles since right-handed leptonic tachyons can exist in principle as free particles (modulo possible Higgs terms). (In this regard the situation is similar to that of time-like photons, except that the set of tachyon physical states cannot be defined in a manner analogous to Gupta-Bleuler electrodynamics where the timelike and longitudinal photons "cancel" each other so that only transverse photons have physical effects.) *Thus there cannot be right-handed leptonic tachyon interactions.*[94]

The doublet nature of the left-handed sector implies at least local SU(2) symmetries implemented with a covariant derivative.

The restricted nature of the right-handed leptonic sector indicates that *only* the Dirac particle in the "right-handed doublet" can have an interaction. Also the appearance of σ_3 in the right-handed term in expression 9.4 breaks SU(2) invariance if the left-handed covariant derivative (eq. 9.5 below) were substituted for $\partial/\partial x^\mu$ in the right-handed term. Thus a U(1) local gauge field interaction, restricted to the Dirac field member of the right-handed doublet and coupling to both members of the left-handed doublet, is the only allowed possibility. Without the U(1) interaction, the "right-handed doublet" would have no trilinear or higher order elementary particle interactions and would be physically irrelevant (except gravitationally) in unbroken gauge theory before spontaneous breakdown.

Putting these symmetries together we obtain a left-handed covariant derivative implementing local SU(2)⊗U(1) invariance found earlier:

$$D_{L\mu} = \partial/\partial x^\mu + \tfrac{1}{2} i g_2 \boldsymbol{\sigma}\cdot\mathbf{W}_\mu + \tfrac{1}{2} i g' B_\mu \tag{9.5}$$

and a right-handed covariant derivative[95]

$$\begin{aligned} D_{R\mu} &= \partial/\partial x^\mu \sigma_3 + \tfrac{1}{2} i g' B_\mu |Q| \sigma_3 \\ &= \partial/\partial x^\mu \sigma_3 + \tfrac{1}{2} i g' B_\mu |Q| \end{aligned} \tag{9.6}$$

where Q is the charge operator using the relation $|Q|\sigma_3 = |Q|$ for leptons. We use the absolute value of Q in order to achieve consistency in form with the right-handed quark sector described in the next chapter. As a result expression 9.4 becomes

$$\Psi_L^\dagger(x)\gamma^0 i\gamma^\mu D_{L\mu}\Psi_L(x) - \Psi_R^\dagger(x)\gamma^0 i\gamma^\mu D_{R\mu}\Psi_R(x) \tag{9.7}$$

Thus the leptonic sector of the lagrangian[96] (modulo mass/Higgs terms) is

$$\mathcal{L}_{\text{lep1}} = \Psi_L^\dagger\gamma^0 i\gamma^\mu D_{L\mu}\Psi_L - \Psi_R^\dagger\gamma^0 i\gamma^\mu D_{R\mu}\Psi_R \tag{9.8}$$

---

94 Right-handed neutrinos must interact with gravitons due to their mass-energy and the universality of the gravitational interaction. The extreme weakness of the gravitational interaction mitigates this effect.

95 The coupling constants are defined by $e = -g'\cos\theta_W = g_2\sin\theta_W$.

96 Note that the gauge fields do not have a tachyon equivalent since they are initially massless prior to spontaneous symmetry breaking.

$$= \Psi_L^{\dagger}\gamma^0 i\gamma^\mu D_{L\mu}\Psi_L + \bar{\psi}_{eR}\gamma^0 i\gamma^\mu D_{R\mu}\psi_{eR} - \bar{\psi}_{\upsilon R}\gamma^0 i\gamma^\mu \partial/\partial x^\mu \psi_{\upsilon R}$$
$$= \Psi_L^{\dagger}\gamma^0 i\gamma^\mu D_{L\mu}\Psi_L + \bar{\psi}_{eR}\gamma^0 i\gamma^\mu(\partial/\partial x^\mu + \tfrac{1}{2}ig'B_\mu)\psi_{eR} - \bar{\psi}_{\upsilon R}\gamma^0 i\gamma^\mu \partial/\partial x^\mu \psi_{\upsilon R}$$

where we identify the tachyon as a neutrino and the Dirac particle as a charged lepton such as the electron. Our leptonic sector lagrangian is now the usual leptonic sector, Weak and Electromagnetic interactions lagrangian with a tachyonic neutrino (neglecting mass related terms).

E. We have local gauge invariance prior to symmetry breaking: The gauge field sector has the usual Yang-Mills lagrangian terms and the B field has lagrangian terms similar to that of the QED lagrangian.

F. Spontaneous symmetry breaking of gauge symmetry, and of L_c group covariance, via the Higgs mechanism (or an alternative mechanism) can be implemented in such a way as to give the electron its known mass as well as the massive vector bosons. Since spontaneous symmetry breaking breaks L_c covariance to Lorentz covariance it is a moot point whether the Higgs sector exhibits a similar covariance.

That concludes the derivation of leptonic sector of the Standard Model (except for mass/Higgs terms. We have shown that the form of the leptonic sector of the Standard Model is fundamental in nature and based on L_c and Reality group covariance of the equations of motion. Thus it is not correct to view the Standard Model as the result of the breakdown of a larger internal symmetry group to SU(2)⊗U(1). On the contrary, the SU(2)⊗U(1) leptonic interaction form is a consequence of the L_c and Reality group covariance of the free dynamical equations. This comment also applies to the Dark Weak and Dark Electromagnetic interactions.

10. Derivation of One Generation Quark Sector Standard Model Features

10.1 Quark Sector of the Standard Model

The derivation of the form of the quark sector of the Standard Model is very similar to the preceding derivation of the form of the leptonic sector – but with some important points of difference. The primary difference is that we identify quarks with complexons. (This is a plausible assumption in view of the appearance of SU(3) symmetry within complexon fields. Since complexons have very different spin characteristics compared to Dirac fields, complexon quarks may resolve difficulties in parton spin studies.) Complexon quarks have complex 3-momentum as we showed earlier. Complexon gluons also have complex 4-momentum (required for consistency with quark terms in dynamic equations.)

10.2 Quark Sector Assumptions

We will make some assumptions that will provide a basis for a construction/derivation of most aspects of the single generation, quark sector Standard Model. These assumptions, based primarily on earlier discussions, connect our theory with physical reality.

1. One generation of quarks is assumed in this chapter. The multi-generation extended Standard Model is developed in a subsequent chapter.
2. The equations of motion of the unbroken form of the Standard Model are determined by covariance under the complex group L_C, and the Reality group (the source of the gauge fields.)
3. Quarks are composed of spin ½ "normal" complexons and tachyonic complexons.
4. The Reality group covariance of the dynamical equations are spontaneously broken through the appearance of mass terms generated by a mechanism such as the Higgs mechanism.

10.3 Derivation of the Form of the Standard Model Quark Sector

The construction/derivation:

A. L(C) covariance of the dynamical equations of motion requires that spin ½ particles be described by equations[97] generalized to an 8 × 8 matrix form (eqs 4.35 – 4.37) based on a doublet consisting of a Dirac-like particle and a tachyon. (They may or may not be complexons.) In in the case of quarks we assume the Dirac-type particle is the top component in the doublet and the tachyon is the bottom component.

B. Thus the 8 × 8 quark matrix formalism is:

$$đ_q(x)\Psi_C(x) = 0 \tag{10.1}$$

where[98]

$$đ_q(x) = \begin{bmatrix} (i\gamma^{\mu}D_{q\mu} - m_0) & 0 \\ 0 & (\gamma^{\mu}D_{q\mu} - m_0) \end{bmatrix} \tag{10.2}$$

with $D_{q\mu}$ given by eq. 7.27 and

$$\Psi_C(x) = \begin{bmatrix} \psi_{Cu}(x) \\ \psi_{Cd}(x) \end{bmatrix} \tag{10.3}$$

The upper 4-component field is a u-type field and the lower 4-component field is a d-type tachyonic field. The generalized equation eq. 10.1 is covariant under 8×8 L(C) transformations similar to eqs. 4.38 – 4.41. The generalized quark fermion equation eq. 10.1 is also covariant under conventional Lorentz transformations in the 8×8 representation.

The free quark sector lagrangian density that corresponds to the 8-dimensional fermion equation eq. 10.1 is

---

[97] The global SU(3) symmetry of complexons makes their identification with quarks reasonable. However, the complexon theory that we develop and use for quark dynamics in the Standard Model is <u>not</u> required but assuming complexon quarks explains their color SU(3) symmetry. Our Standard Model could use Dirac fermion dynamics for the up-type quarks and tachyon dynamics for down-type quarks. We choose to use complexon dynamics for quarks because they have an SU(3)-like structure suggestive of color SU(3). More importantly, their spin dynamics is different and this difference may resolve the differences between theory and experiment for the deep inelastic parton spin-dependent structure functions. Nevertheless, quarks could be similar to leptons in this regard and form a doublet of a Dirac fermion and an ordinary tachyon. Whether quarks are complexons or not is an experimental question!

[98] If we wish to obtain the Standard Model without complexon quarks (which is the conventional Standard Model) then $D_{q\mu} = \partial/\partial x^{\mu}$. If we wish to obtain the Standard Model with complexon quarks (which is a new form of the Standard Model) then $D_{q\mu}$ is given by eq. 9.27.

$$\mathcal{L}_{freeQuark} = \overline{\Psi}_C(x)\not{d}_q(x)\Psi_C(x) \tag{10.4}$$

where

$$\overline{\Psi}_C(x) = \Psi_C{}^{\dagger}\Gamma_C{}^0 \tag{10.5}$$

with

$$\Psi_C{}^{\dagger} = (\Psi_C{}^{\dagger})|_{\mathbf{x_i} = -\mathbf{x_i}} \tag{10.6}$$

for complexon quarks, and with

$$\Psi_C{}^{\dagger} = \text{the hermitean conjugate of } \Psi_C \tag{10.7}$$

in the case of non-complexon quarks. The † in the parentheses on the right side of eq. 10.6 indicates hermitean conjugation. Also

$$\Gamma_C{}^0 = \begin{bmatrix} \gamma^0 & 0 \\ 0 & i\gamma^0\gamma^5 \end{bmatrix} \tag{10.8}$$

C. The free quark field lagrangian is explicitly

$$\mathcal{L}_{freeQuark} = \Psi_C{}^{\dagger} \begin{bmatrix} \gamma^0(i\gamma^{\mu}D_{q\mu} - m_0) & 0 \\ 0 & \gamma^0\gamma^5 i(\gamma^{\mu}D_{q\mu} - m_0) \end{bmatrix} \Psi_C \tag{10.9}$$

Focussing on the derivative term we see that it can be put in the form

$$\Psi_C{}^{\dagger}\gamma^0 \begin{bmatrix} I_4 & 0 \\ 0 & \gamma^5 \end{bmatrix} i\gamma^{\mu}D_{q\mu}\Psi_C = \Psi_C{}^{\dagger}\gamma^0[C^+I + C^-\sigma_3]i\gamma^{\mu}D_{q\mu}\Psi_C \tag{10.10}$$

where I_4 is a 4×4 identity matrix, where I, and the Pauli matrix σ_3, are 2×2 matrices, and C^+ and C^- are defined earlier. Expression 10.10 can be expressed in terms of left-handed and right-handed fields as

$$\Psi_{CL}{}^{\dagger}\gamma^0 i\gamma^\mu D_{q\mu}\Psi_{CL} + \Psi_{CR}{}^{\dagger}\gamma^0 i\gamma^\mu D_{q\mu}\sigma_3\Psi_{CR} \qquad (10.11)$$

D. At this point we are in a position to introduce couplings to gauge fields. In view of the doublet nature of the fields $\Psi_{CL}(x)$ and $\Psi_{CR}(x)$ it would again appear, at first glance, that the symmetry group of the gauge fields would be $SU(2)_L \otimes SU(2)_R$. However the right-handed tachyonic field has the wrong sign in the lagrangian, as has been noted in the earlier discussion of the free tachyon lagrangian and anti-commutator. Consequently, **if quarks were *not* confined,** the right-handed quark tachyon field could not have trilinear or higher order couplings. **If free tachyon quarks existed, and had interactions, then they would rapidly degrade to lower and lower energy by the emission of particles.**

However, because of quark confinement in bound states with discrete energy levels, a bound tachyon quark can only emit particles if a lower energy bound state exists. As a result right-handed tachyon quarks can have interactions, such as the electromagnetic interaction, because quark confinement "tames" their propensity to emit particles due to the "wrong sign" in the lagrangian.

Again there is an analogy to Gupta-Bleuler QED quantization. In Gupta-Bleuler quantization physical states are required to have equal numbers of time-like and longitudinal photons thus canceling their physical effects. Similarly, right-handed tachyon quarks are required to be bound to other quarks by quark confinement to avoid continuous emission of particles.[99] Since interactions are allowed for right-handed tachyon quarks the Higgs mechanism can be used to change their mass.

The doublet nature of the left-handed sector implies at least local SU(2) symmetries. The appearance of σ_3 in the right-handed term in expression 10.11 explicitly breaks SU(2) invariance if the left-handed covariant derivative (eq. 10.12 below) were substituted for $D_{q\mu}$ in the right-handed term. Thus the right-handed fields can only have a U(1) local gauge field interaction, and are SU(2) singlets. We thus obtain a left-handed covariant derivative implementing local SU(2)⊗U(1) covariance:

$$D_{qL\mu} = D_{q\mu} + \tfrac{1}{2} i g_2 \boldsymbol{\sigma}\cdot\mathbf{W}_\mu + i g_1 B_\mu/6 \qquad (10.12)$$

[99] Therefore quark confinement is required in order to have a properly formulated quark sector. Another interaction – the strong interaction – is required for quark confinement. Presently there is only one accepted mechanism for quark confinement – through a non-abelian gauge coupling. (Higher derivative theories with quark confinement are in disfavor.) An additional non-abelian symmetry must be introduced for quarks. As discussed in chapter 7, SU(3) is the natural choice.

where $D_{q\mu}$ is given by eq. 7.27 and a right-handed covariant derivative[100]

$$D_{qR\mu} = D_{q\mu}\sigma_3 + ig_1B_\mu|Q| \tag{10.13}$$

where |Q| is the absolute value of the charge operator (with u eigenvalue 2/3 and d eigenvalue 1/3). The absolute value is used in order to compensate for the minus sign in front of the right-handed tachyon (d quark) term. As a result expression 10.11 becomes

$$\Psi_{CL}^{\dagger}(x)\gamma^0 i\gamma^\mu D_{qL\mu}\Psi_{CL}(x) + \Psi_{CR}^{\dagger}(x)\gamma^0 i\gamma^\mu D_{qR\mu}\Psi_{CR}(x)$$

Thus the quark sector of the lagrangian[101] is

$$\mathcal{L}_{\text{quark1}} = \Psi_{CL}^{\dagger}\gamma^0 i\gamma^\mu D_{qL\mu}\Psi_{CL} + \Psi_{CR}^{\dagger}\gamma^0 i\gamma^\mu D_{qR\mu}\Psi_{CR} \tag{10.14}$$

$$= \Psi_{CL}^{\dagger}\gamma^0 i\gamma^\mu D_{qL\mu}\Psi_{CL} + \overline{\psi}_{CuR} i\gamma^\mu D_{qR\mu}\psi_{CuR} - \overline{\psi}_{CdR} i\gamma^\mu D_{qR\mu}\psi_{CdR}$$

$$= \Psi_{CL}^{\dagger}\gamma^0 i\gamma^\mu D_{qL\mu}\Psi_{CL} + \overline{\psi}_{CuR} i\gamma^\mu(D_{q\mu} + \tfrac{2}{3}ig_1B_\mu)\psi_{CuR} - \overline{\psi}_{CdR} i\gamma^\mu(D_{q\mu} + \tfrac{1}{3}ig_1B_\mu)\psi_{CdR}$$

where we *provisionally* identify the tachyon as a d-type quark and the Dirac-like particle as a u-type quark. Our quark sector lagrangian is now the usual Standard Model quark sector lagrangian modulo complex 3-momenta, and modulo the strong interaction, terms, except that d-type quarks are tachyons and both types of quarks are complexons.

E. The SU(2) gauge field sector has the usual Yang-Mills lagrangian terms, and the B field is a U(1) abelian gauge field.

F. Spontaneous symmetry breaking of gauge symmetry, and of L(C) covariance, via the Higgs mechanism can be implemented in such a way as to give the quarks, and massive vector bosons, their "known" masses. Since spontaneous symmetry breaking breaks L(C) covariance of the dynamical equations of motion to Lorentz covariance it is a moot point whether the Higgs sector (if there is one) is manifestly L(C) covariant or not.

---

[100] The quark SU(2) coupling constant is, by gauge invariance, required to have the same value as the leptonic SU(2) coupling constant. The U(1) coupling constants are not required to be the same in both sectors and, in fact, are different. The coupling constants here are defined by $e = g_1\cos\theta_W = g_2\sin\theta_W$.

[101] Note that the gauge fields do not have a tachyon equivalent since they are initially massless prior to spontaneous symmetry breaking.

Thus L_C covariance of the quark equations of motion generates most of the "unusual" features of the quark sector of the Standard Model.

10.4 Color SU(3)

Our derivation yields the general form of the Electromagnetic and Weak quark sector of the Standard Model with a dynamics modified by the presence of complex spatial 3-momenta. This difference is a positive for the theory. In the discussion of eqs. 3.143-3.154 we showed that complexons have an SU(3) symmetry. The same discussion applies to tachyonic complexons.

Thus, although the discussion in this chapter has hitherto dealt with color scalar quarks,[102] we need only introduce a color index on the quarks to obtain quark fields in the fundamental SU(3) representation $\underline{3}$. Eqs. 3.153, 3.240 and 3.251 contain free color quark field expansions of the $\underline{3}$ representation.

The remaining step is to introduce local SU(3) Yang-Mills gauge fields that interact with the quarks through gauge covariant derivatives. Thus our quark sector has the Electromagnetic, Weak and Strong interactions of the Standard Model flowing naturally from L_C and Reality group covariance of the initially free, quark dynamical equations.

It has long been known that the strong interaction must have a non-abelian symmetry group in order to have quark confinement – a feature required by our derivation and in apparent complete agreement with experimental data. If we require that spin ½ bound quark states exist as they do, then strict quark confinement would rule out an SU(2) color strong interaction. Therefore the appearance of SU(3) symmetry in complexon fields, and the requirement of a "most minimal" non-abelian symmetry group for the color interaction, provide a logical basis for SU(3) as the color symmetry group.

As shown earlier quarks have a complex spatial 3-momentum – they are "normal" complexons and tachyonic complexons. If we think of quark-gluon interactions, and in particular, perturbative diagrams with quark-gluon loops such as the simplest quark self-energy diagram, then imaginary momentum should "flow" around a loop as well as real momentum.

Thus we postulate that gluons are massless spin 1 complexons (although this is not required by our derivation). Therefore we assume an SU(3) non-abelian, massless, version of vector complexons. The gluon lagrangian term is

$$\mathcal{L}_{CG} = -\tfrac{1}{4}\, F_C{}^{a\mu\nu}(x) F_C{}^{a}{}_{\mu\nu}(x) \tag{10.15}$$

where x has a complex 3-momentum and

$$F_C{}^{a}{}_{\mu\nu} = D_\nu A_C{}^{a}{}_{\mu} - D_\mu A_C{}^{a}{}_{\nu} + g f^{abc} A_C{}^{b}{}_{\mu} A_C{}^{c}{}_{\nu} \tag{10.16}$$

---

[102] Dark quarks are SU(3) singlets in our theory.

with D_μ is defined by eq. 3.162, g the coupling constant, and the constants f^{abc} are the SU(3) structure constants.

The theory of the strong color interaction should be developed within a path integral framework that takes account of Faddeev-Popov gauge fixing. Before doing that we must consider the effect of the complex spatial 3-vector on the non-abelian gauge formalism.

10.5 Pure Complexon Gauge Groups

Consider a local Lie group G with the generators of its algebra satisfying the commutation relations:

$$[t^a, t^b] = if^{abc}t^c \tag{10.17}$$

If a complexon field ψ_{Ca} transforms as some representation of G then a transformed field has the form:

$$\psi_{Ca}'(x) = \Theta_{ab}(x)\psi_{Cb}(x) \tag{10.18}$$

where $x = (x^0, \mathbf{x} = \mathbf{x_r} + i\mathbf{x_i})$ and $\Theta_{ab}(x)$ is an element of G. Since G is a local group we can represent $\Theta_{ab}(x)$ in the form:

$$\Theta_{ab}(x) = [\exp(-i\varphi_c(x)t^c)]_{ab} \tag{10.19}$$

where $\varphi_c(x)$ is a local parameter dependent on x, and t^c is the algebraic matrix in the representation of G under consideration.

The gauge covariant derivative in the case of complexon fermions is

$$\mathcal{D}_v = D_v - igA_{Cv} \tag{10.20}$$

where

$$A_{Cv} = A_C{}^a{}_v t^a \tag{10.21}$$

Applying a gauge transformation to the gauge covariant derivative of a complexon fermion field $\mathcal{D}_v\psi_C(x)$:

$$\Theta\mathcal{D}_v\psi_C(x) = \Theta D_v\psi_C(x) - ig\Theta A_{Cv}\Theta^{-1}\Theta\psi_C(x) \tag{10.22}$$

$$= D_v\psi_C'(x) - igA_C{}'_v\psi_C'(x) = (D_v\psi_C(x))'$$

where

$$\psi_C'(x) = \Theta(x)\psi_C(x) \tag{10.23}$$

we find

$$A_C{}'_\nu = (-i/g)(D_\nu\Theta(x))\Theta^{-1}(x) + \Theta(x)A_{C\nu}(x)\Theta^{-1}(x) \qquad (10.24)$$

The reader will note that the form of eqs. 10.15 – 10.24 is identical to those of a conventional non-abelian gauge field with the replacement:

$$\partial/\partial x^\nu \rightarrow D_\nu \qquad (10.25)$$

with D_ν given by eq. 3.162. Thus we see that $\varphi_c(x)$, the local parameter in eq. 10.19 is dependent in general, on time, and the real and imaginary parts of the complex spatial 3-vector. Thus the formalism differs only in small ways from the conventional non-abelian gauge theory. However, perturbation theory and non-perturbative phenomena such as instantons exhibit significant differences.

The commutator of covariant derivatives

$$F_{C\mu\nu} = F_C{}^a{}_{\mu\nu}t^a = (i/g)[\mathcal{D}_\mu, \mathcal{D}_\nu] \qquad (10.26)$$

is itself covariant under gauge transformations:

$$F_C{}'_{\mu\nu} = \Theta F_{C\mu\nu}\Theta^{-1} \qquad (10.27)$$

The strong interaction part of the action has terms of the form

$$I = \int d^7x[\mathcal{L}_{CG} + \bar{\Psi}_q i\gamma^\mu \mathcal{D}_\mu \Psi_q + \ldots] \qquad (10.28)$$

where the "extra" three integrations are over the imaginary spatial coordinates, and Ψ_q represents a generic quark (complexon) field. The action must be supplemented with a constraint similar to those seen earlier for complexons, which ensures the reality of the lagrangian term $\mathcal{L}_{CG}$ (and thus the corresponding Hamiltonian terms and ultimately the unitarity of the theory.) The constraint simply specifies the imaginary part of $\mathcal{L}_{CG}$ is zero:

$$[\partial A_C{}^{a\mu}/\partial x_{ik}]\text{Re } F_C{}^a{}_{\mu k} = [\partial A_C{}^{a\mu}/\partial x_{ik}](\partial A_C{}^a{}_\mu/\partial x_r{}^k - \partial A_C{}^a{}_k/\partial x_r{}^\mu + gf^{abc}A_C{}^b{}_\mu A_C{}^c{}_k) = 0 \qquad (10.29)$$

where $\mathbf{x}_r$ indicates the real part of the spatial coordinates $\mathbf{x} = \mathbf{x}_r + i\mathbf{x}_i$, and $\partial/\partial x_i{}^k$ is the derivative with respect to the kth component of the imaginary coordinate 3-vector $\mathbf{x}_i$. It is not a choice of gauge but rather a restriction on the dependence of the $A_C{}^\nu$ field on the real and imaginary spatial 3-vectors.

If we assume that we can integrate by parts in the imaginary coordinates then we can re-express the constraint as

$$A_C^{a\mu} \text{ Re } \partial F_{C\,\mu k}^{\;a}/\partial x_i^{\,k} = 0 \tag{10.30}$$

which is explicitly

$$A_C^{a\mu}[\partial^2 A_{C\,\mu}^{\;a}/\partial x_r^{\,k}\partial x_i^{\,k} - \partial^2 A_{C\,k}^{\;a}/\partial x_r^{\,\mu}\partial x_i^{\,k} + g f^{abc}\partial(A_{C\,\mu}^{\;b} A_{C\,k}^{\;c})/\partial x_i^{\,k}] = 0 \tag{10.31}$$

by eq. 10.16 where $x_r^{\,0} = x^0$. This restriction can be implemented within the framework of the path integral formalism by using the Faddeev-Popov Method with the introduction of ghosts in a manner similar to that of the Faddeev-Popov Method to implement a gauge condition. The form of the restriction stated in eqs. 10.30 and 10.31 leads to second order derivative Faddeev-Popov ghost terms in the path integral while eq. 10.29 would lead to a third order derivative ghost terms in the path integral with attendant unitarity (and possibly other) issues.

10.6 Pure Gauge Complexon Path Integral Formulation and Faddeev-Popov Method

The path integral formalism for complexon non-abelian, pure, Yang-Mills fields differs significantly from the conventional gauge field path integral formalism. The path integral for a complexon gauge field can be written symbolically as:

$$Z(J^\mu) = N\int DA_C \Delta_{FP}(A_C)\delta(F(A_C))\Delta_C(A_C)\delta(F_C(A_C))\exp\{i\int d^7y[\mathscr{L} + J^\mu(y)A_{C\mu}(y)]\} \tag{10.32}$$

where $\delta(F(A_C))$ specifies the gauge, $\Delta_{FP}(A_C)$ is its Faddeev-Popov determinant; and $\delta(F_C(A_C))$ specifies the complexon condition (eq. 10.31) with $\Delta_C(A_C)$ the Faddeev-Popov determinant for the complexon condition. In both cases the Faddeev-Popov determinant can be calculated in the standard way.[103]

First we consider the gauge fixing delta function. Note that it can be written as a delta function in the gauge times a determinant:

$$\delta(F(A_C^{\,\omega})) = \delta(\omega - \omega_0)|\det \delta F(A_{C\mu}^{\;\;\omega}(x))/\delta\omega(x)|^{-1}|_{F(A_C)=0} \tag{10.33}$$

where ω_0 is a reference gauge, where

$$\begin{aligned} A_{C\,\mu}^{\;a\,\omega}(x) &= A_{C\,\mu}^{\;a}(x) - g^{-1}D_\mu\omega^a + f^{abc}\,\omega^b(x)A_{C\,\mu}^{\;c}(x) \\ &= A_{C\,\mu}^{\;a}(x) + \delta A_{C\,\mu}^{\;a\,\omega}(x) \end{aligned} \tag{10.34}$$

---

[103] See for example Huang (1992).

and

$$\text{Re } F_C{}^a{}_{\mu k}{}^{\omega} = \text{Re } F_C{}^a{}_{\mu k} + f^{abc}\,\omega^b(x)F_C{}^c{}_{\mu k}$$
$$= \text{Re } F_C{}^a{}_{\mu k} + \delta(\text{Re } F_C{}^a{}_{\mu k}{}^{\omega}) \quad (10.35)$$

under an infinitesimal gauge transformation, and where

$$\Delta_{FP}(A_C) = \left|\det \delta F(A_{C\mu}{}^{\omega}(x))/\delta\omega(x)\right|\Big|_{F(A_C)=0,\ \omega=0} \quad (10.36)$$

We will choose the complexon Lorentz gauge to evaluate the Faddeev-Popov determinant:

$$F^a(A_C) = D_\mu A_C{}^{a\mu}(x) = 0 \quad (10.37)$$

We find

$$F^a(A_{C\mu}{}^{\omega}(x)) = D^\mu(A_C{}^a{}_\mu(x) - g^{-1}D_\mu\omega^a(x) + f^{abc}\,\omega^b(x)A_C{}^c{}_\mu(x))$$

$$= -g^{-1}D^\mu D_\mu\omega^a(x) + f^{abc}A_C{}^c{}_\mu(x)D^\mu\omega^b(x) \quad (10.38)$$

Thus

$$\delta F^a(A_{C\mu}{}^{\omega}(x))/\delta\omega^b(x) = -g^{-1}\,\delta^{ab}D^\mu D_\mu + f^{abc}A_C{}^{c\mu}(x)D_\mu \quad (10.39)$$

and

$$\Delta_{FP}(A_C) = \left|\det\,(g^{-1}\delta^{ab}D^\mu D_\mu - f^{abc}A_C{}^{c\mu}(x)D_\mu)\right| \quad (10.40)$$

where | … | represent absolute value.

We can rewrite the Faddeev-Popov determinant as a path integral over anti-commuting c-number fields with a ghost Lagrangian term:

$$\Delta_{FP}(A_C) = \int D\chi^* D\chi\ \exp[\,i\int d^7x\ \mathscr{L}^{ghost}(x)] \quad (10.41)$$

where

$$\mathscr{L}^{ghost}(x) = \chi^{a*}(x)[\delta^{ab}D^\mu D_\mu - gf^{abc}A_C{}^{c\mu}(x)D_\mu]\chi^b(x) \quad (10.42)$$

Faddeev-Popov Application to Complexon Condition

The complexon condition can also be implemented within the path integral formalism using the Faddeev-Popov Mechanism. Using the identity

$$1 = \int DA_C\Delta_C(A_C)\delta(F_C(A_C)) \quad (10.43)$$

we see that an infinitesimal gauge transformation yields eqs. 10.34 and 10.35. This enables us to relate $\Delta_C(A)$ to the determinant

$$\delta(F_C(A_C^{\omega})) = |\det \delta F_C(A_{C\mu}{}^{\omega}(x))/\delta\omega(x)|^{-1}|_{F_C(A_C)=0,\ \omega=0} \tag{10.44}$$

and

$$\Delta_C(A_C) = |\det \delta F_C(A_{C\mu}{}^{\omega}(x))/\delta\omega(x)||_{F_C(A_C)=0,\ \omega=0} \tag{10.45}$$

From eq. 10.31 we see

$$F_C(A_{C\mu}(x)) = A_C{}^{a\mu}[\partial^2 A_C{}^a{}_{\mu}/\partial x_r{}^k \partial x_i{}^k - \partial^2 A_C{}^a{}_k/\partial x_r{}^{\mu}\partial x_i{}^k + gf^{abc}\partial(A_C{}^b{}_{\mu}A_C{}^c{}_k)/\partial x_i{}^k] \tag{10.46}$$

with

$$F_C{}^a(A_{C\mu}) = 0 \tag{10.47}$$

Inserting eq. 10.34 and 10.35 we find

$$[\delta F_C(A_{C\mu}{}^{\omega}(x))/\delta\omega^a(x)]|_{F_C(A_C)=0,\ \omega=0} = \delta[\delta A_C{}^{b\mu\omega}\mathrm{Re}\ \partial F_C{}^b{}_{\mu k}/\partial x_{ik} + A_C{}^{b\mu}\ \partial\delta(\mathrm{Re}\ F_C{}^b{}_{\mu k}{}^{\omega})/\partial x_{ik}\ ]/\delta\omega^a(x)|_{\omega=0}$$

$$= -g^{-1}(\mathrm{Re}\ \partial F_C{}^a{}_{\mu k}/\partial x_{ik})D^{\mu} - f^{abc}A_C{}^{b\mu}(\mathrm{Re}\ F_C{}^c{}_{\mu k})\partial/\partial x_{ik} \tag{10.48}$$

Thus

$$\Delta_C(A_C) = |\det (g^{-1}(\mathrm{Re}\ \partial F_C{}^a{}_{\mu k}/\partial x_{ik})D^{\mu} + f^{abc}A_C{}^{b\mu}(\mathrm{Re}\ F_C{}^c{}_{\mu k})\partial/\partial x_{ik}| \tag{10.49}$$

where | … | represent absolute value.

We can rewrite this Faddeev-Popov determinant as a path integral over anti-commuting c-number fields with a ghost Lagrangian:

$$\Delta_C(A_C) = \lim_{r\to\infty} \int D\chi_C{}^* D\chi_C \exp[ir^{-2}\int d^7x\ \mathscr{L}_C{}^{ghost}(x)] \tag{10.50}$$

where r is a constant that is taken to the limit ∞, and where

$$\mathscr{L}_C{}^{ghost}(x) = \chi_C{}^*(x)\{D^{\mu}D_{\mu} + r^2t^a[(\mathrm{Re}\ \partial F_C{}^a{}_{\mu k}/\partial x_{ik})D^{\mu} + gf^{abc}A_C{}^{b\mu}(\mathrm{Re}\ F_C{}^c{}_{\mu k})\partial/\partial x_{ik}]\}\chi_C(x) \tag{10.51}$$

where t^a is a 3 × 3 matrix of the 3 representation of color SU(3) and $\chi_C(x)$ is a three row field in the 3 representation. *The introduction of $D^{\mu}D_{\mu}$ is based on consistency with the complexon formalism. It is needed to establish a perturbative expansion of the path integral. Its effect vanishes in the limit r → ∞ reducing ghost loops of this type to point interactions.* The reader will note that second order and third order derivative terms appear in the interaction in $\mathscr{L}_C{}^{ghost}(x^{\mu})$ and raise the issue of non-renormalizable divergences. If one uses the Two-Tier approach to quantum field theory developed by Blaha (2005a) then all potential divergences disappear. *The Two-Tier formulation of the pure, complexon, Yang-Mills theory that we are discussing is finite. See Appendix 10-A for a summary of Two-Tier quantum field theory.*

The complete pure complexon, Yang-Mills path integral is

$$Z(J^\mu) = N \int DA_C D\chi^* D\chi D\chi_C{}^* D\chi_C \Delta_{FP}(A_C)\delta(F(A_C))\Delta_C(A_C)\delta(F_C(A_C))\exp\{i\int d^7y\ [\mathscr{L} + J^\mu A_{C\mu}]\} \tag{10.52}$$

where

$$\mathscr{L} = \mathscr{L}_{CG} + \mathscr{L}^{ghost} + \mathscr{L}_C{}^{ghost} \tag{10.53}$$

with the lagrangian terms specified by eqs. 10.15, 10.42, and 10.51.

10.7 Complexon Quark-Gluon Perturbation Theory

In this section we will give the flavor of complexon strong interaction theory by considering a simple diagram. Fig. 10.1 is a self-energy diagram for a quark in which a gluon is emitted and absorbed. If one considers momentum then complexon particles have complex spatial momentum. From the form of the propagators and wave functions previously considered it is clear that the real part of the 3-momentum and the imaginary part of the 3-momentum will be separately conserved at each vertex. The general form of the self-energy integral corresponding to Fig. 10.1 is

$$I_{ab}(p) = \int d^7q\ P_{ab}(q, p)\exp(\text{Gaussian})/Q(q, p) \tag{10.54}$$

where a and b are SU(3) color indices, $P_{ab}(q, p)$ is a polynomial in q and p together with gamma matrix factors and color SU(3) matrix factors, and Q(q, p) is the product of a quark and a gluon propagator denominator factor. A Gaussian exponential factor appears if we use a Two-Tier formulation[104] of the complexon quark and gluon theory. This exponential factor guarantees the convergence of the integral resulting in a finite result.

Thus while perturbation theory is only partly useful in calculating strong interaction phenomena due to the apparent 'largeness' of the coupling constant in general, we see complexon perturbation theory is well defined, finite, and sensible if one uses Two-Tier quantum field theory.

[104] See Blaha (2005a) and the discussion of the Two-Tier formalism in Appendix 10-A.

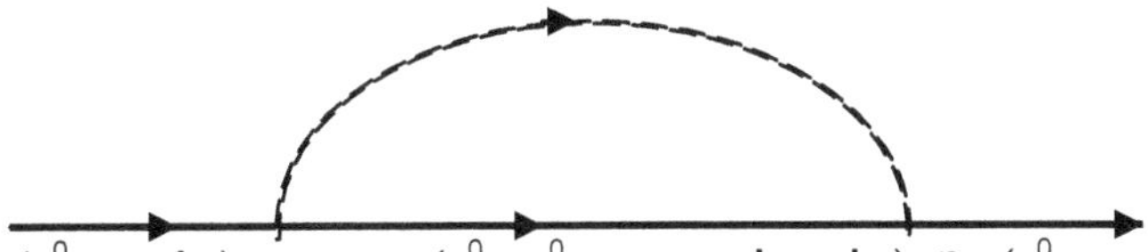

Figure 10.1. Quark self-energy diagram in which a quark emits and subsequently absorbs a gluon. The quark momentum is $p = (p^0, p_r + ip_i)$ and the gluon momentum is $q = (q^0, q_r + iq_i)$ with conservation of energy, and separate conservation of real spatial momentum and imaginary spatial momentum at each vertex.

The integral corresponding to Fig. 10.1 has the form:

$$I_{1ab}(p) = \int dq^0 d^3q_r d^3q_i \, P_{ab}(q, p)\exp(\text{Gaussian})/Q(q, p) \tag{10.55}$$

and includes integrations over both real and imaginary 3-momenta.

10.8 Complexon Quark ElectroWeak Interactions

In our theory quarks are complexons with complex spatial 3-momenta. On the other hand, the Electromagnetic and Weak bosons: $W^{\pm}_{\mu}$, Z_{μ} and A_{μ} are not complexons and so have totally real momenta. In this section we address the issue of quark Electromagnetic and Weak perurbation theory. *The general perturbation theory rule in this situation is that real and imaginary momenta are separately conserved at each vertex.*

Fig. 10.2 shows a quark self-energy diagram in which a quark emits a photon and subsequently reabsorbs it. Since ElectroWeak bosons have real energies and momenta and since real and imaginary momenta are separately conserved at each vertex by the above stated rule, the photon momentum is real while the quark spatial 3-momentum is complex.

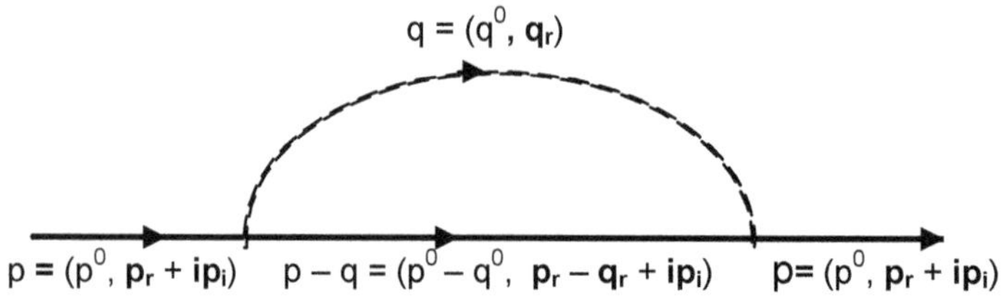

Figure 10.2. Quark self-energy diagram in which a quark emits and subsequently absorbs a photon. The quark momentum is p = (p^0, p_r + ip_i) and the photon momentum is q = (q^0, q_r) with conservation of energy, and separate conservation of real spatial momentum and imaginary spatial momentum at each vertex.

The integral corresponding to Fig. 10.2 has the form:

$$I_{2ab}(p) = \int dq^0 d^3q_r d^3q_i \delta^3(\mathbf{q}_i) P_{ab}(q, p)\exp(\text{Gaussian})/Q(q, p) \qquad (10.56)$$
$$= \int dq^0 d^3q_r P_{ab}(q, p)\exp(\text{Gaussian})/Q(q, p)$$

However, in cases where there are complexon quark loops the imaginary quark 3-momenta affects the result. For example, Fig. 10.3 shows a complexon quark loop contribution to the photon self-energy in which the imaginary quark 3-momenta integration contributes.

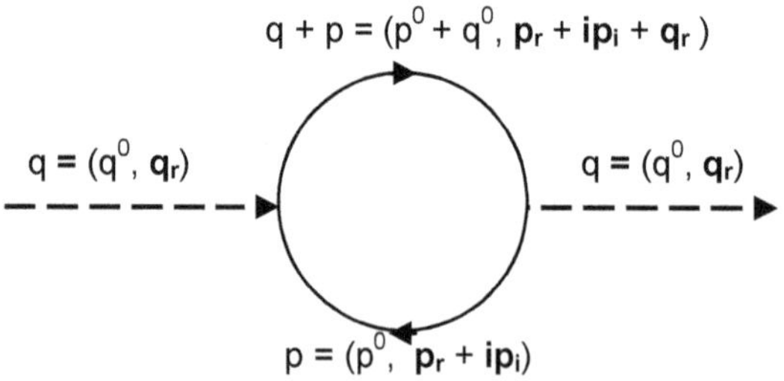

Figure 10.3. Photon self-energy diagram with a quark loop. The photon momentum is q = (q^0, q_r) and quark loop momentum is p = (p^0, p_r + ip_i).

10.9 Complexon Quarks with Color SU(3)

In this chapter we have seen that the complex Lorentz and Reality groups lead to the form of quark ElectroWeak theory and also leads naturally to color SU(3) with complexon gluons.

10-A. Two-Tier Quantum Field Theory Applied to The Standard Model

We derived a set of Standard Models in Blaha (2011c). Since then experiment has revealed new features of Dark Matter that lead us to extend the models to include a Dark sector. The extended Complexon Standard Model which we will simply call The Extended Standard Model embodies the (broken) symmetry group SU(3)⊗SU(2)⊗U(1)⊗SU(2)⊗U(1).

In this chapter we will outline a form of Quantum Field Theory called Two-Tier Quantum Field Theory that eliminates infinities and anomalies that crop up in conventional Quantum Field Theory. Two-Tier Quantum Field Theory was described in detail[105] in Blaha (2005a). It begins by extending real-valued coordinates by a quantum field: $X^\mu(z) = z^\mu + i Y^\mu(z)/M_c^2$ where M_c is a very large mass. These coordinates are then used to develop Two-Tier quantum field theory. and show that it eliminates infinities in perturbation calculations to any order.

In our Conplexon Standard Model all fields will be functions of $X^\mu(z)$ and consequently all perturbation theory calculations will yield finite results including finite fermion triangle "anomaly" diagrams. Thus the theory is completely finite and does not need regularization or other forms of renormalization to eliminate potential infinities that appear in standard quantum field theories.

10-A.1 Two-Tier Formulation of One Generation Standard Model

Two-Tier Quantum Field Theory was first presented in Blaha (2003) and Blaha (2005a). In this chapter we will develop a Two-Tier formulation of the one generation Standard Model described in the previous section. This formulation goes beyond our earlier books by using complex spatial coordinates which necessitate a slightly more complicated formulation.

The basic ansatz of the Two-Tier formalism is to replace every appearance of a real-valued coordinate x with a variable that is, in part, a quantum field Y:

$$x^\mu \rightarrow X^\mu = (y^0, \mathbf{y} + \mathbf{Y}(y^0, \mathbf{y})) \qquad (10\text{-A}.1)$$

where $\mathbf{Y}(y^0, \mathbf{y})$ is a free massless vector field identical in form to the free QED field.

[105] Appendix A at the end of this book has a detailed discussion of Two-Tier Quantum Field Theory.

Then one finds that the momentum space free field Feynman propagators G(k) of all particles acquires a Gaussian factor exp(h(k)):

$$G(k) \rightarrow G(k)\exp(h(k)) \quad (10\text{-A.2})$$

so that all perturbation theory diagrams are finite. The result is a finite perturbative result in all calculations to any order in perturbation theory. Blaha (2005a) shows that Two-Tier theories are finite, Poincare covariant, and unitary.

Simple Two-Tier Formalism

In this subsection we will describe the basic Two-Tier formalism.[106] Replace a lagrangian $\mathscr{L}_F(y)$ with the lagrangian:

$$\mathscr{L}(y) = \mathscr{L}_F(X_\mu(y))J + \mathscr{L}_C(X^\mu(y), \partial X^\mu(y)/\partial y^\nu, y) \quad (10\text{-A.3})$$

where

$$X_\mu(y) = y_\mu + i\, Y_\mu(y)/M_c^2 \quad (10\text{-A.4})$$

with M_c being a very large mass scale and $Y_\mu(y)$ a vector quantum field similar to the electromagnetic field, and where J is the absolute value of the Jacobian of the transformation from X to y coordinates:

$$J = |\partial(X)/\partial(y)| \quad (10\text{-A.5})$$

The lagragian term $\mathscr{L}_C$ is

$$\mathscr{L}_C = +\tfrac{1}{4} M_c^4 F^{\mu\nu}F_{\mu\nu} \quad (10\text{-A.6})$$

with

$$\begin{aligned} F_{\mu\nu} &= \partial X_\mu/\partial y^\nu - \partial X_\nu/\partial y^\mu \quad (10\text{-A.7})\\ &\equiv i\,(\partial Y_\mu/\partial y^\nu - \partial Y_\nu/\partial y^\mu)/M_c^2 \end{aligned}$$

The sign in eq. 10-A.3 is not negative – superficially contrary to the conventional electromagnetic Lagrangian. The reason for this difference is that the quantum field part of X^μ is imaginary. Thus $\mathscr{L}_C$ winds up having the correct sign after taking account of the factor of i in the field strength $F_{\mu\nu}$.

Defining

$$F_{Y\mu\nu} = (\partial Y_\mu/\partial y^\nu - \partial Y_\nu/\partial y^\mu) \quad (10\text{-A.8})$$

---

[106] We note that the lagragian formlation presented in Blaha (2005a) relies on a new form of the Calculus of Variations – not one of the three forms previously known and documented in the literature.

we see the Lagrangian assumes the form of the conventional electromagnetic Lagrangian:

$$\mathscr{L}_C = -\tfrac{1}{4} F_Y^{\mu\nu} F_{Y\mu\nu} \qquad (10\text{-A}.9)$$

The action of this theory has the form

$$I = \int d^4y\, \mathscr{L}(y) \qquad (10\text{-A}.10)$$

The further development of Two-Tier Quantum Field Theory is described in Blaha (2005a).

Two-Tier L_C-based Standard Model Theory

In the present case we will need two variables X_r^μ and X_i^μ for the Standard Model based on L_C covariance since quarks have complex spatial 3-coordinates. We define them similarly to the previous case:

$$X_{r\mu}(y_r) = y_{r\mu} + i\, Y_{r\mu}(y_r)/M_c^2 \qquad (10\text{-A}.11)$$

$$X_{i\mu}(y_i) = y_{i\mu} + i\, Y_{i\mu}(y_i)/M_c^2 \qquad (10\text{-A}.12)$$

where we choose the same mass scale for both the "real" and "imaginary" variables. The Two-Tier, single generation, version of the Standard Model presented in section 10-A.2 has an action of the form

$$I_{SM1tt} = \int dy^0 d^3y_r d^3y_i\, (\mathscr{L}_{SM1}(X_r^\mu(y_r), \mathbf{X}_i^k(y_i)) J_2)|_{y_i^0=0,\; Y_r^0=Y_i^0=0} + \int dy_r^0 d^3y_r\, \mathscr{L}_C(X_r^\mu(y_r), \partial X_r^\mu(y_r)/\partial y_r^\nu, y_r) + $$
$$+ \int dy_i^0 d^3y_i\, \mathscr{L}_C(X_i^\mu(y_i), \partial X_i^\mu(y_i)/\partial y_i^\nu, y_i) \qquad (10\text{-A}.13)$$

where the replacements

$$x^\mu \equiv x_r^\mu \rightarrow X_r^\mu(y_r) \qquad (10\text{-A}.14)$$
$$x_i^k \rightarrow X_i^k(y_i) \qquad (10\text{-A}.15)$$

for $\mu = 0, 1, 2, 3$ and $k = 1, 2, 3$ are made in $\mathscr{L}_{SM1}$ (described in section 10-A.2 below) followed by defining $y_r^0 = y^0$ and making an L_C transformation to a frame where $y_i^0 = 0$, and where J_2 is the absolute value of the Jacobian of the transformation from (X_r, X_i) to (y_r, y_i) coordinates:

$$J_2 = |\partial(X_r, X_i)/\partial(y_r, y_i)| \qquad (10\text{-A}.16)$$

We also choose gauges where $Y_r^0 = Y_i^0 = 0$. These transformations and gauge choices are discussed in detail later. The lagrangian terms $\mathscr{L}_C(X_r^\mu(y_r), \partial X_r^\mu(y_r)/\partial y_r^\nu, y_r)$ and $\mathscr{L}_C(X_i^\mu(y_i), \partial X_i^\mu(y_i)/\partial y_i^\nu, y_i)$ have the form:

$$\mathscr{L}_C = +\tfrac{1}{4} M_c^4 F^{\mu\nu} F_{\mu\nu} \qquad (10\text{-A.17})$$

with

$$F_{\mu\nu} = \partial X_\mu/\partial y^\nu - \partial X_\nu/\partial y^\mu \qquad (10\text{-A.18})$$
$$\equiv i\,(\partial Y_\mu/\partial y^\nu - \partial Y_\nu/\partial y^\mu)/M_c^2$$

or defining

$$F_{Y\mu\nu} = (\partial Y_\mu/\partial y^\nu - \partial Y_\nu/\partial y^\mu) \qquad (10\text{-A.19})$$

we see each lagrangian assumes the form of the conventional electromagnetic Lagrangian:

$$\mathscr{L}_C = -\tfrac{1}{4} F_Y^{\mu\nu} F_{Y\mu\nu} \qquad (10\text{-A.20})$$

The lagrangian is supplemented with the following condition on all complexon fields $\Phi_{\ldots}$:

$$(\partial/\partial X_r^k(y_r))\,(\partial/\partial X_i^k(y_i))\Phi\ldots = 0 \qquad (10\text{-A.21})$$

summed over k = 1, 2, 3. Non-complexon fields Ω... in our left-handed formulation satisfy the subsidiary condition:

$$\{(\partial/\partial X_r^k(y_r))\,(\partial/\partial X_i^k(y_i)) - ((\partial/\partial X_r^k(y_r))^2\,(\partial/\partial X_i^m(y_i))^2)^{1/2}\}\Omega\ldots = 0 \qquad (10\text{-A.22})$$

summed over k = 1, 2, 3 and over m = 1, 2, 3.

Feynman Propagators

The momentum space free field Feynman propagators G...(k) of all particles and ghosts in our Two-Tier Standard Model acquires a Gaussian factor exp(h(k)):

$$G\ldots(k) \rightarrow G\ldots(k)\,\exp(h(k)) \qquad (10\text{-A.23})$$

so that all perturbation theory diagrams are finite. The result is a finite perturbative result in all calculations to any order in perturbation theory. Blaha (2005a) shows that Two-Tier theories are finite, Poincare covariant, and unitary.

An example of the Two-Tier effect on propagators is the case of the Two-Tier photon propagator. The Two-Tier photon propagator[107] is:

---

[107] Blaha (2005a).

$$iD_F^{TT}(y_1 - y_2)_{\mu\nu} = -i \int \frac{d^4p\, e^{-ip\cdot z}\, g_{\mu\nu} R(\mathbf{p}, z)}{(2\pi)^4 (p^2 + i\varepsilon)} \tag{10-A.24}$$

(since the imaginary parts can be taken to be zero: $y_{1i}^{\mu} - y_{2i}^{\mu} = 0$) where

$$z^{\mu} = y_{1r}^{\mu} - y_{2r}^{\mu} \tag{10-A.25}$$

$$R(\mathbf{p}, \mathbf{z}) = \exp[-p^i p^j \Delta_{Tij}(z)/M_c^4] \tag{10-A.26}$$

$$= \exp\{ -\mathbf{p}^2[A(v) + B(v)\cos^2\theta] / [4\pi^2 M_c^4 |\mathbf{z}|^2 \tag{10-A.27}$$

with i, j = 1, 2, 3, and with $\Delta_{Tij}(z)$ being the commutator of the positive frequency part $Y^+_k(y)$ and the negative frequency part $Y^-_k(y)$ of $Y_k(y)$:

$$\Delta_{Tij}(z) = [Y^+_j(y_{1r}), Y^-_k(y_{2r})] = \int d^3k\, e^{ik\cdot(y_{1r} - y_{2r})} (\delta_{jk} - k_j k_k/\mathbf{k}^2)/[(2\pi)^3 2\omega_k] \tag{10-A.28}$$

and

$$v = |z^0|/|\mathbf{z}| \tag{10-A.29}$$

$$A(v) = (1 - v^2)^{-1} + .5v \ln[(v - 1)/(v + 1)] \tag{10-A.30}$$

$$B(v) = v^2(1 - v^2)^{-1} - 1.5v \ln[(v - 1)/(v + 1)] \tag{10-A.31}$$

$$\mathbf{p\cdot z} = |\mathbf{p}|\, |\mathbf{z}| \cos\theta \tag{10-A.32}$$

with $|\mathbf{p}|$ denoting the length of a spatial vector $\mathbf{p}$, $|\mathbf{z}|$ denoting the length of a spatial vector $\mathbf{z}$, and with $|z^0|$ being the absolute value of z^0.

The gaussian factors R(**p**, z) which appear in all Two-Tier propagators damp the large momentum behavior of all perturbation theory integrals producing a completely finite perturbation theory and yet give the usual results of perturbation theory at energies that are small compared to the mass scale M_c.

Complexon Feynman Propagator

In the case of complexons the Two-Tier Feynman propagator differs from the non-complexon case by having an integration over imaginary spatial 3-momenta, a derivative of a delta function embodying the orthogonality of the real and imaginary 3-momenta, and two factors of R(**p**, z): one factor being R($\mathbf{p}_r$, z_r) and the other factor being R($\mathbf{p}_i$, z_i) (where the time components $z_r^0 = z^0$ and $z_i^0 = 0$ since there is only one real time coordinate[108]) thus providing large momentum convergence for both real and imaginary 3-momentum integrations.

For a scalar complexon particle we previously found the Feynman propagator:

$$i\Delta_{CTF}(x-y) = \theta(x^+ - y^+)<0|\phi_{CT}(x)\, \phi_{CT}(y)|0> + \theta(y^+ - x^+)<0|\phi_{CT}(y)\phi_{CT}(x)|0>$$
$$= i\int d^4p_r d^3p_i (2\pi)^{-7}\delta'(\mathbf{p_r}\cdot\mathbf{p_i}/m^2)e^{-ip^+(x^- - y^-) - ip^-(x^+ - y^+) + ip_\perp\cdot(x_\perp - y_\perp) - ip_i\cdot(x_i - y_i)}/(p^2 + m^2 + i\varepsilon)$$

in conventional quantum field theory.

In the case of Two-Tier quantum field a scalar complexon particle has the the Feynman propagator

$$i\Delta_{CTFtt}(x-y) = i\int d^4p_r d^3p_i (2\pi)^{-7}\delta'(\mathbf{p_r}\cdot\mathbf{p_i}/m^2)\, R(\mathbf{p}_r, z_r)R(\mathbf{p}_i, z_i)\cdot$$
$$\cdot e^{-ip^+(x^- - y^-) - ip^-(x^+ - y^+) + ip_\perp\cdot(x_\perp - y_\perp) - ip_i\cdot(x_i - y_i)}/(p^2 - m^2 + i\varepsilon) \qquad (10\text{-A}.33)$$

where the time components $z_r^0 = z^0$ and $z_i^0 = 0$ since there is only one time coordinate and where $p^2. = p^{0\,2} - p_r^2 + p_i^2$.

Propagators for other types of particles are similarly modified in the Two-Tier formalism (See Blaha 2005a).

10-A.3 Renormalization and Divergence Issues

The theory derived from $\mathcal{L}_{SM1}$ when calculated perturbatively in conventional quantum field theory has divergences that are not renormalizable due to its 7-dimensional nature. This issue does not appear to be curable by known renormalization methods. However if the theory is reformulated in the Two-Tier formalism developed by Blaha[109] then a finite theory results. This is also true for the four (three) generation Standard Model to be described later.

The integrals of the terms of the lagrangian can be divided into two sets: integrals of terms that do not contain complexon field factors can be integrated $\int d^4y$ to create one type of contribution to the total action, and integrals of terms that do contain complexon field factors can be integrated $\int dy^0 d^3y_1 \int d^3y_2$ to create the other contributions to the total action. Eq. 10-A.13 exemplifies the definition of a Two-Tier action in general.

[108] We can arrange for $z_i^0 = 0$ by making a L_C transformation to an inertial frame where z is real.
[109] Blaha (2005a) and Blaha (2003).

11. Dark Matter Sector of the SU(3)⊗SU(2)⊗U(1)⊗SU(2)⊗U(1) Extended Standard Model

11.1 Introduction

This chapter describes the dynamic equations of the Dark matter sector of the Extended Standard Model. Because we have a derivation of the Reality group fom the form of Complex Lorentz transformations, we find that Dark matter must have a Dark SU(2)⊗U(1) symmetry and interactions since only one Weak and Electromagnetic SU(2)⊗U(1) is found in the normal matter sector. We distinguish the Dark SU(2)⊗U(1) symmetry by denoting it as DSU(2)⊗DU(1).

Since there is no significant interaction between normal and Dark matter we see that Dark matter does not have color SU(3) symmetry or interactions. Thus DSU(2)⊗DU(1) is the only possible Dark interaction symmetry since there are no further Reality group subgroup factors due to the form of Complex Lorentz transformations.[110]

11.2 The New Extended Standard Model Lagrangian

In previous chapters we derived the form of a one generation Extended Standard Model that included the known parts of the Standard Model (excepting the Higgs sector) and an SU(2)⊗U(1) part for Dark Matter. In the following sections we will discuss the Extended Standard Model Dark matter sector.

11.2.1 Two-Tier Lepton Sector with Dark Leptons

We begin with the definition of a quadruplet of leptons – a pair of doublets, one normal and one Dark, instead of a single doublet. We define left and right lepton quadruplets with[111]

[110] It is possible that further interactions might exist derived from other considerations. Indeed we posit such additional symmetries later: the U(4) Generation group and the U(4) Layer group. These groups and their attendant interactions indirectly emerge from the fermion spectrum generated by the Complex Lorentz group with its four fermion species.

[111] The X's are Two-Tier coordinates.

$$\Psi_{L,R}(X) = \begin{bmatrix} \psi_{DL,R}(X) \\ \psi_{NL,R}(X) \end{bmatrix} \tag{11.1}$$

where $\psi_{NL,R}(X)$ is a "normal" lepton doublet, and where $\psi_{DL,R}(X)$ is a Dark lepton doublet consisting of a Dark electron-like fermion and a Dark neutrino-like fermion.

We define covariant derivative terms which we express in matrix form as

$$D_{L,R}(X) = \begin{bmatrix} \gamma^{\mu}D_{DL,R\mu} & 0 \\ 0 & \gamma^{\mu}D_{NL,R\mu} \end{bmatrix} \tag{11.2}$$

where the normal matter left-handed covariant derivative is

$$D_{NL\mu} = \partial/\partial X^{\mu} - \tfrac{1}{2}ig\boldsymbol{\sigma}\cdot\mathbf{W}_{\mu} + \tfrac{1}{2}ig'B_{\mu} \tag{11.3}$$

and where the Dark matter left-handed covariant derivative is

$$D_{DL\mu} = \partial/\partial X^{\mu} - \tfrac{1}{2}ig_D\boldsymbol{\sigma}\cdot\mathbf{W}'_{\mu} + \tfrac{1}{2}ig_D'B'_{\mu} \tag{11.4}$$

with $\boldsymbol{\sigma}$ a vector composed of the Pauli matrices. The right-handed covariant derivatives have a simpler form. The normal matter right-handed covariant derivative is

$$D_{NR\mu} = \partial/\partial X^{\mu} + ig'B_{\mu} \tag{11.5}$$

and the Dark matter right-handed covariant derivative is

$$D_{DR\mu} = \partial/\partial X^{\mu} + \tfrac{1}{2}ig_D'B'_{\mu} \tag{11.6}$$

The normal and Dark fields above are functions of Two-Tier coordinates X. The Faddeev-Popov mechanism operative for these types of fields is described in appendix 19-A of Blaha (2011c) and in chapter 10.

11.2.2 Quark Sector with Dark Quarks

In the *quark* sector we define left and right quark quadruplets with

$$\Psi_{qL,R}(X_c) = \begin{bmatrix} \psi_{DqL,R}(X_c) \\ \psi_{NqL,R}(X_c) \end{bmatrix} \tag{11.7}$$

where $\psi_{NqL,R}(X_c)$ is a "normal" quark doublet, and where $\psi_{DqL,R}(X_c)$ is a Dark quark doublet consisting of a SU(3) singlet Dark up-quark of unit Dark charge and a SU(3) singlet Dark down-quark of zero Dark charge.

The covariant derivative terms are contained in $D_q(X_c)$ which we express in matrix form as

$$D_{qL,R}(X_c) = \begin{bmatrix} \gamma^\mu D_{qDL,R\mu}(X_c) & 0 \\ 0 & \gamma^\mu D_{qNL,R\mu}(X_c) \end{bmatrix} \tag{11.8}$$

where the normal quark matter left-handed covariant derivative is

$$D_{qNL\mu} = \partial/\partial X_c{}^\mu - ½ig\boldsymbol{\sigma}\boldsymbol{\cdot}\mathbf{W}_\mu - ig'B_\mu/6 + ig_C\tau\boldsymbol{\cdot}A_{C\mu} \tag{11.9}$$

where $A_{C\mu}$ is the color gauge field and where the Dark quark left-handed covariant derivative is

$$D_{qDL\mu} = \partial/\partial X_c{}^\mu - ½ig_D\boldsymbol{\sigma}\boldsymbol{\cdot}\mathbf{W}'_\mu + ½ig_D'B'_\mu \tag{11.10}$$

since Dark quarks are SU(3) singlets with unit or zero Dark charge. The right-handed quark covariant derivatives have a simpler form. The normal quark right-handed covariant derivative is

$$D_{qNR\mu} = \partial/\partial X_c{}^\mu + ig'B_\mu/3 + ig_C\tau\boldsymbol{\cdot}A_{C\mu} \tag{11.11}$$

and the Dark quark right-handed covariant derivative is

$$D_{qDR\mu} = \partial/\partial X_c{}^\mu + ½ig_D'B'_\mu \tag{11.12}$$

The normal and Dark gauge boson fields are functions of $X_{c\cdot} = (X_{r\mu}(y_r), X_{i\mu}(y_i))$ of eqs. 11.11 and 11.12. The Faddeev-Popov mechanism is operative for gauge boson fields and is described in chapter 10 and appendix 19-A of Blaha (2011c).[112] The *complexon* quark Standard Model Sector covariant derivatives in quadruplet matrix form are

$$D_{qL,R}(X_c) = \begin{bmatrix} \gamma^{\mu} D_{qDL,R\mu} & 0 \\ 0 & \gamma^{\mu} D_{qNL,R\mu} \end{bmatrix} \tag{11.13}$$

The remaining parts of the complexon Standard Model are described in chapter 23 of Blaha (2011) and summarized below. The addition of singlet Dark quark Higgs terms is also required.

The lagrangian density and action is

$$\mathcal{L}_{CSM} = \Psi_L^{a\dagger}\gamma^0 i\gamma^{\mu} D_{L\mu}\Psi_L^{\ a} - \Psi_R^{a\dagger}\gamma^0 i\gamma^{\mu} D_{R\mu}\Psi_{3R}^{\ a} + \Psi_{CL}^{a\dagger}\gamma^0 i\gamma^{\mu}\mathcal{D}_{qL\mu}\Psi_{CL}^{\ a} + \Psi_{CR}^{a\dagger}\gamma^0 i\gamma^{\mu}\mathcal{D}_{qR\mu}\Psi_{CR}^{\ a} - $$
$$- \mathcal{L}_{BareMasses} + \mathcal{L}_{Gauge} + \mathcal{L}_{Mass} \tag{11.14}$$

where a is the generation index. $\mathcal{L}_{BareMasses}$ contains the fermion bare mass terms. Also,

$$\mathcal{L}_{Gauge} = \mathcal{L}_{GaugeEW} + \mathcal{L}_{GaugeC} + \mathcal{L}_{GaugeEWD} \tag{11.15}$$

with

$$\mathcal{L}_{GaugeEW} = -\tfrac{1}{4} F_W^{a\mu\nu} F_{W\ \mu\nu}^{\ a} - \tfrac{1}{4} F_B^{\mu\nu} F_{B\mu\nu} + \mathcal{L}_{EW}^{ghost} \tag{11.16}$$

$$\mathcal{L}_{GaugeEWD} = -\tfrac{1}{4} F'^{a\mu\nu}_{W} F'^{\ a}_{W\ \mu\nu} - \tfrac{1}{4} F_{B'}^{\mu\nu} F_{B'\mu\nu} + \mathcal{L}_{W'}^{ghost} \tag{11.17}$$

and

$$\mathcal{L}_{GaugeC} = \mathcal{L}_{CCG} + \mathcal{L}_C^{ghost} + \mathcal{L}_{CC}^{ghost} \tag{11.18}$$

The gauge bosons $W_{\mu}^{\ a}$, and B_{μ} field tensors are:

$$F_{W\ \mu\nu}^{\ a} = \partial W^a_{\ \mu}/\partial x^{\nu} - \partial W^a_{\ \nu}/\partial x^{\mu} + g_2 f^{abc} W^b_{\ \mu} W^c_{\ \nu} \tag{11.19}$$

$$F_{B\mu\nu} = \partial B_{\mu}/\partial x^{\nu} - \partial B_{\nu}/\partial x^{\mu} \tag{11.20}$$

[112] Those who might be concerned about the propagator term <$W_i(X)$, $W_j(X_c)$> and similar propagators where one field is a function of X and the other field is a function of X_c should note that such terms are to very good approximation equal to <$W_i(X)$, $W_j(X)$> for energies much less than M_c (which could be as large as the Planck energy.)

and the Dark gauge bosons W'^{a}_{μ} and B'_{μ} field tensors are:

$$F_{B'\mu\nu} = \partial B'_{\mu}/\partial x^{\nu} - \partial B'_{\nu}/\partial x^{\mu}$$
$$F'^{a}_{W\,\mu\nu} = \partial W'^{a}_{\mu}/\partial x^{\nu} - \partial W'^{a}_{\nu}/\partial x^{\mu} + g_2 f^{abc} W'^{b}_{\mu} W'^{c}_{\nu} \tag{11.21}$$

$\mathcal{L}_{EW}{}^{ghost}$ contains the Faddeev-Popov ghost terms for the W_{μ}^{a} gauge bosons. The complexon color gluon lagrangian $\mathcal{L}_{CCG}$ is defined by

$$\mathcal{L}_{CCG} = -\tfrac{1}{4} F_{CC}{}^{a\mu\nu}(x) F_{CC}{}^{a}{}_{\mu\nu}(x) \tag{11.22}$$

where

$$F_{CC}{}^{a}{}_{\mu\nu} = \partial/\partial X_c{}^{\nu}\, A_C{}^{a}{}_{\mu} - \partial/\partial X_c{}^{\mu}\, A_C{}^{a}{}_{\nu} + g f_{su(3)}{}^{abc} A_C{}^{b}{}_{\mu} A_C{}^{c}{}_{\nu} \tag{11.23}$$

where $A_C{}^{a}{}_{\nu}$ is the color gluon gauge field, g is the color coupling constant, and the $f_{su(3)}{}^{abc}$ are the SU(3) structure constants.

In addition $\mathcal{L}_C{}^{ghost}$ is the color SU(3) Faddeev-Popov ghost terms defined in appendix 19-A of Blaha (2011c) for the complexon Lorentz gauge and $\mathcal{L}_{CC}{}^{ghost}$ is the complexon color SU(3) constraint ghost terms defined through the Faddeev-Popov mechanism. The mass sector $\mathcal{L}_{Mass}$ is presumably based on the Higgs Mechanism.which creates the fermion and vector boson masses, and the generation mixing.

The lagrangian is supplemented with the following condition on all complexon fields $\Phi_{\ldots}$:[113]

$$\nabla_r \cdot \nabla_i \Phi_{\ldots} = 0 \tag{11.24}$$

Non-complexon fields $\Omega\ldots$ in the left-handed formulation under consideration satisfy the subsidiary condition:

$$[\nabla_r \cdot \nabla_i - (\nabla_r^2 \nabla_i^2)^{1/2}]\Omega\ldots = 0 \tag{11.25}$$

which guarantees a complexon's real momentum is parallel to its imaginary momentum.

*We note that **W**$'_{\mu}$ and B'_{μ} are the Dark gauge fields having the Dark symmetry group SU(2)⊗U(1).*

---

[113] These conditions implement the orthogonality of the real and imaginary parts of complexon 3-momentum.

12. Baryon, Lepton, Dark Baryon, and Dark Lepton Conservation Laws, and the U(4) Generation Group

The preceding chapters have derived the interactions of the one generation Extended Standard Model with the symmetry: SU(3)⊗SU(2)⊗U(1)⊗SU(2)⊗U(1) from the form of Complex Lorentz group transformations. In this chapter we consider the extension of the one generation theory to four generations based on a new symmetry group U(4) that we call the Generation group. We will consider for four conservation laws for Baryon, Lepton, Dark Baryon, and Dark Lepton numbers. We will show that the existence of a long range baryonic force supports baryon number conservation in a manner similar to electric charge conservation due to the electromagnetic force. Lepton number conservation suggests a very weak long range force as well. By analogy we postulate similar Dark particle number conservation laws and forces.

Having four conserved (or almost conserved) particle numbers with their attendant forces (interactions) leads naturaly to a U(4) symmetry with four 'diagonal' generators that we call the U(4) Generation group.

12.1 Baryon Number Conservation and a Possible Baryonic Force

We have considered baryon number conservation and a possible baryonic force in Blaha (2014a) and (2014b). Most of this chapter contains selections from these books.

The primary forces involved in the interactions and collisions of baryons include the forces of The Standard Model, the force of gravity, and a fifth force which we take to be the baryonic force, a much discussed force that depends on the baryon numbers of ubjects experiencing it. Gravitation and this force (neglecting Standard Model interactions) between two clumps of baryonic matter containing baryons and other particles: clump1 being of mass m_1 and baryon number n_1, and clump2 being of mass m_2 and baryon number n_2 is

$$F = -Gm_1m_2/r^2 + (\beta^2/4\pi)n_1n_2/r^2 \qquad (12.1)$$

where G is the gravitational constant, β is the baryonic coupling constant and r is the distance between the 'widely' separated clumps. Experimentally a baryonic force between baryons has not been detected with any degree of certainty. Sakurai (1964) discusses early efforts to determine the force of gravity in detail. Eőtvős experiments on the ratio of the observed gravitational mass to the inertial mass showed that that the force is constant to within one part in 100,000,000 as far back as 1922 indicating the baryonic force, if it exists, as we believe it does, is extremely weak compared to the gravitational force. Eőtvős et al[114] found

$$(\beta^2/4\pi)/(Gm_p^2) < 10^{-5}$$

where m_p is the proton mass.

Since then, the experiment has been redone with improved accuracy by Dicke and collaborators.[115] They have improved the accuracy to one part in 100 billion. A further analysis showed a very small discrepancy that suggested the ratio, while small, was non-zero, implying the equivalence principle might not be exact and that the discrepancy changed with the material used in the experiment – just what one might expect if a very small baryonic force was present – often called the "fifth force." At present the existence and amount of the discrepancy is unclear. Nevertheless, we will assume a fifth force.

The primary rationale for the fifth force is the apparent conservation of baryon number. The conservation of baryon number has been repeatedly investigated by experimenters and found to be true to extremely high accuracy. For decades theorists have suggested that a baryon conservation law[116] follows from the existence of a gauge field in a manner much like electric charge conservation follows from the properties of the electromagnetic gauge field.

12.1.1 Estimate of the Baryonic Coupling Constant

The baryonic force, and coupling constant, if it exists, is known to be very small in comparison to gravity and the other known forces. However, measurements of the gravitational constant G are significantly different.[117,118] The reason(s) for these discrepancies is not known. We will assume that both the 2010 and 2013 measurements of G are experimentally correct but disagree because of the baryonic force term in eq. 12.1 that would create a difference in effective G values if the experiments used different masses and thus baryon numbers. Quinn et al found a value for the gravitational constant of $G_1 = 6.67545\times10^{-11}$ $m^3kg^{-1}s^{-2}$. The combined 2010 CODATA value for the gravitational constant was $G_2 = 6.67384\times10^{-11}$ $m^3kg^{-1}s^{-2}$. Both values are subject to estimated uncertainties.

Suppose these values are correct and due to a difference in the chemical composition (metals) of the test masses used in the experiment. Quinn et all use 1.2 kg test masses composed of Cu-0.7% Te free machining alloy. The CODATA value being a composite of many

---

[114] Eőtvős, R. V., Pekár, D., Fekete, E., Ann. d. Physik **68**, 11 (1922).
[115] P. G. Roll, R. Krotkov, R. H. Dicke, Annals of Physics, 26, 442, 1964.
[116] See Gell-Mann, M. and Levy, M. *Nuovo Cimento* 16, 705 (1960) for a proof and Sakurai (1964) for a discussion of the relation of the baryonic gauge field to gravity experimentally.
[117] T. Quinn et al, Phys. Rev. Lett. **111**, 101102 (2013).
[118] P. J. Mohr, B.N. Taylor, and D. B. Newell, Rev. Mod. Phys. 84, 1527 (2012).

experiments does not have an effective equivalent test mass value or composition specified.[119] Suppose the test mass value is $N_1^2m_1^2 + N_{1e}^2m_e^2$ for the G_1 result giving

$$-(N_1^2m_1^2 + N_{1e}^2m_e^2)G_1 = [-G(m_1^2N_1^2 + N_{1e}^2m_e^2) + (\beta^2/4\pi)N_1^2] \qquad (12.2)$$

where G is the *real* value of the gravitational constant. The total test mass is $(m_1^2N_1^2 + N_{1e}^2m_e^2)$ with N_1 baryons of average mass m in each test mass and N_{1e} leptons of average mass m_e.

Suppose further the test mass value is $N_2^2m_2^2 + N_{2e}^2m_e^2$ for the G_2 result giving

$$-(N_2^2m_2^2 + N_{2e}^2m_e^2)G_2 = [-G(m_2^2N_2^2 + N_{2e}^2m_e^2) + (\beta^2/4\pi)N_2^2] \qquad (12.3)$$

where G is again the *real* value of the gravitational constant. The total test mass is $(m_2^2N_2^2 + N_{2e}^2m_e^2)$ with N_2 baryons of average mass m_2 in each test mass and N_{2e} leptons of average mass m_e. Since the test masses are electrically neutral and there are approximately equal numbers of protons and neutrons in a test mass it follows approximately that

$$N_{1e} = ½N_1 \quad \text{and} \quad N_{2e} = ½N_2 \qquad (12.4)$$

Subtracting eq. 12.2 from eq. 12.3 after some algebra[120] we find

$$\Delta G = -G_2 + G_1 = (\beta^2/4\pi)/(m_2^2 + m_e^2/2) - (\beta^2/4\pi)/(m_1^2 + m_e^2/2)$$
$$\simeq (\beta^2/4\pi)(1/m_2^2 - 1/m_1^2) \qquad (12.5)$$

The masses m_1 and m_2 can differ. For example, if m_H is mass of the hydrogen atom, then $m^{-1} = 1.0m_H^{-1}$ for hydrogen, for carbon $m^{-1} = 1.00782m_H^{-1}$, for copper $m^{-1} = 1.00895m_H^{-1}$, and for lead $m^{-1} = 1.00794m_H^{-1}$.[121] Thus using the Quinn et al results, and CODATA results, and assuming copper and lead test masses, we find the order of magnitude *estimate*:

$$\alpha_B = \beta^2/4\pi \simeq \Delta G/[(1.00895^2 - 1.00794^2)\, m_H^2]$$
$$\simeq \Delta G/G\; G\, m_H^2/.002037$$
$$\simeq (0.000241/0.002037)Gm_H^2$$
$$\simeq .118\; Gm_H^2 \qquad (12.6)$$

---

[119] The Eötvös' experiment used a 0.1 gm test mass of $RaBr_2$. R. v. Eötvös, D. Pekár, E. Fekete, Annalen der Physik (Leipzig) 68, 11, 1922.

[120] The reduction of the calculation to algebra reminds the author of Nobelist Hans Bethe's remark that he only felt he understood a physical phenomenon when he could reduce it to algebra. This was quite evident when the author collaborated with Professor Bethe on a study of pion condensation in neutron stars some years ago.

[121] "One Hundred Years of the Eötvös Experiment", I. Bod, E. Fischbach, G. Marx and Maria Náray-Ziegler, August, 1990.

indicating a very weak baryonic force consistent with our general view of the Megaverse. The baryon fine structure constant is minute in comparison to the electromagnetic fine structure constant $\alpha \simeq 1/137$.

Due to our assumptions in the calculation of α_B, which makes it merely an order of magnitude estimate at best, we suggest that an experimental group measure G with differing test masses in the same apparatus to obtain a better value for α_B.

12.1.2 A Baryonic Gauge Field

The Baryonic force and conservation law suggest it has a spin 1 gauge field like the electromagnetic field. However, since baryon number, and any associated gauge field extend beyond our universe (if there is a 'beyond' as the author believes) we will consider a 16-dimensional *Megaverse*[122] implantation of the baryonic gauge field below. The reader may skip this approach and simply assume an electromagnetic-like baryonic gauge field if he/she so desires. The same comments apply to the Dark baryonic force and gauge field, since Dark baryon number as well as baryon number are both not bound by the boundaries of our universe.

We now proceed under the assumption of a Megaverse realizing that the corresponding case of only our universe is completely analgous.

Based on a reasonable value for α_B, we assumed in Blaha (2014) that a baryonic gauge field exists that is similar to the electromagnetic field. This gauge field couples extremely weakly[123] to individual baryons as well as to aggregates of baryons due to their non-zero baryon number. We called the baryonic gauge field particle a *planckton*. Its electromagnetic analogue is the photon.

Plancktons propagate in the Megaverse, both within universes, and exterior to universes. So the planckton field must be defined in 16-dimensional Megaverse coordinates. They will interact with baryons within a universe with Megaverse coordinates mapped to the curved coordinates of the universe. Since a planckton field in 16-dimensional conventional coordinates would lead to divergences we use 16-dimensional quantum coordinates:[124]

$$Y^i(y) = y^i + i\, Y_u{}^i(y)/M_u{}^8 \qquad (12.7)$$

with quantum coordinate derivatives defined by

$$\partial_i = \partial/\partial Y^i(y) = \partial/\partial(y^i - Y_u{}^i(y)/M_u{}^8) \qquad (12.8)$$

[122] The Megaverse is described in Blaha (2015a) and earlier books by the author.
[123] Compared to gravity.
[124] See Appendix A or Blaha (2005a) for a discussion of this new method to eliminate infinities in quantum field theory calculations. **The subscript 'u' simply represents the word 'universe' to distinguish it from the 'B' occasionally used to denote ElectroWeak gauge fields.**

to obtain a completely finite theory of planckton interactions with elementary particles and universe particles.

Plancktons and the $Y_u^i(y)$ field of quantum coordinates are the only fields in the space between universes in the Megaverse. Since the mass-energy and charge of universes is zero, Standard Model fields are zero in the space between universes.[125] It is reasonable to assume that the vacuum between universes does have fermion and boson seas for Standard Model particles. And we will propose new forces: Dark baryon force, lepton force and Dark lepton force corresponding to the three other particle numbers below. These four forces will have gravitational effects between universes in the Megaverse since their particle numbers are also not confined to our universe.

12.1.3 Planckton Second Quantization

The second quantization of the free planckton field $B_u^i(y)$ is similar to the second quantization of the electromagnetic field, and also of the quantum part of the Megaverse quantum coordinates $Y_u^i(y)$. The purpose and role of these fields is quite different: the planckton field generates an interaction between baryons while the $Y_u^i(y)$ field serves as the quantum part of 4-dimensional quantum coordinates giving us a finite quantum field theory of The New Standard Model and gravitation as well as a finite Big Bang for our universe.

We begin by noting that Megaverse quantum coordinates are defined by eqs. 12.7 and 12.8 above. The lagrangian density terms for the free $B_u^i(Y(y))$ fields is

$$\mathscr{L}_{Bu} = -\tfrac{1}{4}\, F_{Bu}{}^{\mu\nu}(Y(y))F_{Bu\mu\nu}(Y(y)) \tag{12.9}$$

with Y(y) given by eq. 12.7. The lagrangian is

$$L_{Bu} = \int d^{15}y\, \mathscr{L}_{Bu}(Y(y)) \tag{12.10}$$

with

$$F_{Bu\mu\nu} = \partial B_{u\mu}(Y(y))/\partial Y^{\nu}(y) - \partial B_{u\nu}(Y(y))/\partial Y^{\mu}(y) \tag{12.11}$$

where the values of μ and ν range from 1 to 16 in this section.

The equal 'time' commutation relations, derived in the usual way, are:

$$[B_u^{\mu}(Y(\mathbf{y}, y^0)), B_u^{\nu}(Y(\mathbf{y}', y^0))] = [\pi_u^{\mu}(Y(\mathbf{y}, y^0)), \pi_u^{\nu}(Y(\mathbf{y}', y^0))] = 0 \tag{12.12}$$

$$[\pi_{uj}(Y(\mathbf{y}, y^0)), B_{uk}(Y(\mathbf{y}', y^0))] = -i\,\delta^{15tr}{}_{jk}(Y(\mathbf{y},0) - Y(\mathbf{y}',0)) \tag{12.13}$$

where

[125] The vacuum energy of the baryonic field and the $Y_u^i(y)$ fields being uniform throughout the Megaverse do not exert forces or cause gravitational effects except possibly through baryonic Casimir forces between universes.

$$\pi_u{}^k = \partial L_u\,(B_u(Y(y)))/\partial B_{uk}{}'(Y(y)) \qquad (12.14)$$
$$\pi_u{}^0 = 0 \qquad (12.15)$$

and

$$\delta^{tr}{}_{jk}(\mathbf{y} - \mathbf{y}') = \int d^{15}k\; e^{i\,\mathbf{k}\bullet(Y(y,0) - Y(y',0))}\,(\delta_{jk} - k_j k_k/\mathbf{k}^2)/(2\pi)^{15} \qquad (12.16)$$
$$B_{uk}{}'(Y(y)) = \partial B_{uk}(Y(y))/\partial y^{16} \qquad (12.17)$$

for j, k = 1, 2, … , 15.

If we choose the Coulomb gauge for $B_{uk}(Y(y))$:

$$B_u{}^{16}(Y(y)) = 0$$
$$\partial B_u{}^j(Y(y))/\partial Y^j(y) = 0$$

for j = 1, 2, … , 15 then fourteen degrees of freedom (polarizations) are present in the vector potential.[126] The Fourier expansion of the vector potential $B_u{}^i(Y(y))$ is:

$$B_u{}^i(Y(y)) = \int d^{15}k\; N_{0B}(k) \sum_{\lambda=1}^{14} \varepsilon^i(k, \lambda)[a_B(k,\lambda) :e^{-ik\cdot Y(y)}: + a_B{}^\dagger(k,\lambda) :e^{ik\cdot Y(y)}:] \qquad (12.18)$$

for i = 1, … , 15 where

$$N_{0B}(k) = [(2\pi)^{15} 2\omega_k]^{-\frac{1}{2}} \qquad (12.19)$$

and (since the field is massless)

$$k^{16} = \omega_k = (\mathbf{k}^2)^{\frac{1}{2}} \qquad (12.20)$$

where k^{16} is the energy, and where the $\varepsilon^i(k, \lambda)$ are the polarization unit vectors for $\lambda = 1, \ldots, 14$ and $k^\mu k_\mu = k^{16\,2} - \mathbf{k}^2 = 0$.

The commutation relations of the Fourier coefficient operators are:

$$[a_B(k,\lambda),\, a_B{}^\dagger(k',\lambda')] = \delta_{\lambda\lambda'}\delta^{15}(\mathbf{k} - \mathbf{k}') \qquad (12.21)$$
$$[a_B{}^\dagger(k,\lambda),\, a_B{}^\dagger(k',\lambda')] = [a_B(k,\lambda),\, a_B(k',\lambda')] = 0 \qquad (12.22)$$

and the polarization vectors satisfy

$$\sum_{\lambda=1}^{14} \varepsilon_i(k, \lambda)\varepsilon_j(k, \lambda) = (\delta_{ij} - k_i k_j/\mathbf{k}^2) \qquad (12.23)$$

The $B_u{}^\mu$ Feynman propagator is

$$iD_F{}^{trTT}(y_1 - y_2)_{jk} = \langle 0|T(B_{uj}(Y(y_1))B_{uk}(Y(y_2)))|0\rangle \qquad (12.24)$$

126 Note we use the Coulomb gauge for Y(y) also.

$$= -\, ig_{jk} \int \frac{d^{16}k\; e^{-ik\cdot(y_1 - y_2)}\, R(\mathbf{k}, y_1 - y_2)}{(2\pi)^{16}\,(k^2 + i\varepsilon)} \qquad (12.25)$$

where g_{jk} is the 16-dimensional Lorentz metric and where $R(\mathbf{k}, y_1 - y_2)$ is given by

$$\begin{aligned} R(\mathbf{k}, y_1 - y_2) &= \exp[\,-k^ik^j\Delta_{Tij}(y_1 - y_2)/M_u^{16}] \\ &= \exp\{\,-k^2[A(v) + B(v)\cos^2\theta] \,/\, [(2\pi)^{14}M_u^{4}z^2]\} \end{aligned} \qquad (12.26)$$

where k^2 is the sum of the squares of the 15 spatial components with

$$\begin{aligned} z^\mu &= y_1^\mu - y_2^\mu \\ z &= |\mathbf{z}| = |\mathbf{y}_1 - \mathbf{y}_2| \\ k &= |\mathbf{k}| \\ v &= |z^0|/z \\ A(v) &= (1 - v^2)^{-1} + .5v\ \ln[(v-1)/(v+1)] \\ B(v) &= v^2(1 - v^2)^{-1} - 1.5v\ \ln[(v-1)/(v+1)] \\ \mathbf{k}\cdot\mathbf{z} &= kz\cos\theta \end{aligned}$$

and $|\mathbf{k}|$ denoting the length of a spatial 15-vector $\mathbf{k}$ while $|z^0|$ is the absolute value of $z^0 \equiv z^{16}$.

As eq. 12.26 indicates, the Gaussian damping factor R(k, z) for all large spatial momentum k^j is the same for both the positive and negative frequency parts of the (Two-Tier) B_u Feynman propagator. We are assuming the spatial momentum is real-valued in this discussion. It is also important to note that R(k, z) does not depend on $k^0 = k^{16}$ (in the B_u and Y_u Coulomb gauges) and thus the integration over k^0 proceeds in the usual way to produce time-ordered positive and negative frequency parts.

The Gaussian exponential factor in *all* spatial coordinates causes the Feynman propagator to be finite and, together with the Gaussian factor in universe particle propagators, causes all perturbation theory calculations when interactions are introduced to be finite as we have seen earlier in The Extended Standard Model.

For small momentum much less than M_u then $R(\mathbf{k}, y_1 - y_2) \to 1$ and the Feynman propagator is the "normal" propagator of conventional 16-dimensional quantum field theory. For large momentum the corresponding potential approaches r^{13} in contrast to the electromagnetic Coulomb potential r^{-1}. The B_u potential is highly non-singular at large energies.

12.1.4 Bary-Electric Fields and Bary-Magnetic Fields

As in electromagnetism there is an antisymmetric tensor of the second rank that appears in the free part of the baryonic field $F_{Bu\mu\nu}(y)$ lagrangian:[127]

$$\mathscr{L}_{Bu} = -\tfrac{1}{4} F_{Bu}{}^{ij}(y)F_{Buij}(y) \qquad (12.27)$$

where

$$F_{Buij}(y) = \partial B_{ui}(y)/\partial y^j - \partial B_{uj}(y)/\partial y^i \qquad (12.28)$$

and i, j = 1, 2, ... , 16. The 16th coordinate corresponds to the time coordinate. While the coordinates are complex in general we will treat the 15 spatial coordinates as real and the 16th coordinate as pure imaginary with the resulting invariant interval

$$ds^2 = dy_1{}^2 + dy_2{}^2 + \ldots + dy_{15}{}^2 - c^2 dy_{16}{}^2 \qquad (12.29)$$

which is invariant under 16 dimensional Lorentz transformations. The coordinates can be transformed into complex-valued coordinates using the Reality group defined in Blaha (2014) and earlier books.

The tensor F_{Buij} is conveniently separated into an baryon electric part and a baryon magnetic part in a manner similar to the separation of the electromagnetic fields into electric and magnetic fields. However the 15 spatial dimensions changes the forms of the baryon fields. Analogously, to electromagnetism the baryon force is given by

$$f_i = F_{Buij}(y)J_B{}^j/c \qquad (12.30)$$

where $J_B{}^j$ is the jth baryonic current.

The baryon "electric" field is

$$E_{Bui} = -F_{Bui0}(y)/c \qquad (12.31)$$

while the baryon "magnetic" field is

$$B_{Bui} = \varepsilon_{ijk}F_{Bu}{}^{jk}(y) \qquad (12.32)$$

where i, j, k = 1, 2, ... , 15 and where ε_{ijk} is a totally anti-symmetric tensor with component values ±1. If $i < j < k$ then ε_{ijk} is +1. Even permutations of these three indices yield a value of +1

[127] Parts of the following appear in Blaha (2014a). They are somewhat modified since we are dealing with the classical, low energy, large distance baryonic field where the quantum coordinate fields Y(y) are well approximated by the classical (non-quantum) Megaverse coordinates y.

for the tensor components. Odd permutations of these three indices yield a value of –1. For example, $\varepsilon_{246} = +1$, $\varepsilon_{426} = -1$, $\varepsilon_{642} = -1$, $\varepsilon_{264} = -1$, $\varepsilon_{462} = +1$, $\varepsilon_{624} = +1$.

With these definitions of the $\mathbf{E_{Bu}}$ and $\mathbf{B_{Bu}}$ fields we can easily derive the 16-dimensional generalization of the *Lorentz force law* for a baryon of charge q and 15-velocity v_j:

$$F_i = qE_{Bui} + q\varepsilon_{ijk}v_jB_{Buk}/c \tag{12.33}$$

for i = 1, 2, … , 15. One important difference from the 4-dimensional case is the forms of the $\mathbf{E_{Bu}}$ and $\mathbf{B_{Bu}}$ fields

$$E_{Bui} = -F_{Bui0}(y)/c = [-\partial B_{u0}(y)/\partial y^i - \partial B_{ui}(y)/\partial y^0] \tag{12.34}$$

or, expressed as a 15-vector,

$$\mathbf{E}_{Bu} = [-\nabla_{15}\phi(y) - \dot{\mathbf{B}}_u(y)]/c \tag{12.35}$$

where ϕ is the baryonic Coulomb potential $B_{u16}(y)$, ∇_{15} is the 15-dimensional grad operator, and $B_u(y)$ is the baryonic 15-vector potential with the "dot" above it signifying a time (y_{16}) derivative.

The 15-dimensional baryon magnetic field has the form of eqn. 12.32. A specific illustrative case shows the baryon magnetic field exhibits more complexity than the 3-dimensional magnetic field of electromagnetism:

$$B_{Bu1} = \varepsilon_{1jk}F_{Bu}{}^{jk}(y)/c = [F_{Bu}{}^{23}(y) + F_{Bu}{}^{24}(y) + \ldots + F_{Bu}{}^{215}(y) + F_{Bu}{}^{34}(y) + F_{Bu}{}^{35}(y) + \ldots + F_{Bu}{}^{315}(y) + F_{Bu}{}^{45}(y) + \ldots + F_{Bu}{}^{14,15}(y)]/c \tag{12.36}$$

Thus each component of the baryon magnetic field impacts on all fifteen spatial directions of the Megaverse.[128]

12.1.5 The Baryonic "Coulombic" Gauge Field

The baryonic gauge field has a "Coulombic" potential part $\phi(y)$, just as the electromagnetic field does. Consequently the total potential between two electromagnetically neutral masses of mass M_1 and M_2, and baryon numbers N_1 and N_2 is[129]

---

[128] For this reason we can use spinning rings and mass configurations to generate baryon magnetic field interactions to enable starships to escape from our universe's three spatial dimensions. We called them *uniships* in Blaha (2014b) and (2014c).

[129] The baryonic r^{-1} factor follows since all other dimensions beyond those of our universe have almost zero distance in our universe.

$$V_{tot} = -GM_1M_2/r + (\beta^2/4\pi)\, N_1N_2/r \qquad (12.37)$$

where G is the gravitational constant, and β is analogous to the electric charge e in the electromagnetic Coulomb potential. If both masses are composed of the same substance and have the same mass, then we can set $M_1 = M_2 = M = Nm$ where m is the average mass of the baryons in the masses.[130] In addition we can set $N_1 = N_2 = N$. Then eq. 12.37 becomes

$$V = [-Gm^2 + (\beta^2/4\pi)]N^2/r \qquad (12.38)$$

Note that the gravitational potential term is attractive, and the baryonic potential term is repulsive between baryons.

In considering eq. 12.1 we have approximated the baryonic potential with only our universe's spatial coordinates. In reality we should be using the spatial separation in all Megaverse coordinates. However, since our universe is close to flat, the distance between two objects that are not too far apart is approximately the same in both coordinate systems. In general, the baryonic potential in Megaverse coordinates is actually

$$\phi(y_1, y_2, \ldots, y_{15}) = (\beta^2/4\pi)N_1N_2/(y_1^{\,2} + y_2^{\,2} + \ldots + y_{15}^{\,2})^{1/2} \qquad (12.39)$$

12.1.6 The Baryonic Force on Baryonic Objects

The baryonic force on a moving baryon mass is given by the baryon Lorentz force for a baryon of baryon charge q and 15-velocity v_j:

$$F_i = qE_{Bui} + q\varepsilon_{ijk}v_jB_{Buk}/c \qquad (12.40)$$

for i = 1, 2, … , 15. The 16-dimensional baryonic Coulombic potential is

$$V = N\phi(y_1, y_2, \ldots, y_{15}) = (\beta^2/4\pi)N/(y_1^{\,2} + y_2^{\,2} + \ldots + y_{15}^{\,2})^{1/2} \qquad (12.41)$$

where N is the baryon number of the baryon mass. The baryon Coulombic force is

$$F_i = N\nabla_{15i}\phi(y) \qquad (12.42)$$

where ∇_{15i} is the ith component of the 15-dimensional grad operator ∇_{15}.

---

[130] We neglect lepton masses since they are negligible relative to the baryon masses.

12.2 Lepton Number, Dark Baryon Number, and Dark Lepton Number Conservation Laws

In the previous section we raised the possibility of an ultra-weak Baryonic force and an associated Baryon number conservation law. This conservation law has been repeatedly tested and found to be satisfied to great accuracy. One possible cause for concern is the Adler-Bell-Jackiw fermion triangle anomaly which follows from a three fermion loop that diverges in conventional quantum field theory. This type of anomaly raises the possibility of baryon number non-conservation. In Two-Tier Quantum Field Theory the triangle graph is convergent – no infinities – and anomalies are not present. Thus our Two-Tier implementation of the Extended Standard Model is anomaly-free and the issue of baryon number non-conservation disappears. (See Appendix A and Blaha (2005a) for more detailed discussions.)

We will assume Baryon number conservation holds.

12.2.1 Other Conserved Particle Numbers

Another conserved number is Lepton Number, denoted L. Again, repeated attempts to find lepton number violation have failed. On that basis *we will assume lepton number conservation.*

If Baryon Number and Lepton Number are both conserved quantities then any linear combination of them is also conserved. Therefore

$$B' = aB + bL \tag{12.43}$$

is also conserved.

If we consider the Dark Matter sector of the Extended Standard Model it is reasonable to assume that *Dark Baryon Number B_D and Dark Lepton Number L_D are conserved* also (although there is no experimental evidence available as yet to confirm (or deny) these assumptions.)

Thus we have four conserved particle Numbers. Linear combinations of these numbers are also conserved:

$$\begin{aligned} B' &= aB + bL + cB_D + dL_D \\ L' &= eB + fL + gB_D + hL_D \\ B_D' &= iB + jL + kB_D + lL_D \\ L_D' &= mB + nL + oB_D + pL_D \end{aligned} \tag{12.44}$$

or

$$N' = AN \tag{12.45}$$

Where N and N' are 4-vectors composed of particle numbers and A is a 4×4 matrix. The number of fermion particle types, 4, is determined by the two families of normal fermions, leptons and quarks, and the number of families of Dark fermions, which is ultimately determined by the Reality group and thus the Complex Lorentz group as we described earlier.

The constants appearing in linear equations of the form of eq. 12.45 seem arbitrary. However if we want the new 'primed' set of conserved Numbers to be an independent set of numbers then the determinant of the constants must be non-zero. Thus the matrix, A, is invertible.

The set of 4×4 matrices of the type of eq. 12.44 form an U(4) group[131] *if we wish to perform these transformations within lagrangians of the type of the Extended Standard Model. The choice of U(4) rather than SU(4) is required since there are four independent particle Numbers. U(4) has four diagonal matrices in its algebra while SU(4) only has three diagonal matrices. U(4) preserves the independence of the four independent particle Numbers. It also allows complex rotations of the form of eq. 12.44.*

12.3 U(4) Number Symmetry

At this point we note the observations of Yang and Mills that Numbers can be local and so we generalize the U(4) Number symmetry to a Yang-Mills symmetry. The transformations then become functions of position A(X) where X represents the space-time coordinates in Two-Tier Quantum Field Theory.

The U(4) rotations of the four Numbers alters the physical interpretation of the number operators without changing their physical implications. Thus we have a symmetry operation induced on particle fields, which in the absence of symmetry breaking terms, becomes a symmetry of the lagrangian.

To implement this symmetry in the particles' lagrangian, all covariant derivatives must acquire another interaction term with 16 U(4) fields corresponding to the 16 generators of U(4). In addition we must define each fermion species in the fundamental representation $\underline{4}$ of U(4). Thus our one generation theory now is a four generation family of four fermion species. We must add another index to each fermion field specifying its generation. Lastly a set of initially massless U(4) gauge field dynamic terms must be added to the Extended Standard Model lagrangian.

The initially massless U(4) gauge transformation symmetry is broken by the Higgs Mechanism with the gauge fields acquiring masses (with two exceptions) and the fermions of the four generations acquiring masses which are generation dependent. The symmetry breaking is described later.

---

[131] If we wish to further limit the values of the 'primed' Numbers to integers assuming the unprimed Numbers are integers then the group of the transformation is the set of permutations of four entities – the Symmetric group S_4. However the 'primed' numbers can be integer or not. There is no apparent physical principle requiring integer Numbers. Quarks are usually assigned Baryon Number 1/3. Also Numbers are not necessarily positive valued.

13. Fermion Generations and the Broken U(4) Generation Group Interaction

In sections 12.2 and 12.3 we showed that a local U(4) symmetry based on conserved fermion particle numbers existed in Nature and added a new assumption to our construction of the Extended Standard Model. U(4) has 16 generators, which we denote G_i for i = 1, 2, … , 16. Its fundamental representation has 4×4 matrices. When we introduce this U(4) symmetry directly into the one generation Extended Standard Model each fermion acquires a new index and becomes a four generation set of fermions.[132] Symmetry breaking via a Higgs Mechanism for the U(4) gauge fields gives a different mass to each of the members of each species. Higgs particle lagrangian terms for U(4) breaking will be described in section 13.3.

13.1 Four Generation Extended Standard Model

Previously we derived the form of a one generation Extended Standard Model that included the known parts of the Standard Model (excepting the Higgs sector) and an SU(2)⊗U(1) part for Dark Matter.

In this section we generalize to the four generation Extended Standard Model that results.[133] Covariant derivatives acquire another interaction term with 16 U(4) fields U_i^{μ}. In addition we add another index to each fermion field specifying its generation. Lastly a set of initially massless gauge field dynamics terms is added to the Extended Standard Model lagrangian to specify U(4) gauge field evolution and interactions.

13.1.1 Two-Tier Lepton Sector

We begin with the definition of a quadruplet of leptons – a pair of doublets, one normal and one Dark, instead of a single doublet. We define left and right lepton quadruplets with[134]

---

[132] At present there are only three known generations. We suggest a fourth generation exists that is not accessible at current accelerator energies – given the almost exponential increase of fermion masses with each new generation.

[133] It is based on the three principles based on Ockham's Razor ("The simplest choice is often the best."): 1) The only connecting interaction is a weak interaction, 2) The form of ElectroWeak theory remains unchanged, and 3) Dark Matter parallels normal matter in its general characteristics: four generations, SU(3) singlets, an SU(2)⊗U(1) symmetry analogous to ElectroWeak symmetry, SU(2)⊗U(1) dark lepton and dark quark doublets.

[134] The X's are Two-Tier coordinates.

$$\Psi_{L,Ra}(X) = \begin{bmatrix} \psi_{DL,Ra}(X) \\ \psi_{NL,Ra}(X) \end{bmatrix} \tag{13.1}$$

where a is a generation index ranging from 1 to 4, $\psi_{NL,R}(X)$ is a "normal" ElectroWeak-like lepton doublet, and where $\psi_{DL,R}(X)$ is a Dark ElectroWeak-like lepton doublet consisting of a Dark electron-like fermion and a Dark neutrino-like fermion.

We define covariant derivative terms which we express in matrix form are

$$D_{L,R}(X) = \begin{bmatrix} \gamma^{\mu} D_{DL,R\mu} & 0 \\ 0 & \gamma^{\mu} D_{NL,R\mu} \end{bmatrix} \tag{13.2}$$

where the normal matter left-handed covariant derivative is

$$D_{NL\mu} = \partial/\partial X^{\mu} - \tfrac{1}{2} i g \boldsymbol{\sigma}\cdot\mathbf{W}_{\mu} + \tfrac{1}{2} i g' B_{\mu} - \tfrac{1}{2} i g_G \mathbf{G}\cdot\mathbf{U}_{\mu} \tag{13.3}$$

where g_G is an ultra-weak generational coupling constant, $\mathbf{G}\cdot\mathbf{U}_{\mu}$ is the sum of the inner product of 16 U(4) generators G_i and gauge fields $U_i(X)$, and where the Dark matter left-handed covariant derivative is

$$D_{DL\mu} = \partial/\partial X^{\mu} - \tfrac{1}{2} i g_D \boldsymbol{\sigma}\cdot\mathbf{W}'_{\mu} + \tfrac{1}{2} i g_D' B'_{\mu} - \tfrac{1}{2} i g_G \mathbf{G}\cdot\mathbf{U}_{\mu} \tag{13.4}$$

with $\boldsymbol{\sigma}$ a vector composed of the Pauli matrices. The right-handed covariant derivatives have a simpler form. The normal matter right-handed covariant derivative is

$$D_{NR\mu} = \partial/\partial X^{\mu} + \tfrac{1}{2} i g' B_{\mu} - \tfrac{1}{2} i g_G \mathbf{G}\cdot\mathbf{U}_{\mu} \tag{13.5}$$

and the Dark matter right-handed covariant derivative is

$$D_{DR\mu} = \partial/\partial X^{\mu} + \tfrac{1}{2} i g_D' B'_{\mu} - \tfrac{1}{2} i g_G \mathbf{G}\cdot\mathbf{U}_{\mu} \tag{13.6}$$

The normal and Dark electroweak fields above are functions of a Two-Tier X. The Faddeev-Popov mechanism operative for these types of fields is described in appendix 19-A of Blaha (2011c) and in chapter 10.

13.1.2 Quark Sector

In the *quark* sector we define left and right quark quadruplets with

$$\Psi_{qL,Ra}(X_c) = \begin{bmatrix} \psi_{DqL,Ra}(X_c) \\ \psi_{NqL,Ra}(X_c) \end{bmatrix} \tag{13..7}$$

where $\psi_{NqL,Ra}(X_c)$ is a "normal" ElectroWeak-like quark doublet, and where $\psi_{DqL,Ra}(X_c)$ is a Dark ElectroWeak-like quark doublet consisting of a SU(3) singlet Dark up-quark of unit Dark charge and a SU(3) singlet Dark down-quark of zero Dark charge in the ath generation.

The covariant derivative terms are contained in $D_q(X_c)$ which we express in matrix form as

$$D_{qL,R}(X_c) = \begin{bmatrix} \gamma^\mu D_{qDL,R\mu}(X_c) & 0 \\ 0 & \gamma^\mu D_{qNL,R\mu}(X_c) \end{bmatrix} \tag{13.8}$$

where the normal quark matter left-handed covariant derivative is

$$D_{qNL\mu} = \partial/\partial X_c^{\ \mu} - \tfrac{1}{2}ig\boldsymbol{\sigma}\cdot\mathbf{W}_\mu - ig'B_\mu/6 - \tfrac{1}{2}ig_G\mathbf{G}\cdot\mathbf{U}_\mu + ig_C\tau\cdot A_{C\mu} \tag{13.9}$$

and where the Dark quark left-handed covariant derivative is

$$D_{qDL\mu} = \partial/\partial X_c^{\ \mu} - \tfrac{1}{2}ig_D\boldsymbol{\sigma}\cdot\mathbf{W}'_\mu + \tfrac{1}{2}ig_D B'_\mu - \tfrac{1}{2}ig_G\mathbf{G}\cdot\mathbf{U}_\mu \tag{13.10}$$

since Dark quarks are SU(3) singlets with unit or zero Dark charge. The right-handed quark covariant derivatives have a simpler form. The normal quark right-handed covariant derivative is

$$D_{qNR\mu} = \partial/\partial X_c^{\ \mu} + \tfrac{1}{2}ig'B_\mu/3 - \tfrac{1}{2}ig_G\mathbf{G}\cdot\mathbf{U}_\mu + ig_C\tau\cdot A_{C\mu} \tag{13.11}$$

and the Dark quark right-handed covariant derivative is

$$D_{qDR\mu} = \partial/\partial X_c^{\ \mu} + \tfrac{1}{2}ig_D B'_\mu - \tfrac{1}{2}ig_G\mathbf{G}\cdot\mathbf{U}_\mu \tag{13.12}$$

The normal and Dark gauge boson fields are functions of $X_c. = (X_{r\mu}(y_r), X_{i\mu}(y_i))$. The Faddeev-Popov mechanism is operative for gauge boson fields and is described in earlier in this book and in Appendix 19-A of Blaha (2011c).[135] The *complexon* quark Extended Standard Model ElectroWeak Sector covariant derivatives in quadruplet matrix form are

$$D_{qL,R}(X_c) = \begin{bmatrix} \gamma^{\mu} D_{qDL,R\mu} & 0 \\ 0 & \gamma^{\mu} D_{qNL,R\mu} \end{bmatrix} \tag{13.13}$$

The remaining parts of the complexon Standard Model are described in chapter 23 of Blaha (2011) and summarized below. The addition of singlet Dark quark Higgs terms is also required.

The lagrangian density and action is

$$\mathcal{L}_{CSM} = \Psi_{La}^{\dagger}\gamma^0 i\gamma^{\mu} D_{L\mu}\Psi_{La} - \Psi_{Ra}^{\dagger}\gamma^0 i\gamma^{\mu} D_{R\mu}\Psi_{3Ra} + \Psi_{CLa}^{\dagger}\gamma^0 i\gamma^{\mu}\mathcal{D}_{qL\mu}\Psi_{CLa} + \Psi_{CRa}^{\dagger}\gamma^0 i\gamma^{\mu}\mathcal{D}_{qR\mu}\Psi_{CRa} - $$
$$- \mathcal{L}_{BareMasses} + \mathcal{L}_{Gauge} + \mathcal{L}_{Mass} + \mathcal{L}_{Ufields} \tag{13.14}$$

where a is the generation index. $\mathcal{L}_{BareMasses}$ contains the fermion bare mass terms. Also,

$$\mathcal{L}_{Gauge} = \mathcal{L}_{GaugeEW} + \mathcal{L}_{GaugeC} + \mathcal{L}_{GaugeEWD} \tag{13.15}$$

with

$$\mathcal{L}_{GaugeEW} = -\tfrac{1}{4} F_W{}^{a\mu\nu} F_W{}^{a}{}_{\mu\nu} - \tfrac{1}{4} F_B{}^{\mu\nu} F_{B\mu\nu} + \mathcal{L}_{EW}{}^{ghost} \tag{13.16}$$

$$\mathcal{L}_{GaugeEWD} = -\tfrac{1}{4} F'_W{}^{a\mu\nu} F'_W{}^{a}{}_{\mu\nu} - \tfrac{1}{4} F_{B'}{}^{\mu\nu} F_{B'\mu\nu} + \mathcal{L}_{W'}{}^{ghost} \tag{13.17}$$

and

$$\mathcal{L}_{GaugeC} = \mathcal{L}_{CCG} + \mathcal{L}_C{}^{ghost} + \mathcal{L}_{CC}{}^{ghost} \tag{13.18}$$

$$\mathcal{L}_{Ufields} = -\tfrac{1}{4} F_U{}^{a\mu\nu} F_{U\mu\nu\nu} + \mathcal{L}_U{}^{ghost} + \mathcal{L}_U{}^{UHiggs} \tag{13.19}$$

where $\mathcal{L}_U{}^{UHiggs}$ is discussed in section 13.4. The ElectroWeak gauge bosons $W_{\mu}{}^{a}$, B_{μ} and B'_{μ} field tensors are:

$$F_W{}^{a}{}_{\mu\nu} = \partial W^a{}_{\mu}/\partial X^{\nu} - \partial W^a{}_{\nu}/\partial X^{\mu} + g_2 f^{abc} W^b{}_{\mu} W^c{}_{\nu} \tag{13.20}$$

$$F_{B\mu\nu} = \partial B_{\mu}/\partial X^{\nu} - \partial B_{\nu}/\partial X^{\mu} \tag{13.21}$$

---

[135] Those who might be concerned about the propagator term $<W_i(X), W_j(X_c)>$ and similar propagators where one field is a function of X and the other field is a function of X_c should note that such terms are to very good approximation equal to $<W_i(X), W_j(X)>$ for energies much less than M_c (which could be as large as the Planck energy.)

and the Dark ElectroWeak gauge bosons $W'^{\;a}_{\mu}$ and B'_{μ} field tensors are:

$$F_{B'\mu\nu} = \partial B'_{\mu}/\partial X^{\nu} - \partial B'_{\nu}/\partial X^{\mu}$$

$$F'^{\;a}_{W\,\mu\nu} = \partial W'^{a}_{\mu}/\partial X^{\nu} - \partial W'^{a}_{\nu}/\partial X^{\mu} + g_2 f^{abc} W'^{b}_{\mu} W'^{c}_{\nu} \qquad (13.22)$$

The U fields' tensor is:

$$F_{U\,\mu\nu}^{\;a} = \partial U^{a}_{\mu}/\partial X^{\nu} - \partial U^{a}_{\nu}/\partial X^{\mu} + g_G f_4^{\;abc} U^{b}_{\mu} U^{c}_{\nu} \qquad (13.23)$$

where $f_4^{\;abc}$ are the U(4) algebra commutator constants.

$\mathcal{L}_{EW}^{ghost}$ contains the Faddeev-Popov ghost terms for the ElectroWeak $W_{\mu}^{\;a}$ gauge bosons. The complexon color gluon lagrangian $\mathcal{L}_{CCG}$ is defined by

$$\mathcal{L}_{CCG} = -\tfrac{1}{4} F_{CC}^{\;a\mu\nu}(X) F_{CC\;\mu\nu}^{\;a}(X) \qquad (13.24)$$

where

$$F_{CC\;\mu\nu}^{\;a} = \partial/\partial X_c^{\;\nu} A_{C\;\mu}^{\;a} - \partial/\partial X_c^{\;\mu} A_{C\;\nu}^{\;a} + g f_{su(3)}^{\;abc} A_{C\;\mu}^{\;b} A_{C\;\nu}^{\;c} \qquad (13.25)$$

where $A_{C\;\nu}^{\;a}$ is the color gluon gauge field, g is the color coupling constant, and the $f_{su(3)}^{\;abc}$ are the SU(3) structure constants.

In addition $\mathcal{L}_{C}^{ghost}$ is the color SU(3) Faddeev-Popov ghost terms defined earlier and in appendix 19-A of Blaha (2011c) for the complexon Lorentz gauge and $\mathcal{L}_{CC}^{ghost}$ is the complexon color SU(3) constraint ghost terms defined through the Faddeev-Popov mechanism. The mass sector $\mathcal{L}_{Mass}$ is presumably based on the Higgs Mechanism.which creates the fermion and ElectroWeak vector boson masses, and generation mixing.

The lagrangian is supplemented with the following condition on all complexon fields $\Phi_{\ldots}$:[136]

$$\nabla_r \cdot \nabla_i \Phi\ldots = 0 \qquad (13.26)$$

Non-complexon fields $\Omega\ldots$ in the left-handed formulation under consideration satisfy the subsidiary condition:

$$[\nabla_r \cdot \nabla_i - (\nabla_r^{\,2} \nabla_i^{\,2})^{1/2}]\Omega\ldots = 0 \qquad (13.27)$$

---

[136] These conditions implement the orthogonality of the real and imaginary parts of complexon 3-momentum.

which guarantees a complexon's real momentum is parallel to its imaginary momentum.

13.2 U(4) Gauge Symmetry Breaking and Long Range Forces

In chapter 12 we showed that there was good experimental evidence for a conserved Baryon Number B and we proceeded to develop a simple U(1) gauge theory that would imply Baryon Number conservation in a manner analogous to QED's implying electric charge conservation. In section 13.1 we used a new symmetry group local U(4) to generalize the one generation Extended Standard Model to a four generation Extended Standard Model based on four conserved particle numbers: B, L, B_D, and L_D.[137]

We now assume in our construction that the four generation Extended Standard Model has a local U(4) symmetry that is broken by mass terms gewnerated by the Higgs Mechanism.

Further, we will assume that the Higgs breakdown yields two massless (long range) fields which we associate with Baryon Number B and Dark Baryon Number B_D. The remaining fields acquire masses and generate short range forces.

We use the following U(4) diagonal matrices:

$$\begin{aligned} G_1 &= \text{diag}(1, 1, 1, 1) \\ G_2 &= \text{diag}(0, 1, 0, 0) \\ G_3 &= \text{diag}(0, 0, 1, 0) \\ G_4 &= \text{diag}(0, 0, 0, 1) \end{aligned} \tag{13.28}$$

The U(4) algebra has 16 hermitean matrices that satisfy

$$G_i^\dagger = G_i \tag{13.29}$$

The particle numbers can be expressed in terms of the diagonal generators as

$$\begin{aligned} B &= G_1 - G_2 - G_3 - G_4 \\ B_D &= G_2 \\ L &= G_3 \\ L_D &= G_4 \end{aligned} \tag{13.30}$$

The covariant derivatives have the general form:

$$D_{\dots\mu} = \partial/\partial X^\mu + \dots - \tfrac{1}{2} i g_G \mathbf{G}\cdot\mathbf{U}_\mu \tag{13.31}$$

[137] Charge, although a conserved number, is a part of the ElectroWeak sector, account of which has already been taken.

where the ellipses indicate the other details of the particular covariant derivative. We now wish to express the four gauge fields $U_i(X)$ for i = 1, 2, 3, 4 corresponding to the diagonal generators in terms of the fields of the four particle number gauge fields: B_μ, L_μ, $B_{D\mu}$, and $L_{D\mu}$.

$$U_{i\mu} = A_{ik} N_{k\mu} \qquad (13.32)$$

where A_{ik} are the elements of a matrix of constants and

$$N_\mu = \begin{bmatrix} B_\mu(X) \\ L_\mu(X) \\ B_{D\mu}(X) \\ L_{D\mu}(X) \end{bmatrix} \qquad (13.33)$$

is a column vector consisting of the gauge fields corresponding to each of the conserved particle numbers.

The matrix A must have non-zero determinant so that eq. 13.32 can be inverted to express the particle number fields in terms of the four $U_i(X)$ gauge fields:

$$N_\mu = A^{-1}U_\mu \qquad (13.34)$$

resulting in

$$\begin{aligned} B_\mu(X) &= U_{1\mu} \\ L_\mu(X) &= U_{1\mu} + U_{2\mu} \\ B_{D\mu}(X) &= U_{1\mu} + U_{3\mu} \\ L_{D\mu}(X) &= U_{1\mu} + U_{4\mu} \end{aligned} \qquad (13.35)$$

Then

$$D_{\ldots\mu} = \partial/\partial X^\mu + \ldots - \tfrac{1}{2} i g_G[\sum_{i=5}^{16} \mathbf{G}_i\mathbf{U}_{i\mu} + BB_\mu(X) + LL_\mu(X) + B_D B_{D\mu}(X) + L_D L_{D\mu}(X)] \qquad (13.36)$$

where the particle numbers, which are analogous to the charges Q and Q' in ElectroWeak theory, are B, L, B_D, and L_D. They are expressed in terms of U(4) generators by eqs. 13.30.

13.3 Higgs Mass Mechanism for U(4) Gauge Fields

We now require that there are two massless fields, one coupled to Baryon number and one coupled to Dark Baryon number. The Dark sector is assumed to be analogous to the normal particle sector in this respect. There are fourteen remaining fields that acquire a mass and longitudinal components. These fields become short range, ultra-weak generational forces. The masses they acquire through the Higgs Mechanism are presumably very large as these gauge particles have not been found experimentally.

We assume that a scalar Higgs field exists which is a U(4) vector with four components corresponding to the fermion generations. It is an SU(2)⊗U(1)⊗SU(3) scalar. Its lagrangian density is

$$\mathcal{L}_U^{\,UHiggs} = (\partial\eta^\dagger/\partial X^\mu)(\partial\eta/\partial X^\mu) - \lambda(\eta^\dagger\eta - \rho^2)^2 + \mathcal{L}_U^{\,UHiggs}{}_{FermionMasses}$$

where $\mathcal{L}_U^{\,UHiggs}{}_{FermionMasses}$ are the fermion masses produced by the Generation group, U, Higgs Mechanism and where we choose a unitary gauge in which

$$\rho = \begin{bmatrix} 0 \\ \rho_1 \\ 0 \\ \rho_2 \end{bmatrix} \tag{13.37}$$

where ρ_1 and ρ_2 are real fields. Then the covariant derivative of ρ is

$$D_{\ldots\mu}\,\rho = \{\partial/\partial X^\mu + \ldots - \tfrac{1}{2}ig_G[\Sigma\mathbf{G}_i\mathbf{U}_{i\mu} + BB_\mu(X) + LL_\mu(X) + B_DB_{D\mu}(X) + L_DL_{D\mu}(X)]\}\begin{bmatrix} 0 \\ \rho_1 \\ 0 \\ \rho_2 \end{bmatrix} \tag{13.38}$$

The sum over i is from 5 through 16, and $[\mathbf{G}_i]_{jk}$ is the jk element of $\mathbf{G}_i$. Then

$$D_{\ldots\mu}\,\rho = \begin{bmatrix} -\tfrac{1}{2}ig_G\{\rho_1\Sigma[\mathbf{G}_i]_{12}\mathbf{U}_{i\mu} + \rho_2\Sigma[\mathbf{G}_i]_{14}\mathbf{U}_{i\mu}\} \\ \partial\rho_1/\partial X^\mu - \tfrac{1}{2}ig_G\rho_1L_\mu - \tfrac{1}{2}ig_G\{\rho_1\Sigma[\mathbf{G}_i]_{22}\mathbf{U}_{i\mu} + \rho_2\Sigma[\mathbf{G}_i]_{24}\mathbf{U}_{i\mu}\} \\ -\tfrac{1}{2}ig_G\{\rho_1\Sigma[\mathbf{G}_i]_{32}\mathbf{U}_{i\mu} + \rho_2\Sigma[\mathbf{G}_i]_{34}\mathbf{U}_{i\mu}\} \\ \partial\rho_2/\partial X^\mu - \tfrac{1}{2}ig_G\rho_2L_{D\mu} - \tfrac{1}{2}ig_G\{\rho_1\Sigma[\mathbf{G}_i]_{42}\mathbf{U}_{i\mu} + \rho_2\Sigma[\mathbf{G}_i]_{44}\mathbf{U}_{i\mu}\} \end{bmatrix} \tag{13.39}$$

$$= \begin{bmatrix} -\tfrac{1}{2}ig_G\Sigma\{\rho_1[\mathbf{G}_i]_{12} + \rho_2[\mathbf{G}_i]_{14}\}\mathbf{U}_{i\mu} \\ \partial\rho_1/\partial X^\mu - \tfrac{1}{2}ig_G\rho_1L_\mu - \tfrac{1}{2}ig_G\rho_2\Sigma[\mathbf{G}_i]_{24}\mathbf{U}_{i\mu} \\ -\tfrac{1}{2}ig_G\Sigma\{\rho_1[\mathbf{G}_i]_{32} + \rho_2[\mathbf{G}_i]_{34}\}\mathbf{U}_{i\mu} \\ \partial\rho_2/\partial X^\mu - \tfrac{1}{2}ig_G\rho_2L_{D\mu} - \tfrac{1}{2}ig_G\rho_1\Sigma[\mathbf{G}_i]_{42}\mathbf{U}_{i\mu} \end{bmatrix} \tag{13.40}$$

since the generators $\mathbf{G}_i$ have zeroes along their diagonals for i = 5, … , 16.

From eq. 13.39 we find the corresponding Higgs field kinetic terms in the lagrangian are

$$(D_{\ldots\mu}\,\rho)^\dagger D_{\ldots}^{\ \mu}\,\rho = \partial\rho_1/\partial X^\mu \partial\rho_1/\partial X_\mu + \partial\rho_2/\partial X^\mu\, \partial\rho_2/\partial X_\mu + g_G^{\,2}\rho_1^{\,2}L_\mu\, L^\mu/4 + g_G^{\,2}\rho_2^{\,2}L_{D\mu}\, L_D^{\,\mu}/4 + \ldots \tag{13.41}$$

Note there are differing mass squared terms for the Lepton ($g_G^{\,2}\rho_1^{\,2}/4$) and Dark Lepton ($g_G^{\,2}\rho_2^{\,2}/4$) gauge fields making them short range fields with the likelihood of very large masses much beyond ElectroWeak gauge field masses, and with an ultra weak coupling constant g_G as suggested by the "experimental" coupling for the Baryonic force given in eq. 12.6.

The Baryonic and Dark Baryonic gauge fields are massless and thus long range although their coupling constant appears to be ultra weak – much below the gravitational coupling constant G.

We now turn to calculating the remaining terms in eq. 13.41 that determine the masses of the remaining 14 gauge fields. We begin by assigning matrix elements for the remaining hermitean U(4) generators:

$$\begin{aligned}
[G_5]_{ik} &= \delta_{i1}\delta_{k2} + \delta_{i2}\delta_{k1} \\
[G_6]_{ik} &= -i\delta_{i1}\delta_{k2} + i\delta_{i2}\delta_{k1} \\
[G_7]_{ik} &= \delta_{i1}\delta_{k3} + \delta_{i3}\delta_{k1} \\
[G_8]_{ik} &= -i\delta_{i1}\delta_{k3} + i\delta_{i3}\delta_{k1} \\
[G_9]_{ik} &= \delta_{i1}\delta_{k4} + \delta_{i4}\delta_{k1} \\
[G_{10}]_{ik} &= -i\delta_{i1}\delta_{k4} + i\delta_{i4}\delta_{k1} \\
[G_{11}]_{ik} &= \delta_{i2}\delta_{k3} + \delta_{i3}\delta_{k2} \\
[G_{12}]_{ik} &= -i\delta_{i2}\delta_{k3} + i\delta_{i3}\delta_{k2} \\
[G_{13}]_{ik} &= \delta_{i2}\delta_{k4} + \delta_{i4}\delta_{k2} \\
[G_{14}]_{ik} &= -i\delta_{i2}\delta_{k4} + i\delta_{i4}\delta_{k2} \\
[G_{15}]_{ik} &= \delta_{i3}\delta_{k4} + \delta_{i4}\delta_{k3} \\
[G_{16}]_{ik} &= -i\delta_{i3}\delta_{k4} + i\delta_{i4}\delta_{k3}
\end{aligned} \tag{13.42}$$

Then completing eq. 13.41 using eq. 13.40 we find

$$\begin{aligned}
(D_{\ldots\mu}\,\rho)^\dagger D_{\ldots}^{\ \mu}\,\rho = {} & \partial\rho_1/\partial X^\mu \partial\rho_1/\partial X_\mu + \partial\rho_2/\partial X^\mu\, \partial\rho_2/\partial X_\mu + g_G^{\,2}\rho_1^{\,2}L_\mu\, L^\mu/4 + g_G^{\,2}\rho_2^{\,2}L_{D\mu}\, L_D^{\,\mu}/4 + \\
& + (g_G/2)^2\rho_1^{\,2}(U_5^{\,2} + U_6^{\,2}) + (g_G/2)^2\rho_2^{\,2}(U_9^{\,2} + U_{10}^{\,2}) + (g_G/2)^2\rho_1^{\,2}(U_{11}^{\,2} + U_{12}^{\,2}) + \\
& + (g_G/2)^2(\rho_1^{\,2} + \rho_2^{\,2})(U_{13}^{\,2} + U_{14}^{\,2}) + \ + (g_G/2)^2\rho_2^{\,2}(U_{15}^{\,2} + U_{16}^{\,2})
\end{aligned} \tag{13.43}$$

up to total divergences which generate surface terms which we discard, and assuming that all fields satisfy the gauge condition

$$\partial U_i^{\,\mu}/\partial X^\mu = 0 \tag{13.44}$$

Note that there are no mass terms for $U_{7\mu}(X)$ and $U_{8\mu}(X)$ as well as $B_\mu(X)$ and $B_{D\mu}(X)$ due to our choice of unitary gauge eq. 13.37.[138] Consequently there are four massless long range fields and 12 gauge fields that acquire masses of three different values: $(g_G/2)\rho_{10}$, $(g_G/2)\rho_{20}$, and $(g_G/2)(\rho_{10}^{\,2} + \rho_{20}^{\,2})^{1/2}$ where ρ_{10} and ρ_{20} ar the vacuum expectation values of ρ_1 and ρ_2 respectively. The fields $U_7(X)$ and $U_8(X)$ are not "diagonal" and thus appear in the fermion sector as terms connecting fermions in different generations within the four species of normal fermions and within the four species of Dark fermions.[139] Therefore they do not change the values of any of the four species of particle numbers.

Based on the estimate of eq. 12.6 the ultra weak value of the coupling constant is

$$g_G = (4\pi\alpha_B)^{1/2} \approx 1.218\,(Gm_H^{\,2})^{1/2} \tag{13.45}$$

The ultra weak value of the coupling constant implies that the baryonic force with gauge field $B_\mu(X)$ which is now part of a quadruplet of fields. It is a massless, long range field that looks to our discussion of the 16-dimensional *Megaverse* in which our universe resides where the baryonic force and the Dark Baryon force exist. Beyond our universe they cause interactions with other possible "island" universes. (The leptonic and Dark leptonic forces are short range and thus do not extend beyond our universe.)

The two non-diagonal long range forces, being between different generations of a species, and having an ultra weak coupling are not of great consequence, because of the short lifetime of the higher generations of each species. Therefore,,despite their long range, they have only the "shortest" time to exert an inter-generation force before a higher generation particle decays.

Since we expect the other massive fields to have very large masses (and thus very large Higgs field vacuum expectation values) and ultra weak coupling they are not likely to be experimentally found for the foreseeable future.

13.4 Impact of this U(4) Higgs Mechanism on Fermion Generation Masses

The fermion masses of the charged lepton, and the up-type quark, and down-type quark species' generations all show a rapid increase of mass with the generation. For example the u quark mass is a few MeV while the t quark (third generation) has a mass of about 170 GeV/c. The ratio of these masses is about 170,000. While one can account for this great difference by

---

[138] If we modify eq. 13.37 making $\rho = (0, \rho_1, \rho_3, \rho_2)$ then $U_{7\mu}(X)$, $U_{8\mu}$ (X), and $B_{D\mu}(X)$ acquire mass terms. Then Dark baryon number would be short range. Baryon number $B_\mu(X)$ would remain massless and thus long range. Whether the choice above or this alternate change is the correct one can only be answered by experiment. These considerations should be compared to Layer group symmetry breaking in chapter 15.

[139] Neutral lepton, charged lepton, up-type quark and down-type quark plus the four corresponding Dark species.

the judicious choices of Higgs' parameter values, when one considers the Generation group and its associated numerical quantities: ultra weak coupling, very large U particle masses – perhaps of the order of hundreds or thousands of GeV/c, and the corresponding very large Higgs particle vacuum expectation values in this U gauge field sector[140] then the differences in fermion masses within a species become more understandable and natural.

Thus the popular view that the ElectroWeak gauge field symmetry breaking is solely via ElectroWeak Higgs fields is not part of our Extended Standard Model unless the U(4) sector is removed. In our model there are two sets of contributions (at present) to fermion symmetry breaking: ElectroWeak Higgs particles symmetry breaking, and Generation group U(4) Higgs particles symmetry breaking. The Generation group causes each species to break into four generations.

In the conventional Standard Model the breakup of each species into generations is inserted "by hand." It is not a consequence of the existence of SU(2)⊗U(1) symmetry or symmetry breaking. In our approach the U(4) Generation group causes the appearance of generations. We base the existence of the Generation group[141] on the four conserved particle numbers.

13.5 Generation Group Higgs Mechanism for Fermion Masses

We now consider the Generation group Higgs Mechanism for the eight species of fermions (four species of "normal" matter and four species of Dark Matter). We shall consider the mass terms for the four normal species which is the same as that of the four Dark species except for the values in the various species mass matrices. We define the generations 4-vector:

$$\Psi_s = \begin{bmatrix} \Psi_{11} \\ \Psi_{12} \\ \Psi_{13} \\ \Psi_{14} \\ \cdots \\ \Psi_{41} \\ \Psi_{42} \\ \Psi_{43} \\ \Psi_{44} \end{bmatrix} \tag{13.46}$$

[140] They are not the Higgs particles of the SU(2)⊗U(1) ElectroWeak sector.

[141] In earlier books we suggested the fermion generations might be the result of a wormhole to another 4-dimensional universe. The new approach is simpler and more consistent with known facts.

where ψ_{ki} is the generation index for the ith generation of the kth species. ψ_{k1} is the wave function for the 1st generation, ψ_{k4} is the 4th generation member of the kth species, and we omit other indices in the interests of clarity. The normal fermion species are ordered charged lepton (k = 1), up-type quark, neutral lepton, and down-type quark (k = 4). Other indices of these wave functions are surpressed in the interests of clarity. A 4th generation fermion of any species is yet to be found experimentally. The lagrangian density mass term for the four normal fermion species is

$$\mathcal{L}_U{}^{UHiggs}{}_{FermionMassesk} = \Sigma_{\alpha,\beta}\, \bar{\psi}_{kL\alpha}\, \eta m_{k\alpha\beta}\, \psi_{kR\beta} + \text{c.c.} \qquad (13.47)$$

where $m_{k\alpha\beta}$ is complex constant matrix for the kth species, and where α, β = 1, … , 4. The total fermion lagrangian mass terms[142] for the kth species are

$$\mathcal{L}^{Higgs}{}_{FermionMassesk} = \mathcal{L}_U{}^{UHiggs}{}_{FermionMassesk} + \mathcal{L}_{EW}{}^{Higgs}{}_{FermionMassesk} \qquad (13.48)$$

where $\mathcal{L}_{EW}{}^{Higgs}{}_{FermionMassesk}$ is the contribution of ElectroWeak Higgs Mechanism to the fermion masses. Using the vacuum expectation value of η in eq. 13.37, and summing over k, we find the total

$$\mathcal{L}_U{}^{UHiggs}{}_{FermionMasses} = \Sigma_{\alpha,\beta}\, \{\bar{\psi}_{2L\alpha}\, \rho_1 m_{2\alpha\beta}\psi_{2R\beta} + \bar{\psi}_{4L\alpha}\, \rho_2 m_{4\alpha\beta}\psi_{4R\beta}\} + \text{c.c.} \qquad (13.49)$$

giving mass terms for the up-type and down-type quark species but not for lepton species. There is an implicit color summation over the color quarks in each generation and quark species. *Qualitatively eq. 13.49 could be viewed as corresponding to the experimentally determined largeness of quark masses relative to lepton masses in each generation of normal matter.*

The mass matrices $m_2 = [m_{2\alpha\beta}]$ and $m_4 = [m_{4\alpha\beta}]$ are both complex, constant mass matrices. They can be brought to diagonal form with non-negative values by U(4) matrices A_k and B_k:

$$A_2 m_2 B_2^{-1} = D_2 \qquad (13.50)$$
$$A_4 m_4 B_4^{-1} = D_4$$

or

$$m_2 = A_2^{-1} D_2\, B_2 \qquad (13.51)$$
$$m_4 = A_4^{-1} D_4\, B_4$$

---

[142] At this point. There are also Layer group contributions that we will consider later.

We now note, that although, both D_2 and D_4 have non-negative real values, down-type quarks are all tachyonic and up-type quarks are all non-tachyonic due to their lagrangian kinetic terms as seen earlier.

We further note that $m_2^\dagger m_2$ and $m_4^\dagger m_4$ are hermitean, and A_k and B_k are members of U(4) as is D_k for k = 2,4, with the result that m_2 and m_4 are also both members of the U(4) group. Thus

$$m_2^{-1} = m_2^\dagger \qquad (13.52)$$
$$m_4^{-1} = m_4^\dagger$$

We can express the mass matrices in terms of U(4) generators

$$m_2 = \Sigma G_i m_{2i} \qquad (13.53)$$
$$m_4 = \Sigma G_i m_{4i}$$
$$m_2^{-1} = m_2^\dagger = \Sigma G_i m_{2i}{}^* \qquad (13.54)$$
$$m_4^{-1} = m_4^\dagger = \Sigma G_i m_{4i}{}^*$$

since the matrices G_i are all hermitean, where $\{m_{2i}\}$ and $\{m_{4i}\}$ are each a set of sixteen complex constants.

While we do not as yet know the 4th generation fermions or their masses, the third generation quarks have masses that are far greater than the 1st and 2nd generation quarks or their sum suggesting that the trace of m_2 and m_4.is dominated by the 4th generation mass of the two quark species with a similar situation holding, perhaps, for the two Dark quark species. Therefore if we take the trace of m_2 and m_4 then it seems probable based on the trend of the generations that the 4th generation mass dominates the trace:

$$D_{24} \approx \text{tr } D_2 \qquad (13.55)$$
$$D_{44} \approx \text{tr } D_4$$

We can use these A_k and B_k U(4) transformations to define the eight "physical" (up to further ElectroWeak Higgs effects) up-type and down-type quark generations fields:

$$\overline{\psi}_{2L\alpha}\, \rho_1 m_{2\alpha\beta}\psi_{2R\beta} + \overline{\psi}_{4L\alpha}\, \rho_2 m_{4\alpha\beta}\psi_{4R\beta} = (\overline{\psi}_{2L}\, A_2^{-1})_\alpha\, \rho_1 D_{2\alpha\beta}(B_2\psi_{2R})_\beta + (\overline{\psi}_{4L}\, A_4^{-1})_\alpha\, \rho_2 D_{4\alpha\beta}(B_4\psi_{4R})_\beta$$
$$= \overline{\psi}_{2Lphys\alpha}\, \rho_1 D_{2\alpha\beta}\psi_{2Rphys\beta} + \overline{\psi}_{4Lphs\alpha}\, \rho_2 D_{4\alpha\beta}\psi_{4Rphys\beta} \qquad (13.56)$$

Matching Species: up-type quarks down-type quarks

13.5.1 Dark Matter Generation Symmetry Breaking

The preceding discussion with changes in the values of constants and constant matrices holds for Dark Matter also where the Dark quarks acquire mass terms but the Dark leptons do not. The Dark Matter species mass terms, with the subscript D signifying Dark Matter, are

$$= \bar{\psi}_{D2Lphys\alpha}\, \rho_{D1} D_{D2\alpha\beta} \psi_{D2Rphys\beta} + \bar{\psi}_{D4Lphs\alpha}\, \rho_{D2} D_{D4\alpha\beta} \psi_{D4Rphys\beta} \tag{13.57}$$

Matching Dark Species: up-type quarks down-type quarks

13.6 Four Generation CKM Matrix

The four generation generalization of the CKM matrix[143]

$$C_4 = A_2 B_4^{-1} \tag{13.58}$$

is a constant U(4) Generation group matrix. Redefining field phases it can be reduced to an SU(4) matrix. It appears in the charge-changing quark current

$$J^{\mu}_{charged} = \bar{\psi}_{4Lphs} \gamma^{\mu} C_4 \psi_{2Lphys} \tag{13.59}$$

Thus a specific Generation group matrix determines the mixing between the generations. This matrix may be expected to generate CP violation. C_4 can be constrcted as a product of SU(4) factors with nine arbitrary parameters.

13.7 Interactions of Generation Gauge Fields with Fermions

In addition to the role of the Generation group in mixing between generations, its also gives interactions between generations. The interactions of Generation group gauge fields with fermions is discussed in chapter 16 with illustrative Feynman diagrams.

[143] M. Ko
bayashi and K. Maskawa, Prog. Theo. Phys. **49**, 652 (1975) and references therein.

14. Extended Standard Model Multi-Quark Particles, and Dark Matter 'Chemistry'

14.1 New Multi-quark Particles such as Penta-Quark Particles

The CERN LHC has found evidence for penta-quark particles – particles consisting of five quarks.[144] This discovery is a step beyond the tetraquark (four quark) particle named $Z_C(3900)$ found, and announced,[145] in 2013. The Z_C particle could be viewed as two D-mesons bound together by the strong color force in a color singlet "hadron molecule."

Similarly, the pentaquark could be viewed as a particle consisting of a two quark meson and a three quark baryon formed into a type of molecule by the strong interaction.

14.2 Multi-Quark Molecules - Quark Chemistry

These recent discoveries are the initial steps to a spectrum of multi-quark particles consisting of

$$2m + 3b$$

quarks where m is the number of "meson-like" constituents and b is the number of "baryon-like" constituents. The pentaquark has m = b = 1 constituents.

We are now confronted by a new type of "molecule" where the binding force is not electromagnetic but, instead, the strong color force. The tight binding of these "molecules" by the strong force makes them effectively into particles. But they are perhaps more comparable to the binding of protons and neutrons into atomic nuclei by the nuclear force.

The development of a quark molecule chemistry would, if molecules were produced in quantity, lead to quark materials with extremely important physical characteristics – quark matter – with both new and important "chemical" properties. However, quark molecules have an extraordinarily short lifetime and so the creation of a number of quark molecules, and their fabrication into materials, does not appear to be feasible. Nevertheless, their theoretical and experimental study might serve to drive conventional chemistry and materials science forward.

---

[144] Announced July 15, 2015. See arXiv.org/abs/1507.03414.

[145] June 20, 1013. Discovery by the Belle Collaboration, High Energy Accelerator Research Organization, in Tsukuba, Japan; and BESIII Collaboration at the Beijing Electron Positron Collider in China. Phys. Rev. Lett. **110**, 252001, 2013 and **110**, 252002, 2013 .

14.3 Large Multi-Quark "Molecules" and Eventually Quark Stars

One can envision the eventual creation of very "large" multi-quark particles/molecules. Taken to the limit of extremely large "molecules" of extraordinary size it appears possible to develop a dynamics of quark stars where the dominant force is not the nuclear force as we usually know it and electromagnetism but rather the strong force. Astronomy has yet to discover an unambiguous quark star.

14.4 Dark Matter Features: Particle Spectrum and Basic Chemistry of Dark Matter, Dark Matter Bodies

In chapter 39 of Blaha (2012a) we initially described Dark Matter particles, Dark atoms, the Dark Periodic Table, and Dark basic chemistry – all based on the additional SU(2)⊗U(1) symmetry in the Extended Standard Model that more or less mirrors normal ElectroWeak symmetry. The major differences were 1) that Dark quarks are assumed to be SU(3) singlets as suggested by the known weakness of Dark Matter interactions with normal matter and 2) that *physical* hadron charges – both electric charge and Dark electric charge – are quantized with whole number values. (Dark quarks are the emasculated "hadrons" of this theory since they do not experience the Strong Interaction.) This chapter contains material from chapter 39 of Blaha (2012a) for completeness as well as some additional thoughts.

Another important issue is the effect of gravitation on Dark Matter—Can Dark matter aggregate under the force of gravity to form galactic clusters, galaxies, and smaller objects such as Dark suns and perhaps Dark planets? We discuss these possibilities below.

This chapter is based on the assumption of four generations.[146] Previous chapters present a strong theoretical case for four generations of fermions.

14.5 Fundamental Dark Particles

We now consider the Dark particles that are associated with our new Dark SU(2)⊗U(1) symmetry. Since Dark Matter only interacts weakly with known matter, Dark particles must be color singlets. Thus there will be 12 (in the three generation case) or 16 (in the four generation case) Dark particles (plus their antiparticles.) Recent experiments suggest Dark particles have extremely large masses of the order of 8.6 GeV/c or larger. Dark quarks are color singlets with complex 3-momenta in our Extended Standard Model.

---

[146] This chapter is based on the assumption of one layer of four generations. The introduction of the Layer group and four layers of four fermion generations dramatically increases the number of Dark fermion states.

"Periodic Table" of 'Known' Fermions

NORMAL ELECTROWEAK FERMION DOUBLETS

Generation	Leptons		Color Triplet Quarks	
	Real-Valued 3-Momenta		Complex-Valued 3-Momenta	
1	e	ν_e	u_1 u_2 u_3	d_1 d_2 d_3
2	μ	ν_μ	c_1 c_2 c_3	s_1 s_2 s_3
3	τ	ν_τ	t_1 t_2 t_3	b_1 b_2 b_3
4?	ω	ν_ω	v_1 v_2 v_3	w_1 w_2 w_3

DARK SU(2)⊗U(1) FERMION DOUBLETS

Generation Quarks	Dark Leptons		Color Singlet Dark	
	Real-Valued 3-Momenta		Complex-Valued 3-Momenta	
1	e_D	ν_{eD}	u_D	d_D
2	μ_D	$\nu_{\mu D}$	c_D	s_D
3	τ_D	$\nu_{\tau D}$	t_D	b_D
4?	ω_D	$\nu_{\omega D}$	v_D	w_D

Figure 14.1 "Periodic Table" of four generations of normal and Dark fundamental fermions assuming only one layer of fermions.

In addition to Dark fermions there will be four SU(2)⊗U(1) Dark gauge bosons – also with large masses:

Dark Particle Gauge Bosons

U(1): $W'_0{}^{\mu}$ SU(2): $W'_1{}^{\mu}$ $W'_2{}^{\mu}$ $W'_3{}^{\mu}$

14.6 Dark Particle Chemistry

The chemistry of Dark Matter will be different from the chemistry of known matter due to the absence of the color interaction. Dark particles cannot combine through a strong interaction to form a hadronic spectrum, or atomic nuclei, such as we see in normal matter. Thus Dark particle atoms are like hydrogen atoms, and consist of a Dark quark particle bound to a Dark lepton particle by the Dark electric force assuming Dark charge is quantized and has equal integer absolute values for Dark quarks and leptons. More complex Dark molecules are also possible as we will see later.

Suppose all Dark charged particles have charge $\pm 1e_D$ (e_D is the Dark unit of charge which may possibly be equal to e, the electromagnetic charge), and each Dark doublet has a Dark particle with unit Dark charge and a neutral Dark particle. Then we would expect that (in the case of four generations) Dark Matter would consist of:

1. Lepton-like fundamental particles: Four Dark charged and four Dark charge neutral particles and their anti-particles.

2. Quark-like fundamental particles: Four Dark charged and four Dark charge neutral particles and their anti-particles.

3. There would be a total of eight neutral Dark quarks and leptons.

4. Atoms are composed of oppositely charged Dark particles of different types. There are 16 Dark "atoms" of the form leptoDark particle - quarkDark particle, $e_D u_D$, $e_D c_D$, $e_D t_D$, $\mu_D u_D$, $\mu_D c_D$, $\mu_D t_D$, $\tau_D u_D$, $\tau_D c_D$, $\tau_D t_D$, $e_D v_D$, $\mu_D v_D$, $\tau_D v_D$, $\omega_D v_D$, $\omega_D u_D$, $\omega_D c_D$, and $\omega_D t_D$ plus their anti-matter equivalents. There are 12 "quasi-stable" particle-anti-particle combinations: six leptoDark - antileptoDark particle combinations, and six quarkDark - antiquarkDark particle combinations. (There is no attractive nuclear force.) All of these atoms are bound by the Dark electric force.

5. Simple molecules of the type of Figs. 14.2 – 14.4 below, and so on, based on Dark dipole interactions, Dark van der Waals forces and other Dark electromagnetic interactions are possible.

After a sufficiently long time, collisions would lead perhaps to the dominance of Dark particles and the "disappearance" of antiDark particles if the number of Dark particles is overwhelmingly dominant in a fashion similar to normal matter.[147] (The other possibility is not excluded.) The Dark Periodic Table is presented in Fig. 14.2.

---

[147] The study of electron-positron production through Weak Interaction with Dark Matter by M. Aguilar et al, Phys. Rev. Letters **110**, 141102 (2013) does not seem to clarify this issue.

Periodic Table of Simple Dark Particle Atoms
(Assuming Dark anti-particles are Annihilated)

$e_D u_D$	$\mu_D u_D$	$\tau_D u_D$	$\omega_D u_D$	d_D
$e_D c_D$	$\mu_D c_D$	$\tau_D c_D$	$\omega_D c_D$	s_D
$e_D t_D$	$\mu_D t_D$	$\tau_D t_D$	$\omega_D t_D$	b_D
$e_D v_D$	$\mu_D v_D$	$\tau_D v_D$	$\omega_D v_D$	w_D

Figure 14.2 "Periodic Table" of Simple Dark atoms assuming only one layer of fermions.

There also are similar antiparticle atoms. Bound states are assumed bound, perhaps into hydrogen-like atoms, through a Dark electromagnetic force. The last column consists of Dark charge neutral quarks. Antiparticle atoms of these states might also exist or be created through the Dark ElectroWeak interactions. One could extend the table with a column of Dark neutrinos. The decays, and mixing between generations, remains to be determined in the unknown Dark Higgs sector.

The periodic table (Fig. 14.2) that we constructed is based on an analogy with the features of normal matter: quarks are much heavier than leptons and leptons "revolve" around the Dark quark nuclei. If this view is correct then one can conceive of a chemistry of Dark Matter with molecules bound by Dark electromagnetic forces. Pair bonding of Dark leptons would be possible and so one could conceive of a fairly complex Dark chemistry bounded by the fact that a Dark particle atom has only one Dark lepton. Thus there would only be 64 bound pairs of atoms[148] similar to H_2.

e_D

u_D $\qquad\qquad$ u_D

e_D

Figure 14.3. Example of two Dark atoms binding in a manner similar to the binding of hydrogen molecules in H_2. Considering the possible combinations of quarks and leptons there are 64 bound varieties of this type. Chains of these molecules are also possible in principle.

Dark particles can be combined into chains and more complex molecules. A simple chain of Dark particles appears in Fig. 14.4.

---

[148] Plus anti-particle equivalents.

e_D u_D

u_D u_D e_D e_D

e_D u_D

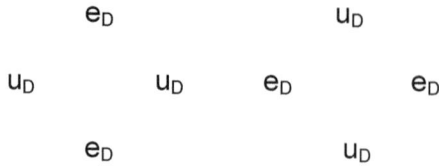

Figure 14.4. Example of a simple Dark Matter chain.

e_D u_D u_D e_D

u_D u_D e_D e_D e_D e_D u_D u_D

e_D u_D u_D e_D

Figure 14.5. A segment of two strands interacting with each other. Other combinations are possible using other Dark charged fermions. This diagram is suggestive of a Dark form of DNA. Dark Life?

More complex chains as well as two and three dimensional bound aggregates are also possible. These considerations suggest something approaching the complexity of simple life forms may be possible.

Dark dipole effects could lead to the (weaker) binding of larger assemblages of Dark atoms. For example, if the masses of the leptonic and quark Dark particles are not too dissimilar, crystalline Dark Matter appears possible.

We thus arrive at a Dark Matter sector with much less variety than normal matter. It may, or may not, preclude the existence of Dark life forms, and possibly of Dark solids composed of Dark Matter. Solid Dark planets may exist. The detection of planet-like and star-like Dark Matter objects within the Dark Matter "cloud" surrounding a galaxy would be of great importance.

14.7 Gravitation and Dark Matter Bodies: Galactic Clusters, Galaxies

Dark Matter is thought to equal approximately 83% of the total matter in the universe. It is known to form halos around galaxies and to form filaments between galaxies in galactic clusters. It appears to influence the orbits of stars in galaxies possibly causing the rotational speed of stars at large radial distances from galactic centers to be roughly constant and independent of the radius.

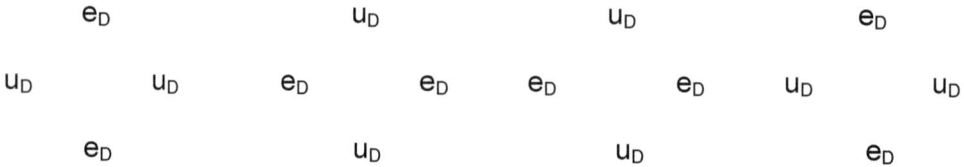

When two galaxies collide it appears that the Dark Matter drags between the galaxies as they separate. Thus Dark Matter has forces as we suggested earlier.

We see these effect in the Dark Matter halos around galaxies. Galaxies are permeated with Dark Matter. Our galaxy, for example, appears to have ten times as much Dark Matter as normal matter.[149] Galaxies that are composed of predominantly Dark matter have been found.

14.8 Dark Stars, Dark Planets, Dark Globular Clusters—Or Just Darkened?

However we have not seen (as yet?) clumps of Dark Matter within our galaxy similar in size to stars and planets or globular clusters. Nor do we see the earth, or other Solar System planets, having a discrepancy in mass due to Dark Matter clumps within these bodies. Nor is there a Dark planet in our Solar System.

Yet there is no apparent reason why Dark Matter should not clump within planets (and our sun) causing a discrepancy in solar system dynamics and ostensibly the force of gravity. The ten to one dominance of Dark Matter in our galaxy creates a striking discrepancy with the apparent absence of Dark Matter gravitational effects in the Solar System.

14.8.1 Open Questions on Dark Matter in Our Solar System

1. Why does not the earth's mass have a significant contribution from the Dark Matter within the earth since our galaxy has 10 times more Dark Matter than normal matter? Does not Dark Matter clump somewhat around planet size objects?

2. Same question as 1 but applied to the sun. Where is the effect of the Dark Matter within the sun?

3. Why is there no Dark Jupiter in our Solar System with normal moons circling it?

4. Why are not the planetary dynamics of the Solar System affected by the presence of Dark Matter?

---

[149] Therefore one would expect more gravitational phenomena reflecting the presence of Dark Matter locally in our galaxy than its effect on galactic rotation. Since Dark Matter hardly interacts with normal matter it would intersperse with normal matter throughout the galaxy forming a hidden part of the galaxy – present but almost completely not detectable. The hidden part would occupy the same space as normal galactic matter rather like an evanescent ghost in a horror movie.

14.8.2 Open Questions on Dark(ened) Stars and Dark(ened) Globular Clusters of Stars

There are many stars in the galaxy composed of normal matter apparently. However the ten to one ratio of Dark Matter to normal matter in our galaxy should be reflected in stellar dynamics which is a supposedly well understood field. Stellar dynamics depends significantly on the mass of stars.

1. Why does not stellar dynamics take account of the Dark Matter contribution to the mass of stars?

2. When a star contracts generating gravitational energy why does not the contraction of the Dark Matter clumped within the star impact gravitationally on the contraction of the normal matter?

3. Why is not the evolution and structure of globular clusters of stars affected by their presumably large Dark Matter content?

4. Will we find a Dark star circled by some large normal matter planets?

5. Can we detect a Dark cluster by its gravitational effects?

14.8.3 Comments on these Questions

The only currently apparent resolution of these questions is a lack of significant Dark chemical interactions/reactions between Dark Matter atoms that may prevent the formation of Dark solids and liquids.[150] Dark nuclear decays similar to the normal nuclear reactions that power stars may also not exist. Thus no Dark stars. But Dark aggregates of the mass of the sun or more should be possible.

Dark Matter in galaxies appears to be rather like a distributed gas within a gravitational well formed by the combined masses of a galaxy. It also forms a filament between galaxies.[151] Within a galaxy Dark Matter appears dispersed in a somewhat uniform way with only very minor gravitational clumping by clusters, stars and planets.

---

[150] Thus Dark atoms may not have Dark dipole forces or other multipole forces that would induce the formation of Dark Matter liquids and solids.

[151] A galaxy composed overwhelmingly of Dark Matter has recently been found.

15. The New U(4) Layer Group Leading to Four Layers of Fermion Generations

This chapter[152] adds a new symmetry group, the U(4) Layer group, with its interactions to the extended Standard Model. The form of Complex Lorentz group transformations led to the Reality group: R = SU(3)⊗SU(2)⊗U(1)⊗SU(2)⊗U(1), which is the symmetry group of the 'original' Standard Model. The U(4) Generation group, which followed from the four particle number conservation laws, increase the group to SU(3)⊗SU(2)⊗U(1)⊗SU(2)⊗U(1)⊗U(4). We now note that the four levels within the four fermion generations of each species can also be numbered and thus embody a broken U(4) symmetry that we call the Layer group. This group, with its interactions, and with the assumption that each generation level is in the Layer group fundamental representation $\underline{4}$, yields four layers of four generation fermions pictured in Fig. 15.1.

Figure 15.1. The four layers of fermions. Each layer (oval) has four generations of fermions. Level 1 is the level that we have found. The 4th generation of level 1, the Dark part of level 1, and the remaining three more massive levels constitute Dark matter.

[152] This chapter is extracted from Blaha (2016a) and (2016b).

The rationale for the Layer group, which is based on four generation numbers: L_1, L_2, L_3, and L_4, counting the number of particles in each generation, is the same as the rationale for the U(4) Generation group, which was based on the four particle numbers B, L, B_D, and L_D. We discuss the Layer group numbers in detail below.

15.1 The Origin of the Layer Group

The Generation group was based on U(4) rotations of the four number operators B, L, B_D, and L_D in the one generation Extended Standard Model. We can visualize these rotations horizontally as

$$\longleftarrow B \quad L \quad B_D \quad L_D \longrightarrow$$

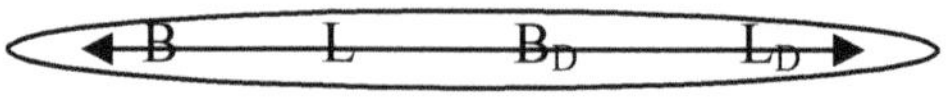

Given U(4) in the one generation case, it is natural to assume that the symmetry applies in a larger case – the case of a four generation Extended Standard Model. So we attached a U(4) generation index to each fermion field ranging from 1 through 4 and introduced gauge field interactions between the generations. The result is the four generation Extended Standard Model presented in Blaha (2015a) and earlier books.

After generations are introduced, then it becomes possible to consider vertical rotations amongst the four generations:

$$\begin{array}{c} L_1 \\ L_2 \\ L_3 \\ L_4 \end{array} \updownarrow$$

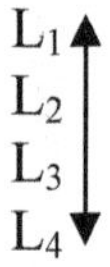

These rotations are based on four 'conserved' numbers L_1, L_2, L_3, and L_4 that count the number of fermions of each generation in a state.[153] *L_i counts the number of fundamental fermions in generation i for i = 1, 2, 3, 4.*

Fermions have positive $L_i = +1$ values and anti-fermions have negative $L_i = -1$ values. For example, if a state has 3 u quarks, 1 d quark, 1 anti-s quark, 2 electrons, 2 anti-τ leptons, 2 Dark muon neutrinos, and one Dark electron neutrino ν_{De} then $L_1 = 3+1+2+1 = 7$, $L_2 = -1+2 = 1$, $L_3 = -2$ and $L_4 = 0$.

It is important to note that the Layer numbers are independent of the baryon and lepton particle numbers that form the basis of the Generation group, and so the physics embodied in the Generation group is not the same as the physics of the Layer group defined here.

Layer numbers are conserved under strong and electromagnetic interactions but broken by the Electromagnetic and Weak interactions as well as their Dark counterparts.

---

[153] The generations are numbered from 1 to 4 with the lowest masses generation (e, ν_e, u, d), and the Dark analogues, being generation 1.

The partial conservation of the L_i Numbers enables us to define a new broken U(4) symmetry called the Layer group. Thus we have four new (partially) conserved particle numbers. Linear combinations of these numbers are also (partly) conserved:[154]

$$L_1' = aL_1 + bL_2 + cL_3 + dL_4$$
$$L_2' = eL_1 + fL_2 + gL_3 + hL_4$$
$$L_3' = iL_1 + jL_2 + kL_3 + lL_4$$
$$L_4' = mL_1 + nL_2 + oL_3 + pL_4$$

using constants labeled from "a" through "p" or

$$L' = AL$$

where A is a 4×4 U(4) matrix.

If we want the new 'primed' set of conserved numbers to be an independent set of numbers then the determinant of the constants must be non-zero. Thus the matrix, A, is invertible.

The set of 4×4 matrices of the above type form a U(4) group.[155] *The choice of U(4) rather than SU(4) is required since there are four independent layer particle numbers and U(4) has four diagonal matrices in its algebra while SU(4) only has three diagonal matrices. U(4) preserves the independence of the four independent particle numbers. Thus we have the Layer group.*

Since the coefficients in Layer group transformations can be local functions, the Layer group is implemented as a Yang-Mills theory.

Just as we extended the reach of the Generation group from one generation to four generations, we can extend the Layer group representation to 4 similarly by assuming that there are four layers of fermions obtained by adding a U(4) vector layer index to each fermion field.

Further, *the gauge fields for all interactions must be the same.*[156]

To implement the Layer symmetry in the Extended Standard Model lagrangian the following steps are required:

---

[154] This discussion parallels the discussion of particle numbers in the Generation group.

[155] If we wish to further limit the values of the 'primed' Numbers to integers assuming the unprimed Numbers are integers then the group of the transformation is the set of permutations of four entities – the Symmetric group S_4. However the 'primed' numbers can be integer or not. There is no apparent physical principle requiring integer Numbers. Quarks are usually assigned Baryon Number 1/3. Also Numbers are not necessarily positive valued.

[156] Later we will see that Weak interaction mass mixing occurs between particles in different layers. However the ultra-weakness of the Layer gauge field interactions makes the gauge field mixing negligible – although it is present in principle.

1, All covariant derivatives must acquire another interaction term with 16 U(4) fields corresponding to the 16 generators of the U(4) Layer group.

2. Expand the Extended Standard Model to embody the Layer group symmetry by adding another index ranging from 1 through 4 to each fermion field making four layers of four generations of fermions. Thus the fermion fields are in the fundamental representation of the Layer group.

3. Each layer should have its own set of Higgs particles (modulo mixing) contributing to fermion masses. One expects that the masses of fermions should be substantially larger for the three 'upper' layers beyond our layer. Otherwise we would have found particles from these upper layers.

4. Insert an interaction term of the form $g_V V^a{}_\mu G_L{}^a$ in the covariant derivative of each fermion using the 16 U(4) Layer gauge fields $V^a{}_\mu$ with a = 1, 2, …, 16 and the 16 U(4) generators $G_L{}^a$ that couple within and between layers using the new layer index where g_V is an ultra-weak coupling constant much smaller than the ElectroWeak coupling constants. This interaction will be between normal matter, between Dark Matter, between normal and Dark matter, and between fermions in the same and/or different layers. This interaction will be the only interaction between particles in different layers. Its small coupling constant, and its presumably very large gauge field masses will essentially make the four layers almost independent of each other except for gravitation.

5. Insert Layer group symmetry breaking Higgs fields (independently for all layers) that generate fermion mass term contributions independently for each layer.

The above modifications to the extended Standard Model lagrangian will be mathematically similar to the development of the features of the Generation group in Blaha (2015a) presented earlier. The form of the "periodic table" of fermions that results appears in Fig. 15.2 below.

15.2 Layer Group Interactions

The new Layer group interactions play two roles:

1. The four diagonal generator terms of $g_V V^a{}_\mu G_{La}$ create transitions amongst the Normal and Dark fermions, including transitions between Normal and Dark, between Normal and Normal, and between Dark and Dark in each layer separately.

2. The non-diagonal generator terms of $V^{a}_{\mu}G_{La}$ create transitions amongst the Normal and Dark fermions between *differing* layers. The non-diagonal interactions are the source of decays from higher layers to lower layers. Since no experimental evidence of such decays has been found as yet we conclude that the interaction constant g_V is ultra-small and/or the Layer gauge boson masses are extraordinarily large. Thus the distribution of fermions in layers may be approximately the same as that at the time of the Big Bang. We considered this issue in Blaha (2016a) and (2016b), and showed the Dark Matter fraction implied by the fermion spectrum (Fig. 15.2) is 83.33% – a result consistent with cosmological data. The fermions of the higher layers must be considered constituents of "Dark Matter" (as well as the Dark part of our layer) since only the ultra-weak Layer interaction and graviton interactions can connect layers.

THE FERMION PERIODIC TABLE

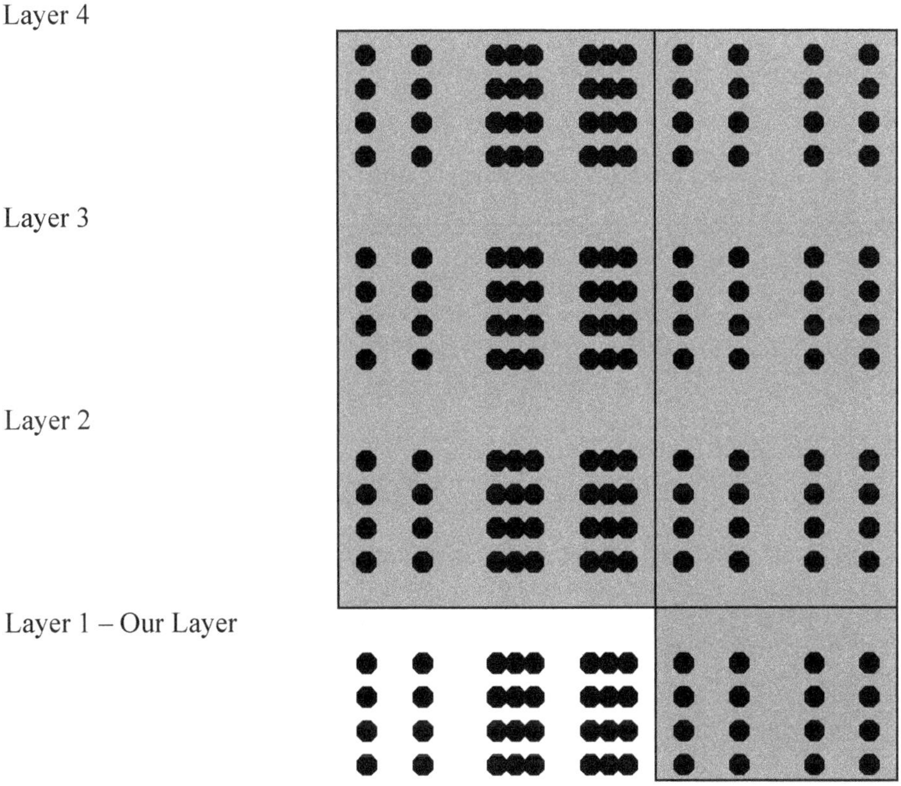

Figure 15.2. Dark parts of the periodic table are gray. Light parts are the known fermions with an additional, as yet not found, 4th generation shown.

15.3 Layer Group Interaction

In this section we describe Layer group interactions in detail.[157] The Theory of Everything (Extended Standard Model) symmetry group that we have developed in earlier books was:

$$SU(3)\otimes SU(2)\otimes U(1)\otimes SU(2)\otimes U(1)\otimes U(4)\otimes U(4) \quad (15.1)$$

with Layer interaction

$$g_V V^a{}_\mu G_{La} \quad (15.2)$$

We now modify the leptonic left-handed and right-handed covariant derivatives in the normal and Dark ElectroWeak sectors, which were:[158]

$$D_{NL\mu} = \partial/\partial X^\mu - \tfrac{1}{2}ig\boldsymbol{\sigma}\cdot\mathbf{W}_\mu + \tfrac{1}{2}ig'B_\mu - \tfrac{1}{2}ig_G\mathbf{G}\cdot\mathbf{U}_\mu \quad (15.3)$$
$$D_{DL\mu} = \partial/\partial X^\mu - \tfrac{1}{2}ig_D\boldsymbol{\sigma}\cdot\mathbf{W}'_\mu + \tfrac{1}{2}ig_D'B'_\mu - \tfrac{1}{2}ig_G\mathbf{G}\cdot\mathbf{U}_\mu \quad (15.4)$$
$$D_{NR\mu} = \partial/\partial X^\mu + ig'B_\mu - \tfrac{1}{2}ig_G\mathbf{G}\cdot\mathbf{U}_\mu \quad (15.5)$$
$$D_{DR\mu} = \partial/\partial X^\mu + \tfrac{1}{2}ig_D'B'_\mu - \tfrac{1}{2}ig_G\mathbf{G}\cdot\mathbf{U}_\mu \quad (15.6)$$

We also now modify eqs. 15.8 and 15.10 in the quark ElectroWeak covariant derivatives, which were:

$$D_{qNL\mu} = \partial/\partial X_c{}^\mu - \tfrac{1}{2}ig\boldsymbol{\sigma}\cdot\mathbf{W}_\mu - ig'B_\mu/6 + ig_C\tau\cdot A_{C\mu} - \tfrac{1}{2}ig_G\mathbf{G}\cdot\mathbf{U}_\mu \quad (15.7)$$
$$D_{qDL\mu} = \partial/\partial X_c{}^\mu - \tfrac{1}{2}ig_D\boldsymbol{\sigma}\cdot\mathbf{W}'_\mu + \tfrac{1}{2}ig_D'B'_\mu - \tfrac{1}{2}ig_G\mathbf{G}\cdot\mathbf{U}_\mu \quad (15.8)$$
$$D_{qNR\mu} = \partial/\partial X_c{}^\mu + ig'B_\mu/3 + ig_C\tau\cdot A_{C\mu} - \tfrac{1}{2}ig_G\mathbf{G}\cdot\mathbf{U}_\mu \quad (15.9)$$
$$D_{qDR\mu} = \partial/\partial X_c{}^\mu + \tfrac{1}{2}ig_D'B'_\mu - \tfrac{1}{2}ig_G\mathbf{G}\cdot\mathbf{U}_\mu \quad (15.10)$$

where $A_{C\mu}$ is the color gauge field.

The new covariant derivatives that contain the U(4) Layer group interaction are:

1) Leptonic covariant derivatives in the normal and Dark ElectroWeak sectors:

$$D_{NL\mu} = \partial/\partial X^\mu - \tfrac{1}{2}ig\boldsymbol{\sigma}\cdot\mathbf{W}_\mu + \tfrac{1}{2}ig'B_\mu - \tfrac{1}{2}i\, g_V\mathbf{G}_L\cdot\mathbf{V}_\mu - \tfrac{1}{2}ig_G\mathbf{G}\cdot\mathbf{U}_\mu \quad (15.3')$$
$$D_{DL\mu} = \partial/\partial X^\mu - \tfrac{1}{2}ig_D\boldsymbol{\sigma}\cdot\mathbf{W}'_\mu + \tfrac{1}{2}ig_D'B'_\mu - \tfrac{1}{2}i\, g_V\mathbf{G}_L\cdot\mathbf{V}_\mu - \tfrac{1}{2}ig_G\mathbf{G}\cdot\mathbf{U}_\mu \quad (15.4')$$
$$D_{NR\mu} = \partial/\partial X^\mu + ig'B_\mu - \tfrac{1}{2}i\, g_V\mathbf{G}_L\cdot\mathbf{V}_\mu - \tfrac{1}{2}ig_G\mathbf{G}\cdot\mathbf{U}_\mu \quad (15.5')$$
$$D_{DR\mu} = \partial/\partial X^\mu + \tfrac{1}{2}ig_D'B'_\mu - \tfrac{1}{2}i\, g_V\mathbf{G}_L\cdot\mathbf{V}_\mu - \tfrac{1}{2}ig_G\mathbf{G}\cdot\mathbf{U}_\mu \quad (15.6')$$

---
[157] Much of this section is extracted from Blaha (2016b).
[158] Equations from Blaha (2015a).

2) Quark ElectroWeak covariant derivatives in the normal and Dark ElectroWeak sectors:

$$D_{qNL\mu} = \partial/\partial X_c^{\mu} - \tfrac{1}{2}ig\boldsymbol{\sigma}\cdot\mathbf{W}_{\mu} - ig'B_{\mu}/6 + ig_C\tau\cdot A_{C\mu} - \tfrac{1}{2}i\, g_V\mathbf{G}_L\cdot\mathbf{V}_{\mu} - \tfrac{1}{2}ig_G\mathbf{G}\cdot\mathbf{U}_{\mu} \quad (15.7')$$
$$D_{qDL\mu} = \partial/\partial X_c^{\mu} - \tfrac{1}{2}ig_D\boldsymbol{\sigma}\cdot\mathbf{W}'_{\mu} + \tfrac{1}{2}ig_D'B'_{\mu} - \tfrac{1}{2}i\, g_V\mathbf{G}_L\cdot\mathbf{V}_{\mu} - \tfrac{1}{2}ig_G\mathbf{G}\cdot\mathbf{U}_{\mu} \quad (15.8')$$
$$D_{qNR\mu} = \partial/\partial X_c^{\mu} + ig'B_{\mu}/3 + ig_C\tau\cdot A_{C\mu} - \tfrac{1}{2}i\, g_V\mathbf{G}_L\cdot\mathbf{V}_{\mu} - \tfrac{1}{2}ig_G\mathbf{G}\cdot\mathbf{U}_{\mu} \quad (15.9')$$
$$D_{qDR\mu} = \partial/\partial X_c^{\mu} + \tfrac{1}{2}ig_D'B'_{\mu} - \tfrac{1}{2}i\, g_V\mathbf{G}_L\cdot\mathbf{V}_{\mu} - \tfrac{1}{2}ig_G\mathbf{G}\cdot\mathbf{U}_{\mu} \quad (15.10')$$

We add a new Layer group index to each fermion field[159] with index number values ranging from 1 through 4. The new Layer group interaction term $g_V G_{La} V^a{}_{\mu}$ uses 16 U(4) gauge fields $V^a{}_{\mu}$ with a = 1, 2, …, 16, and 16 U(4) generators, denoted G_{La}, *that couple to the new Layer group indexes.*[160] The $V^a{}_{\mu}$ gauge fields have a standard U(4) kinetic energy lagrangian term.

The Layer interaction constant g_V is an ultra-weak coupling constant assumed to be much smaller than the ElectroWeak and Generation group coupling constants.

15.4 Layer Group Higgs Mechanism Contributions to Layer Gauge Field Masses

In this section we will determine the Layer group Higgs contributions to gauge field masses. We will see that all[161] layers have Layer group Higgs contributions to their fermion masses.

We begin by assuming that a scalar Higgs field η exists, which is a U(4) Layer group 4-vector with four components corresponding to the four conserved generation number operators L_1, L_2, L_3, and L_4 of section 15.1. η is an SU(2)⊗U(1)⊗SU(3) Electromagnetic, Weak and Strong Interaction scalar. Its lagrangian density terms are[162]

$$\mathcal{L}_V{}^{Higgs} = (\partial\eta^{\dagger}/\partial X^{\mu})(\partial\eta/\partial X^{\mu}) - \lambda(\eta^{\dagger}\eta - \rho^2)^2 + \mathcal{L}_V{}^{Higgs}{}_{FermionMasses} \quad (15.11)$$

where $\mathcal{L}_V{}^{Higgs}{}_{FermionMasses}$ are the fermion masses produced by the Layer Higgs Mechanism and where we set the η Layer 4-vector with Higgs field components to

---

[159] Higgs bosons also acquire Layer group indexes.

[160] The Generation group U(4) matrices G^a couple to Generation group indices.

[161] All Layers have Layer group Higgs contributions is required to avoid massless Layer group gauge fields.

[162] We use the standard formulation of the Higgs Mechanism because of its familiarity.

$$\eta = \begin{bmatrix} \rho_1 \\ \rho_2 \\ \rho_3 \\ \rho_4 \end{bmatrix} \qquad \begin{matrix} \underline{\text{Corresponding Conserved Number}} \\ L_1 \\ L_2 \\ L_3 \\ L_4 \end{matrix} \tag{15.12}$$

where ρ_1, ρ_2, ρ_3 and ρ_4 are real fields.[163] Then the covariant derivative of η is

$$D_{\ldots\mu}\eta = \{\partial/\partial X^\mu + \ldots - \tfrac{1}{2}ig_V[\Sigma \mathbf{G}_{Li}\mathbf{V}_{i\mu} + \mathbf{G}_{L1}\mathbf{V}_{1\mu} + \mathbf{G}_{L2}\mathbf{V}_{2\mu} + \mathbf{G}_{L3}\mathbf{V}_{3\mu} + \mathbf{G}_{L4}\mathbf{V}_{4\mu}]\} \begin{bmatrix} \rho_1 \\ \rho_2 \\ \rho_3 \\ \rho_4 \end{bmatrix} \tag{15.13}$$

The sum over i is from 5 through 16 (non-diagonal matrices), and $[\mathbf{G}_{Li}]_{jk}$ is the jk[th] element of $\mathbf{G}_{Li}$. Then

$$D_{\ldots\mu}\eta = \begin{bmatrix} \partial\rho_1/\partial X^\mu - \tfrac{1}{2}ig_V\{\rho_1\mathbf{G}_{L1}\mathbf{V}_{1\mu} + \rho_2\Sigma[\mathbf{G}_{Li}]_{11}\mathbf{V}_{i\mu} + \rho_2\Sigma[\mathbf{G}_{Li}]_{12}\mathbf{V}_{i\mu} + \rho_3\Sigma[\mathbf{G}_{Li}]_{13}\mathbf{V}_{i\mu} + \rho_4\Sigma[\mathbf{G}_{Li}]_{14}\mathbf{V}_{i\mu}\} \\ \partial\rho_2/\partial X^\mu - \tfrac{1}{2}ig_V\{\rho_2\mathbf{G}_{L2}\mathbf{V}_{2\mu} + \rho_1\Sigma[\mathbf{G}_{Li}]_{21}\mathbf{V}_{i\mu} + \rho_2\Sigma[\mathbf{G}_{Li}]_{22}\mathbf{V}_{i\mu} + \rho_3\Sigma[\mathbf{G}_{Li}]_{23}\mathbf{V}_{i\mu} + \rho_4\Sigma[\mathbf{G}_{Li}]_{24}\mathbf{V}_{i\mu}\} \\ \partial\rho_3/\partial X^\mu - \tfrac{1}{2}ig_V\{\rho_3\mathbf{G}_{L3}\mathbf{V}_{3\mu} + \rho_1\Sigma[\mathbf{G}_{Li}]_{31}\mathbf{V}_{i\mu} + \rho_2\Sigma[\mathbf{G}_{Li}]_{32}\mathbf{V}_{i\mu} + \rho_3\Sigma[\mathbf{G}_{Li}]_{33}\mathbf{V}_{i\mu} + \rho_4\Sigma[\mathbf{G}_{Li}]_{34}\mathbf{V}_{i\mu}\} \\ \partial\rho_4/\partial X^\mu - \tfrac{1}{2}ig_V\{\rho_4\mathbf{G}_{L4}\mathbf{V}_{4\mu} + \rho_1\Sigma[\mathbf{G}_{Li}]_{41}\mathbf{V}_{i\mu} + \rho_2\Sigma[\mathbf{G}_{Li}]_{42}\mathbf{V}_{i\mu} + \rho_3\Sigma[\mathbf{G}_{Li}]_{43}\mathbf{V}_{i\mu} + \rho_4\Sigma[\mathbf{G}_{Li}]_{44}\mathbf{V}_{i\mu}\} \end{bmatrix} \tag{15.14}$$

$$= \begin{bmatrix} \partial\rho_1/\partial X^\mu - \tfrac{1}{2}ig_V\{\rho_1\mathbf{G}_{L1}\mathbf{V}_{1\mu} + \rho_2\Sigma[\mathbf{G}_{Li}]_{12}\mathbf{V}_{i\mu} + \rho_3\Sigma[\mathbf{G}_{Li}]_{13}\mathbf{V}_{i\mu} + \rho_4\Sigma[\mathbf{G}_{Li}]_{14}\mathbf{V}_{i\mu}\} \\ \partial\rho_2/\partial X^\mu - \tfrac{1}{2}ig_V\{\rho_2\mathbf{G}_{L2}\mathbf{V}_{2\mu} + \rho_1\Sigma[\mathbf{G}_{Li}]_{21}\mathbf{V}_{i\mu} + \rho_3\Sigma[\mathbf{G}_{Li}]_{23}\mathbf{V}_{i\mu} + \rho_4\Sigma[\mathbf{G}_{Li}]_{24}\mathbf{V}_{i\mu}\} \\ \partial\rho_3/\partial X^\mu - \tfrac{1}{2}ig_V\{\rho_3\mathbf{G}_{L3}\mathbf{V}_{3\mu} + \rho_1\Sigma[\mathbf{G}_{Li}]_{31}\mathbf{V}_{i\mu} + \rho_2\Sigma[\mathbf{G}_{Li}]_{32}\mathbf{V}_{i\mu} + \rho_4\Sigma[\mathbf{G}_{Li}]_{34}\mathbf{V}_{i\mu}\} \\ \partial\rho_4/\partial X^\mu - \tfrac{1}{2}ig_V\{\rho_4\mathbf{G}_{L4}\mathbf{V}_{4\mu} + \rho_1\Sigma[\mathbf{G}_{Li}]_{41}\mathbf{V}_{i\mu} + \rho_2\Sigma[\mathbf{G}_{Li}]_{42}\mathbf{V}_{i\mu} + \rho_3\Sigma[\mathbf{G}_{Li}]_{43}\mathbf{V}_{i\mu}\} \end{bmatrix} \tag{15.15}$$

since the generators $\mathbf{G}_i$ have zeroes along their diagonals for i = 5, … , 16.

From eq. 15.15 we find the corresponding Higgs field kinetic terms in the lagrangian are

$$(D_{\ldots\mu}\eta)^\dagger D_{\ldots}^{\ \mu}\eta = \partial\rho_1/\partial X^\mu\, \partial\rho_1/\partial X_\mu + \partial\rho_2/\partial X^\mu\, \partial\rho_2/\partial X_\mu + \partial\rho_3/\partial X^\mu\, \partial\rho_3/\partial X_\mu + \partial\rho_4/\partial X^\mu\, \partial\rho_4/\partial X_\mu +$$

---

[163] Each field ρ_i can be expressed as a Pseudoquantum field: $\rho_i = \varphi_{1i} + \varphi_{2i}$ where φ_{1i} has the vacuum expectation value $\rho_{i0.\ \text{for } i = 1, \ldots, 4.}$ Thus our Pseudoquantum field theory version is implemented easily. We discuss Pseudoquantization later.

$$+ g_V^2\rho_1^2\mathbf{V}_{1\mu}\mathbf{V}_1^{\mu}/4 + g_V^2\rho_2^2\mathbf{V}_{2\mu}\mathbf{V}_2^{\mu}/4 + g_V^2\rho_3^2\,\mathbf{V}_{3\mu}\mathbf{V}_3^{\mu}/4 + g_V^2\rho_4^2\,\mathbf{V}_{4\mu}\mathbf{V}_4^{\mu}/4 + \ldots \tag{15.16}$$

We now turn to calculating the remaining terms in eq. 15.16 that determine the masses of the remaining 14 gauge fields. We begin by assigning matrix elements for the remaining hermitean U(4) generators:

$$\begin{aligned}
[G_{L5}]_{ik} &= \delta_{i1}\delta_{k2} + \delta_{i2}\delta_{k1} \\
[G_{L6}]_{ik} &= -i\delta_{i1}\delta_{k2} + i\delta_{i2}\delta_{k1} \\
[G_{L7}]_{ik} &= \delta_{i1}\delta_{k3} + \delta_{i3}\delta_{k1} \\
[G_{L8}]_{ik} &= -i\delta_{i1}\delta_{k3} + i\delta_{i3}\delta_{k1} \\
[G_{L9}]_{ik} &= \delta_{i1}\delta_{k4} + \delta_{i4}\delta_{k1} \\
[G_{L10}]_{ik} &= -i\delta_{i1}\delta_{k4} + i\delta_{i4}\delta_{k1} \\
[G_{L11}]_{ik} &= \delta_{i2}\delta_{k3} + \delta_{i3}\delta_{k2} \\
[G_{L12}]_{ik} &= -i\delta_{i2}\delta_{k3} + i\delta_{i3}\delta_{k2} \\
[G_{L13}]_{ik} &= \delta_{i2}\delta_{k4} + \delta_{i4}\delta_{k2} \\
[G_{L14}]_{ik} &= -i\delta_{i2}\delta_{k4} + i\delta_{i4}\delta_{k2} \\
[G_{L15}]_{ik} &= \delta_{i3}\delta_{k4} + \delta_{i4}\delta_{k3} \\
[G_{L16}]_{ik} &= -i\delta_{i3}\delta_{k4} + i\delta_{i4}\delta_{k3}
\end{aligned} \tag{15.17}$$

Then completing eq. 15.16 using eq. 15.15 we find

$$\begin{aligned}
(D_{\ldots\mu}\eta)^{\dagger} D_{\ldots}^{\ \mu}\eta = {}& \partial\rho_1/\partial X^{\mu}\,\partial\rho_1/\partial X_{\mu} + \partial\rho_2/\partial X^{\mu}\,\partial\rho_2/\partial X_{\mu} + \partial\rho_3/\partial X^{\mu}\,\partial\rho_3/\partial X_{\mu} + \partial\rho_4/\partial X^{\mu}\,\partial\rho_4/\partial X_{\mu} + \\
& + g_V^2\rho_1^2\mathbf{V}_{1\mu}\mathbf{V}_1^{\mu}/4 + g_V^2\rho_2^2\,V_2^2/4 + g_V^2\rho_3^2\,V_3^2/4 + g_V^2\rho_4^2\,V_4^2/4 + \\
& + (g_V/2)^2(\rho_1^2 + \rho_2^2)(V_5^2 + V_6^2) + (g_V/2)^2(\rho_1^2 + \rho_3^2)(V_7^2 + V_8^2) + \\
& + (g_V/2)^2(\rho_1^2 + \rho_4^2)(V_9^2 + V_{10}^2) + (g_V/2)^2(\rho_2^2 + \rho_3^2)(V_{11}^2 + V_{12}^2) + \\
& + (g_V/2)^2(\rho_2^2 + \rho_4^2)(V_{13}^2 + V_{14}^2) + (g_V/2)^2(\rho_3^2 + \rho_4^2)(V_{15}^2 + V_{16}^2)
\end{aligned} \tag{15.18}$$

up to total divergences, which generate surface terms which we discard. We also assume that all fields satisfy the gauge condition

$$\partial V_i^{\mu}/\partial X^{\mu} = 0 \tag{15.19}$$

Eq. 15.18 shows all Layer group gauge fields have masses. Thus Layer symmetry is completely broken. The combination of an ultra-weak coupling constant and very large gauge field masses results in extremely weak interactions between the fields in each layer, which leads to almost independent layers of fermions. Thus the Darkness! They result in very rare decays between layers, and very weak interactions between fermions in different layers. The higher layers with presumably much more massive fermions are thus well "insulated" from our layer.

We estimate Layer group gauge field masses to be very large – of the order of many TeV or they would have been detected at CERN by now. Their detection must await the construction of much more powerful accelerators. *The "non-diagonal" Layer gauge fields are the means by which we may hope to eventually find fermions of the higher layers.*

15.5 Layer Group Higgs Mechanism Contributions to Fermion Masses

The fermion masses of the charged lepton, and the up-type quark, and down-type quark species' generations all show a rapid increase of mass with the generation. For example the u quark mass is a few MeV while the t quark (third generation) has a mass of about 170 GeV/c. The ratio of these masses is about 170,000. While one can account for this great difference by the judicious choices of Higgs' parameter values, when one considers the Layer group and its associated numerical quantities: ultra-weak coupling, its very large Layer gauge field masses – perhaps of the order of hundreds or thousands of GeV/c, then a large difference in particle masses between layers is understandable and natural.

The form of the layers of fermion mass terms is[164]

$$\mathcal{L}^{\text{Higgs}}_{\text{FermionMasses}} = \Sigma_{k,a,\alpha,\beta}\bar{\psi}_{kaL\alpha\delta}\,\eta_k m_{EW_{ka\alpha\beta}}\psi_{kaR\beta} + \Sigma_{k,a,\alpha,\beta}\bar{\psi}_{D_{kaL\alpha}}\eta_{Dk} m_{DEW_{ka\alpha\beta}}\psi_{D_{kaR\beta}} + \qquad \text{ElectroWeak}$$

$$+ \Sigma_{k,a,\alpha,\beta}\bar{\psi}_{UkaL\alpha}\,\eta_{Uka} m_{Uka\alpha\beta}\psi_{UkaR\beta} + \qquad \text{Generation}$$
$$+ \Sigma_{k,a,\alpha,\beta}\bar{\psi}_{DUkaL\alpha}\eta_{DUka} m_{DUka\alpha\beta}\psi_{DUkaR\beta} + \qquad \text{Group U}$$

$$+ \Sigma_{k,g,\delta,\gamma}\bar{\psi}_{LkgL\delta}\,\eta_{Lg} m_{Lg\delta\gamma}\psi_{LkgR\gamma} + \qquad \text{Layer}$$
$$+ \Sigma_{k,g,\delta,\gamma}\bar{\psi}_{DLkgL\delta}\,\eta_{DLg} m_{DLg\delta\gamma}\psi_{DLkgR\gamma} + \qquad \text{Group L}$$

$$+ \Sigma_{k,a}\bar{\psi}_{GkaL}\eta_{Ga} m_{Gka}\psi_{GkaR} + \Sigma_{k,a}\bar{\psi}_{DGkaL}\eta_{DGa} m_{DGka}\psi_{DGkaR} + \qquad \text{Gravitational}^{165}$$

$$+ \text{c.c.} \qquad (15.20)$$

where the subscripts EW, D, U, L and G label ElectroWeak origin, D Dark type, U Generation group origin, L Layer group origin, and G Gravitational origin respectively. The fields labeled η (with subscripts) are Higgs fields that have non-zero vacuum expectation values.[166] The indices k label species – normal and Dark separately, g labels the (four) generations, and a labels the layers. The indices δ and γ label *layer* rows and columns (with implicit sums over generations in

[164] Layer group contributions have been added to the original eq. 5.56 in Blaha (2015b) in accord with Blaha (2016a).
[165] The U(4) gravitational gauge interaction will be discussed later. We present it here for completeness.
[166] The Higgs fields $\eta\ldots$ in our Pseudoquantum formulation are $\eta\ldots = \varphi_{1\ldots}(x) + \varphi_{2\ldots}(x)$ as described earlier.

the Layer group terms.) The Layer group mass contribution is the same for each fermion in each generation for each species in each layer. The matrices labeled m (with subscripts) are the complex constant mass matrices of species. The indices α, β = 1, … , 4 label *generation* rows and columns.

15.6 Layer Mixing of Fermions

Eq. 15.20 contains the mass terms for the four layers of fermions in our Theory of Everything. *For each species and generation the Layer group terms mix the Layer mass contributions.*

$$\mathcal{L}_V{}^{Higgs}{}_{FermionMasses} = \Sigma_{k,g,\delta,\gamma}\bar{\psi}_{LkgL\delta}\,\eta_{Lg}\,m_{Lg\delta\gamma}\psi_{LkgR\gamma} + \Sigma_{k,g,\delta,\gamma}\bar{\psi}_{DLkgL\delta}\,\eta_{DLg}\,m_{DLg\delta\gamma}\psi_{DLkgR\gamma} + \text{c.c.} \quad (15.21)$$

where $m_{Lg\delta\gamma}$ and $m_{DLg\delta\gamma}$ are complex constant matrices, and where δ, γ = 1, … , 4 label generations. The total of fermion lagrangian mass terms is

$$\mathcal{L}^{Higgs}{}_{FermionMasses} = \mathcal{L}_{EWFermionMasses} + \mathcal{L}_{UFermionMasses} + \mathcal{L}_{VFermionMasses} + \mathcal{L}_{GravFermionMasses} + \text{c.c.} \quad (15.22)$$

where $\mathcal{L}_{EW}{}^{Higgs}$ is the contribution of Electromagnetic-Weak Higgs Mechanism to the fermion masses (discussed later). Using the vacuum expectation value of η in eq. 15.12, and assuming $\eta_{Lg} = \eta_{DLg}$ we find

$$\begin{aligned}\mathcal{L}_{VFermionMasses} = \Sigma_{k,\delta,\gamma}\{&\bar{\psi}_{Lk1L\delta}\rho_1 m_{L1\delta\gamma}\psi_{Lk1R\gamma} + \bar{\psi}_{DLk1L\delta}\rho_1 m_{DL1\delta\gamma}\psi_{DLk1R\gamma} +\\ &+ \bar{\psi}_{Lk2L\delta}\rho_2 m_{L2\delta\gamma}\psi_{Lk2R\gamma} + \bar{\psi}_{DLk2L\delta}\rho_2 m_{DL2\delta\gamma}\psi_{DLk2R\gamma} +\\ &+ \bar{\psi}_{Lk3L\delta}\rho_3 m_{L3\delta\gamma}\psi_{Lk3R\gamma} + \bar{\psi}_{DLk3L\delta}\rho_3 m_{DL3\delta\gamma}\psi_{DLk3R\gamma} +\\ &+ \bar{\psi}_{Lk4L\delta}\rho_4 m_{L4\delta\gamma}\psi_{Lk4R\gamma} + \bar{\psi}_{DLk4L\delta}\rho_4 m_{DL4\delta\gamma}\psi_{DLk4R\gamma}\} + \text{c.c.}\end{aligned} \quad (15.23)$$

where the indices k label species – normal and Dark separately, and the indices δ and γ label *layer* matrix rows and columns. The integers 1, … , 4 label generations.

The mass matrices in eq. 15.22 are complex, constant mass matrices that can be totaled and brought to diagonal form with non-negative values by U(4) matrices. The resulting diagonalized mass matrices are the mass matrices of the physical fermions. See section 13.4 for an example of the procedure.

The fermion masses in the resulting three "upper" layers have terms with similar forms but with different mass values. These values are presumably very large. We expect that they are in the multi-TeV and may extend to tens of TeVs ranges – probably putting most of them out of range of the current CERN LHC.

Due to the weakness of this ultra-weak interaction, but the anticipated large vacuum expectation values of $\rho_1, \ldots, \rho_4$, the size of the mass cross terms in the Layer group mass matrices of the different layers is problematic. The mass cross terms (mixing) appear likely to be small.

15.7 Four Layer CKM-like Matrix

Similarly to the case of the Generation group (Chapter 13) there are sixteen, four layer, CKM-like Layer group matrices,[167] which we denote C_{LayerTgs} where g = 1, 2, 3, 4 labels the generation and s = 1, 2, 3, 4 labels the species. We use T = 1 to denote normal matter and T = 2 to denote Dark matter. Thus there are thirty-two CKM-like matrices associated with the Layer group if we take account of Dark as well as normal matter. Each CKM-like matrix is a 4×4 constant, complex U(4) Layer group matrix in general Each matrix has nine independent parameters.

The Layer changing current has the form

$$J^{\mu}_{\text{Tgs}} = \overline{\psi}_{\text{physTgsa}}\gamma^{\mu}[C_{\text{LTgs}}]_{\text{ab}}\psi_{\text{physTgsb}} \qquad (15.24)$$

for T = 1, 2 with sums over the Layer indices a and b. Thus, these constant Layer group matrices determine the mixing between the layers. They may be expected to generate CP violation.

Eq. 15.24 shows the mixing in the layer changing current is between each four corresponding fermions in the four layers – fermion by fermion. Thus the Layer group is seen to be truly a direct product with the other Standard Model group factors.

15.8 Interactions of Layer Gauge Fields with Fermions

The interactions of Layer group gauge fields with fermions is discussed in chapter 16 with illustrative Feynman diagrams.

---

[167] The four generation CKM matrix and the sixteen four layer CKM-like matrices associated with the Layer group may be expected to have similarities.

16. The 192 Fermions of the Extended Standard Model

The 192 fermions that form the fermions of the 'Fermion Periodic Table' are displayed in Fig. 15.2. There are four species of 'normal' fermions and four species of Dark fermions. Normal matter up-type quarks and down-type quarks each consist of three subspecies. Dark matter quarks do not have subspecies since they are color SU(3) singlets. There are four generations of each species (and subspecies) due to the Generation group. There are four layers of each set of four generations due to the Layer group. Fig. 16.1 has the Periodic Table of Fermions with rows and columns labeled with quadruplets of numbers.

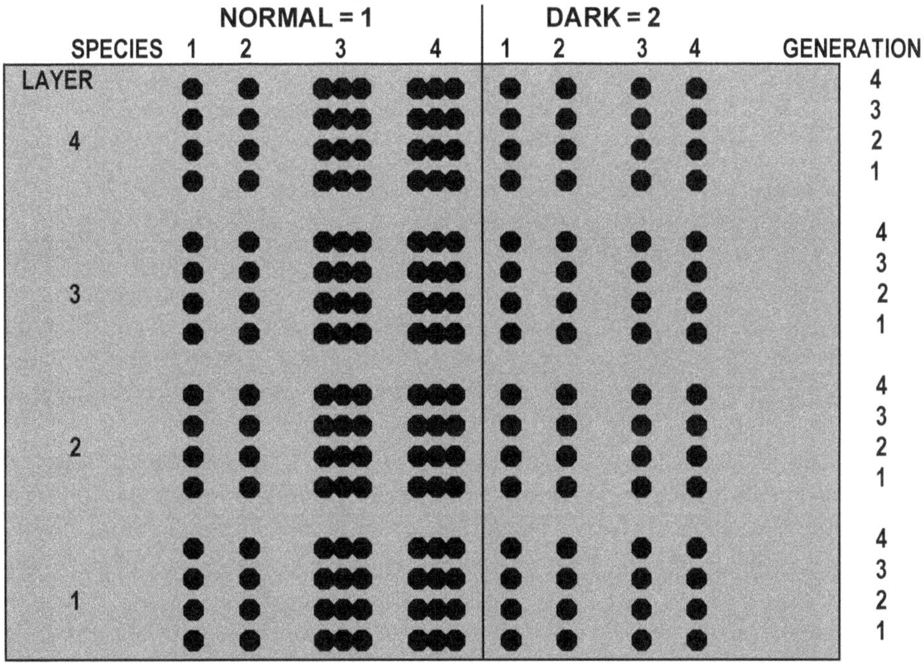

Figure 16.1. The Periodic Table of Fermions with each fermion identified by a quadruplet of integers: (T, S, L, G) where T identifies Normal or Dark, S identifies the species, G identifies the generation, and L identifies the layer. For example the electron has the triplet (1, 1, 1, 1), and the second generation, fourth layer, Dark up-type quark is (2, 3, 4, 2).

In the case of normal quark species, which each consist of a color triplet, a fifth integer would be needed to identify each color quark within a triplet: (T, S, L, G, C) where C = 1, 2, 3 identifies the color ("red, white or blue"). For example, the second generation, fourth layer, normal up-type red quark is (1, 3, 4, 2, 1) where we treat 'red' as having the value 1.

However, only the color SU(3) interaction distinguishes between the colors in interactions. Thus we can use quadruplets for all practical purposes – for electromagnetic, Weak, Generation group, and Layer interactions of fermions.

16.1 Identifying Fermions with Quadruplets of Numbers

Species are numbered from 1 through 4 – separately for normal and Dark species: charged lepton, neutral lepton, three up-type quarks, three down-type quarks, Dark charged lepton, Dark neutral lepton, one Dark up-type quark species, and one Dark down-type quark species. Layers are numbered from 1 through 4 with our layer being layer 1. Generations are numbered from 1 through 4 from lightest to the heaviest.[168] (e, ν_e, u and d constitute the known part of generation 1 of our layer.)

We will call the quadruplet (or quintet) of a fermion its *ID number*.

16.2 Group Theoretic Origin of Fermion ID Numbers

Since the Periodic table of Fermions is 2-dimensional – like the Chemical Periodic Table of Elements one might ask: Wh not use two numbers to identify fermions in a manner similar to the Chemistry table? The answer is related to the group structure of the Extended Standatd Model interactions. Except for T (the type of matter), the other three digits in a quadruplet are each related to a group. Thus G specifies the generation and thus the position of a fermion in the $\underline{4}$ of the U(4) Generation group. The integer L specifies the layer and thus the position of the fermion in the $\underline{4}$ of the U(4) Layer group.

The integer S specifies the position of a fermion amongst the species. Later we will show that the four species of each type of matter follow from a U(4) group, that we call the General Relativistic Reality Group, derived from a consideration of the structure of Complex General Coordinate transformations. The U(4) gauge fields are denoted $A_S^{\mu}(x)$ in discussions below.

Thus the fermion quadruplet is physically motivated by the group structure of the 'further' Extended Standard Model: $SU(3)\otimes SU(2)\otimes U(1)\otimes SU(2)\otimes U(1)\otimes U(4)\otimes U(4)\otimes U(4)$. *This group structure has been derived from the space-time considerations in this book. It is not simply postulated based on esthetic considerations.*

[168] The ordering by mass may not hold in the currently Dark part of the fermion spectrum.

16.3 Extended Standard Model Gauge Field – Fermion Interactions

We now turn to consider the ElectroWeak, Generation group and Layer group interactions.

16.3.1 Strong Interactions

For completeness we point out the interactions of color quarks due to color SU(3).

16.3.2 ElectroWeak Interactions

The ElectroWeak interactions both for normal and for Dark particles occur within, and between, generations and species of one layer. The solid lines below represent fermions; the dashed lines represent ElectroWeak gauge bosons. The bar above a quadruplet indicates an anti-particle.

In *general* the form of basic Electromagnetic and Weak interactions is:

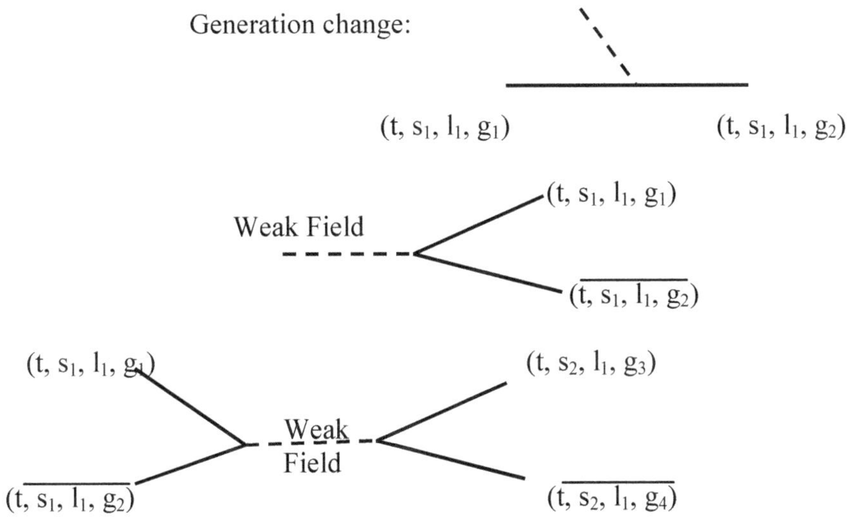

16.3.3 Generation Group Interactions

The Generation group interactions for both normal and for Dark particles occur within generations of one layer, and/or within individual species. The 16 gauge field interactions of the Generation group 'overlap' with the four Electromagnetic and Weak gauge field interactions since U(4) has an SU(2)⊗U(1) subgroup. Thus they partly "duplicate" Electromagnetic and Weak interactions but "add more." The Electromagnetic and Weak, and Generation group interactions are distinctively different. The anticipated ultra-weak Generation group coupling constant:

$$g_G = (4\pi\alpha_B)^{1/2} \approx 1.218\ (Gm_H^{2})^{1/2}$$

makes their contributions much smaller than ElectroWeak contributions to interactions. The solid lines below represent fermions; the dashed lines represent Generation group gauge bosons. In *general* the form of the Generation group interactions is:

Change of
Generation

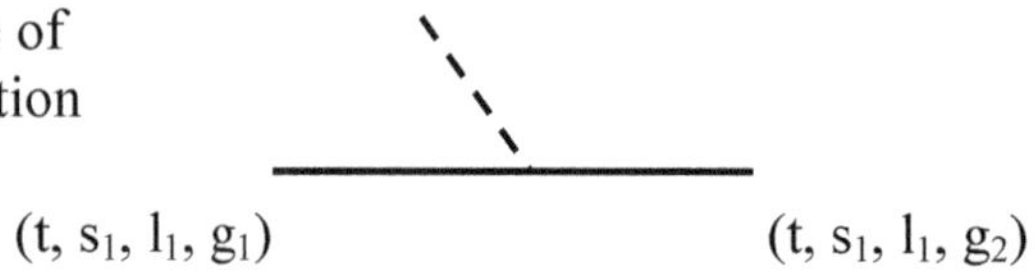

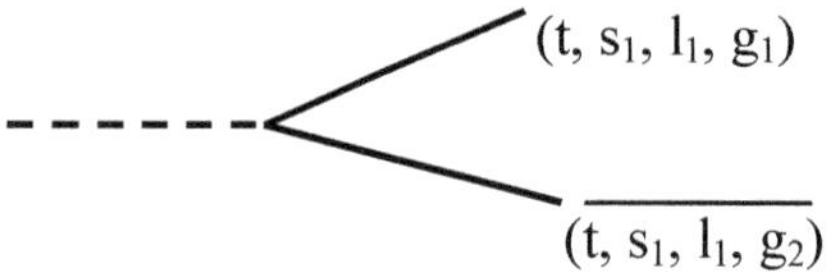

16.3.4 U(4) Layer Group Interactions

The U(4) Layer group interactions for both normal and for Dark particles can occur within one layer, and/or between layers, of each species individually for each fermion of the four generations and four species. It can provide transitions between normal fermions and Dark fermions in our layer (and in other layers) through the four diagonal U(4) Layer interactions such as the annihilation diagram:

$$N\bar{N} \rightarrow \gamma_L \rightarrow D\bar{D}$$

where γ_L is a Layer group gauge boson. (See also the third diagram below.)

The ultra-weak Generation group coupling constant g_V, which is also most likely of the order of $(Gm_H^{2})^{1/2}$, causes only minimal coupling between the layers – making the layers more or less independent of each other except for rare interactions. Thus it gives a reason for the Darkness of the Dark sector of our layer and of the other layers.

The solid lines below represent fermions; the dashed lines represent Layer group gauge bosons. In *general* the form of the Layer group interactions is:

Change of Layer

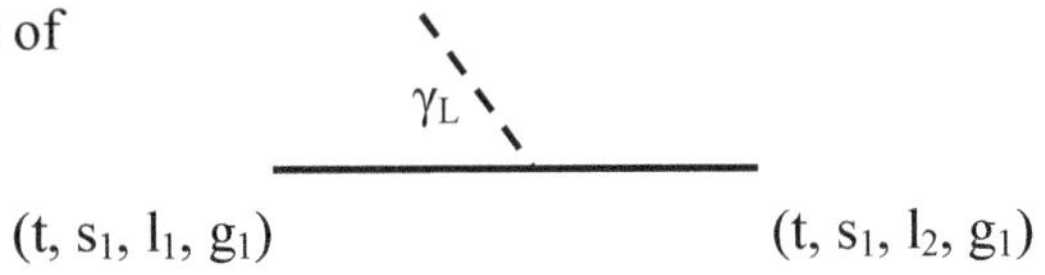

Creation

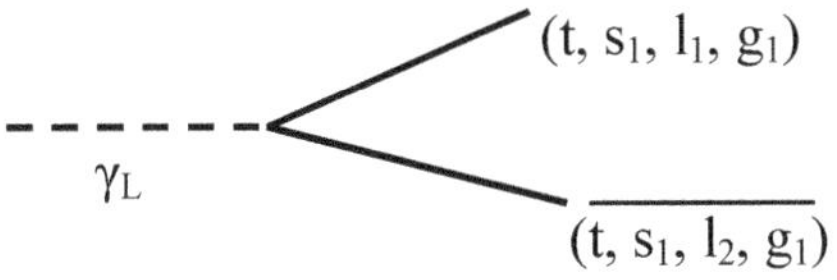

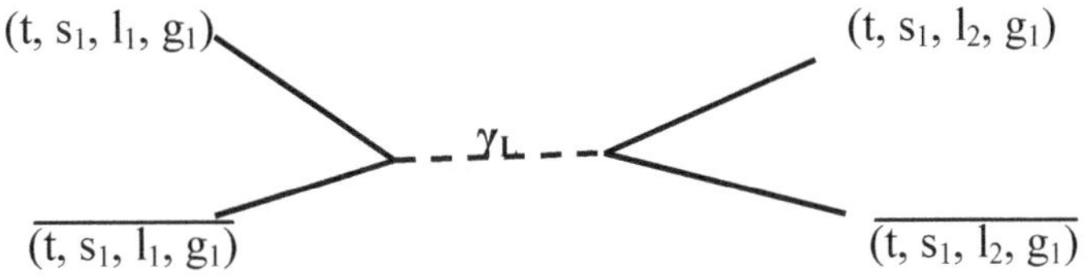

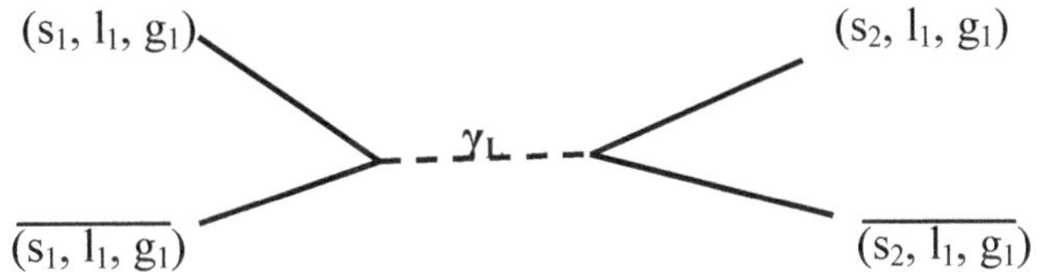

16.3.5 U(4) General Relativistic Reality Gauge Group Interactions

This gauge group (discussed later) causes transitions (rotations) between species and generates the interactions of the form:

Change of Species

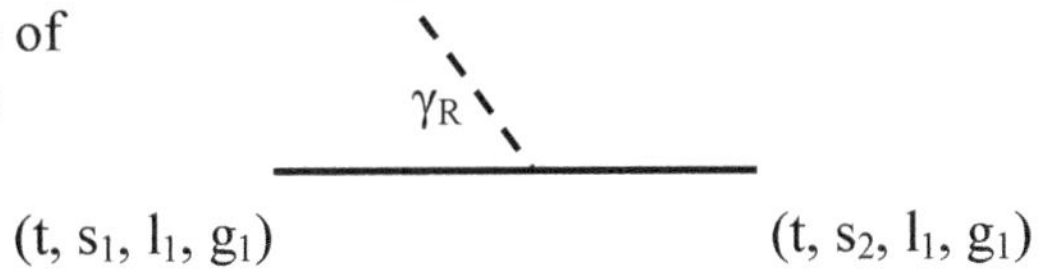

The change of species of a fermion is from a specific species to another of the four species although momentum and spin may change.

17. Normal vs. Dark Matter: Origin of 83⅓% Dark Matter

In this chapter[169] we suggest a rationale for the dominant abundance of Dark matter: Dark matter has been found experimentally to constitute 81% - 85% of the total matter in the universe.

17.1 Equipartition Principle for Fermion Degrees of Freedom

In a closed system at equilibrium the thermal energy of a system is equally partitioned (distributed) among its degrees of freedom. This Equipartition Principle is well known. The application of this principle to the beginning of the universe *when all fermions were massless* and all symmetries were unbroken suggests that the distribution of fermions should be the same for all fermionic degrees of freedom at that time. Thus there should be approximately equal numbers of fermions of each of the 192 types, that we found, with the same fraction of the total thermal energy.

We now estimate the relative proportion of Normal and Dark matter in the universe at its beginning based on this Equipartition Principle.

17.2 Proportion of Dark Matter in the Universe

First we note that 8 of the 12 species (counting quark of each color as a separate species) in layer 1 – the layer with which we are familiar are Normal matter fermions. (Our discussion is based on our Extended Standard Model.) Four of the 12 first layer species are Dark.

The other three layers are all Dark from our point of view since they have not been detected. Thus we find that 40 of the 48 species are Dark yielding a percentage of Dark matter equal to 40/48 = 83.33%. The same counting could have been done by counting fermions with the same results.

Recent studies of the proportion of Dark matter in the universe have yielded two estimates: 84.5% by Aghanim et al in Astronomy and Astrophysics 1303;5062 and 81.5% from a NASA fit to various models.

Thus our estimate based on our fermion Equipartition Principle is midway between these experimental estimates.

Two possibilities emerge with respect to the present proportion of Dark Matter:

1. The percentage has not changed from the Beginning and the approximate estimates are

[169] The material in this chapter appeared in Blaha (2016a).

slightly off. The lack of change could be due to the extremely small decay rates of the fermions in the higher levels.

2. The percentage of matter in the upper layers has decreased due to decay and so the current proportion may be somewhat below 83.33%.

17.3 Penetrating the veil of the Big Bang

Studies of the Big Bang are hampered by infinities that often appear at t = 0. There appear to be two possible initial states for the expansion of our universe. One possibility is that the universe existed in a metastable state for some time before expansion began. The other possibility is instantaneous creation. We will not consider this possibility although it could support the Equipartition Principle. We have previously suggested a model (that includes inflation) that has an extremely short-lived initial metastable state.[12]

If there was an initial period of quasi-stability then it is possible that a state of equilibrium was achieved and the fermion Equipartition Principle applied. Then the estimate of the proportion of Dark matter would be correct.

Today experiments at CERN and other accelerators are creating quark-gluon plasma states that approximate the conditions near the Big Bang point. It would be interesting if it were possible to determine the proportions of all fermions generated in an ion-ion collision – both leptons and quarks – to see if the numbers of each type of produced fermion are approximately equal. If a rough equality were found then there would be significant support for the fermion Equipartition Principle.

If the Equipartition Principle holds at the time of the Big Bang then modeling of the Big Bang state would be constrained to be more realistic and we will have begun to penetrate the veil of the Big Bang.

18. Higgs Mechanism in the ElectroWeak Sector

This chapter[170] presents the Higgs Mechanism in the ElectroWeak Sector. The discussion is presented using the standard approach to the Higgs Mechanism because of its familiarity. Later we will 'separate' the ElectroWeak formulation into Electromagnetic and Weak interaction sectors for good reason.

In chapter 19 we present a rationale for the existence of Higgs particles for all interactions except color SU(3). Subsequently we present a new formulation of Higgs particles using Pseudoquantum field theory. It produces the same results as the conventional approach. But it offers a superior form of quantization, and it makes significant predictions not present in the usual Higgs theory.

18.1 ElectroWeak Higgs Mechanism Generation of ElectroWeak Gauge Field Masses

We require that there is one massless field, the electromagnetic field coupled to electric charge and three massive fields that receive masses via the Higgs Mechanism which breaks SU(2)⊗U(1) symmetry. *The Dark sector is assumed to be analogous to the normal particle sector in this respect since it has a SU(2)⊗U(1) symmetry in our Extended Standard Model.* The massive fields become the short range Weak interactions.

We assume that a doublet Higgs field exists with two components:

$$\eta = \begin{bmatrix} \varphi_+ \\ \varphi_0 \end{bmatrix} \tag{18.1}$$

with conjugate Higgs doublet

$$\eta' = \begin{bmatrix} \varphi_0 \\ -\varphi_- \end{bmatrix} \tag{18.2}$$

The Higgs sector lagrangian has the form:

[170] Most of this chapter appears in Blaha (2015a) and earlier books by the author.

$$\mathcal{L}_{EW}{}^{Higgs} = (\partial\eta^\dagger/\partial X^\mu)(\partial\eta/\partial X^\mu) - \lambda(\eta^\dagger\eta - \rho^2)^2 + \mathcal{L}_{EW}{}^{Higgs}{}_{EWMasses} \quad (18.3)$$

where the symmetry breaking follows from the choice of unitary gauge

$$\rho = \begin{bmatrix} 0 \\ \rho \end{bmatrix} \quad (18.4)$$

where ρ is a real field. Then the covariant derivative of η is

$$D_{\ldots\mu}\eta = \{\partial/\partial X^\mu + \ldots + ig\mathbf{t}\cdot\mathbf{W}_\mu + + ig't_0W_{0\mu}\} \begin{bmatrix} 0 \\ \rho \end{bmatrix} \quad (18.5)$$

where g and g' are coupling constants, t_0.is a ½ the identity matrix, and the **t** matrices are ½ the vector of Pauli matrices.The ellipses indicate additional indices and additional terms respectively.

Then

$$D_{\ldots\mu}\eta = \begin{bmatrix} \tfrac{1}{2}ig\rho(W_{1\mu} - iW_{2\mu}) \\ \partial\rho/\partial X^\mu - \tfrac{1}{2}ig\rho(\cos\theta_W)^{-1}Z_\mu \end{bmatrix} \quad (18.6)$$

where θ_W is the Weinberg angle and

$$W_3{}^\mu = Z^\mu\cos\theta_W + A^\mu\sin\theta_W$$
$$W_0{}^\mu = -Z^\mu\sin\theta_W + A^\mu\cos\theta_W \quad (18.7)$$

From eq. 18.6 we find the corresponding Higgs field kinetic terms in the lagrangian are

$$(D_{\ldots\mu}\eta)^\dagger D_{\ldots}{}^\mu\eta = \partial\rho/\partial X^\mu\partial\rho/\partial X_\mu + g^2\rho^2[W_1{}^\mu W_{1\mu} + W_2{}^\mu W_{2\mu}]/4 + g^2\rho^2 Z^\mu Z_\mu/(2\cos\theta_W)^2 \quad (18.8)$$

with the $W_1{}^\mu$ and $W_2{}^\mu$ and Z^μ gauge bosons acquiring masses and the electromagnetic field A^μ massless.

18.2 ElectroWeak Higgs Mechanism Generation of Fermion Masses

We now consider the ElectroWeak Higgs Mechanism for the eight species of fermions (four species of "normal" matter (counting three normal matter colored quarks as one species), and the four species of Dark Matter, which have quark color singlets). We shall first consider the *mass terms for the four normal species* which is the same as that of the four Dark species except for the values in the various species mass matrices. We define the *normal* matter 4-vector:

$$\Psi_s = \begin{bmatrix} \psi_{11} \\ \psi_{12} \\ \psi_{13} \\ \psi_{14} \\ \cdots \\ \psi_{41} \\ \psi_{42} \\ \psi_{43} \\ \psi_{44} \end{bmatrix} \tag{18.9}$$

where ψ_{ki} is the generation index for the ith generation of the kth species: ψ_{k1} is the wave function for the 1st generation, .. , ψ_{k4} is the 4th generation member of the kth species. We omit other indices in the interests of clarity. The normal fermion species are ordered: charged lepton (k = 1), up-type quark, neutral lepton, and down-type quark (k = 4). A 4th generation fermion of any species is yet to be found experimentally. Nevertheless we will assume it exists due to our earlier arguments for the U(4) Generation group.

We now assume that a "double" doublet Higgs field (with a conjugate doublet) *exists:*

$$\eta = \begin{bmatrix} \varphi_{1+} \\ \varphi_{10} \\ \varphi_{2+} \\ \varphi_{20} \end{bmatrix} \tag{18.10}$$

$$\eta' = \begin{bmatrix} \varphi_{10} \\ -\varphi_{1-} \\ \varphi_{20} \\ -\varphi_{2-} \end{bmatrix} \tag{18.11}$$

The Higgs sector lagrangian then has the form:

$$\mathcal{L}_{EW}{}^{Higgs} = (\partial\eta^\dagger/\partial X^\mu)(\partial\eta/\partial X^\mu) - \lambda(\eta^\dagger\eta - \rho^2)^2 + \mathcal{L}_{EW}{}^{Higgs}{}_{FermionMasses} \tag{18.12}$$

where the symmetry breaking follows from the choice of unitary gauge (similar in form to eq. 13.37) vacuum expectation values:

$$\rho = \begin{bmatrix} 0 \\ \rho_1 \\ 0 \\ \rho_2 \end{bmatrix} \qquad \rho' = \begin{bmatrix} \rho_1 \\ 0 \\ \rho_2 \\ 0 \end{bmatrix} \tag{18.13}$$

and where ρ is a real field quadruplet.

The lagrangian density mass term for the four normal fermion species is

$$\mathcal{L}_{EW}{}^{Higgs}{}_{FermionMasses} = \Sigma_{\alpha,\beta}\,\{\bar\psi_{kL\alpha}\eta m_{k\alpha\beta}\,\psi_{kR\beta} + \bar\psi_{kL\alpha}\eta' m'_{k\alpha\beta}\psi_{kR\beta}\} + \text{c.c.} \tag{18.14}$$

where $m_{k\alpha\beta}$ and $m'_{k\alpha\beta}$ are complex constant matrices, where α, β = 1, ... , 4, and where the second term is the double conjugation doublet used to produce a total mass term invariant under weak hypercharge. The total fermion lagrangian mass terms are

$$\mathcal{L}^{Higgs}{}_{FermionMasses} = \mathcal{L}_G{}^{GHiggs}{}_{FermionMasses} + \mathcal{L}_{EW}{}^{Higgs}{}_{FermionMasses} \tag{13.48}$$

$\mathcal{L}_{EW}{}^{Higgs}{}_{FermionMasses}$ is the contribution of ElectroWeak Higgs Mechanism to the fermion masses. $\mathcal{L}_G{}^{GHiggs}{}_{FermionMasses}$ is the Generation group Higgs Mechanism contribution to particle masses. Using the vacuum expectation values from eq. 18.13 we find

$$\begin{aligned}\mathcal{L}_{EW}{}^{Higgs}{}_{FermionMasses} = \Sigma_{\alpha,\beta}\,\{&\bar\psi_{2L\alpha}\,\rho_1 m_{2\alpha\beta}\psi_{2R\beta} + \bar\psi_{4L\alpha}\,\rho_2 m_{4\alpha\beta}\psi_{4R\beta} + \\ &+ \bar\psi_{1L\alpha}\,\rho_1 m'_{1\alpha\beta}\psi_{1R\beta} + \bar\psi_{3L\alpha}\,\rho_2 m'_{3\alpha\beta}\psi_{3R\beta}\} + \text{c.c.}\end{aligned} \tag{18.15}$$

giving mass terms to all four species. **There is an implicit color summation over the color quarks in each generation and quark species.** The values of ρ_1 and ρ_2 are unlikely to be the same as those appearing in the Generation group mass terms of chapter 13.

The four mass matrices m_1 , ... , m_4 are all complex, constant mass matrices. They can be brought to diagonal form D_k with non-negative values by U(4) matrices A_k and B_k:

$$A_k m_k B_k^{-1} = D_k \tag{18.16}$$

or

$$m_k = A_k^{-1} D_k B_k \tag{18.17}$$

for k= 1, …, 4.

We now note, that although, D_k has non-negative real values, down-type quarks are all tachyonic and up-type quarks are all non-tachyonic. Neutral leptons are all tachyonic and charged leptons are all Dirac non-tachyonic leptons, due to their lagrangian kinetic terms as seen earlier.

We further note that $m_k^\dagger m_k$ is hermitean, and A_k and B_k are members of U(4) as is D_k for k = 1, 2, 3, 4, with the result that all matrices m_k are members of a U(4) group.[171]

We can use these U(4) transformations A_k and B_k to define the sixteen "physical" fermion fields:

$$\begin{aligned}
&\bar\psi_{2L\alpha}\,\rho_1 m_{2\alpha\beta}\psi_{2R\beta} + \bar\psi_{4L\alpha}\,\rho_2 m_{4\alpha\beta}\psi_{4R\beta} + \bar\psi_{1L\alpha}\,\rho_1 m_{1\alpha\beta}\psi_{1R\beta} + \bar\psi_{3L\alpha}\,\rho_2 m_{3\alpha\beta}\psi_{3R\beta} \\
&= (\bar\psi_{2L}A_2^{-1})_\alpha \rho_1 D_{2\alpha\beta}(B_2\psi_{2R})_\beta + (\bar\psi_{4L}A_4^{-1})_\alpha\, \rho_2 D_{4\alpha\beta}(B_4\psi_{4R})_\beta + \\
&\qquad + (\bar\psi_{1L}A_1^{-1})_\alpha\, \rho_1 D_{1\alpha\beta}(B_1\psi_{1R})_\beta + (\bar\psi_{3L}A_3^{-1})_\alpha \rho_2 D_{3\alpha\beta}(B_3\psi_{3R})_\beta
\end{aligned} \tag{18.18}$$

$$= \bar\psi_{2Lphys\alpha}\rho_1 D_{2\alpha\beta}\psi_{2Rphys\beta} + \bar\psi_{4Lphs\alpha}\rho_2 D_{4\alpha\beta}\psi_{4Rphys\beta} + \bar\psi_{1Lphys\alpha}\rho_1 D_{1\alpha\beta}\psi_{1Rphys\beta} + \bar\psi_{3Lphs\alpha}\rho_2 D_{3\alpha\beta}\psi_{3Rphys\beta}$$

Species: Up-type quarks down-type quarks charged leptons neutral leptons

18.2.1 Dark Matter Masses

The preceding discussion with changes in the values of constants and constant matrices holds for Dark Matter also where the Dark quarks and leptons acquire Dark ElectroWeak mass terms. The Dark Matter species Dark ElectroWeak mass terms, with the subscript D signifying Dark Matter, are[172]

$$\bar\psi_{D2Lphys\alpha}\rho_{D1}D_{D2\alpha\beta}\psi_{D2Rphys\beta} + \bar\psi_{D4Lphs\alpha}\rho_{D2}D_{D4\alpha\beta}\psi_{D4Rphys\beta} + \bar\psi_{D1Lphys\alpha}\rho_{D1}D_{D1\alpha\beta}\psi_{D1Rphys\beta} + \bar\psi_{D3Lphs\alpha}\rho_{D2}D_{D3\alpha\beta}\psi_{D3Rphys\beta} \tag{18.19}$$

Dark
Species: Up-type quarks down-type quarks charged leptons neutral leptons

---

[171] This group is not the Generation group.
[172] Dark quarks are Color SU(3) singlets.

18.3 Combined Effect of Generation Group and ElectroWeak Higgs Mechanisms on Fermion Masses

The Generation group Higgs Mechanism and the ElectroWeak Higgs Mechanism combine to give masses to the fermion species and the Dark fermion species. The combined mass terms for normal fermions is

$$\overline{\psi}_{2Lphys\alpha}(\rho_{1G}D_{G2\alpha\beta} + \rho_1 D_{2\alpha\beta})\psi_{2Rphys\beta} + \overline{\psi}_{4Lphs\alpha}(\rho_{2G}D_{G4\alpha\beta} + \rho_2 D_{4\alpha\beta})\psi_{4Rphys\beta} + \overline{\psi}_{1Lphys\alpha}\rho_1 D_{1\alpha\beta}\psi_{1Rphys\beta} + \\ + \overline{\psi}_{3Lphs\alpha}\rho_2 D_{3\alpha\beta}\psi_{3Rphys\beta} \tag{18.20}$$

(where 'G' denotes a Generation group quantity) showing that the quark species acquire Higgs contributions from both the Generation group and ElectroWeak group symmetry breaking. A similar situation occurs for Dark Matter. Its combined mass terms have the same form as eq. 18.20 with a subscript "D" inserted to indicate it is for Dark Matter.

There is one possible source of ambiguity. The order of the rotations of the generation 4-vectors of the two quark species and the two Dark quark species to produce the 4-vectors of physical quark species due to Generation and ElectroWeak symmetry breaking is an issue. For example in eq. 13.57 we see Generation symmetry breaking necessitates the rotation $\psi_{G2Rphys} = D_{G2}\psi_{G2R}$ to diagonalize the up-type quark 4-vector while in eq. 18.18 we see that ElectroWeak symmetry breaking necessitates the rotation $\psi_{EW2Rphys} = B_{EW2}\psi_{EW2R}$ to diagonalize the up-type quark 4-vector.[173] *In principle we should diagonalize the sum of the ElectroWeak and Generation group mass matrices.*

However because the mass matrices of each group are quite different in the values of their elements we can *approximately* choose an order of rotations since they do not commute. We choose to do Generation group rotations first followed by ElectroWeak symmetry breaking rotations because the Generation group is more primary due to its role as the origin of the generations of the various species. Thus

$$\psi_{2Rphys} \cong D_{EW2}D_{G2}\psi_{G2R} \tag{18.21}$$

with a corresponding expression for the Left-handed species Generation 4-vector.

This issue does not arise for lepton and Dark lepton species as their 4-vectors only undergo ElectroWeak symmetry breaking rotations.

---

[173] We introduce the subscripts 'G' and 'EW' to distinguish the Generation group rotation from the ElectroWeak rotation.

The discussion of this section does not include Layer group symmetry breaking which will also cause a mixing of fermions between the four layers together with fermion mass contributions.The form of the extension to Layer group effects is similar to that of the above.

18.4 Generalization of the ElectroWeak Higgs Mechanism for Gauge Field Masses to Include Dark Gauge Fields

In section 18.2 we introduced a double doublet (a quadruplet) to derive the symmetry breaking fermion mass spectrum of the four normal fermion families and saw that a similar derivation should be operative for Dark fermions since it is the simplest possible approach to handling the broken SU(2)⊗U(1) symmetry that the Dark ElectroWeak sector has in the Extended Standard Model.

Now we generalize the SU(2)⊗U(1) ElectroWeak symmetry breaking via the Higgs Mechanism that gives mass to three SU(2)⊗U(1) gauge bosons to include Dark SU(2)⊗U(1) symmetry breaking via the Higgs Mechanism that will give mass to three Dark SU(2)⊗U(1) gauge bosons and also yield a massless gauge boson analogous to the electromagnetic field. This assumption is the simplest choice within the context of the Extended Standard Model and achieves a maximal effect with a minimal extension of the formalism. We will therefore be constructing an SU(2)⊗U(1)⊗SU(2)⊗U(1) Higgs Mechanism symmetry breaking derivation using the double doublets of section 18.2.

We assume that a "double" doublet Higgs field exists with four components:

$$\eta = \begin{bmatrix} \varphi_{1+} \\ \varphi_{10} \\ \varphi_{2+} \\ \varphi_{20} \end{bmatrix} \tag{18.22}$$

with conjugate Higgs doublet

$$\eta' = \begin{bmatrix} \varphi_{10} \\ -\varphi_{1-} \\ \varphi_{20} \\ -\varphi_{2-} \end{bmatrix} \tag{18.23}$$

The ElectroWeak Higgs sector lagrangian has the form:

$$\mathcal{L}_{EW}{}^{Higgs} = (\partial\eta^\dagger/\partial X^\mu)(\partial\eta/\partial X^\mu) - \lambda(\eta^\dagger\eta - \rho^2)^2 + \mathcal{L}_{EW}{}^{Higgs}{}_{EWMasses} \tag{18.24}$$

where the symmetry breaking follows from the choice of unitary gauge (similar in form to eq. 18.13)

$$\rho = \begin{bmatrix} 0 \\ \rho_1 \\ 0 \\ \rho_2 \end{bmatrix} \qquad \rho' = \begin{bmatrix} \rho_1 \\ 0 \\ \rho_2 \\ 0 \end{bmatrix} \tag{18.25}$$

where ρ is a real field quadruplet.

The covariant derivative of η in the unitary gauge is

$$D_{\ldots\mu}\,\eta = \begin{bmatrix} \{\partial/\partial X^\mu + \ldots + ig\mathbf{t}\cdot\mathbf{W}_\mu + + ig' t_0 W_{0\mu}\} & 0 \\ 0 & \{\partial/\partial X^\mu + \ldots + ig_D\mathbf{t}\cdot\mathbf{W}_{D\mu} + + ig_D' t_0 W_{D0\mu}\} \end{bmatrix} \begin{bmatrix} 0 \\ \rho_1 \\ 0 \\ \rho_2 \end{bmatrix} \tag{18.26}$$

where g and g' are coupling constants with g_D and g_D' their Dark equivalents, t_0.is a ½ the identity matrix, the **t** matrices are ½ the vector of Pauli matrices, and the zeros and derivative terms are all 2×2 submatrices.The ellipses indicate additional indices and additional terms respectively.

Then

$$D_{\ldots\mu}\,\eta = \begin{bmatrix} \tfrac{1}{2} ig\rho_1(W_{1\mu} - iW_{2\mu}) \\ \partial\rho_1/\partial X^\mu - \tfrac{1}{2} ig\rho_1(\cos\theta_W)^{-1} Z_\mu \\ \tfrac{1}{2} ig_D\rho_2(W_{D1\mu} - iW_{D2\mu}) \\ \partial\rho_2/\partial X^\mu - \tfrac{1}{2} ig_D\rho_2(\cos\theta_{WD})^{-1} Z_{D\mu} \end{bmatrix} \tag{18.27}$$

where θ_{WD} is the Dark Weinberg angle and

$$\begin{aligned} W_3^{\mu} &= Z^\mu\cos\theta_W + A^\mu\sin\theta_W \\ W_0^{\mu} &= -Z^\mu \sin\theta_W + A^\mu\cos\theta_W \\ W_{D3}^{\mu} &= Z_D^{\mu}\cos\theta_{WD} + A_D^{\mu}\sin\theta_{WD} \\ W_{D0}^{\mu} &= -Z_D^{\mu} \sin\theta_{WD} + A_D^{\mu}\cos\theta_{WD} \end{aligned} \tag{18.28}$$

From eq. 18.27 we find the corresponding Higgs field kinetic terms in the lagrangian are

$$(D_{\ldots\mu}\eta)^\dagger D_{\ldots}^{\ \mu}\eta = \partial\rho_1/\partial X^\mu \partial\rho_1/\partial X_\mu + g^2\rho_1^{\ 2}[W_1^{\ \mu}W_{1\mu} + W_2^{\ \mu}W_{2\mu}]/4 + g^2\rho_1^{\ 2}\, Z^\mu Z_\mu/(2\cos\theta_W)^2 + $$
$$+ \partial\rho_2/\partial X^\mu \partial\rho_2/\partial X_\mu + g_D^{\ 2}\rho_2^{\ 2}[W_{D1}^{\ \ \mu}\, W_{D1\mu} + W_{D2}^{\ \ \mu}W_{D2\mu}]/4 + g_D^{\ 2}\rho_2^{\ 2}\, Z_D^{\ \mu}Z_{D\mu}/(2\cos\theta_{WD})^2 \qquad (18.29)$$

with the $W_1^{\ \mu}$ and $W_2^{\ \mu}$ and Z^μ gauge bosons acquiring masses, and the electromagnetic field A^μ massless, and with the Dark $W_{D1}^{\ \ \mu}$ and $W_{D2}^{\ \ \mu}$ and $Z_D^{\ \mu}$ gauge bosons acquiring masses, and the Dark electromagnetic field $A_D^{\ \mu}$ massless.

Thus the SU(2)⊗U(1)⊗SU(2)⊗U(1) Higgs Mechanism symmetry breaking yields the desired results. Chapter 14 discusses Dark Matter Chemistry. Its basis is consistent with a massless Dark electromagnetic field.

18.5 The Reality Group Compared to the Generations Group U(4)

It appears to be a somewhat remarkable coincidence that the Reality group, which maps complex coordinates to real-valued coordinates, consists of the tensor product of U(4) subgroups, and the generations group consists of (broken) U(4). However the "fourness" of the groups can be traced back to the dimension of space-time, and the consequent four boosts that generate the four families of fermions: charged leptons, neutral leptons, up-type quarks and down-type quarks and the corresponding four Dark fermion families, which in turn yield four conserved particle numbers B, L, B_D, and L_D that can undergo U(4) transformations and yield conserved linear combinations.

Thus the dimensionality of complex space-time *ultimately* is the origin of the "fourness" of the Reality group and the Generations group.

18.6 Completion of Construction/Derivation of the Form of The Extended Standard Model and Quantum Gravity - Unification

This concludes the first stage of our construction/derivation of the form of the Extended Standard Model. Its lies primarily in the assumption of a complex-valued space-time that is mapped to the physical reality of real space-time coordinates by the Reality group, to multi-generation fermions by the Generation group, to fermion layers by the Layer group, and to a Two-Tier quantum field thery formulation that eliminates infinities in calculations. The basis of the form of The Extended Standard Model in complex space-time is clear. The basis of General Relativity also is complex space-time geometry as we will see later.

Our construction/derivation does not lead to the determination of the values of constants[174] that appear in The Extended Standard Model: the many coupling constants and masses; and does not determine the value of the gravitational constant G. These issues remain to be resoved.

18.7 The Higgs Mechanism is Not Needed in Pseudoquantum Field Theory

We note that our approach to the Extended Standard Model using Two-Tier field theory (described in detail in appendices A and B) makes the use of the Higgs Mechanism unnecessary to achieve renormalizability. However, as we shall see in the next chapter scalar bosons do appear naturally in our formulation. These bosons may well be the scalar particle(s) found at the CERN LHC.

---

[174] Later we will suggest that a Higgs Mechanism may exist that determines the 'arbitrary' constants appearing in the Extended Standard Model. This possibility may give a new purpose to the Higgs Mechanism.

19. The Genesis of Scalar (Higgs?) Particle Fields from Complex Gauge Fields

This chapter shows that scalar particles can be 'extracted' from all spin 1 gauge fields except for color SU(3).[175] We began with the derivation of four species of fermions using Complex Lorentz group transformations. We found the up-type and down-type quark species each had three subspecies due to color SU(3) symmetry. We then found that the Reality group was necessary to map complex coordinates to real0valued coordinates. This group was the source of the SU(3)⊗SU(2)⊗U(1)⊗SU(2)⊗U(1) symmetry group of the Extended Standard Model which included normal and Dark matter. Consideration of the four number conservation laws (baryon number, lepton number, and their Dark analogues) led us to a U(4) group that we called the Generation group. We attributed four fermion generations to each species putting each species in the fundamental representation 4 of a local U(4) Generation group. Further consideration of the generations of each species suggested a broken U(4) local Layer group with each fermion in the 4 of the Layer group. Both the Generation group and the Layer group had spin 1 gauge bosons generating interactions between fermions. The Extended Standard Model symmetry group then became SU(3)⊗SU(2)⊗U(1)⊗SU(2)⊗U(1)⊗U(4)⊗U(4). When we consider Complex General Coordinate transformations later we will find that another U(4) gauge group based on the gauge field $\mathbf{A}_S{}^{\mu}(x)\cdot\mathbf{G}_S$ appears. As a result the complete gauge symmetry of the Extended Standard Model is

$$SU(3)\otimes SU(2)\otimes U(1)\otimes SU(2)\otimes U(1)\otimes U(4)\otimes U(4)\otimes U(4) \qquad (19.1)$$

Since our Extended Standard Model is ultimately based on the Complex General Coordinate Transformations and the Complex Lorentz group (thus complex-valued coordinate systems), it appears reasonable to consider all spin 1 gauge fields to be initially similarly complex-valued. Most gauge fields can be rotated to real values. However we shall see that color SU(3) gauge fields are *necessarily* complex-valued. All other gauge fields can be rotated to real values. The price of rotation is the introduction of scalar fields. Some of these fields may be Higgs particle fields and generate gauge boson masses (symmetry breaking) and fermion masses.

Thus we view scalar particles including Higgs particles as inherently associated with gauge fields.

---

[175] This chapter first appeared in Blaha (2015c) and (2016c).

19.1 The Difference between the Strong Gauge Field and the Other Gauge Fields in the Extended Standard Model

In our Extended Standard Model the only gauge field without an associated Higgs particle is the strong interaction gluon gauge field. *We view this exception as a particularly important clue as to the nature of the relation between gauge fields and Higgs particles.*[176]

How does the strong interaction gauge field differ from all other gauge fields in the Extended Standard Model and our Theory of Everything? An examination of the gauge fields dynamic equations (and other lagrangian terms) of our Extended Standard Model reveals that all gauge field dynamic equation kinetic terms *except the strong interaction gauge field* have the form:

$$\partial/\partial x_\mu F^a_{\ \mu\nu} + gf^{abc} A^{b\mu} F^c_{\ \mu\nu} = j^a_{\ \nu} \tag{19.1}$$

where

$$F^a_{\ \mu\nu} = \partial/\partial x^\nu A^a_{\ \mu} - \partial/\partial x^\mu A^a_{\ \nu} + gf^{abc} A^b_{\ \mu} A^c_{\ \nu} \tag{19.2}$$

where the coordinates x^ν *are real-valued*, where a, b, c are structure constant indices, where g is a coupling constant, and where $j^a_{\ \nu}$ is the corresponding current. The gauge field $A^a_{\ \mu}$ is real for ElectroWeak gauge fields, Generation group gauge fields, and Layer group gauge fields. Thus eqns. 19.1 and 19.2 are real-valued.

The strong interaction gauge field[177] in our Extended Standard Model differs from the other gauge fields by being *necessarily* complex[178] due to the complex 3-space complexon derivatives that appear in the corresponding equations:

$$D^\mu F_C{}^a_{\ \mu\nu} + gf^{abc} A_C{}^{b\mu} F_C{}^c_{\ \mu\nu} = j^a_{\ \nu} \tag{19.3a}$$

with

$$F_C{}^a_{\ \mu\nu} = D_\nu A_C{}^a_{\ \mu} - D_\mu A_C{}^a_{\ \nu} + gf^{abc} A_C{}^b_{\ \mu} A_C{}^c_{\ \nu} \tag{19.3b}$$

where

$$D_k = \partial/\partial x_r^{\ k} + i\, \partial/\partial x_i^{\ k} \tag{19.4}$$
$$D_0 = \partial/\partial x^0$$

for k = 1, 2, 3 where $A_C{}^a_{\ \mu}$ is the complexon color Strong interaction gauge field. The complexon spatial coordinates have the form $x_r^{\ k} + i\, x_i$. The time coordinate is real-valued. These equations

---

[176] Most of the material in this chapter appeared in Blaha (2015c).
[177] This field is called a complexon gauge field in Blaha (2015a) and earlier books.
[178] One cannot cleanly separate the real and imaginary parts of its dynamic equations.

are eqs. 10.16 and 5.162 of Blaha (2015a) for complexon gauge fields,[179] which carry the strong interaction in the Extended Standard Model.

This difference enables us to differentiate the strong gauge field from all other gauge fields in The Extended Standard Model. Thereby we can develop a unified formalism for the non-strong gauge fields and their corresponding Higgs particles.

The necessarily complex nature of the colr SU(3) field is the reason that the Strong Interaction gauge fields do not acquire a mass via the Higgs Mechanism. As shown below, the necessary complexity of Strong Interaction gauge fields precludes the generation of Higgs fields from Yang-Mills gauge fields.

19.2 Generation of Higgs Fields From Non-Abelian Gauge Fields

In the prior section we considered the difference between the strong gauge field and the other gauge fields of The Extended Standard Model. Unlike strong gauge fields the other gauge fields (ElectroWeak and so on) could be real or complex. In a manner similar to what we did in the preceding *Physics is Logic* books (and earlier books) we can assume the gauge fields are initially complex, and then transform them to real-valued fields using a phase transformation that introduces scalar fields, some of which we will take to be Higgs fields.

We define a complex phase transformation for a gauge field $A^{b\mu}$ with

$$A'^{a\mu}(x) = \Phi(x)^a{}_b A^{b\mu}(x) \qquad (19.5)$$

where $\Phi(x) = \text{diag}(\exp[i\varphi_1(x)], \exp[i\varphi_2(x)], \ldots, \exp[i\varphi_n(x)])$, and n is the number of symmetry components of $A^{b\mu}$. Inserting $A'^{a\mu}(x)$ in eq. 19.1 we find that eq. 19.1 becomes:

$$\partial/\partial x_\mu F'^{a}{}_{\mu\nu} + gf^{abc}A'^{b\mu} F'^{c}{}_{\mu\nu} = j^a{}_\nu \qquad (19.6)$$

where

$$F'^{a}{}_{\mu\nu} = \partial/\partial x^\nu\{\exp[i\varphi_a(x)]A^a{}_\mu\} - \partial/\partial x^\mu\{\exp[i\varphi_a(x)]A^a{}_\nu\} + gf^{abc}\exp[i\varphi_b(x)]\, A^b{}_\mu \exp[i\varphi_c(x)]A^c{}_\nu \qquad (19.7)$$

If we now assume that $\varphi_a(x)$ is small for all a then

$$\exp[i\varphi_a(x)] \simeq 1 + i\varphi_a(x) \qquad (19.8)$$

[179] In The Extended Standard Model we also identify quark species particles as having complex 3-momentum. We call them complexon fermions.

to first order. Substituting in eqs. 19.6 and 19.7, and keeping terms to leading order yields the real part:

$$\partial/\partial x_\mu F^a_{\ \mu\nu} + gf^{abc}A^{b\mu} F^c_{\ \mu\nu} = j^a_{\ \nu} \tag{19.9}$$

where $F^a_{\ \mu\nu}$ is given by eq. 19.2, and the imaginary part is:

$$\partial/\partial x_\mu F_i{}^a_{\ \mu\nu} + gf^{abc}A^{b\mu} F_i{}^c_{\ \mu\nu} = 0 \tag{19.10}$$

to leading order where

$$F_i{}^a_{\ \mu\nu} = \partial/\partial x^\nu \varphi_a(x)A^a_{\ \mu} - \partial/\partial x^\mu \varphi_a(x)A^a_{\ \nu} \tag{19.11}$$

Substituting eq. 19.11 in eq. 19.10 we find

$$A^a_{\ \nu}\Box\varphi_a(x) - A^a_{\ \mu}\, \partial/\partial x_\mu \partial/\partial x^\nu\, \varphi_a(x) - gf^{abc}A^{b\mu}\, [A^c_{\ \mu}\, \partial/\partial x^\nu\, \varphi_a(x) - A^c_{\ \nu}\, \partial/\partial x^\mu\, \varphi_a(x)] = 0 \tag{19.12}$$

in the Landau gauge, with no sum over a. Eq. 19.12 is a form of Klein-Gordon equation having interaction terms with the gauge field. If the gauge field is weak then only the first two terms are important.

Note that only derivatives of $\varphi_a(x)$ appear in eq. 19.12. Consequently shifts of the $\varphi_a(x)$ field by a constant still yield solutions of eq. 19.12. This feature makes $\varphi_a(x)$ a candidate to be a Higgs particle.

Note also that complexon gauge fields cannot have such a phase change, with a subdivision into real and imaginary dynamic equations, due to the complexity of the spatial coordinates. This difference appears to be the reason why the strong interaction gauge field does not have an associated Higgs particle.

The $\varphi_a(x)$ particles can be made into Higgs particles by adding an appropriate potential:

$$V = A\, \varphi_a^{\ 2}(x) + B\, \varphi_a^{\ 4}(x) \tag{19.13}$$

where A and B are constants. Approximating eq. 19.12 with its first two terms and inserting the potential term we find the Higgs-like equation:

$$A^a_{\ \nu}\Box\varphi_a(x) - A^a_{\ \mu}\, \partial/\partial x_\mu \partial/\partial x^\nu\, \varphi_a(x) + \partial V/\partial\varphi_a = 0 \tag{19.14}$$

$\varphi_a(x)$ has a minimum at the minimum of the potential in the corresponding lagrangian.

The second and third terms in eq. 19.14 constitute the interaction. Neglecting these terms we see that eq. 19.14 becomes the free, massless, field Klein-Gordon equation

$$\Box\varphi_a(x) = 0 \tag{19.15}$$

The pairing of Higgs particles with real-valued gauge fields is thus established.[180] The non-existence of a matching Higgs field for the strong interaction is due to the inherently complex nature of the strong interaction (complexon) gauge field in the Extended Standard Model also follows.

The derivation presented here is analogous to the derivation of Higgs fields in Complex General Relativity – also a gauge theory – in *Physics is Logic Part II.*

One of the remarkable aspects of The Extended Standard Model is its ability to directly prove qualitative properties of elementary particles: four fermion species, Parity violation, the distinction between leptons and quarks, the match of the Standard Models (broken) symmetries with the Reality group consisting of subgroups of U(4), and now the existence of Higgs gauge fields in the ElectroWeak sector but not for the strong interactions. We take these successes to be indicators of the correctness of The Extended Standard Model.

19.3 General Higgs Formulation of Gauge and Fermion Particle Masses

We have seen seven of the interactions present in our Extended Standard Model. Four more interactions will be presented later. One of them are the gauge fields $A_S{}^{\mu}$ generated from the Reality group of complex General Relativistic transformations. There are two more interactions associated with General Relativistic transformations and an all-encompassing interaction $A_\Omega{}^{\mu}(x)$. We shall consider the eight spin 1 gauge field interactions that appear in fermion covariant derivatives. They can be put in a vector form:[181]

$$\mathbf{A}_I{}^{\mu} = (g_1\mathbf{A}_{SU(3)}{}^{\mu}(x_C), g_2\mathbf{W}^{\mu}(x)\,, g_3\mathbf{A}_E{}^{\mu}(x), g_4\mathbf{W}_D{}^{\mu}(x), g_5\mathbf{A}_{DE}{}^{\mu}(x), g_6\mathbf{U}^{\mu}(x), g_7\mathbf{V}^{\mu}(x), g_8\mathbf{A}_S{}^{\mu}(x)) \tag{19.16}$$

where each element is a vector of the gauge fields in the group of the gauge field and the respective coupling constants are labeled $g_1, g_2, \ldots, g_8$. The subscript 'D' labels Dark matter interactions. 'W' labels Weak fields, 'E' labels Electromagnetic fields. $V^{\mu}(x)$ labels U(4)

---

[180] Some of the Higgs fields so generated may not have vacuum expectation values and so may only play a role in interactions.

[181] Later we will reformulate this discussion in terms of Pseudoquantum field theory.

Generation group fields., 'V' labels U(4) Layer group fields. A_S labels the U(4) General Relativistic transformations Reality field.

The interactions' symmetry is SU(3)⊗SU(2)⊗U(1)⊗SU(2)⊗U(1)⊗U(4)⊗U(4)⊗U(4) respectively. The number of fields for each of the elements of the vectors is 8. 3, 1, 3, 1, 16, 16, and 16 respectively – totaling 64 fields.

Similarly we define an 8-vector of 64 generators

$$\mathbf{T}_I = (\mathbf{T}_{SU(3)}, \boldsymbol{\tau}_{SU(2)}, \mathbf{I}_{U(1)}, \boldsymbol{\tau}_{DSU(2)}, \mathbf{I}_{DU(1)}, \mathbf{G}_{U(4)}, G_{LU(4)}, \mathbf{G}_S) \qquad (19.17)$$

Then the total gauge fields interaction term within a covariant derivative corresponding to the eight interactions, the spinor interaction, and the Ω-interaction can be expressed as

$$\mathbf{A}_I{}^\mu{}_k \mathbf{T}_{Ik} + g_B B^\mu + g_\Omega A_\Omega{}^\mu(x) \qquad (19.18)$$

summed separately over k for each interaction. The remaining additional interactions are real-valued gravitational connections that we will describe later. The covariant derivative of a fermion field is

$$\{\partial^\mu + i\,[\mathbf{A}_I{}^\mu + g_B B^\mu + g_\Omega A_\Omega{}^\mu(x)]\}\gamma_\mu \psi = 0 \qquad (19.19)$$

Note the complexon nature of the SU(3) gauge field makes us use the covariant derivative

$$\{\partial / \partial x_{C\mu} + i\,[\mathbf{A}_I{}^\mu + g_B B^\mu + g_\Omega A_\Omega{}^\mu(x)]\}\gamma^\mu \psi = 0 \qquad (19.19a)$$

for the SU(3) quark dynamic equations where the other gauge fields are functions of $x_r = \text{Re}\ x_C$.

We now consider the combined effects of the eight interactions, $\mathbf{A}_I{}^\mu$, on generating gauge boson masses (symmetry breaking) and fermion masses. We begin by defining a composite Higgs field for all 8 interactions:

$$\eta = \prod_{k=1}^{8} \eta_{kTSLg} \qquad (19.20)$$

where k labels the group, T labels the type of matter, S labels the species, L labels the layer and g labels the generation. We now consider

$$D^\mu \eta = \{\partial^\mu + i\,[\mathbf{A}_I{}^\mu + g_B B^\mu + g_\Omega A_\Omega{}^\mu(x)]\}\eta \qquad (19.21)$$

Letting η be a real field ρ whose elements are composed of zeroes and non-zero real fields

$$\eta = \prod_{k=1}^{8} \rho_{kTSLg} \tag{19.22}$$

Then we find that

$$(D_\mu\eta)^\dagger D^\mu\eta = \sum_{kTSLg} \partial_\mu\rho_{kTSLg}\, \partial^\mu\rho_{kTSLg} + \sum_{kTSL} g_k^{\,2}\beta_{kTSL}U_{kTSL}^{\;2} \tag{19.23}$$

where β_{kTSL} is a sum of terms, each of which is quadratic in the vacuum expectation value of a Higgs field. The second term above yields the masses of the gauge fields of non-zero mass.

The lagrangian terms that generate fermion masses have the form

$$\mathscr{L}_{FermionMasses} = \sum_{kTSLgh} \bar{\psi}_{LkTSLg}\, \rho_{0kTSL} m_{kTSLgh} \psi_{RkTSLh} \tag{19.24}$$

where the sums over g and h are over the generations of a specific T, S, and L of a group labeled k, and ρ_{0kTSL} is the vacuum expectation value of a Higgs particle. The initial 'L' and 'R' subscripts represent Left and Right.

The passages of chapter 13 with eqs 13.37 through 13.49, and of chapter 15 with eqs. 15.11 through 15.21 illustrate the above general Higgs procedure.

We note that the fermion mass matrices can be diagonalized using a matrix A_{TSL}:

$$m_{TSLphys} = A_{LTSL} \sum_k \rho_{0kTSL} m_{kTSL} A_{RTSL}^{\;\;-1} \tag{19.25}$$

where $m_{TSLphys}$ is the diagonal mass matrix for the generations specified by T, S, and L.[182]

Thus the lagrangian fermion mass terms for physical fermions become

$$\mathscr{L}_{FermionMasses} = \sum_{TSL} \bar{\psi}_{LphysTSL}\, m_{TSLphys} \psi_{RphysTSL} + \text{c.c.} \tag{19.26}$$

with diagonal mass matrices $m_{TSLphys}$.

19.4 The Mixing Pattern in the Fermion Periodic Table

The preceding discussions describe the pattern of mixing resulting from the ElectroWeak, Generation group, and Layer group. Fig. 19.1 pictorially presents an example of the mixing pattern within the Periodic Table of Fermions.

---

[182] This passage corrects the approximate discussion of section 18.3 concerning the interplay of the ElectroWeak and Generation group in determining the fermion mass matrices.

THE FERMION PERIODIC TABLE

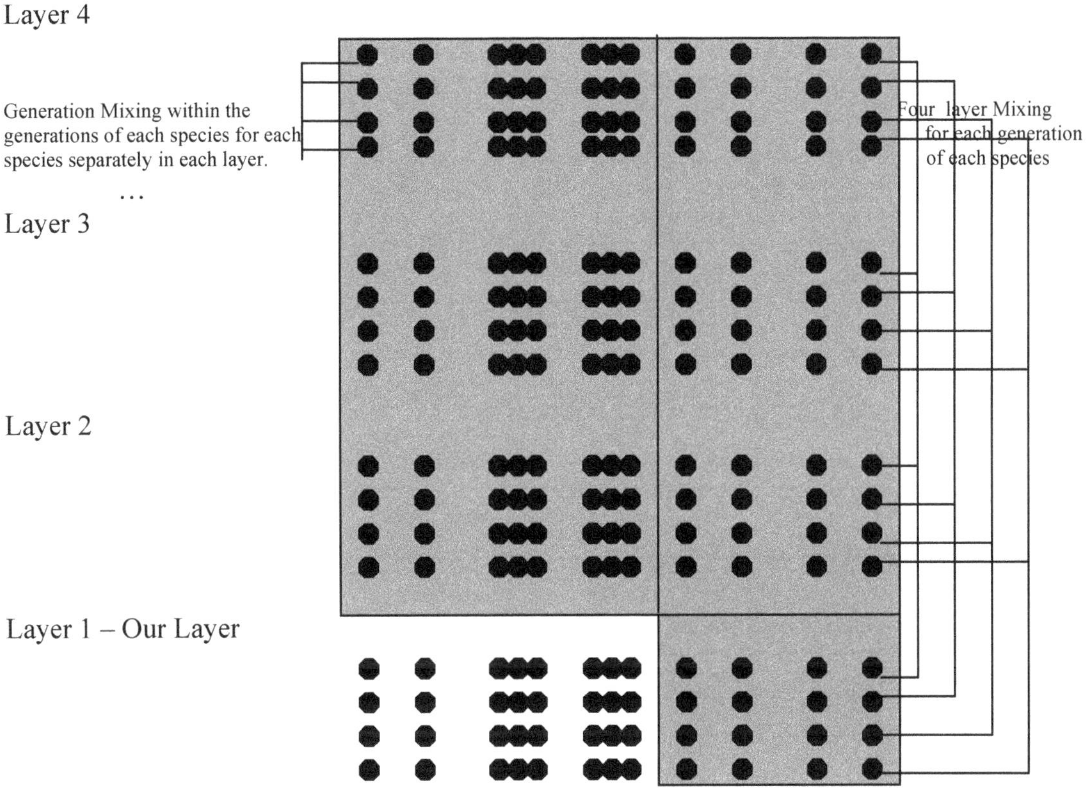

Figure 19.1. Partial example of pattern of mass mixing of the Generation group and of the Layer group. Dark parts of the periodic table are gray. Light parts are the known fermions with an additional, as yet not found, 4th generation shown. The lines on the left side show an example of the Generation mixing within one species. The Generation mixing applies to each species in each layer. The lines on the right side show an example of Layer mixing within one species with the mixing amongst all four layers of the species for each generation individually.

19.5 The Full Extended Standard Model Fermion Mass Matrices

Eqs. 19.24 – 19.26 lead to the total mass matrices for each of the four layers listed below. The masses for each particle are different although we use the same symbol for each type of mass. We include the mass contributions, m_{Gi}, from the General Relativistic Reality group with gauge fields A_S^{μ}. We will discuss this group, and its physical effects, in detail in chapter 22.

The below list is for one layer. The other three layers have a similar pattern. The masses listed in the list are symbolic and are not of the same value for each particle type. We denote them generally as: m_{Wi} for the Weak group contribution, m_{Li} for the Layer group contribution, m_{Geni} for the Generation group contribution, and m_{Gi} for the Gravitational Reality group contribution with gauge fields $\mathbf{A}_S^{\mu}$.

Charged Lepton Species Total Mass Matrix

$$m_{etot} = m_{We} + m_{Le} + m_{Ge}$$

Neutral Lepton Species Mass Matrix

$$m_{\upsilon tot} = m_{W\upsilon} + m_{L\nu} + m_{G\upsilon}$$

Up-Type Quark Species Mass Matrix (for each color)

$$m_{utot} = m_{Wu} + m_{Lu} + m_{Genu} + m_{Gu}$$

Down-Type Quark Species Mass Matrix (for each color)

$$m_{dtot} = m_{Wd} + m_{Ld} + m_{Gend} + m_{Gd}$$

Dark Charged Lepton Species Total Mass Matrix

$$m_{Detot} = m_{DWe} + m_{DLe} + m_{Ge}$$

Dark Neutral Lepton Species Mass Matrix

$$m_{D\upsilon tot} = m_{DW\upsilon} + m_{DL\upsilon} + m_{G\upsilon}$$

Dark Up-Type Quark Species Mass Matrix

$$m_{Dutot} = m_{DWu} + m_{DLu} + m_{DGenu} + m_{Gu}$$

Dark Down-Type Quark Species Mass Matrix

$$m_{Ddtot} = m_{DWd} + m_{DLd} + m_{DGend} + m_{Gd}$$

The gravitational contribution to each fermion mass, m_{Gi} for each fermion type i, sets the scale for all fermion masses (and secondarily of massive gauge bosons' masses) yielding the "principle" of Newton, Einstein and others that *inertial mass equals gravitational mass*. See chapter 22 for a discussion of the Higgs Mechanism for General Relativistic Reality group symmetry breaking.

NOTE: The generation group contributions, in the spontaneous breakdown that we described, appear only in quark and Dark quark mass matrices providing, possibly, a reason why quark masses are so much larger than lepton masses.

20. Pseudoquantum Field Theory Formalism for the Higgs Mechanism

20.1 The Enigma of Higgs Particles and the Higgs Mechanism

In our previous work on the Standard Model, and its generalization to The Extended Standard Model described in a series of books entitled *Physics is Logic ...*, we showed that the fermion spectrum results from Complex Special Relativity, the gauge interactions result from the Reality group, the fermion generations result from the Generation group, and the Theory of Everything results from a combination with Complex General Relativity. The Higgs particles and the Higgs Mechanism were inserted to generate particle masses and symmetry breaking effects.

Whence comes Higgs particles? There does not appear to be a more fundamental cause. And so the Higgs sector appears to be an expedient mechanism to insert much needed symmetry breaking and masses into the theory.

However, there are a number of peculiarities in the implementation of the Higgs Mechanism:

1. First, it is selective in the sense that some gauge fields have associated Higgs particles and utilize the Higgs Mechanism, and some gauge fields do not have associated Higgs particles. In particular, the ElectroWeak gauge fields, the Generation group gauge fields, the Layer group fields, and the complex gravity field have associated Higgs particles. The strong interaction (gluon) gauge fields do not.

2. The Higgs potentials have a quadratic mass term of the "wrong" sign plus a quartic interaction term, which together, generate non-zero vacuum expectation values. They obviously accomplish their goal. But the source of these potentials, and why they have the same form, is unknown. One expects a fundamental principle should be operative here.

3. One can imagine creating a Higgs microscope at some super-accelerator. Using this microscope in the presence of a (classical) condensate could enable the Uncertainty Principle to be violated. This possibility, in the case of a microscope

using electromagnetic fields, was the source of a heuristic argument for the need to quantize the electromagnetic field.[183]

4. The formulation of the Higgs Mechanism uses classical fields under the assumption that a path integral formulation justifies their use. While this may be true, the path integral formulation relies on implicit, unstated boundary conditions that obscure the physics of the quantum field theoretic nature of the mechanism. A direct quantum field theoretic study of the Higgs Mechanism is needed and would further elucidate its character.

5. Scalar fields have a cloud hanging over them that spin ½ fields do not. A spin ½ particle cannot transition to negative energy because there is a filled sea of negative energy particles. No additional particles can fall into the sea due to the Pauli Exclusion Principle that forbids two fermions with the same 4-momentum and quantum numbers. In the case of scalar particles the Pauli Exclusion Principle does not apply and so a *filled* negative energy sea of scalar particles is not possible and positive energy scalar particles can transition to negative energy without hindrance. This problem has been "resolved" by an appropriate definition of the scalar particle vacuum to exclude transitions to negative energy. But the rationale for the definition is lacking. Dirac was asked about this issue many years ago. He said he had a solution to the problem. However he did not present it – in keeping with his well-known taciturn nature. So the issue remains an open question.

For the above reasons we will show that a more satisfactory method of achieving the goals of mass generation and symmetry breaking exists.[184] This method relies on a larger Fock space that enables the appearance of a vacuum expectation value for Higgs particles to be understood within a truly quantum framework. One major consequence of this approach is the appearance of a local Arrow of Time – a concept that has been a subject of interest for over one hundred years. Another consequence is a rationale for ElectroWeak Higgs bosons and for their absence for the strong (gluon) interaction.

---

[183] Heitler (1954) p. 86 provides a good discussion of the need to quantize the electromagnetic field.

[184] In the Extended Standard Model of Blaha (2015a) we have shown that the basic iota particles have a mass, the Landauer mass, so that the theory is symmetry violating from the very start. We have also shown that our Two-Tier formalism (Appendices A and B) for quantum field theories always yields finite results in perturbation theory calculations – making the renormalization approach of t'Hooft and others, which relied on initially massless gauge fields, unnecessary.

20.2 Pseudoquantization of Scalar Particles

We will now consider the pseudoquantization of a scalar particle field that will become a Higgs particle with a non-zero vacuum expectation value.[185] We begin by defining two fields that correspond to the scalar particle: $\varphi_1(x)$ and $\varphi_2(x)$.[186] These fields will be assumed to have the equal time commutators

$$[\varphi_i(x), \pi_j(y)] = i(1 - \delta_{ij})\delta^3(\mathbf{x} - \mathbf{y}) \qquad (20.1)$$
$$[\varphi_i(x), \varphi_j(y)] = 0$$
$$[\pi_i(x), \pi_j(y)] = 0$$

where δ_{ij} is the Kronecker δ and where $\pi_i(x)$ is the canonically conjugate momentum to $\varphi_i(x)$. The fields $\varphi_1(x)$ and $\pi_1(y)$ will be observable classical fields. The fields $\varphi_2(x)$ and $\pi_2(y)$ will not be observables so that $\varphi_1(x)$ and $\pi_1(y)$ can both be sharp on the set of physical states.

We now specify the lagrangian density for a scalar Klein-Gordon particle:

$$\mathcal{L} = \partial\varphi_1/\partial x_\mu \partial\varphi_2/\partial x^\mu \qquad (20.2a)$$

with hamiltonian density

$$\mathcal{H} = \pi_1\, \pi_2 + \partial\varphi_1/\partial x_i \partial\varphi_2/\partial x^i \qquad (20.2b)$$

where i labels spatial coordinates, and $\pi_1 = \partial\varphi_2/\partial t$ and $\pi_2 = \partial\varphi_1/\partial t$. Eqs. 20.2 are without a potential or mass term.

The lagrangian and hamiltonian for a massive scalar particle are

$$\mathcal{L} = \partial\varphi_1/\partial x_\mu \partial\varphi_2/\partial x^\mu - m^2\, \varphi_1\varphi_2 \qquad (20.2c)$$

with hamiltonian density

$$\mathcal{H} = \pi_1\, \pi_2 + \partial\varphi_1/\partial x_i \partial\varphi_2/\partial x^i + m^2\, \varphi_1\varphi_2 \qquad (20.2d)$$

The fields can be fourier expanded in terms of creation and annihilation operators:

$$\varphi_i(\mathbf{x}, t) = \int d^3k\, [a_i(k)f_k(x) + a_i^\dagger(k)f_k^*(x)] \qquad (20.3)$$

for i = 1, 2 where

$$f_k(x) = e^{-ik\cdot x}/(2\omega_k(2\pi)^3)^{1/2}$$

with $\omega_k = |\mathbf{k}|$.

---

[185] Much of this chapter appears in Blaha (2016c), and earlier books, as well as in S. Blaha, Phys. Rev. **D17**, 994 (1978). The case of fermion Pseudoquantization is also discussed in S. Blaha, Il Nuovo Cimento **49A**, 35 (1979).
[186] The subscripts on the fields are not gauge symmetry indices but simply identifiers distinguishing the fields from each other.

The creation and annihilation operators satisfy the commutation relations:

$$[a_i(k), a_j^\dagger(k')] = (1 - \delta_{ij})\delta^3(\mathbf{k} - \mathbf{k'}) \qquad (20.4)$$
$$[a_i(k), a_j(k')] = 0$$
$$[a_i^\dagger(k), a_j^\dagger(k')] = 0$$

for i, j = 1, 2.

In this formulation the defining properties of a physical state are:

$$\varphi_1(x)|\Phi, \Pi> = \Phi(x)|\Phi, \Pi> \qquad (20.5)$$
$$\pi_1(x)|\Phi, \Pi> = \Pi(x)|\Phi, \Pi>$$

where $\Phi(x)$ and $\Pi(x)$ are sharp on the states and thus classical fields with

$$\Phi(\mathbf{x}, t) = \int d^3k\, [\alpha(k)f_k(x) + \alpha^*(k)f_k{}^*(x)] \qquad (20.6)$$

and correspondingly for $\Pi(x)$.

20.3 Vacuum States for Scalar (Higgs) Particles with Non-Zero Vacuum Expectation Values

When we implement the mass mechanism Φ becomes constant. We can define a set of states

$$a_1(k)|\alpha> = \alpha(k)|\alpha>$$
$$a_1^\dagger(k)|\alpha> = \alpha^*(k)|\alpha>$$

and correspondingly a set of coherent states

$$|\alpha> = C\exp\left\{\int d^3k\, [\alpha(k)a_2^\dagger(k) + \alpha^*(k)a_2(k)]\right\}|0> \qquad (20.7)$$

where C is a normalization constant and where the vacuum state |0> satisfies

$$a_1(k)|0> = a_1^\dagger(k)|0> = 0 \qquad (20.8a)$$

$$a_2(k)|0> \neq 0 \qquad\qquad a_2^\dagger(k)|0> \neq 0 \qquad (20.8b)$$

The dual vacuum state satisfies

$$<0|a_2(k) = <0|a_2^\dagger(k) = 0 \qquad (20.9a)$$

$$<0|a_1(k) \neq 0 \qquad\qquad <0|a_1^\dagger(k) \neq 0 \tag{20.9b}$$

With this coherent state formalism, which gives purely classical fields and yet also has quantum fields through the use of φ_2 and its creation and annihilation operators, we now have the machinery to define a mass mechanism without the introduction of a potential whose origin can only be described as dubious.

For we can define a coherent state for some k as

$$|\Phi, \Pi> = C\exp\{[(2\pi)^3\omega_k/2]^{1/2}\Phi[a_2^\dagger(k) + a_2(k)]\}|0> \tag{20.10}$$

where C is a normalization constant, that yields a non-zero vacuum expectation value:

$$\varphi_1(x)|\Phi, \Pi> = \Phi| \Phi, \Pi> \tag{20.11}$$

where Φ is a constant. Evaluating a fermion interaction term we find a mass term emerges[187]

$$\overline{\psi}(\varphi_1 + \varphi_2)\psi \ \rightarrow \overline{\psi}(\Phi + \varphi_2)\psi \tag{20.12}$$

It generates a mass for an interaction with a gauge field of the form

$$A^\mu(\varphi_1 + \varphi_2)^2A_\mu \ \rightarrow A^\mu(\Phi + \varphi_2)^2A_\mu \tag{20.13}$$

It also yields a quantum field theoretic interaction that would result in the production of ElectroWeak particles from these scalar fields. The production of Higgs particles that decay into ElectroWeak gauge particles has recently been found at CERN.

The present formalism provides a clean way to separate the vacuum expectation value of a scalar particle from its quantum field part in contrast to the Higgs Mechanism where one has to separate a Higgs field into parts manually.

20.4 Interpretation of Negative Energy Scalar Particle States

As we noted earlier scalar particle physics has the problem of no barrier to the decay of positive energy states to negative energy states due to the absence of a Pauli Exclusion Principle for bosons. The pseudoquantization procedure that we developed in 1978 and describe here allows negative energy states as one would physically expect and raises the possibility of disastrous particle decays to negative energy. Eqs. 20.7 and 20.8 show that negative energy states are possible in this theory.

---

[187] When matrix elements with a "vacuum state" such as eq. 3.10 are taken.

However eq. 20.7 also shows that combined positive and negative energy boson states can be interpreted as classical field states. In addition the ability of any number of boson particles to have the same 4-momentum and quantum numbers shows that a *macroscopic* classical scalar field state can be constructed.

Thus we can view states containing negative energy particles as classical field states and thus solve[188] the issue of interpreting negative energy particle states – a more satisfactory approach than the standard quantization procedure does – with due respect to Professor Dirac.

We note that macroscopic many particle fermion states can only have one particle in any mode unlike bosons. Therefore we cannot use this formalism to create macroscopic classical fermion field states.[189] And the filled Dirac sea of negative energy fermions precludes the transition of a positive energy Dirac fermion to a negative energy state. Thus there is a certain complementarity between fermions that cannot become classical fields but have a filled sea precluding decays to negative energy states, and bosons that can become classical fields but support decays to negative energy states.

20.5 Contrast with Conventional Second Quantization of Scalar Particles

The pseudoquantization procedure followed in this chapter uses different boundary conditions than the usual scalar particle quantization procedure. The essence of the difference is embodied in a comparison of the definition of the vacuum in eqs. 20.8 and 20.9 and the definition of the conventional second quantized field vacuum:

$$a|0\rangle = 0 \qquad \text{Conventional Approach} \qquad (20.14)$$
$$a^\dagger|0\rangle \neq 0$$

In the conventional approach the creation of negative energy boson states is eliminated *ab initio* whereas in our approach it is allowed in order to support classical field states with non-zero vacuum expectation values that are a form of classical field. While one cannot discredit the conventional choice for conventional scalar fields, one can see that our approach yields a physically more important result – particularly for Higgs fields – because it leads to the arrow of time *locally* – an important feature of physical phenomena that has been a subject of much discussion and dispute. One can say that the conventional approach sweeps the issue "under the rug" rather than seeking a deeper justification – differing from Dirac's implied notion that the

[188] Also a boson that has no interactions cannot transition from to a positive energy state to a negative energy state due to conservation of energy.

[189] However we can create Pseudoquantum fermion states. See S. Blaha, Phys. Rev. **D17**, 994 (1978) and references therein to earlier papers by the author.

issue merited attention. We will discuss the "arrow of time" within the framework of our pseudoquantization approach later.

20.6 Why Inertial Reference Frames are Special

The great physicists of the early 20th century raised numerous questions about Special Relativity after Einstein and Poincarè's discovery. Prominent among them was the question of why inertial reference frames are of especial importance in Special Relativity, and afterwards in General Relativity.

It appears that our formulation of the mass generation mechanism sheds significant light on the reason for the special prominence of inertial frames. Earlier we considered the case of a massless pseudoquantized scalar. We now consider massive scalars since experiments at CERN have apparently discovered a Higgs particle with a 125 GeV/c mass. Eqs. 20.2c and 20.2d describe a massive scalar particle. If the scalar is massive, then the "vacuum" state, eq. 20.10, that yields a non-zero expectation value must change to

$$|\Phi, \Pi\rangle = C\exp\{(2\pi)^3 m/2]^{\frac{1}{2}}[a_2^\dagger(\mathbf{0},m) + a_2(\mathbf{0},m)]\}|0\rangle \qquad (20.10')$$

to have operators for a particle of mass m in its rest frame. Then, having established this preferred frame for a Higgs particle, in The Extended Standard Model, and requiring that invariant intervals

$$ds^2 = dt^2 - d\mathbf{x}^2 \quad \text{(in rectangular coordinates)} \qquad (20.15)$$

are unchanged by a (complex or real) Lorentz transformation, we find that inertial reference frames are singled out as "special" in the sense that they are the only accessible reference frames that can be generated by a Lorentz boost/transformation from the Higgs particle rest frame. *The Higgs particle vacuum state singles out the class of inertial reference frames.*

Thus Higgs particles play a central role in establishing the basis of physical reality.

20.7 Pseudoquantization Reveals More Physical Consequences than the Higgs Mechanism of Scalar Particles

Earlier we pointed out that our pseudoquantization theory of Higgs particles reveals more physical consequences than the conventional approach, which implements the Higgs Mechanism by simply using a potential term that has a minimum at a non-zero vacuum expectation value. This chapter and the following chapters show the major results of a properly implemented mechanism. We find a better explanation of the negative energy state problem of boson field theories. We find a local arrow of time that explains the direction of time that we, and all of nature, experiences. We find the reason why inertial reference frames have a special physical significance – a result long sought by physicists.

In addition we saw that real gauge fields should have an associated Higgs particle, while necessarily complex gauge fields (the strong interaction gauge field in The Extended Standard Model) do not have an associated gauge field. These results correspond to experimental reality.

20.8 The T Invariance Issues of Our Pseudoquantized Scalar Particle Theory

The pseudoquantized scalar particle hamiltonian equations are invariant under time reversal t → t' = –t. The vacuum states defined by eqs. Eqs. 20.8 and 20.9 break the time reversal invariance of the theory resulting in retarded particle propagators.

The hamiltonian equations

$$[H, \varphi_1(\mathbf{x}, t)] = -i\partial\varphi_1/\partial t \qquad (20.16)$$
$$[H, \varphi_2(\mathbf{x}, t)] = -i\partial\varphi_2/\partial t$$

are invariant under time reversal. If we define a time reversal operator transformation U then the time reversed equations are

$$[UHU^{-1}, \varphi_1(\mathbf{x}, -t)] = +i\partial\varphi_1(\mathbf{x}, -t)/\partial(-t) \qquad (20.17)$$
$$[UHU^{-1}, \varphi_2(\mathbf{x}, -t)] = +i\partial\varphi_2(\mathbf{x}, -t)/\partial\,(-t)$$

The operator U, which is unitary, transforms H into –H. This operation is legal because the hamiltonian – in this case – is not positive definite and admits negative energy states.[190] Thus

$$[H, \varphi_1(\mathbf{x}, -t)] = -i\partial\varphi_1(\mathbf{x}, -t)/\partial\,(-t) \qquad (20.18)$$
$$[H, \varphi_2(\mathbf{x}, -t)] = -i\partial\varphi_2(\mathbf{x}, -t)/\partial\,(-t)$$

and the time reversal invariance of the equations of motion is established for this case.

Time reversal invariance is broken by our choice of vacuum states. This choice is necessary to obtain classical field states as we showed earlier. A demonstration of the time reversal symmetry breaking is presented later where we show theory has retarded propagators for particle propagation to and from asymptotic states.

Within the interaction region the particle propagators are the sum of retarded and advanced parts that combine to yield principle value propagators – not Feynman propagators. Many years ago Feynman and Wheeler championed principle value propagators for electrodynamics to obtain an action-at-a distance theory of Quantum Electrodynamics. While their theory, and ours, differ from the standard quantum field theory approach there is no reason

---

190 Unlike the usual case of second quantized Klein-Gordon quantum field theory.

to view them as faulty, or having serious physical defects. The only question is whether nature chooses conventional quantum field theory or pseudoquantized quantum field theory. In our case the need for a classical scalar particle non-zero vacuum expectation value strongly motivates our choice of psedoquantized Higgs particles.

20.9 Retarded Propagators for Our Quantized Higgs Particles

In the previous section we pointed out that our pseudoquantization Higgs theory has an arrow of time due to is boundary conditions as expressed by its definition of the vacuum state and its dual. In this section we will show that the theory uses retarded propagators for propagation to and from the interaction region to asymptotic in-states and out-states. Within an interaction region the theory uses half-retarded – half-advanced propagators. We discuss aspects of the perturbation theory and propagators of our scalar particles in this chapter.

First we note that in-states at $t = -\infty$ are composed of superpositions of $a_2(k)$ and $a_2^\dagger(k)$ creation and annihilation operators by eq. 20.8b:

$$a_2(k)|0\rangle \neq 0 \qquad\qquad a_2^\dagger(k)|0\rangle \neq 0 \qquad (20.8b)$$

while the out-states composed of superpositions of $a_1(k)$ and $a_1^\dagger(k)$ creation and annihilation operators by eq. 20.9b:

$$\langle 0|a_1(k) \neq 0 \qquad\qquad \langle 0|a_1^\dagger(k) \neq 0 \qquad (20.9b)$$

Consequently when in-state particles (x_1) propagate into the interaction region (x_2) the relevant propagators are retarded propagators with the form

$$\begin{aligned} G_{in}(x_2, x_1) &= \langle 0|T(\varphi_{1\,in}(x_2), \varphi_{2\,in}(x_1))|0\rangle \\ &= \theta(x_{20} - x_{10})\langle 0|[\varphi_{1\,in}(x_2), \varphi_{2\,in}(x_1)]\,|0\rangle \end{aligned} \qquad (20.19)$$

Eq. 20,19 is a manifestly retarded propagator. The choice of vacuums clearly results in a time asymmetry giving a retarded propagation reflecting the familiar Arrow of Time.

A similar situation prevails for propagation to out-states (x_3) from the interaction (x_2) region:

$$\begin{aligned} G_{out}(x_3, x_2) &= \langle 0|T(\varphi_{1\,out}(x_3), \varphi_{2\,out}(x_2))|0\rangle \\ &= \theta(x_{30} - x_{20})\langle 0|[\varphi_{1\,out}(x_3), \varphi_{2\,out}(x_2)]\,|0\rangle \end{aligned} \qquad (20.20)$$

Within the interaction region the Higgs particles have principle value propagators.

Thus we find Pseudoquantized Higgs particles embody a local Arrow of Time. The locality of the Arrow of Time is embodied in all the particles that interact with the Higgs particle. Since the mass of *every* particle – bosons and fermions – has a Higgs contribution, and thus *every* particle interacts with the Higgs particles, the Arrow of Time permeates The Extended Standard Model as well as the more familiar Standard Model known from experiment.

20.10 The *Local* Arrow of Time

In the *Physics is Logic* series of monographs we saw that complex coordinates led to the form of the fermion spectrum, that the mapping of complex coordinates to real-valued coordinates yielded the Reality group and The Extended Standard Model gauge interactions, that Complex General Relativity led to Higgs particles that were directly united with elementary particle masses and gave us the equality of inertial mass and gravitational mass. In the present volume, we saw the reduction of complex gauge fields to real gauge fields explained the appearance of Higgs fields in The Standard Model and The Extended Standard Model.

Now we see that the pseudoquantization procedure leads to retarded Higgs field propagators and thence to a *local* arrow of time. Many arguments have been put forward over the past hundred plus years for the Arrow of Time. Many arguments based on Statistical Mechanics, Entropy, and Boltzmann's statistical atomic theory have suggested the Arrow of Time is a global statistical consequence. This view seems to contradict the results of elementary particle experiments where a *local* Arrow of Time is evident.

Our rationale for the Arrow of Time begins with retarded Higgs fields. Then we note that Higgs field quantum interactions appear for all fermions and gauge particles. Thus all particle interactions are imbued with an Arrow of Time. Particles united to form macroscopic matter inherit their combined Arrows of Time producing the global Arrow of Time we experience.

Thus our pseudoquantization approach offers a more satisfactory solution of the origin of the Arrow of Time.

It is remarkable that complex quantities – coordinates and fields – through the Higgs phenomena that we have considered, lead to the equality of inertial mass and gravitational mass, and an Arrow of Time. This unity of mass and time phenomena may reflect the deeper fact that we can have no practical Arrow of Time if all particles were massless, for particle dynamics at light speed would then be pointless. This view has been expressed by DeWitt, Unruh, and others who have pointed out that, physically, time is meaningful and measurable only if masses exist; the larger the mass, the more accurate the time measurement in principle.[191]

[191] No mass, no clock; no clock, no physical time. See Blaha (2015a) pp. 368-371 for a discussion including comments by DeWitt and Unruh.

20.11 Space-Time Dependent Particle Masses

It is possible that the Theory of Everything has masses that evolve with time and may also be spatially varying – different values in different parts of the universe. Presently there is no decisive evidence for this possibility although astrophysical studies continue. In this section we will describe the mechanism for space-time dependent masses.

Consider a classical field (time and spatially varying):

$$\Phi(\mathbf{x}, t) = \int d^3k\, [\alpha(k)f_k(x) + \alpha^*(k)f_k{}^*(x)] \quad (20.21)$$

If we define the coherent vacuum state

$$|\alpha\rangle = C \exp\left\{\int d^3k\, [\alpha(k)a_2{}^\dagger(k) + \alpha^*(k)a_2(k)]\right\}|0\rangle \quad (20.22)$$

then

$$\varphi_1(x)|\Phi, \Pi\rangle = \Phi(x)|\Phi, \Pi\rangle \quad (20.23)$$
$$\pi_1(x)|\Phi, \Pi\rangle = \Pi(x)|\Phi, \Pi\rangle$$

where

$$\varphi_i(\mathbf{x}, t) = \int d^3k\, [a_i(k)f_k(x) + a_i{}^\dagger(k)f_k{}^*(x)] \quad (20.24)$$

for i = 1, 2 and where

$$f_k(x) = e^{-ik\bullet x}/(2\omega_k(2\pi)^3)^{1/2}$$

with ω_k equal to the energy.

20.12 Inertial Mass Equals Gravitational Mass

From the days of Newton through Einstein[192] to the present the equality of gravitational mass and inertial mass has been a topic of interest. Mach, who played an important role, in this ongoing discussion, thought distant masses in the universe were the source of the equality. However the origin of the equality, which has been shown experimentally to very high accuracy, remained uncertain until the *Physics is Logic* series of books, in which we showed the interconnection of the Extended Standard Model and Complex Gravitation via Higgs generated masses that united gravitational and inertial mass.

In Blaha (2016h) we showed that a Complex General Relativity transformation can be factored into the product of a complex-valued transformation and a real-valued General Coordinate transformation. The set of complex valued transformations form a U(4) group that

[192] For example, Einstein and Grossman in 1913 stated, "The theory herein described originates in the conviction that the proportionality between the inertial and gravitational mass of a body is an exact law of nature that must be expressed as a foundation principle of theoretical physics."

we called the General Coordinate Reality group. This group has gauge fields that undergo spomtaneous symmetry breaking and generate contributions to all fermion masses.

Since fermion field masses are now sums of ElectroWeak Higgs contributions, Generation group Higgs contributions, Layer group Higgs contributions, and General Coordinate Reality group contributions, and since the gravitational Higgs fields appear in all fermion masses, the equality of inertial and gravitational mass is proven. The gravitational Higgs particles' equations depend, in part, on the gravitational field by Blaha (2016h) and so set the mass scale of gravitational mass, and thereby of all Higgs mass contributions. They set the scale of inertial masses equal to the scale of gravitational masses. ***Since an expression cannot mix mass scales, the gravitational mass scale must be the same as the inertial mass scale. Inertial Mass equals gravitational mass.***

We have established the equality of inertial and gravitational mass at the short distance quantum level. In our view, this explanation is far more satisfying than basing the equality on a combination of large distance phenomena and quantum phenomena. As Einstein and Weyl have pointed out, all fundamental physics phenomena should be based on a local theory. Complex Gravity as we have constructed it, combined with the Extended Standard Model, furnishes a completely local basic Theory of Everything.

Eq. 20.22 contains a coherent state $|\alpha>$ for a time and spatially varying mass. The above equations can be generalized to the case of multiple space-time varying masses.[193]

$$|\Phi_1,\Phi_2, \ldots ,\Phi_n;\Pi_1,\Pi_2, \ldots ,\Pi_n> = C \prod_{i=1}^{n} \exp\left\{\int d^3k \, [\alpha_i(k)a_{2i}^\dagger(k) + \alpha_i^*(k)a_{2i}(k)]\right\}|0> \quad (20.25)$$

Then all n mass vacuum expectation values are space-time dependent:

$$\varphi_{1i}(x) \mid \Phi_1, \Phi_2, \ldots , \Phi_n; \Pi_1, \Pi_2, \ldots , \Pi_n> = \Phi_i(x) \mid \Phi_1, \Phi_2, \ldots , \Phi_n; \Pi_1, \Pi_2, \ldots , \Pi_n> \quad (20.26)$$

Thus our formalism can accommodate space-time varying masses should they be found in the Cosmos.

20.13 Benefits of the Pseudoquantization Method

In this book and earlier work we showed that a more physically satisfactory method for avoiding the negative energy state problem exists. This method relies on the use of a larger Fock space in which negative energy states (or partially negative energy states) are interpreted as states containing classical fields or a mix of classical fields and individual boson particles.

[193] The "vacuum" state $|0>$ in eq. 20.25 also implicitly has factors for the vacuum expectation values used for fields that give masses to fermions and vector bosons as described in Blaha (2016h).

This approach resolves the negative energy boson issue and provides a common framework for boson particles and classical boson fields.

One consequence of the pseudoquantization method is that it enables the appearance of a vacuum expectation value for Higgs particles (a constant classical field) to be understood within a truly quantum framework. Another major consequence of this approach is the appearance of a *local* Arrow of Time due to the Higgs mass generation mechanism – a concept that has been a subject of interest for over one hundred years. A macroscopic arrow of time is often described as a statistical result. But our approach yields an arrow of time at the single particle level.

The conventional approach to boson field quantization sweeps these issues "under the rug" rather than seeking a deeper justification. It differs from Dirac's implied notion that the issue merited attention.

Another important consequence of the pseudoquantization method is that it singles out inertial reference frames when applied to the case of Higgs particles.

Yet another more subtle consequence of boson pseudoquantization is that it provides a rationale/explanation for the presence of ElectroWeak Higgs bosons, *and for their absence for the strong (gluon) interactions. The question of why there are no strong interaction Higgs bosons has not been previously considered to the best of this author's knowledge.*

21. Higgs Mechanism for Coupling Constants

In this chapter[194] we will specify the role of the coupling constants of The Extended Standard Model and Gravitation, which we can view as a Theory of Everything.

21.1 The Interaction Coupling Constants and their Pseudoquantum Field Vacuum Expectation Values

Ten of our Theory of Everything coupling constants are:[195]

- The Strong interaction coupling constant field g_S.
- The Weak SU(2) coupling constant g_W.
- The Electromagnetic U(1) coupling constant g_E.
- The Dark Weak SU(2) coupling constant g_{DW}.
- The Dark Electromagnetic U(1) coupling constant g_{DE}.
- The U(4) Generation group coupling constant g_G.
- The U(4) Layer group coupling constant g_V.
- The U(4) General Coordinate Reality group coupling constant g_S.
- The SU(3)⊗U(64) Ω-interaction group coupling constant g_Ω.
- The complex gravitational coupling constant $g_{GR} = \kappa^{-1} = (4\pi G)^{-½}$.

Based on the discussions of chapter 20 we can define Higgs vacuum expectation values for these coupling constants using a mass factor to obtain the correct coupling constant dimensions. We use the Pseudoquantum formalism of chapter 20:

- The Strong interaction coupling constant field $\Phi_1 = m_1 g_S$.
- The Weak SU(2) coupling constant $\Phi_2 = m_2 g_W$.
- The Electromagnetic U(1) coupling constant $\Phi_3 = m_3 g_E$.
- The Dark Weak SU(2) coupling constant $\Phi_4 = m_4 g_{DW}$.
- The Dark Electromagnetic U(1) coupling constant $\Phi_5 = m_5 g_{DE}$.
- The U(4) Generation group coupling constant $\Phi_6 = m_6 g_G$.
- The U(4) Layer group coupling constant $\Phi_7 = m_7 g_V$.
- The U(4) General Coordinate Reality group coupling constant $\Phi_8 = m_8 g_S$.

---

[194] This chapter is largely extracted from Blaha (2015d).

[195] We place g_{Rflat}, g_{GR}, g_Ω for three interactions: the U(4) General Coordinate Reality group, Gravitational coupling constant, and the Ω-symmetry group. The interactions of these groups will be discussed later.

- The SU(3)⊗U(64) Ω-interaction group coupling constant $\Phi_9 = m_9 g_\Omega$.
- The complex gravitational coupling constant $\Phi_{10} = m_{10} g_{GR} = \kappa^{-1} = (4\pi G)^{-½}$.

The ten masses, m_1 , m_2 , … , m_{10} may be equal or they may have different values. It is also possible they all may be equal to κ^{-1}, which would yield

- The Strong interaction coupling constant field $\Phi_1 = \kappa^{-1} g_S$.
- The Weak SU(2) coupling constant $\Phi_2 = \kappa^{-1} g_W$.
- The Electromagnetic U(1) coupling constant $\Phi_3 = \kappa^{-1} g_E$.
- The Dark Weak SU(2) coupling constant $\Phi_4 = \kappa^{-1} g_{DW}$.
- The Dark Electromagnetic U(1) coupling constant $\Phi_5 = \kappa^{-1} g_{DE}$.
- The U(4) Generation group coupling constant $\Phi_6 = \kappa^{-1} g_G$.
- The U(4) Layer group coupling constant $\Phi_7 = \kappa^{-1} g_V$.
- The U(4) General Coordinate Reality group coupling constant $\Phi_8 = \kappa^{-1} g_S$.
- The SU(3)⊗U(64) Ω-interaction group coupling constant $\Phi_9 = \kappa^{-1} g_\Omega$.
- The complex gravitational coupling constant $\Phi_{10} = \kappa^{-1} g_{GR} = \kappa^{-1} = (4\pi G)^{-½}$.

Then scaling the above vacuum expectation values by κ^{-1} would give:[196]

- The strong interaction coupling constant[197] vacuum expectation value $\Phi_1' = g_S = 1.22$
- The Weak SU(2) coupling constant vacuum expectation value $\Phi_2' = g_W = 0.619$.
- The Electromagnetic U(1) coupling constant vacuum expectation value $\Phi_3' = e = g_E = 0.303$.
- The Dark Weak SU(2) coupling constant vacuum expectation value $\Phi_4' = \Phi_2'$. (?)
- The Dark Electromagnetic U(1) coupling constant vacuum expectation value $\Phi_5' = = \Phi_3'$. (?).
- The U(4) Generation group coupling constant $\Phi_6' = g_G$.
- The U(4) Layer group coupling constant $\Phi_7' = g_V$.
- The U(4) General Coordinate Reality group coupling constant $\Phi_8' = g_S$.
- The SU(3)⊗U(64) Ω-interaction group coupling constant $\Phi_9' = g_\Omega$.
- The complex gravitational coupling constant $\Phi_{10}' = 1$.

These *scaled* vacuum expectation values,[198] which are in fact the coupling constants, have a comparable range of values[199] as opposed to the range of values for the unscaled constants

---

[196] All coupling constant values are based on data extracted from K. A. Olive et al (Particle Data Group), Chinese Physics **C38**, 090001 (2014).
[197] Based on the running coupling constant value $\alpha_s(M_Z^2) = 0.1193 \pm 0.0016$.
[198] The closeness of all the values to one is suggestive: The value $\alpha = 1$ (or $e = (4\pi)^{½} = 3.54$) was the value found in our calculation in the Johnson, Baker, Willey model of QED. Perhaps a larger calculation along the lines of our paper in massless ElectroWeak theory might yield scaled coupling constant values near unity.
[199] The weakness of the Weak interactions is primarily due to the large masses of the Z and W vector bosons – not the values of their coupling constants g and g'.

which range from the ultra-small gravitational vacuum expectation value to values, perhaps, within a few orders of magnitude of unity.

Given the range of known values above, it appears reasonable to conjecture that the unknown values would also be of the order of unity.

The known coupling constant values above are of comparable value, which suggests that our Theory of Everything, at current energies, may be close to the GUT level at which coupling constants are equal.

21.2 Theory of Everything Lagrangian Coupling Constants

We begin with the Theory of Everything lagrangian density $\mathcal{L}_{TE}$ with coupling constants explicitly displayed[200]

$$\mathcal{L}_{TE} = \mathcal{L}_{TE}(g_S, g_W, g_E, g_{DW}, g_{DE}, g_G, g_V, g_S, g_{GR}, g_\Omega) \tag{21.1}$$

and fields and space-time coordinates not displayed.

In terms of vacuum expectation values as discussed in section 21.1 we see we can write[201]

$$\mathcal{L}_{TE} = \mathcal{L}_{TE}(\Phi_1/m_1, \Phi_2/m_2, \ldots, \Phi_{11}/m_{11}) \tag{21.2}$$

where

$$| \Phi_1, \Phi_2, \ldots, \Phi_{11}; \Pi_1, \Pi_2, \ldots, \Pi_{11}> = C \prod_{i=1}^{11}\{\exp[[(2\pi)^3 m_i/2]^{1/2}\Phi_i[a_{i2}^\dagger(\mathbf{0},m_i) + a_{i2}(\mathbf{0},m_i)]]\}|0> \tag{21.3}$$

Assuming all $m_i = \kappa^{-1}$ we obtain

$$| \Phi_1, \Phi_2, \ldots, \Phi_{11}; \Pi_1, \Pi_2, \ldots, \Pi_{11}> = C \prod_{i=1}^{11}\{\exp[[(2\pi/\kappa)^3/2]^{1/2}\Phi_i'[a_{i2}^\dagger(\mathbf{0}, \kappa^{-1}) + a_{i2}(\mathbf{0}, \kappa^{-1})]]\}|0> \tag{21.4}$$

Then eq. 21.2 can be written as

$$\mathcal{L}_{TE} = \mathcal{L}_{TE}(\Phi_1', \Phi_2', \ldots, \Phi_{11}') \tag{21.5}$$

Setting $m_i = g_{CG} = \kappa^{-1}$ = the Planck mass, simplifies the above expressions and *supports the belief that we are close to the unification of all interactions.* However having particles of such large mass makes them undetectable by accelerators. It also seems too large from the

[200] The 11th coupling constant g_Ω and the Ω symmetry group are discussed in chapter 31.

[201] The "vacuum" state |0> in eq. 21.3 also has factors for the vacuum expectation values used for fields that give masses to fermions and vector bosons as described in Blaha (2015b).

viewpoint of physical intuition. Consequently eqs. 21.2 and 21.3 may be the correct expressions with masses perhaps in the TeV range.

We finally note

$$\varphi_{1i}|\ \Phi_1, \Phi_2, \ldots, \Phi_{11}; \Pi_1, \Pi_2, \ldots, \Pi_{11}> = \Phi_{1i}|\ \Phi_1, \Phi_2, \ldots, \Phi_{11}; \Pi_1, \Pi_2, \ldots, \Pi_{11}> \qquad (21.6)$$

21.3 Big Bang Vacuum

At the origin of the universe – the Big Bang – there was a vacuum state in principle. In our earlier books[202] we showed that the universe existed in an ultra-small, but finite, region for an infinitesimal time before it began an explosive inflationary expansion to become the familiar universe. In this time period there were no infinities – a finite temperature and so on.

Thus it is reasonable to assume one of two possibilities for the above ten coupling constants: 1) they have remained unchanged since the beginning, or 2) they have changed with time.

In this section we note, that if our scaling with the Planck mass κ^{-1} in preceding discussions is correct, then it is reasonable to assume that the vacuum state in the beginning is eq. 21.4 with |0> including factors setting fermion and vector boson masses as described in Blaha (2015b).

21.4 Evolving/Space-Time Dependent Coupling Constants

It is possible that the Theory of Everything coupling constants evolve with time and may also be spatially varying – different constants in different parts of the universe. Presently there is no decisive evidence for either possibility. In this section we will describe the mechanism to support either or both possibilities.

Consider a classical field (time and spatially varying):

$$\Phi(\mathbf{x}, t) = \int d^3k\ [\alpha(k)f_k(x) + \alpha^*(k)f_k{}^*(x)] \qquad (21.7)$$

If we define the coherent vacuum state

$$|\alpha> = C \exp\left\{\int d^3k\ [\alpha(k)a_2{}^\dagger(k) + \alpha^*(k)a_2(k)]\right\}|0> \qquad (21.8)$$

then

$$\varphi_1(x)|\Phi, \Pi> = \Phi(x)|\Phi, \Pi> \qquad (21.9)$$
$$\pi_1(x)|\Phi, \Pi> = \Pi(x)|\Phi, \Pi>$$

---

[202] Blaha (2015a) and Blaha (2004).

where

$$\varphi_i(\mathbf{x}, t) = \int d^3k\ [a_i(k)f_k(x) + a_i^{\dagger}(k)f_k^*(x)] \qquad (21.10)$$

for i = 1, 2 with

$$f_k(x) = e^{-ik\cdot x}/(2\omega_k(2\pi)^3)^{½}$$

where $\omega_k = |\mathbf{k}|$.

Eq. 21.8 contains the coherent state $|\alpha>$ for a time and spatially varying vacuum expectation value (classical) field. The above equations can be generalized to the case of the eleven coupling constant vacuum expectation values:[203]

$$|\ \Phi_1, \Phi_2, \ldots, \Phi_{11}; \Pi_1, \Pi_2, \ldots, \Pi_{11}> = C \prod_{i=1}^{11} \exp\left\{\int d^3k\ [\alpha_i(k)a_{2i}^{\dagger}(k) + \alpha_i^{*}(k)a_{2i}(k)]\right\}|0> \quad (21.11)$$

Then all eleven coupling constant vacuum expectation values are space-time dependent:

$$\varphi_{1i}(x)\ |\ \Phi_1, \Phi_2, \ldots, \Phi_{11}; \Pi_1, \Pi_2, \ldots, \Pi_{11}> = \Phi_i(x)\ |\ \Phi_1, \Phi_2, \ldots, \Phi_{11}; \Pi_1, \Pi_2, \ldots, \Pi_{11}> \quad (21.12)$$

and the Theory of Everything lagrangian eq. 21.1 becomes

$$\mathcal{L}_{TE} = \mathcal{L}_{TE}(\Phi_1(x), \Phi_2(x), \ldots, \Phi_{11}(x)) \qquad (21.13)$$

for matrix elements between the vacuum defined by eq. 21.5 and its conjugate,

Thus our formalism can accommodate space-time varying coupling constants should they be found in the Cosmos.

21.5 A Theory of Everything Lagrangian Without Any Constants

The preceding chapters put coupling constants within the same framework as particle masses completing the process of eliminating constants from The Theory of Everything lagrangian. Instead the vacuum contains the values of all coupling constants and particle masses. In one sense this new formulation is a tradeoff. The values of all constants are shifted to the vacuum. However the shift has some advantages technically. One advantage is the ability to have space-time dependent coupling constants as shown above. It would also be straightforward to make masses, and mixing angles, space-time dependent. The possible space-time dependence of coupling constants and particle masses has been an active area of experimental interest for many years although cosmological data seems to indicate these quantities have not changed significantly since the universe began.

---

[203] The "vacuum" state |0> in eq. 21.5 also has factors for the vacuum expectation values used for fields that give masses to fermions and vector bosons as described in Blaha (2015b).

The question of changes in lagrangian constant physical values is of great philosophical importance since it appears that the existence of life, as we know it, depends sensitively on their values. This dependence has been embodied in the Anthropomorphic Principle and studied by a number of physicists and philosophers.

Since our formulation allows space-time varying physical constants the question of the Anthropomorphic Principle attains new importance. As we saw in the case of the theory of Black Holes, which was a theory without evidence for over forty years before Black Holes were discovered, Nature seems to provide phenomena that have been shown to be theoretically possible. Many other cases of this sort have also occurred – the most recent example at the time of this writing is Weyl fermions.

Lastly, our approach opens the possibility of a study of all the many constants in The Theory of Everything lagrangian *on the same footing* rather than in the piecemeal fashion used up to the present. It replaces the scattered hodge-podge of constants in the lagrangian with a centralized location for all constants in the vacuum state permitting a direct study of their interconnection. *The study of the vacuum now becomes of central importance.*

21.6 The Form is Determined But Not the Constants

The derivation of The Extended Standard Model in the Blaha (2015) books and earlier work was based on Asynchronous Logic (to support physical processes spread in space and time); on complex space-time coordinates, the Complex Lorentz group and complex General Coordinate transformations; the Reality group to map complex coordinates to the real-valued coordinates that we observe; the Generations group that yields the four fermion generations and particle number interactions such as the baryon number and lepton number interactions, and the Reality group for complex general coordinate transformations.

This firm basis in fundamental considerations enables us to forge a path to The Extended Standard Model, which included the known features of The Standard Model. We thus were able to avoid the many possible variants and extensions of The Standard Model that have been considered in the Physics literature over the past thirty years.

Two remarkable features of The Extended Standard Model derivation were:

1. A precise fixing of the form of the Extended Standard Model and the Theory of Everything.
2. The absence of any constraints on the values of its coupling constants or masses.

Particle masses were fixed by either the original Higgs Mechanism or by our new mechanism that was based on an extension of Quantum Field Theory to include classical fields such as the vacuum expectation values that cropped up in the original Higgs Mechanism and were handled "by hand." (See Blaha (2015c).)

Thus, up to this point, we have a Theory of Everything (known) except for a basis for the values of the coupling constants that appear in the theory. The coupling constants have a wide range of values. A fundamental basis for their values has been wanting. We will suggest a mechanism for the determination of their values and a (badly broken) U(8) symmetry for The Theory of Everything and the eight coupling constants in The Extended Standard Model.

21.7 How Can Coupling Constants be Determined?

The renormalized coupling constants of The Theory of Everything can be determined experimentally. However a theoretical determination is lacking. This gap in our understanding suggests that there is a major aspect of fundamental physics that is not understood. The fact that we can determine the form – but not the values of coupling constants – so directly from basic principles suggests that a new basic principle(s) is needed to complete The Extended Standard Model. A similar comment applies to fermion and boson masses – both our mechanism,[204] and the vanilla Higgs Mechanism, arbitrarily fixes particle masses. (Attempts to relate particle masses using various symmetries beg the question. As Isidore Rabi (Columbia) once said in a different context, "Who ordered them?" Proposed symmetries are typically "pulled out of a hat.")

The *one* meaningful attempt to determine a coupling constant in a non-trivial 4-dimensional quantum field theory was that of Johnson, Baker and Willey[205] in a 4-dimensional model – massless Quantum Electrodynamics. They developed the theory to the point where if one function, that they called the eigenvalue function, had a zero at the value of the fine structure constant $\alpha \approx 1/137$ then the theory would have no infinities. Adler then showed that the eigenvalue constant zero must be an essential singularity, IF it had a zero,. This author then developed an approximate solution for the eigenvalue function in perhaps the most comprehensive 4-dimensional quantum field theory calculation to all orders in α. The approximate calculation agreed with known exact results to 6th order in e.[206] However the author found a zero at $\alpha = 1$, and the zero was not an essential singularity.

While the Johnson, Baker, Willey model QED was not successful in finding α its method illustrates one possible approach to determining the coupling constants of The Extended Standard Model. It might be possible to use a consistency condition(s) to fix coupling constant values. Since The Extended Standard Model does not have infinities, the motivation of Johnson, Baker and Willey is absent. A fundamental set of consistency conditions is not apparent and so this approach is not currently viable.

What other approaches are possible? There is an anthropomorphic approach which posits the necessity of certain ranges of some coupling constants for human life, and life in

---

[204] Blaha (2015c).
[205] M. Baker and K. Johnson, Phys. Rev. **D8**, 1110 (1973) and references therein.
[206] Equation 1 in our paper S. Blaha, Phys. Rev. **D9**, 2246 (1974).

general, to exist. We are not comfortable with this approach since it seems to "beg the question." The input is equivalent to the output mitigating is character as fundamental.

One could also study the set of coupling constants in a 10-dimensional space looking for the set of values.

Given these considerations we have chosen to pursue a less ambitious approach: to specify the coupling constants as vacuum expectation values of a set of new Higgs-like scalar fields. This approach conceptually parallels the determination of particle masses as vacuum expectation values of scalar Higgs fields.

After reducing coupling constants to vacuum expectation values we considered the possibility that the vacuum state at the Big Bang point determined the coupling constants. We also considered the possibility that coupling constants evolve slowly with time and/or may vary in differing spatial locations.

22. A New Formulation of Complex General Relativity

We have seen that Complex Special Relativity is the basis of flat space-time phenomena. Flat space-time coordinates are complex-valued in general. The real-valued coordinates that we experience in everyday life are the result of our measuring instruments: clocks and rulers. Real-valued coordinates are generated from complex-valued coordinates by Reality group transformations.

If flat space-time is governed by Complex Special Relativity then it is clear that curved 'space-time' is governed by Complex General Relativity. Here again there is a Reality group the General Relativity Reality group – a U(4) group – that maps complex-valued General Relativity coordinates to real-valued curved coodinates.[207]

In Blaha (2004) we considered the General Relativity of complex-valued space-time. We pointed out that Complex General Relativity can be obtained from the General Relativity of real-valued space-time coordinates by piece-wise analytic continuation using complex variable theory. We then constructed Complex General Relativity and considered some new features that appeared.

We now factor complex General Relativistic coordinate transformations into parts that consist of a real-valued General Coordinate transformation and complex-valued coordinate transformations. It will be apparent that the General Relativistic Reality group emerges in this discussion. The General Relativistic Reality group is distinct from the Reality group that led to the symmetry group of the Standard Model: R = SU(3)⊗SU(2)⊗U(1)⊗SU(2)⊗U(1).

22.1 Tetrad (Vierbein) Formalism

The *vierbein* formalism begins with the Equivalence Principle that allows us to define an inertial coordinate system in the neighborhood of any point Z in space-time. We will use the notation $\varsigma^\alpha(Z)$ to denote the inertial coordinates at Z. We define a tetrad or vierbein as

$$v^\alpha{}_\mu(x) = (\partial\varsigma^\alpha(x)/\partial x^\mu)_{x=Z} \tag{22.1}$$

and, in a neighborhood of Z, we can invert the relation between ς and x to define an inverse

$$w^\mu{}_\alpha(x) = (\partial x^\mu(\varsigma)/\partial\varsigma^\alpha)_{x=X} \tag{22.2}$$

such that

---

[207] Much of this chapter appears in Blaha (2016h) and (2017a).

$$w^{\mu}{}_{\alpha}(x)v^{\alpha}{}_{\nu}(x) = \delta^{\mu}{}_{\nu} \tag{22.3}$$
$$w^{\mu}{}_{\beta}(x)v^{\alpha}{}_{\mu}(x) = \delta^{\alpha}{}_{\beta} \tag{22.4}$$

In real General Relativity all *tetrads* are real-valued. In Complex General Relativity a *tetrad* $v^{\alpha}{}_{\mu}(x)$ is complex-valued.

The metric at a curved space-time point X is defined in terms of *tetrads* as

$$g_{\rho\sigma}(x) = \eta_{\alpha\beta}\, v^{\alpha}{}_{\rho}(x)v^{\beta}{}_{\sigma}(x) \tag{22.5}$$
$$g^{\rho\sigma}(x) = \eta^{\alpha\beta}\, w^{\rho}{}_{\alpha}(x)w^{\sigma}{}_{\beta}(x) \tag{22.6}$$

The inverse of a *tetrad* transformation can also be expressed as

$$w_{\beta}{}^{\nu}(x) = v_{\beta}{}^{\nu}(x) = \eta_{\beta\alpha}g^{\nu\mu}(x)v^{\alpha}{}_{\mu}(x) \tag{22.7}$$

Then a *tetrad* and its inverse satisfy the relations

$$v^{\alpha}{}_{\mu}(x)v_{\beta}{}^{\mu}(x) = \delta^{\alpha}{}_{\beta} \tag{22.8}$$

and

$$v^{\alpha}{}_{\mu}(x)v_{\alpha}{}^{\nu}(x) = \delta^{\nu}{}_{\mu} \tag{22.9}$$

There are two general types of space-time transformations that can be performed on a tetrad.

1. A complex-valued (possibly real-valued) General Relativistic coordinate transformation:

$$v'^{\alpha}{}_{\mu}(x) = \partial x^{\nu}/\partial x'^{\mu}\, v^{\alpha}{}_{\nu}(x) \tag{22.10}$$

2. A complex-valued, local *Lorentzian transformation*

$$v'^{\beta}{}_{\mu}(x) = \Lambda(x)^{\beta}{}_{\alpha}\, v^{\alpha}{}_{\mu}(x) \tag{22.11}$$

where $\Lambda(x)^{\beta}{}_{\alpha}$ is an element of a subset of the local Complex Lorentz Group.

The local Lorentzian transformations $\Lambda(x)^{\beta}{}_{\alpha}$ consist of local Lorentz transformations that are real-valued, and complex-valued Lorentz transformations. Both types of transformations satisfy the orthogonality condition:

$$\eta_{\alpha\beta}\Lambda^{\alpha}{}_{\rho}(x)\Lambda^{\beta}{}_{\sigma}(x) = \eta_{\rho\sigma} \tag{22.12}$$

Thus the *tetrad* partakes of both local (position dependent) General Relativistic transformations and local Lorentzian transformations.

22.2 Complex General Relativistic Transformations

The General Relativistic Reality group interaction emerges from complex General Relativistic transformations. We can separate elements of the set of all complex General Coordinate transformations into a product of two factors: a real-valued General Coordinate transformation and a complex-valued General Coordinate transformation. The set of complex factors can be further factored into those that satisfy

$$\Lambda(\omega, \mathbf{u})^T G \Lambda(\omega, \mathbf{u}) = G \tag{22.13}$$

and those that do not. We then showed that the set of those that do not satisfy eq. 22.13 form a curved space representation of the U(4) group under 'multiplication' of transformations.

The elements of the set of real and complex General Coordinate transformations whose flat complex space-time limit satisfies eq. 22.13 form the elements of the Complex Lorentz group.[208]

We thus find the set of all 4-dimensional complex, curved space General coordinate transformations can be visualized as in Fig. 22.1. The next section describes the interplay of the three parts displayed in Fig. 22.1.

22.3 Structure of Complex General Coordinate Transformations

Complex General Coordinate transformations can be uniquely factored into products of two terms, which will later be further factored into three factors. They have the form

$$\partial x''^{\nu}(x)/\partial x^{\mu} = U(x'')^{\nu}{}_{\beta}\, \partial x'^{\beta}(x)/\partial x^{\mu} \tag{22.14}$$

where

$$x''^{\nu}(x) = U(x'')^{\nu}{}_{\beta} x'^{\beta}$$
$$x''^{\mu}(x) = U^{-1\mu}{}_{b}(x'')\, x''^{b}$$

where $U(x')^{\nu}{}_{\beta}$ is complex and where $\partial x'^{\beta}(x)/\partial x^{\mu}$ is a purely real General Coordinate transformation.

We define

$$U\,(x'')^{\mu}{}_{\nu} = w^{\mu}{}_{a}(x'')\big[\exp(i\textstyle\sum_k g_k \Phi_k(x'')\tau_k)\big]^{a}{}_{b}\, v^{b}{}_{\nu}(x'') \tag{22.15}$$

[208] It is this part of curved space-time General Relativity that becomes the flat space-time Complex Lorentz group, which leads to the SU(3)⊗SU(2)⊗U(1)⊗SU(2)⊗U(1) Standard Model Reality group.

$$U^{-1}(x'')^{\mu}{}_{\nu} = w^{\mu}{}_{a}(x'')\left[\exp\left(-i\sum_k g_k\Phi_k(x'')\tau_k\right)\right]^{a}{}_{b}\, v^{b}{}_{\nu}(x'') \qquad (22.16)$$

where the constants g_k are real, and Φ_k and τ_k are hermitean. The uniqueness of the factorization follows from the Reality group (and U(4)) property that any complex 4-vector can be uniquely mapped to any specified real 4-vector.

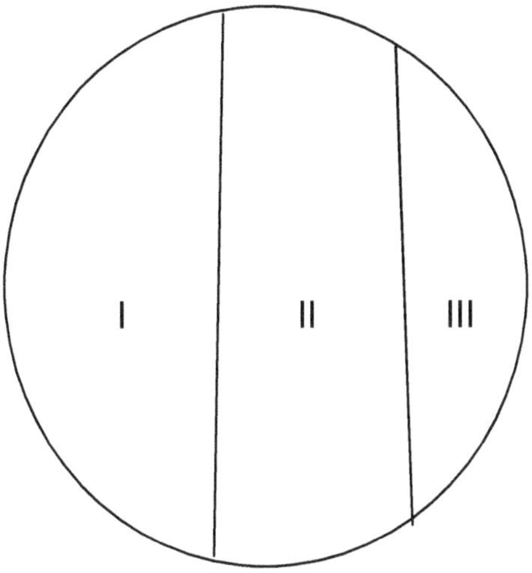

Figure 22.1. A visualization of the set of General Coordinate transformations separated into real-valued General coordinate transformations (part I), complex transformations that satisfy eq. 22.13 (part II), and complex transformations that do not satisfy eq. 22.13 (part III). Part I and part II combine in the limit of flat space-time to form the Complex Lorentz group. Parts II and III elements form a U(4) group that we call the General Relativistic Reality group.

Given the factorization (eq. 22.14) it becomes possible to separate the affine connection correspondingly.

22.4 Complex Affine Connection – General Relativistic Reality Group

The structure of a complex general coordinate transformation (eq. 22.14) enables us to calculate its affine connection for later use in determining the covariant derivative, and the dynamic equations. First the transformation to the real-valued x' coordinates from inertial coordinates is

$$\Gamma^{\sigma}{}_{\lambda\mu}(x') = \partial x'^{\sigma}/\partial\varsigma^{\rho}\ \partial^2\varsigma^{\rho}/\partial x'^{\lambda}\partial x'^{\mu} \qquad (22.17)$$

Next the Reality group transformation has the affine connection

$$\Gamma^{\sigma}{}_{\lambda\mu}(x'') = \partial x''^{\sigma}/\partial\varsigma^{\rho}\, \partial^2\varsigma^{\rho}/\partial x''^{\lambda}\partial x''^{\mu} \qquad (22.18)$$

which can be re-expressed as

$$\Gamma^{\sigma}{}_{\lambda\mu}(x'') = \partial x''^{\sigma}/\partial x'^{\beta}\, \partial x'^{\beta}(\varsigma)/\partial\varsigma^{\rho}\, \partial/\partial x''^{\mu}[\partial\varsigma^{\rho}/\partial x'^{\alpha}\, \partial x'^{\alpha}/\partial x''^{\lambda}] \qquad (22.19)$$

Using eq. 22.17 we find eq. 22.19 has the form

$$\Gamma^{\sigma}{}_{\lambda\mu}(x'') = \partial x''^{\sigma}/\partial x'^{\beta}\, \partial x'^{\alpha}/\partial x''^{\lambda}\, \partial x'^{\gamma}/\partial x''^{\mu}\, \Gamma^{\beta}{}_{\alpha\gamma}(x') + \partial x''^{\sigma}/\partial x'^{\beta}\, \partial^2 x'^{\beta}/\partial x''^{\lambda}\partial x''^{\mu} \qquad (22.20)$$

Next substituting the General Relativistic Reality group transformation

$$\begin{aligned} x''^{\nu}(x) &= U(x'')^{\nu}{}_{\beta}x'^{\beta} \\ x'^{\mu}(x) &= U^{-1}(x'')^{\mu}{}_{\beta}\, x''^{\beta} \end{aligned} \qquad (22.21)$$

together with

$$\partial x''^{\sigma}/\partial x'^{\beta} = \partial[U(x'')^{\sigma}{}_{\alpha}x'^{\alpha}]/\partial x'^{\beta} = U(x'')^{\sigma}{}_{\beta} + x'^{\alpha}\,\partial U(x'')^{\sigma}{}_{\alpha}/\partial x'^{\beta} \qquad (22.22)$$

$$\partial x'^{\sigma}/\partial x''^{\beta} = \partial[U^{-1}(x'')^{\sigma}{}_{\alpha}x''^{\alpha}]/\partial x''^{\beta} = U^{-1}(x'')^{\sigma}{}_{\beta} + x''^{\alpha}\,\partial U^{-1}(x'')^{\sigma}{}_{\alpha}/\partial x''^{\beta} \qquad (22.23)$$

we find the second term in eq. 22.20 is the Reality fields affine connection

$$\Gamma_R{}^{\sigma}{}_{\lambda\mu}(x'') = \partial[U(x'')^{\sigma}{}_{\alpha}x'^{\alpha}]/\partial x'^{\beta}\, \partial\{\partial[U^{-1}(x'')^{\beta}{}_{\alpha}x''^{\alpha}]/\partial x''^{\lambda}\}/\partial x''^{\mu} \qquad (22.24)$$

and so we find the affine connections are approximately additive. Eq. 22.20 is approximately

$$\Gamma^{\sigma}{}_{\lambda\mu}(x'') = \Gamma_{GR}{}^{\sigma}{}_{\lambda\mu}(x') + \Gamma_R{}^{\sigma}{}_{\lambda\mu}(x'') \qquad (22.25)$$

if $x''^{\sigma} \backsimeq x'^{\sigma}$.

A complex transformation of types II and III in Fig. 22.1 has the form:

$$\begin{aligned} U(x'')^{\mu}{}_{\nu} &= w^{\mu}{}_{a}(x'')[\exp(i\textstyle\sum_k \Phi_k(x'')\tau_k)]^{a}{}_{b}\, l^{b}{}_{\nu}(x'') \\ U^{-1}(x'')^{\mu}{}_{\nu} &= w^{\mu}{}_{a}(x'')[\exp(-i\textstyle\sum_k \Phi_k(x'')\tau_k)]^{a}{}_{b}\, l^{b}{}_{\nu}(x'') \end{aligned} \qquad (22.26)$$

Its infinitesimal transformation is approximately

$$U(x'')^{\nu}{}_{\beta} \approx \delta^{\nu}{}_{\beta} + i\textstyle\sum_k \Phi_k(x'')[\tau_k]^{\nu}{}_{\beta} \qquad (22.27)$$

$$U^{-1}(x'')^{\nu}{}_{\beta} \approx \delta^{\nu}{}_{\beta} - i \sum_k \Phi_k(x'')[\tau_k]^{\nu}{}_{\beta} \tag{22.28}$$

using the *vierbein* flat space-time limits

$$w^{\mu}{}_{a}(x'') \approx \delta^{\mu}{}_{a}$$
$$l^{b}{}_{\nu}(x'') \approx \delta^{b}{}_{\nu}$$

where

$$\Phi_k(x) = \int^{x} dy_\lambda\, A_{Rk}{}^{\lambda}(y) \tag{22.29}$$

Then

$$\Gamma_R{}^{\sigma}{}_{\lambda\mu} = -\tfrac{1}{2}i\{\sum_k A_{Rk}(x'')_\mu[\tau_k]^{\sigma}{}_{\lambda} + \sum_k A_{Rk}(x'')_\lambda[\tau_k]^{\sigma}{}_{\mu}\} \tag{22.30}$$
$$= A_R{}^{\sigma}{}_{\mu\lambda} + A_R{}^{\sigma}{}_{\lambda\mu}$$

(summed over k) with the matrix $A_R{}^{\sigma}{}_{\mu\lambda}$ given by

$$A_R{}^{\sigma}{}_{\mu\lambda} = -\tfrac{1}{2}i\sum_k A_{Rk\mu}[\tau_k]^{\sigma}{}_{\lambda} \tag{22.31}$$

with $A_R{}^{\sigma}{}_{\mu\lambda}$ transformable to matrix row and column numbers

$$A_{R\text{flat}}{}^{i\mu a}{}_{b} = A_{R\text{flat}}{}^{i\mu}[\tau_k]^{\sigma}{}_{\lambda}\delta_{\sigma}{}^{a}\delta^{\lambda}{}_{b} \tag{22.32}$$

using the flat space-time vierbein values, and so $A_{R\text{flat}}{}^{ia}{}_{\mu b}$ may be written in matrix form as

$$A_{R\text{flat}}{}^{i}{}_{\mu} = -\tfrac{1}{2}i\sum_k A_{R\text{flat}}{}^{i}{}_{k\mu}\tau_k \tag{22.33}$$

In the flat space-time limit $A_{Rk}{}^{\lambda}(y)$ becomes a U(4) gauge field $A_{R\text{flat}k}{}^{\lambda}(y)\tau_k$.

22.5 Pseudoquantization of Affine Connections

Having obtained the form of the general affine connection (eqs 22.25, 22.30, and 22.33) we now Pseudoquantize them for use in our unification program in chapter 24.

We define

$$R^{1\beta}{}_{\sigma\nu\mu} = \partial_\mu H^{1\beta}{}_{\sigma\nu} - \partial_\nu H^{1\beta}{}_{\sigma\mu} + H^{1\gamma}{}_{\nu\sigma}H^{1\beta}{}_{\gamma\mu} - H^{1\gamma}{}_{\mu\sigma}H^{1\beta}{}_{\gamma\nu} \tag{22.34}$$

$$R^{2\beta}{}_{\sigma\nu\mu\rho} = \partial_\mu H^{2\beta}{}_{\sigma\nu} - \partial_\nu H^{2\beta}{}_{\sigma\mu} + H^{2\gamma}{}_{\nu\sigma}H^{2\beta}{}_{\gamma\mu} - H^{2\gamma}{}_{\mu\sigma}H^{2\beta}{}_{\gamma\nu} +$$
$$+ H^{1\gamma}{}_{\nu\sigma}H^{2\beta}{}_{\gamma\mu} - H^{1\gamma}{}_{\mu\sigma}H^{2\beta}{}_{\gamma\nu} + H^{2\gamma}{}_{\nu\sigma}H^{1\beta}{}_{\gamma\mu} - H^{2\gamma}{}_{\mu\sigma}H^{1\beta}{}_{\gamma\nu} \tag{22.35}$$

where

$$H^{\sigma}{}_{\nu\mu} = \Gamma_{GR}{}^{\sigma}{}_{\nu\mu} + \Gamma_{GR}{}^{2\sigma}{}_{\nu\mu} + \Gamma_{R}{}^{1\sigma}{}_{\nu\mu} + \Gamma_{R}{}^{2\sigma}{}_{\nu\mu} \quad (22.23)$$

where $\Gamma_{GR}{}^{\sigma}{}_{\nu\mu}$ and $\Gamma_{GR}{}^{2\sigma}{}_{\nu\mu}$ are affine connections for real-valued General Relativity, and $\Gamma_{R}{}^{1\sigma}{}_{\nu\mu}$ and $\Gamma_{R}{}^{2\sigma}{}_{\nu\mu}$ are affine connections for a complex-valued set of transformations embodying a U(4) gauge group that combine with real-valued General Relativistic transformations to yield Complex General Relativistic transformations.

The affine connection is most often viewed as a derived quantity—part of the derivation of the curvature tensor in General Relativity. It is typically derived from manipulations of the metric $g_{\mu\nu}$. However, the affine connection can also be viewed as a set of independent fields that become related to the metric via dynamic equations.

Some years ago A. Einstein and H. Weyl[209] pointed out that the metric and the affine connection should be treated as independent quantities and subject to independent arbitrary infinitesimal variations:

> "In contrast to Einstein's original "metric" conception in terms of the $g_{\nu\mu}$ there was later developed, by Eddington, by Einstein himself, and recently by Schrödinger, an affine field theory operating with the components $\Gamma^{\sigma}{}_{\nu\mu}$ of an affine connection. But in 1925 Einstein also advocated a "mixed" formulation by means of a lagrangian in which both the $g_{\nu\mu}$ and the $\Gamma^{\sigma}{}_{\nu\mu}$ are taken as basic field quantities and submitted to independent arbitrary infinitesimal variations.[210] In certain respects this seems to be the most natural procedure."

Following this approach we have introduced the above affine connections for use in the construction of our unification of the eleven particle interactions.

---

[209] H. Weyl, Phys. Rev. **77**, 699 (1950).
[210] A. Einstein, Sitzungsber., Preuss. Akad. Der Wissensch. (1925), p. 414.

23. Higgs Mechanism for General Relativistic U(4) Reality Group – The Species Group

23.1 General Relativity U(4) Reality (Species) Group Gauge Fields

From eq. 22.33 the flat space-time limit of $A_{Rk}{}^{\lambda}(y)$ is a local U(4) gauge field $A_{R_{flatk}}{}^{\lambda}(y)$. The mathematical features of this field is quite similar to to the U(4) Layer group fields. We can thus adapt the formalism developed in sections 15.4 through 15.6 to the General Relativity U(4) Reality Group. The interaction that appears in covariant derivatives is $g_8\mathbf{A}_{Rflat}{}^{\mu}(x) = g_8\mathbf{A}_{Rflatk}{}^{\mu}(x)$ $\mathbf{G}_{RflatU(4)k}$ where k is summed from 1, … , 16.

23.1.1 General Relativity U(4) Reality (Species) Group Transformations

The General Relativity U(4) Reality Group originates as a group of transformations on **space**-time coordinates. However it induces U(4) transformations of fermion spinors. In that role it generates gauge fields similar to the other gauge fields of The Standard Model.

Consider a transformation of the coordinates of a (for the moment free) fermion field. Then we see it induces a transformation on the fermion spinor in the manner:

$$\begin{aligned}\psi'(r') &= \psi'(A_{Rflat\mu}r^{\mu}) \\ &= U(A_{Rflat)})\psi(r)\end{aligned}$$

where, given the four components of spinors and the fact that any 4-vector can be made real-valued by a U(4) transformation, we can write

$$U(A_{Rflat)}) = \mathbf{c}_k(x)G_{RflatU(4)k}$$

summed over k, where $\mathbf{c}_k(x)$ is a local 4-vector and $G_{RflatU(4)k}$ is a 4×4 U(4) matrix in its $\underline{4}$ representation. Thus, we find a U(4) transformation group induced on fermion fields that is generated by General Relativity Reality group transformations of coordinates.

We will see in subsection 23.1.3 that the primary qualitative effect of the General Relativity U(4) Reality Group is to change the species of a fermion. Therefore we will call this group the *Species group*.

23.1.2 Species Group Covariance

A Species group transformation on a Dirac equation must be covariant. Consider the Dirac equation with $\mathbf{A}_{Rflat}{}^{\mu}(x)$ but with other gauge field terms in the covariant derivative omitted for simplicity:

$$\bar{\psi}(x)[i\gamma_\mu(\partial/\partial x_\mu - ig_8 A_{Rflatk}{}^{\mu}(x)G_{RflatU(4)k}) - m]\psi(x) = 0$$

summed over k. If we perform a Species group transformation U:

$$\bar{\psi}(x)[i\gamma_\mu(\partial/\partial x_\mu - ig_8 A_{Rflatk}{}^{\mu}(x)G_{RflatU(4)k}) - m]U^{-1}U\psi(x) = 0$$

or

$$\bar{\psi}(x)U^{-1}U[iU^{-1}U\gamma_\mu U^{-1}U(\partial/\partial x_\mu - ig_8 A_{Rflatk}{}^{\mu}(x)G_{RflatU(4)k}) - m]U^{-1}U\psi(x) = 0$$

we find

$$\bar{\psi}'(x)[i\gamma_\mu' U(\partial/\partial x_\mu - ig_8 A_{Rflatk}{}^{\mu}(x)G_{RflatU(4)k})U^{-1} - m]\psi'(x) = 0$$

where

$$\gamma_\mu'(x) = U\gamma_\mu U^{-1}$$

is locally equivalent to a Dirac matrix by Good's Theorem.[211] If we set

$$A'_{Rflat}{}^{\mu}(x) = -(i/g_8)U[\partial U^{-1}/\partial x^\mu] + UA_{Rflat}{}^{\mu}(x)U^{-1}$$

then the transformed Dirac equation is

$$\bar{\psi}'(x)[i\gamma_\mu'(x)(\partial/\partial x_\mu - ig_8 A'_{Rflatk}{}^{\mu}(x)G_{RflatU(4)k}) - m]\psi'(x) = 0$$

and has the same form as the original equation above and thus is covariant. We note the indices of the matrices $G_{RflatU(4)k}$ are spinor indices and so $G_{RflatU(4)k}\gamma_\mu$ has an implicit spinor matrix summation.

The coordinate dependence of $\gamma_\mu'(x)$ introduces locality into the Dirac matrix. This locality might be viewed with concern except that an inverse Species group transformation exists that removes the locality. Thus the physical impact of this 'new' locality is eliminated.

---

[211] R. H. Good, Jr., Rev. Mod. Phys., **27**, 187 (1955).

23.1.3 Physical Role of the Species Group

The physical role of elements of this group is to 'rotate' fermion fields within a given species (for fixed Type, generation and layer), and also to 'rotate' a fermion field from one species to a field in another species (again for fixed Type, generation and layer). Thus if we consider the four types of species: charged lepton, neutral lepton, up-type quark, and down-type quark, then a General Relativity U(4) Reality Group transformation 'rotates' any fermion field to another fermion field within its own species, or to a fermion field in a different species. These rotations do not change the Type (normal or Dark), the generation, or the layer of the fermion field[212] – only its species at most.

If a Species group transformation is applied to a free fermion field it will yield a new free fermion field in the same or a diferent species. If this fermion field has a complex energy then the Species group transformation must be followed by a Standard Model Reality group transformation that transforms the fermion field into a field with a real-valued energy since free fundamental particles must have real-valued energies (as we pointed out earlier) since they cannot decay. The fermion field so produced, with real-valued energy, must be in one of the four species (six species if you count colored subspecies as distinct) for each type of matter: normal or Dark.

23.2 Spontaneous Symmetry Breaking of the General Relativity U(4) Reality Group – The Species Group

Given the rather long appellation of the group we will shorten it's name to the *Species group. Similarly we let* $\mathbf{A}_{\mathrm{Rflat}}{}^{\mu}(x) = \mathbf{A}_{\mathrm{S}}{}^{\mu}(x)$ *and* $\mathbf{G}_{\mathrm{RflatU(4)}} = \mathbf{G}_{\mathrm{S}}$. We begin the discussion of Species group symmetry breaking by defining a Higgs field η which is a Species group 4-vector

$$\eta = \begin{bmatrix} \rho_1 \\ \rho_2 \\ \rho_3 \\ \rho_4 \end{bmatrix} \qquad (23.1)$$

where ρ_1, ρ_2, ρ_3 and ρ_4 are real fields.[213] Then the covariant derivative of η (taking account only of the Species group) is

---

[212] It also does not change the color of an SU(3) color quark.

[213] Each field ρ_i can be expressed as a Pseudoquantum field: $\rho_i = \varphi_{1i} + \varphi_{2i}$ where φ_{1i} has the vacuum expectation value $\rho_{i0.\ \mathrm{for}\ i\, =\, 1,\, \ldots\, ,\, 4.}$ Thus our Pseudoquantum field theory version is implemented easily.

$$D_{\dots\mu}\eta = \{\partial/\partial X^\mu + \dots - \tfrac{1}{2} i g_8 \Sigma A_{Sk}{}^{\mu}(x) G_{Sk}\} \begin{bmatrix} \rho_1 \\ \rho_2 \\ \rho_3 \\ \rho_4 \end{bmatrix} \tag{23.2}$$

Following steps similar to eqs. 15.4 through 15.17 we find with ρ_i being the vacuum expectation value of the Higgs field:

$$\begin{aligned}(D_{\dots\mu}\eta)^\dagger D_{\dots}{}^{\mu}\eta = {} & \partial\rho_1/\partial X^\mu\, \partial\rho_1/\partial X_\mu + \partial\rho_2/\partial X^\mu\, \partial\rho_2/\partial X_\mu + \partial\rho_3/\partial X^\mu\, \partial\rho_3/\partial X_\mu + \partial\rho_4/\partial X^\mu\, \partial\rho_4/\partial X_\mu + \\ & + \tfrac{1}{4} g_8{}^2\{\rho_1{}^2 A_{S1}{}^2 + \rho_2{}^2 A_{S2}{}^2 + \rho_3{}^2 A_{S3}{}^2 + \rho_4{}^2 A_{S4}{}^2 + \\ & + (\rho_1{}^2 + \rho_2{}^2)(V_5{}^2 + V_6{}^2) + \tfrac{1}{4}(\rho_1{}^2 + \rho_3{}^2)(V_7{}^2 + V_8{}^2) + \\ & + (\rho_1{}^2 + \rho_4{}^2)(V_9{}^2 + V_{10}{}^2) + \tfrac{1}{4}(\rho_2{}^2 + \rho_3{}^2)(V_{11}{}^2 + V_{12}{}^2) + \\ & + (\rho_2{}^2 + \rho_4{}^2)(V_{13}{}^2 + V_{14}{}^2) + \tfrac{1}{4}(\rho_3{}^2 + \rho_4{}^2)(V_{15}{}^2 + V_{16}{}^2)\}\end{aligned} \tag{23.3}$$

up to total divergences, which generate surface terms which we discard. We also assume that all fields satisfy the gauge condition

$$\partial A_{Si}{}^{\mu}/\partial X^\mu = 0 \tag{23.4}$$

Eq. 23.3 shows all Species group gauge fields have masses. Thus Species group symmetry is completely broken. The combination of an ultra-weak coupling constant and very large gauge field masses results in extremely Species interactions.

We assume Species group gauge field masses to be very large – of the order of the Planck mass in view of its origin in Complex General Relativity.

23.3 Species Group Higgs Mechanism Contributions to Fermion Masses

The symmetry breaking of the Species group results in a contribution to each fermion mass of all types, species, generations, and layers. The Species group contributions to normal and Dark fermion mass terms are

$$\mathcal{L}^{Higgs}{}_{FermionMassesSpecies} = \Sigma_{s,g,l}\bar{\psi}_{sglL}\rho_s m_{sgl}\psi_{sglR} + \Sigma_{s,g,l}\bar{\psi}_{DsglL}\rho_s m_{Dsgl}\psi_{DsglR} + \text{c.c.} \tag{23.5}$$

The η field expectation value has components labeled ρ_s.[214] The mass matrices m_{sgl} and m_{Dsgl} are the complex constant Species mass matrix contributions for normal and Dark species.

[214] The Higgs fields $\eta\dots$ in our Pseudoquantum formulation are $\eta\dots = \varphi_{1\dots}(x) + \varphi_{2\dots}(x)$ as described earlier.

24. Types of Unification of Interactions

In this chapter[215] we consider various forms of the unification of particle interactions. We note that particle interactions appear to always involve spin 1 or spin 2 bosons.[216]

24.1 Symbolic ElectroWeak-Type Unification

One type of interaction unification is simply to additively combine the lagrangian terms for each interaction to form a 'total' interaction. The 'unification' is further enhanced by combining gauge fields of separate interactions through rotations such as the ElectroWeak Weinberg angle rotation.[217] Then one can use a covariant derivative, which takes advantage of the spin 1 gauge field nature of the interaction, to insert interactions in the fermion terms of the total lagrangian.

Gauge field particles appear to obtain a mass through the symmetry breaking Higgs Mechanism. The form of the Higgs sector terms is critical to the success of this form of unification. Unfortunately our knowledge of the nature and form of Higgs particle lagrangian terms awaits experimental determination. Thus, in the ElectroWeak[218] case, in particular, the success of unification remains to be proven experimentally. A hopeful factor is the proof that ElectroWeak theory is renormalizable.[219]

One wonders, however, to what extent a symbolic theoretical unification of this type is truly unification or merely a symbolic 'gluing' of different interactions.

In the case of ElectroWeak theory, should it prove to be superficial, one must recognize that it played a constructive historical role in increasing our understanding of the Electromagnetic and Weak interactions. Thus ElectroWeak theory was a success from that viewpoint regardless of its future role in Physics.

---

[215] The contents of this chapter first appeared in Blaha (2016h) and (2017a).

[216] This feature of interactions make them particularly conducive for use in covariant derivatives, and, perhaps more deeply, for use in defining a Riemann-Christoffel curvature tensor that unites the interactions.

[217] Although technically such a roation, in itself, is equivalent to the unrotated form.

[218] We use the ElectroWeak term often because of its familiarity. In this work we point out that the gauge field mixing by the Weinberg angle is simply one application of our 'rotation of interactions' formalism developed later. We will treat the SU(2) Weak interaction separately from the U(1) Electromagnetic interaction in what follows, and show ElectroWeak gauge rotations are but an example of our formalism.

[219] Although this benefit is not of consequence in our *fully finite* Two-Tier quantum field theory formulation of the Extended Standard Modl. Two-Tier QFT is described in Appendices A and B.

24.2 Group-Based Unification

It is also possible to envision a unification of interactions within one symmetry group. This type of unification has been a dream of many theorists particularly because various renormaization studies of the trend of interaction coupling constants at high energy suggests the various interaction strengths are becoming similar.

If, taking a positive attitude, interaction strengths become similar at high energy, one has to wonder if the disparate group structure seen at today's energy can evolve into a union within one large enveloping group. An enveloping group would necessarily have additional generators that cause an interplay between the different symmetry groups seen at low energy.

For example, there might be generators that 'couple' the strong and Weak sectors. These generators, and their associated gauge fields, would have to 'materialize' as particle physics goes to higher energies. There is no evidence at present for the appearance of new interaction terms of this sort. Thus unification within a larger enveloping group is much more than achieving an equality of coupling constants.

24.3 Unification Based on Energy-Momentum Considerations

All gauge boson interactions affect the energy and momentum in a physical system. Consequently they must affect the curvature of space and time to a greater or lesser degree. As a result they must contribute to the total Riemann-Christoffel curvature tensor.

Their common appearance in the form of covariant derivative terms leads to a unification framework that encompasses all possible interactions. It also yields 'interactions between interactions' that do not sem to have received attention although these new interactions appear to have experimental consequences of interest.

24.4 Unification Based on a 'Rotation of Interactions' Symmetry

Interactions are typically constructed separately and treated separately. However it is possible that they may jointly have a unity based on a very general concept of a 'rotation' that transforms, and interchanges, interactions amongst each other. Thus, this new form of unification principle is based on the rotation of interactions just as we view the unity of space as based on spatial rotations.

We suggest that a set of eight particle interactions[220] can be rotated amongst each other by local SU3)⊗U(64) transformations, which we call *Ω-transformations*. Since there are 64 fields in the eight interactions, the transformations will have the form of the <u>1</u> representation of SU(3) and the <u>64</u> representation of U(64). The 192 fermions will form a <u>192</u> representation of SU3)⊗U(64). The Ω-transformation SU3)⊗U(64) fields symmetry will be seen to be a broken symmetry. Its gauge fields will acquire a mass – presumably due to the Higgs Mechanism.

---

[220] See section 19.3 for a list of these interactions.

These eight particle interactions are united with two General Relativistic interactions via the Riemann-Christoffel curvature tensor. The eleventh interaction, which embodies the 'rotation of particle interactions,' will be discussed in detail in chapter 31. We call this SU3)⊗U(64) interaction symmetry – *Ω-Symmetry*.

25. Summary of the Eleven Interactions

In this chapter[221] we define the eleven interactions of our unified theory. We begin by defining the fields of the eleven interactions using the pseudoquantization formalism described in Blaha (2016c) and earlier books. We use the Pseudoquantization formalism described earlier that defines two gauge fields for each gauge interaction. In this chapter we omit coupling constants. In the following chapters we introduce coupling constants.

25.1 The SU(2) Weak Interaction

The Weak interaction SU(2) gauge fields are defined as $W^{1i\mu}(x)$ and $W^{2i\mu}(x)$ for i = 1, 2, 3. Using SU(2) generators we define the matrix form by $W^{k\mu}(x) = W^{ki\mu}(x)\tau_i$ for k= 1, 2. Under an SU(2) gauge transformation C_W the gauge fields transform as

$$W^{1\mu}(x) \to C_W(x)W^{1\mu}(x)C_W^{-1}(x) - i\, C_W(x)\partial^\mu C_W^{-1}(x) \tag{25.1}$$

and

$$W^{2\mu}(x) \to C_W(x)W^{2\mu}(x)C_W^{-1}(x) \tag{25.2}$$

25.2 The U(1) Electromagnetic Interaction

The U(1) electromagnetic gauge fields[222] are defined as $A_E^{1\mu}(x)$ and $A_E^{2\mu}(x)$. Under a local electromagnetic gauge transformation $C_E(x)$ the gauge fields transform as

$$A_E^{1\mu}(x) \to C_E(x)A_E^{1\mu}(x)C_E^{-1}(x) - i\, C_E(x)\partial^\mu C_E^{-1}(x) \tag{25.3}$$

and

$$A_E^{2\mu}(x) \to C_E(x)A_E^{2\mu}(x)C_E^{-1}(x) \tag{25.4}$$

25.3 The SU(3) Strong Interaction

The Strong SU(3) gauge fields are defined as $A_{SU(3)}{}^{1i\mu}(x)$ and $A_{SU(3)}{}^{2i\mu}(x)$ for i = 1, … , 8. Using SU(3) generators we define the matrix form by $A_{SU(3)}{}^{1\mu}(x) = A_{SU(3)}{}^{1i\mu}(x)T_i$ and $A_{SU(3)}{}^{2\mu}(x) = A_{SU(3)}{}^{2i\mu}(x)T_i$. Under an SU(3) gauge transformation C the gauge field transforms as

$$A_{SU(3)}{}^{1\mu}(x) \to C(x)A_{SU(3)}{}^{1\mu}(x)C^{-1}(x) - i\, C(x)\partial^\mu C^{-1}(x) \tag{25.5}$$

---

[221] Most of the material in this chapter appeared in Blaha (2016h) and earlier books.

[222] We introduce two fields as we did in our article S. Blaha, Phys. Rev. D**10**, 4268 (July, 1974). These fields enable us to define a free electromagnetic lagrangian that is linear in the fields for reasons given elsewhere.

and

$$A_{SU(3)}{}^{2\mu}(x) \to C(x)A_{SU(3)}{}^{2\mu}(x)C^{-1}(x) \quad (25.6)$$

25.4 The U(4) Generation Group Interaction

The U(4) Generation group[223] generators are denoted G_i and its gauge fields are denoted $U_{\mu i}(X)$. Thus the Generation group terms in covariant derivatives are

$$\mathbf{U}^i{}_\mu\mathbf{\cdot G}_i \quad (25.7)$$

where i = 1, 2. Under a U(4) Generation transformation C_G the gauge field transforms as

$$U^{1\mu}(x) \to C_G(x)U^{1\mu}(x)C_G{}^{-1}(x) - i\, C_G(x)\partial^\mu C_G{}^{-1}(x) \quad (25.8)$$

and

$$U^{2\mu}(x) \to C_G(x)U^{2\mu}(x)C_G{}^{-1}(x) \quad (25.9)$$

25.5 The U(4) Layer Group Interaction

The U(4) Layer group[224] generators are denoted G_{Lk} and its gauge fields are denoted $V^i{}_{\mu k}(X)$. Thus the Layer group terms in covariant derivatives are

$$\mathbf{V}^i{}_\mu\mathbf{\cdot G}_{Li} \quad (25.10)$$

where i = 1, 2, and k = 1, 2, …, 16. Under a U(4) Generation transformation C_G the gauge field transforms as

$$V^{1\mu}(x) \to C_G(x)V^{1\mu}(x)C_G{}^{-1}(x) - i\, C_G(x)\partial^\mu C_G{}^{-1}(x) \quad (25.11)$$

and

$$V^{2\mu}(x) \to C_G(x)V^{2\mu}(x)C_G{}^{-1}(x) \quad (25.12)$$

25.6 The SU(2) Dark Weak Interaction

We assume Dark Weak interactions have the same form as the known SU(2) Weak interactions. the The Dark Weak interaction SU(2) gauge fields are defined as $W_D{}^{1i\mu}(x)$ and $W_D{}^{2i\mu}(x)$ for i = 1, 2, 3. Using SU(2) generators we define the matrix form by $W_D{}^{k\mu}(x) = W_D{}^{ki\mu}(x)\tau_{Di}$ for k= 1, 2 where the generator matrices τ_{Di} are not in the same subspace as the normal SU(2) generators. Under a Dark SU(2) gauge transformation C_{DW} the gauge field transforms as

---

[223] If there are only three generatons of fermions then the Generation group is U(3).
[224] If there are only three generatons of fermions then the Layer group is also U(3).

$$W_D^{1\mu}(x) \rightarrow C_{DW}(x)W_D^{1\mu}(x)C_{DW}^{-1}(x) - i\, C_{DW}(x)\partial^\mu C_{DW}^{-1}(x) \qquad (25.13)$$

and

$$W_D^{2\mu}(x) \rightarrow C_{DW}(x)W_D^{2\mu}(x)C_{DW}^{-1}(x) \qquad (25.14)$$

25.7 The U(1) Dark Electromagnetic Interaction

The U(1) Dark electromagnetic gauge field[225] are defined as $A_{DE}^{1\mu}(x)$ and $A_{DE}^{2\mu}(x)$. Under a local Dark electromagnetic gauge transformation $C_{DE}(x)$ the gauge fields transform as

$$A_{DE}^{1\mu}(x) \rightarrow C_{DE}(x)A_{DE}^{1\mu}(x)C_{DE}^{-1}(x) - i\, C_{DE}(x)\partial^\mu C_{DE}^{-1}(x) \qquad (25.15)$$

and

$$A_{DE}^{2\mu}(x) \rightarrow C_{DE}(x)A_{DE}^{2\mu}(x)C_{DE}^{-1}(x) \qquad (25.16)$$

25.8 The U(4) General Relativistic Reality Group Interaction – The Species Group

The U(4) Species group (the General Relativistic Reality Group) interaction gauge fields[226] are $A_S^{1\mu}(x) = A_{Rflat}^{1\mu}(x)$ and $A_S^{2\mu}(x) = A_{Rflat}^{2\mu}(x)$. Under a gauge transformation $C_R(x)$ they transform as

$$A_S^{1\mu}(x) \rightarrow C_R(x)A_S^{1\mu}(x)C_R^{-1}(x) - i\, C_R(x)\partial^\mu C_R^{-1}(x) \qquad (25.17)$$
$$A_S^{2\mu}(x) \rightarrow C_R(x)A_S^{2\mu}(x)C_R^{-1}(x)$$

This group rotates fermion fields amongst the four normal species and the four Dark species. Each Color subspecies is rotated to a color species of the same color. Intermediate rotations fall into one species or another. See chapter 23 for more details.

25.9 The 'Interaction Rotation' Interaction - A_Ω

The SU3)⊗U(64) interaction rotation group gauge fields[227] are defined as $A_\Omega^{1ij\mu}(x)$ and $A_\Omega^{2ij\mu}(x)$ for i = 1, …, 8 and j = 1, …, 64. They total 512 gauge field components. Using their 72 generators expressed in the SU(3) <u>3</u> representation and the U(64) <u>64</u> representation, with matrix denoted $T_{\Omega ij}$, we can define the matrix form as

$$A_\Omega^{k\mu}(x) = A_\Omega^{kij\mu}(x)T_{\Omega ij} \qquad (25.18)$$

[225] We introduce two fields as we did in our article S. Blaha, Phys. Rev. D**10**, 4268 (July, 1974). These fields enable us to define a free electromagnetic lagrangian that is linear in the fields for reasons given elsewhere.
[226] See chapters 22 and 23 for a discussion of the origin of this interaction in Complex General Relativity.
[227] The SU(3) factor is *not* color SU(3).

for k = 1, 2, where $T_{\Omega ij}$ is a cross product of SU3)⊗U(64) generators. The tensor product generator matrices are 192×192 matrices (since 3*64 = 192). We choose to have a representation with 192×192 matrices due to the 192 fermions in our Periodic Table of Fermions. (See chapter 16.)

Under a local SU3)⊗U(64) gauge transformation C_Ω a gauge field transforms as

$$A_\Omega{}^{1\mu}(x) \rightarrow C_\Omega(x)A_\Omega{}^{1\mu}(x)C_\Omega{}^{-1}(x) - i\, C_\Omega(x)\partial^\mu C_\Omega{}^{-1}(x) \tag{25.19}$$

$$A_\Omega{}^{2\mu}(x) \rightarrow C_\Omega(x)A_\Omega{}^{2\mu}(x)C_\Omega{}^{-1}(x)$$

This interaction, which we will call the *Ω-interaction,* is described in detail in chapter 31. The gauge fields will correspondingly be called *Ω-fields*.

25.10 The Spinor Connection Interaction

The spinor connection used in formulations of vierbein gravity is $B^1{}_{\mu ab}(x)$ where a and b are tangent space indices. The vector is combined with γ matrices for use in matrix equations:

$$B^{1\mu} = B^{1\mu}{}_{ab}\Sigma^{ab} \tag{25.20}$$

where

$$\Sigma^{ab} = i\,[\gamma^a, \gamma^b]/4 \tag{25.21}$$

Under a local Lorentz transformation S

$$B^{1\mu}(x) \rightarrow S(x)B^{1\mu}(x)S^{-1}(x) - i\, S(x)\partial^\mu S^{-1}(x) \tag{25.22}$$

Similarly we define a secondary spinor connection:

$$B^{2\mu} = B^{2\mu}{}_{ab}\Sigma^{ab} \tag{25.23}$$

where

$$\Sigma^{ab} = i\,[\gamma^a, \gamma^b]/4 \tag{25.24}$$

Under a local Lorentz transformation S

$$B^{2\mu}(x) \rightarrow S(x)B^{2\mu}(x)S^{-1}(x) \tag{25.25}$$

A simple spin ½ field transforms as

$$(\partial^\mu + i\, B^{1\mu} + i\, B^{2\mu})\psi \rightarrow S(\partial^\mu + i\, B^{1\mu} + i\, B^{2\mu})\psi \tag{25.26}$$

25.11 Real-Valued General Coordinate Connection (Interaction)

The usual gravitational metric field $g_{\mu\nu}$ is supplemented with a secondary metric field to enable us to define a higher derivative gravitation theory and still use the canonical Euler-

Lagrange formalism to generate dynamical equations. Thus we define a second metric field and associated affine connection

$$g^{2\mu\nu} = g^{2\nu\mu} \qquad (25.27)$$

$$\Gamma_{GR}{}^{2\lambda}{}_{\mu\nu} = \tfrac{1}{2} g^{2\lambda\alpha}(\partial_\mu g^2{}_{\alpha\nu} + \partial_\nu g^2{}_{\alpha\mu} - \partial_\alpha g^2{}_{\mu\nu}) \qquad (25.28)$$

in addition to the usual real-valued metric and affine connection denoted $\Gamma_{GR}{}^{1\sigma}{}_{\nu\mu} = \Gamma_{GR}{}^{\sigma}{}_{\nu\mu}$.

26. The Total Covariant Derivative and Riemann-Christoffel Curvature Tensor

We begin by defining a 'space' vector which also is a fundamental representation vector of SU(3)⊗SU(2)⊗U(1)⊗SU(2)⊗U(1)⊗U(4)⊗U(4)⊗U(4)⊗SU(3)⊗U(64).[228]

$$V_\sigma = V_\sigma^{\ aijkmno}(x)\gamma_a T_i \tau_j \tau_{Dj} \mathbf{G}_{U(4)k} G_{LU(4)m} \mathbf{G}_{Sn} G_{SU(3)\otimes U(64)o} \qquad (26.1)$$

where $\boldsymbol{\tau}_D$ represents the Dark SU(2) generators and $G_{SU(3)\otimes U(64)}$ is the set of SU(3)⊗U(64) generators.

Then we use the following generalized covariant derivative of this vector:[229,230]

$$\begin{aligned} D_\nu V_\mu &= (\partial_\nu + iF_\nu)V_\mu - H^\sigma{}_{\nu\mu}V_\sigma \\ &= [g^\sigma{}_\mu\partial_\nu + ig^\sigma{}_\mu F_\nu - H^\sigma{}_{\nu\mu}]V_\sigma \\ &= [g^\sigma{}_\mu\partial_\nu + iD^\sigma{}_{\mu\nu}]V_\sigma \end{aligned} \qquad (26.2)$$

where[231]

$$F^\mu = g_\Omega A_\Omega{}^{1\mu}(x) + g_\Omega A_\Omega{}^{2\mu}(x) + \mathbf{A}_I{}^{1\mu}(x) + \mathbf{A}_I{}^{2\mu}(x) + B^{1\mu} + B^{2\mu} \qquad (26.3)$$

by eqs. 19.16 and 19.17, and

$$H^\sigma{}_{\nu\mu} = \Gamma_{GR}{}^\sigma{}_{\nu\mu} + \Gamma_{GR}{}^{2\sigma}{}_{\nu\mu} \qquad (26.4)$$

$$D^\sigma{}_{\mu\nu} = g^\sigma{}_\mu F_\nu + iH^\sigma{}_{\nu\mu} \qquad (26.5)$$

where we have abstracted the complex part of the complex affine connection into the U(4) gauge field $A_S{}^\mu$. Eq. 26.4 is the real-valued part of the complex affine connection.

Commutators of the vector fields in F_μ are implicit when the covariant derivative is applied to vectors and tensors such as V_σ.

---

[228] Most of the material in this chapter appeared in Blaha (2016h) and earlier books.

[229] We use the superscript '1' to distinguish primary connections from secondary connections labeled '2'. The discussion in this subsection and in the following subsection parallels that of chapter 4 of I.

[230] *Commutator 'cross products' are usually implicit in the subsequent equations.*

[231] We will omit the insertion of coupling constants of $B^{1\mu}$ and $B^{2\mu}$ in the interests of simplifying expressions.

26.1 The Riemann-Christoffel Tensor

Eqs. 26.1 – 26.5 enable us to calculate the Riemann-Christoffel curvature tensor, and then its contractions $R_{\mu\nu}$ and R using[232]

$$(D_\nu D_\mu - D_\mu D_\nu)V_\sigma = R^\beta{}_{\sigma\nu\mu}V_\beta \tag{26.6}$$

The second order covariant derivative of V_σ is

$$D_\nu D_\mu V_\sigma = \{g^\alpha{}_\mu(\partial_\nu + iF_\nu) - H^\alpha{}_{\mu\nu}\}\{g^\beta{}_\sigma(\partial_\alpha + iF_\alpha)V_\beta - H^\beta{}_{\sigma\alpha}V_\beta\} - H^\gamma{}_{\nu\sigma}\{g^\alpha{}_\gamma(\partial_\mu + iF_\mu)V_\alpha - H^\alpha{}_{\gamma\mu}V_\alpha\} \tag{26.7}$$

with implicit commutators with the guage field terms.

26.2 The Eleven Interaction Riemann-Christoffel Curvature Tensor

Using the definitions in chapter 4 we find

$$\begin{aligned}
R'^\beta{}_{\sigma\nu\mu}V_\beta = {} & g^\alpha{}_\mu(\partial_\nu + iF_\nu)g^\beta{}_\sigma(\partial_\alpha + iF_\alpha)V_\beta - H^\alpha{}_{\mu\nu}g^\beta{}_\sigma(\partial_\alpha + iF_\alpha)V_\beta + \\
& + H^\alpha{}_{\mu\nu}H^\beta{}_{\sigma\alpha}V_\beta - g^\alpha{}_\mu(\partial_\nu + iF_\nu)H^\beta{}_{\sigma\alpha}V_\beta - H^\gamma{}_{\nu\sigma}\{g^\alpha{}_\gamma(\partial_\mu + iF_\mu)V_\alpha - H^\alpha{}_{\gamma\mu}V_\alpha\} - \\
& - \{\mu\leftrightarrow\nu\} \\
= {} & ig^\beta{}_\sigma(\partial_\nu F_\mu - \partial_\mu F_\nu - i[F_\nu, F_\mu])V_\beta + (\partial_\mu H^\beta{}_{\sigma\nu} - \partial_\nu H^\beta{}_{\sigma\mu} + H^\gamma{}_{\nu\sigma}H^\beta{}_{\gamma\mu} - H^\gamma{}_{\mu\sigma}H^\beta{}_{\gamma\nu})V_\beta \\
= {} & ig^\beta{}_\sigma(F_E{}^1{}_{\nu\mu} + F_E{}^2{}_{\nu\mu} + F_W{}^1{}_{\nu\mu} + F_W{}^2{}_{\nu\mu} + F_{DE}{}^1{}_{\nu\mu} + F_{DE}{}^2{}_{\nu\mu} + F_{DW}{}^1{}_{\nu\mu} + F_{DW}{}^2{}_{\nu\mu} + F_{SU(3)}{}^1{}_{\nu\mu} + \\
& + F_{SU(3)}{}^2{}_{\nu\mu} + F_U{}^1{}_{\nu\mu} + F_U{}^2{}_{\nu\mu} + F_V{}^1{}_{\nu\mu} + F_V{}^2{}_{\nu\mu} + F_\Omega{}^1{}_{\nu\mu} + F_\Omega{}^2{}_{\nu\mu})V_\beta + \\
& + (ig^\beta{}_\sigma B^1{}_{\nu\mu} + ig^\beta{}_\sigma B^2{}_{\nu\mu} + \partial_\mu H^\beta{}_{\sigma\nu} - \partial_\nu H^\beta{}_{\sigma\mu} + H^\gamma{}_{\nu\sigma}H^\beta{}_{\gamma\mu} - H^\gamma{}_{\mu\sigma}H^\beta{}_{\gamma\nu})V_\beta \\
= {} & R'_E{}^\beta{}_{\sigma\nu\mu}V_\beta + R'_{SU(2)}{}^\beta{}_{\sigma\nu\mu}V_\beta + R'_{DE}{}^\beta{}_{\sigma\nu\mu}V_\beta + R'_{DSU(2)}{}^\beta{}_{\sigma\nu\mu}V_\beta + R'_{SU(3)}{}^\beta{}_{\sigma\nu\mu}V_\beta + R'_U{}^\beta{}_{\sigma\nu\mu}V_\beta + \\
& + R'_V{}^\beta{}_{\sigma\nu\mu}V_\beta + R'_S{}^\beta{}_{\sigma\nu}V_\beta + R'_\Omega{}^\beta{}_{\sigma\nu}V_\beta + R'_B{}^\beta{}_{\sigma\nu}V_\beta + R'_G{}^\beta{}_{\sigma\nu\mu}V_\beta
\end{aligned} \tag{26.8}$$

where

$$\begin{aligned}
R'_{SU(3)}{}^\beta{}_{\sigma\nu\mu} &= ig^\beta{}_\sigma(F_{SU(3)}{}^1{}_{\nu\mu} + F_{SU(3)}{}^2{}_{\nu\mu}) \\
R'_{SU(2)}{}^\beta{}_{\sigma\nu\mu} &= ig^\beta{}_\sigma(F_W{}^1{}_{\nu\mu} + F_W{}^2{}_{\nu\mu}) \\
R'_E{}^\beta{}_{\sigma\nu\mu} &= ig^\beta{}_\sigma(F_E{}^1{}_{\nu\mu} + F_E{}^2{}_{\nu\mu}) \\
R'_U{}^\beta{}_{\sigma\nu\mu} &= ig^\beta{}_\sigma(F_U{}^1{}_{\nu\mu} + F_U{}^2{}_{\nu\mu}) \\
R'_V{}^\beta{}_{\sigma\nu\mu} &= ig^\beta{}_\sigma(F_V{}^1{}_{\nu\mu} + F_V{}^2{}_{\nu\mu}) \\
R'_{DSU(2)}{}^\beta{}_{\sigma\nu\mu} &= ig^\beta{}_\sigma(F_{DW}{}^1{}_{\nu\mu} + F_{DW}{}^2{}_{\nu\mu}) \\
R'_{DE}{}^\beta{}_{\sigma\nu\mu} &= ig^\beta{}_\sigma(F_{DE}{}^1{}_{\nu\mu} + F_{DE}{}^2{}_{\nu\mu})
\end{aligned} \tag{26.9}$$

---

[232] Much of the material in this chapter appeared in Blaha (2016h) and earlier books.

$$R'_S{}^{\beta}{}_{\sigma\nu\mu} = ig^{\beta}{}_{\sigma}(F_S{}^1{}_{\nu\mu} + F_S{}^2{}_{\nu\mu})$$
$$R'_\Omega{}^{\beta}{}_{\sigma\nu\mu} = ig^{\beta}{}_{\sigma}(F_\Omega{}^1{}_{\nu\mu} + F_\Omega{}^2{}_{\nu\mu})$$
$$R'_B{}^{\beta}{}_{\sigma\nu\mu} = ig^{\beta}{}_{\sigma}(F_B{}^1{}_{\nu\mu} + F_B{}^2{}_{\nu\mu})$$

and

$$\begin{aligned} R'_G{}^{\beta}{}_{\sigma\nu\mu} &= \partial_\mu H^{1\beta}{}_{\sigma\nu} - \partial_\nu H^{1\beta}{}_{\sigma\mu} + H^{1\gamma}{}_{\nu\sigma}H^{1\beta}{}_{\gamma\mu} - H^{1\gamma}{}_{\mu\sigma}H^{1\beta}{}_{\gamma\nu} + \partial_\mu H^{2\beta}{}_{\sigma\nu} - \partial_\nu H^{2\beta}{}_{\sigma\mu} + \\ &\quad + H^{2\gamma}{}_{\nu\sigma}H^{2\beta}{}_{\gamma\mu} - H^{2\gamma}{}_{\mu\sigma}H^{2\beta}{}_{\gamma\nu} + H^{1\gamma}{}_{\nu\sigma}H^{2\beta}{}_{\gamma\mu} - H^{1\gamma}{}_{\mu\sigma}H^{2\beta}{}_{\gamma\nu} + H^{2\gamma}{}_{\nu\sigma}H^{1\beta}{}_{\gamma\mu} - \Gamma^{2\gamma}{}_{\mu\sigma}\Gamma^{\beta}{}_{\gamma\nu} \\ &= R^{1\beta}{}_{\sigma\nu\mu} + R^{2\beta}{}_{\sigma\nu\mu} \end{aligned} \tag{26.10}$$

with

$$H^{\beta}{}_{\sigma\nu\mu} = \partial_\mu H^{\beta}{}_{\sigma\nu} - \partial_\nu H^{\beta}{}_{\sigma\mu} + H^{\gamma}{}_{\nu\sigma}H^{\beta}{}_{\gamma\mu} - H^{\gamma}{}_{\mu\sigma}H^{\beta}{}_{\gamma\nu} \tag{26.11}$$
$$R^{1\beta}{}_{\sigma\nu\mu} = \partial_\mu H^{1\beta}{}_{\sigma\nu} - \partial_\nu H^{1\beta}{}_{\sigma\mu} + H^{1\gamma}{}_{\nu\sigma}H^{1\beta}{}_{\gamma\mu} - H^{1\gamma}{}_{\mu\sigma}H^{1\beta}{}_{\gamma\nu} \tag{26.12}$$
$$\begin{aligned} R^{2\beta}{}_{\sigma\nu\mu\rho} &= \partial_\mu H^{2\beta}{}_{\sigma\nu} - \partial_\nu H^{2\beta}{}_{\sigma\mu} + H^{2\gamma}{}_{\nu\sigma}H^{2\beta}{}_{\gamma\mu} - H^{2\gamma}{}_{\mu\sigma}H^{2\beta}{}_{\gamma\nu} + \\ &\quad + H^{1\gamma}{}_{\nu\sigma}H^{2\beta}{}_{\gamma\mu} - H^{1\gamma}{}_{\mu\sigma}H^{2\beta}{}_{\gamma\nu} + H^{2\gamma}{}_{\nu\sigma}H^{1\beta}{}_{\gamma\mu} - H^{2\gamma}{}_{\mu\sigma}H^{1\beta}{}_{\gamma\nu} \end{aligned} \tag{26.13}$$

and

$$H^{1\sigma}{}_{\nu\mu} = \Gamma_{GR}{}^{\sigma}{}_{\nu\mu}$$
$$H^{2\sigma}{}_{\nu\mu} = \Gamma_{GR}{}^{2\sigma}{}_{\nu\mu}$$

and where

$$F_{SU(3)}{}^1{}_{\kappa\mu} = \partial A_{SU(3)}{}^1{}_\mu/\partial x^\kappa - \partial A_{SU(3)}{}^1{}_\kappa/\partial x^\mu + ig_1[A_{SU(3)}{}^1{}_\kappa, A_{U(3)}{}^1{}_\mu] \tag{26.14}$$
$$F_W{}^1{}_{\kappa\mu} = \partial W^1{}_\mu/\partial x^\kappa - \partial W^1{}_\kappa/\partial x^\mu + ig_2[W^1{}_\kappa, W^1{}_\mu]$$
$$F_E{}^1{}_{\kappa\mu} = \partial A_E{}^1{}_\mu/\partial x^\kappa - \partial A_E{}^1{}_\kappa/\partial x^\mu$$
$$F_{DW}{}^1{}_{\kappa\mu} = \partial W_D{}^1{}_\mu/\partial x^\kappa - \partial W_D{}^1{}_\kappa/\partial x^\mu + ig_4[W_D{}^1{}_\kappa, W_D{}^1{}_\mu]$$
$$F_{DE}{}^1{}_{\kappa\mu} = \partial A_{DE}{}^1{}_\mu/\partial x^\kappa - \partial A_{DE}{}^1{}_\kappa/\partial x^\mu$$
$$F_U{}^1{}_{\kappa\mu} = \partial U^1{}_\mu/\partial x^\kappa - \partial U^1{}_\kappa/\partial x^\mu + ig_6[U^1{}_\kappa, U^1{}_\mu]$$
$$F_V{}^1{}_{\kappa\mu} = \partial V^1{}_\mu/\partial x^\kappa - \partial V^1{}_\kappa/\partial x^\mu + ig_7[V^1{}_\kappa, V^1{}_\mu]$$
$$F_S{}^1{}_{\kappa\mu} = \partial A_S{}^1{}_\mu/\partial x^\kappa - \partial A_S{}^1{}_\kappa/\partial x^\mu + ig_8[A_S{}^1{}_\kappa, A_S{}^1{}_\mu]$$
$$F_\Omega{}^1{}_{\kappa\mu} = \partial A_\Omega{}^1{}_\mu/\partial x^\kappa - \partial A_\Omega{}^1{}_\kappa/\partial x^\mu + ig_\Omega[A_\Omega{}^1{}_\kappa, A_\Omega{}^1{}_\mu]$$
$$F_B{}^1{}_{\kappa\mu} = \partial B^1{}_\mu/\partial x^\kappa - \partial B^1{}_\kappa/\partial x^\mu + i[B^1{}_\kappa, B^1{}_\mu]$$

$$\begin{aligned} F_{SU(3)}{}^2{}_{\kappa\mu} &= \partial A_{SU(3)}{}^2{}_\mu/\partial x^\kappa - \partial A_{SU(3)}{}^2{}_\kappa/\partial x^\mu + ig_1[A_{SU(3)}{}^2{}_\kappa, A_{SU(3)}{}^2{}_\mu] + ig_1[A_{SU(3)}{}^1{}_\kappa, A_{SU(3)}{}^2{}_\mu] + \\ &\quad + ig_1[A_{SU(3)}{}^2{}_\kappa, A_{SU(3)}{}^1{}_\mu] \end{aligned}$$
$$F_W{}^2{}_{\kappa\mu} = \partial W^2{}_\mu/\partial x^\kappa - \partial W^2{}_\kappa/\partial x^\mu + ig_2[W^2{}_\kappa, W^2{}_\mu] + ig_2[W^1{}_\kappa, W^2{}_\mu] + ig_2[W^2{}_\kappa, W^1{}_\mu]$$
$$F_E{}^2{}_{\kappa\mu} = \partial A_E{}^2{}_\mu/\partial x^\kappa - \partial A_E{}^2{}_\kappa/\partial x^\mu$$
$$F_{DW}{}^2{}_{\kappa\mu} = \partial W_D{}^2{}_\mu/\partial x^\kappa - \partial W_D{}^2{}_\kappa/\partial x^\mu + ig_4[W_D{}^2{}_\kappa, W_D{}^2{}_\mu] + ig_4[W_D{}^1{}_\kappa, W_D{}^2{}_\mu] + ig_4[W_D{}^2{}_\kappa, W_D{}^1{}_\mu]$$
$$F_{DE}{}^2{}_{\kappa\mu} = \partial A_{DE}{}^2{}_\mu/\partial x^\kappa - \partial A_{DE}{}^2{}_\kappa/\partial x^\mu$$
$$F_U{}^2{}_{\kappa\mu} = \partial U^2{}_\mu/\partial x^\kappa - \partial U^2{}_\kappa/\partial x^\mu + ig_6[U^2{}_\kappa, U^2{}_\mu] + ig_6[U^1{}_\kappa, U^2{}_\mu] + ig_6[U^2{}_\kappa, U^1{}_\mu]$$

$$F_{V\ \kappa\mu}^{\ 2} = \partial V^2_{\ \mu}/\partial x^\kappa - \partial V^2_{\ \kappa}/\partial x^\mu + ig_7[V^2_{\ \kappa}, V^2_{\ \mu}] + ig_7[V^1_{\ \kappa}, V^2_{\ \mu}] + ig_7[V^2_{\ \kappa}, V^1_{\ \mu}]$$
$$F_{S\ \kappa\mu}^{\ 2} = \partial A_S{}^2_{\ \mu}/\partial x^\kappa - \partial A_S{}^2_{\ \kappa}/\partial x^\mu + ig_8[A_S{}^2_{\ \kappa}, A_S{}^2_{\ \mu}] + ig_8[A_S{}^1_{\ \kappa}, A_S{}^2_{\ \mu}] + $$
$$+ ig_8[A_S{}^2_{\ \kappa}, A_S{}^1_{\ \mu}]$$
$$F_{\Omega\ \kappa\mu}^{\ 2} = \partial A_\Omega{}^2_{\ \mu}/\partial x^\kappa - \partial A_\Omega{}^2_{\ \kappa}/\partial x^\mu + ig_\Omega[A_\Omega{}^2_{\ \kappa}, A_\Omega{}^2_{\ \mu}] + ig_\Omega[A_\Omega{}^1_{\ \kappa}, A_\Omega{}^2_{\ \mu}] + ig_\Omega[A_\Omega{}^2_{\ \kappa}, A_\Omega{}^1_{\ \mu}]$$
$$F_{B\ \kappa\mu}^{\ 2} = \partial B^2_{\ \mu}/\partial x^\kappa - \partial B^2_{\ \kappa}/\partial x^\mu + i[B^2_{\ \mu}, B^2_{\ \kappa}] + i[B^1_{\ \mu}, B^2_{\ \kappa}] + i[B^2_{\ \mu}, B^1_{\ \kappa}]$$

Note that $R'^{\beta}_{\ \sigma\nu\mu}$ factorizes into U(1)⊗SU(2)⊗U(1)⊗SU(2)⊗SU(3)⊗U(4)⊗U(4)⊗U(4)⊗U(4)⊗U(64) parts and a Riemann-Christoffel Gravitational curvature tensor part. For later use in defining a lagrangian we define

$$R'^{\beta}_{\ \sigma\nu\mu} = R'_E{}^{1\beta}_{\ \sigma\nu\mu} + R'_E{}^{2\beta}_{\ \sigma\nu\mu} + R'_{SU(2)}{}^{1\beta}_{\ \sigma\nu\mu} + R'_{SU(2)}{}^{2\beta}_{\ \sigma\nu\mu} + R'_{DE}{}^{1\beta}_{\ \sigma\nu\mu} + R'_{DE}{}^{2\beta}_{\ \sigma\nu\mu} + R'_{DSU(2)}{}^{1\beta}_{\ \sigma\nu\mu} +$$
$$+ R'_{DSU(2)}{}^{2\beta}_{\ \sigma\nu\mu} + R'_{SU(3)}{}^{1\beta}_{\ \sigma\nu\mu} + R'_{SU(3)}{}^{2\beta}_{\ \sigma\nu\mu} + R'_U{}^{1\beta}_{\ \sigma\nu\mu} + R'_U{}^{2\beta}_{\ \sigma\nu\mu} + R'_V{}^{1\beta}_{\ \sigma\nu\mu} + R'_V{}^{2\beta}_{\ \sigma\nu\mu} +$$
$$+ R'_S{}^{1\beta}_{\ \sigma\nu\mu} + R'_S{}^{2\beta}_{\ \sigma\nu\mu} + R'_\Omega{}^{1\beta}_{\ \sigma\nu\mu} + R'_\Omega{}^{2\beta}_{\ \sigma\nu\mu} + R'_B{}^{1\beta}_{\ \sigma\nu\mu} + R'_B{}^{2\beta}_{\ \sigma\nu\mu} + R^{1\beta}_{\ \sigma\nu\mu} + R^{2\beta}_{\ \sigma\nu\mu} \qquad (26.15)$$

where

$$R'_E{}^{1\beta}_{\ \sigma\nu\mu} = ig^\beta_{\ \sigma}F_E{}^1_{\ \nu\mu}$$
$$R'_E{}^{2\beta}_{\ \sigma\nu\mu} = ig^\beta_{\ \sigma}F_{DE}{}^2_{\ \nu\mu}$$

$$R'_{DE}{}^{1\beta}_{\ \sigma\nu\mu} = ig^\beta_{\ \sigma}F_E{}^1_{\ \nu\mu}$$
$$R'_{DE}{}^{2\beta}_{\ \sigma\nu\mu} = ig^\beta_{\ \sigma}F_{DE}{}^2_{\ \nu\mu}$$

$$R'_{SU(2)}{}^{1\beta}_{\ \sigma\nu\mu} = ig^\beta_{\ \sigma}F_W{}^1_{\ \nu\mu}$$
$$R'_{SU(2)}{}^{2\beta}_{\ \sigma\nu\mu} = ig^\beta_{\ \sigma}F_{DW}{}^2_{\ \nu\mu}$$

$$R'_{DSU(2)}{}^{1\beta}_{\ \sigma\nu\mu} = ig^\beta_{\ \sigma}F_W{}^1_{\ \nu\mu}$$
$$R'_{DSU(2)}{}^{2\beta}_{\ \sigma\nu\mu} = ig^\beta_{\ \sigma}F_{DW}{}^2_{\ \nu\mu}$$

$$R'_{SU(3)}{}^{1\beta}_{\ \sigma\nu\mu} = ig^\beta_{\ \sigma}F_{SU(3)}{}^1_{\ \nu\mu}$$
$$R'_{SU(3)}{}^{2\beta}_{\ \sigma\nu\mu} = ig^\beta_{\ \sigma}F_{SU(3)}{}^2_{\ \nu\mu}$$

$$R'_U{}^{1\beta}_{\ \sigma\nu\mu} = ig^\beta_{\ \sigma}F_U{}^1_{\ \nu\mu}$$
$$R'_U{}^{2\beta}_{\ \sigma\nu\mu} = ig^\beta_{\ \sigma}F_U{}^2_{\ \nu\mu}$$

$$R'_V{}^{1\beta}_{\ \sigma\nu\mu} = ig^\beta_{\ \sigma}F_V{}^1_{\ \nu\mu}$$
$$R'_V{}^{2\beta}_{\ \sigma\nu\mu} = ig^\beta_{\ \sigma}F_V{}^2_{\ \nu\mu}$$

$$R'_S{}^{1\beta}_{\ \sigma\nu\mu} = ig^\beta_{\ \sigma}F_S{}^1_{\ \nu\mu}$$
$$R'_S{}^{2\beta}_{\ \sigma\nu\mu} = ig^\beta_{\ \sigma}F_S{}^2_{\ \nu\mu}$$

$$R'_{\Omega}{}^{1\beta}{}_{\sigma\nu\mu} = ig^{\beta}{}_{\sigma}F_{\Omega}{}^{1}{}_{\nu\mu}$$
$$R'_{\Omega}{}^{2\beta}{}_{\sigma\nu\mu} = ig^{\beta}{}_{\sigma}F_{\Omega}{}^{2}{}_{\nu\mu}$$

$$R'_{B}{}^{1\beta}{}_{\sigma\nu\mu} = ig^{\beta}{}_{\sigma}B^{1}{}_{\nu\mu}$$
$$R'_{B}{}^{2\beta}{}_{\sigma\nu\mu} = ig^{\beta}{}_{\sigma}B^{2}{}_{\nu\mu}$$

The total Ricci tensor is

$$R'_{\sigma\mu} = R'^{\beta}{}_{\sigma\beta\mu} \tag{26.16}$$

$$= iF_{E}{}^{1}{}_{\sigma\mu} + iF_{E}{}^{2}{}_{\sigma\mu} + iF_{W}{}^{1}{}_{\sigma\mu} + iF_{W}{}^{2}{}_{\sigma\mu} + iF_{DE}{}^{1}{}_{\sigma\mu} + iF_{DE}{}^{2}{}_{\sigma\mu} + iF_{DW}{}^{1}{}_{\sigma\mu} + iF_{DW}{}^{2}{}_{\sigma\mu} + iF_{SU(3)}{}^{1}{}_{\sigma\mu} + iF_{SU(3)}{}^{2}{}_{\sigma\mu} +$$
$$+ iF_{U}{}^{1}{}_{\sigma\mu} + iF_{U}{}^{2}{}_{\sigma\mu} + iF_{V}{}^{1}{}_{\sigma\mu} + iF_{V}{}^{2}{}_{\sigma\mu} + iF_{S}{}^{1}{}_{\sigma\mu} + iF_{S}{}^{2}{}_{\sigma\mu} + iF_{\Omega}{}^{1}{}_{\sigma\mu} + iF_{\Omega}{}^{2}{}_{\sigma\mu} + iB^{1}{}_{\sigma\mu} + iB^{2}{}_{\sigma\mu} +$$
$$+ \partial_{\mu}H^{1\beta}{}_{\sigma\beta} - \partial_{\beta}H^{1\beta}{}_{\sigma\mu} + H^{1\gamma}{}_{\beta\sigma}H^{1\beta}{}_{\gamma\mu} - H^{1\gamma}{}_{\mu\sigma}H^{1\beta}{}_{\gamma\beta} +$$
$$+ \partial_{\mu}H^{2\beta}{}_{\sigma\beta} - \partial_{\beta}H^{2\beta}{}_{\sigma\mu} + H^{2\gamma}{}_{\beta\sigma}H^{2\beta}{}_{\gamma\mu} - H^{2\gamma}{}_{\mu\sigma}H^{2\beta}{}_{\gamma\beta} + H^{1\gamma}{}_{\beta\sigma}H^{2\beta}{}_{\gamma\mu} - H^{1\gamma}{}_{\mu\sigma}H^{2\beta}{}_{\gamma\beta} + H^{2\gamma}{}_{\beta\sigma}H^{1\beta}{}_{\gamma\mu} - H^{2\gamma}{}_{\mu\sigma}H^{1\beta}{}_{\gamma\beta}$$

$$= R'_{E}{}^{1}{}_{\sigma\mu} + R'_{E}{}^{2}{}_{\sigma\mu} + R'_{SU(2)}{}^{1}{}_{\sigma\mu} + R'_{SU(2)}{}^{2}{}_{\sigma\mu} + R'_{DE}{}^{1}{}_{\sigma\mu} + R'_{DE}{}^{2}{}_{\sigma\mu} + R'_{DSU(2)}{}^{1}{}_{\sigma\mu} + R'_{DSU(2)}{}^{2}{}_{\sigma\mu} + R'_{SU(3)}{}^{1}{}_{\sigma\mu} +$$
$$+ R'_{SU(3)}{}^{2}{}_{\sigma\mu} + R'_{U}{}^{1}{}_{\sigma\mu} + R'_{U}{}^{2}{}_{\sigma\mu} + R'_{V}{}^{1}{}_{\sigma\mu} + R'_{V}{}^{2}{}_{\sigma\mu} + R'_{S}{}^{1}{}_{\sigma\mu} + R'_{S}{}^{2}{}_{\sigma\mu} + R'_{\Omega}{}^{1}{}_{\sigma\mu} + R'_{\Omega}{}^{2}{}_{\sigma\mu} + R'_{B}{}^{1\beta}{}_{\sigma\beta\mu} +$$
$$+ R'_{B}{}^{2\beta}{}_{\sigma\beta\mu} + R^{1}{}_{\sigma\mu} + R^{2}{}_{\sigma\mu}$$
$$= R'^{1}{}_{\sigma\mu} + R'^{2}{}_{\sigma\mu} \tag{26.17}$$

where

$$R'^{1}{}_{\sigma\mu} = R'_{E}{}^{1}{}_{\sigma\mu} + R'_{SU(2)}{}^{1}{}_{\sigma\mu} + R'_{DE}{}^{1}{}_{\sigma\mu} + R'_{DSU(2)}{}^{1}{}_{\sigma\mu} + R'_{SU(3)}{}^{1}{}_{\sigma\mu} + R'_{U}{}^{1}{}_{\sigma\mu} + R'_{V}{}^{1}{}_{\sigma\mu} + R'_{S}{}^{1}{}_{\sigma\mu} +$$
$$+ R'_{\Omega}{}^{1}{}_{\sigma\mu} + R'_{B}{}^{1\beta}{}_{\sigma\beta\mu} + R^{1}{}_{\sigma\mu} \tag{26.18}$$

$$R'^{2}{}_{\sigma\mu} = R'_{E}{}^{2}{}_{\sigma\mu} + R'_{SU(2)}{}^{2}{}_{\sigma\mu} + R'_{DE}{}^{2}{}_{\sigma\mu} + R'_{DSU(2)}{}^{2}{}_{\sigma\mu} + R'_{SU(3)}{}^{2}{}_{\sigma\mu} + R'_{U}{}^{2}{}_{\sigma\mu} + R'_{V}{}^{2}{}_{\sigma\mu} + R'_{S}{}^{2}{}_{\sigma\mu} +$$
$$+ R'_{\Omega}{}^{2}{}_{\sigma\mu} + R'_{B}{}^{2\beta}{}_{\sigma\beta\mu} + R^{2}{}_{\sigma\mu}$$

with

$$R'_{E}{}^{1}{}_{\sigma\mu} = iF_{E}{}^{1}{}_{\sigma\mu}$$
$$R'_{E}{}^{2}{}_{\sigma\mu} = iF_{E}{}^{2}{}_{\sigma\mu}$$

$$R'_{SU(2)}{}^{1}{}_{\sigma\mu} = iF_{W}{}^{1}{}_{\sigma\mu}$$
$$R'_{SU(2)}{}^{2}{}_{\sigma\mu} = iF_{W}{}^{2}{}_{\sigma\mu}$$

$$R'_{DE}{}^{1}{}_{\sigma\mu} = iF_{DE}{}^{1}{}_{\sigma\mu}$$
$$R'_{DE}{}^{2}{}_{\sigma\mu} = iF_{DE}{}^{2}{}_{\sigma\mu}$$

$$R'_{DSU(2)}{}^{1}{}_{\sigma\mu} = iF_{DW}{}^{1}{}_{\sigma\mu}$$

$$R'_{DSU(2)}{}^{2}{}_{\sigma\mu} = iF_{DW}{}^{2}{}_{\sigma\mu}$$

$$R'_{SU(3)}{}^{1}{}_{\sigma\mu} = iF_{SU(3)}{}^{1}{}_{\sigma\mu}$$
$$R'_{SU(3)}{}^{2}{}_{\sigma\mu} = iF_{SU(3)}{}^{2}{}_{\sigma\mu}$$

$$R'_{U}{}^{1}{}_{\sigma\mu} = iF_{U}{}^{1}{}_{\sigma\mu}$$
$$R'_{U}{}^{2}{}_{\sigma\mu} = iF_{U}{}^{2}{}_{\sigma\mu}$$

$$R'_{V}{}^{1}{}_{\sigma\mu} = iF_{V}{}^{1}{}_{\sigma\mu}$$
$$R'_{V}{}^{2}{}_{\sigma\mu} = iF_{V}{}^{2}{}_{\sigma\mu}$$

$$R'_{S}{}^{1}{}_{\sigma\mu} = iF_{S}{}^{1}{}_{\sigma\mu}$$
$$R'_{S}{}^{2}{}_{\sigma\mu} = iF_{S}{}^{2}{}_{\sigma\mu}$$

$$R'_{\Omega}{}^{1}{}_{\sigma\mu} = iF_{\Omega}{}^{1}{}_{\sigma\mu}$$
$$R'_{\Omega}{}^{2}{}_{\sigma\mu} = iF_{\Omega}{}^{2}{}_{\sigma\mu}$$

$$R'_{B}{}^{1}{}_{\sigma\mu} = iB^{1}{}_{\sigma\mu}$$
$$R'_{B}{}^{2}{}_{\sigma\mu} = iB^{2}{}_{\sigma\mu}$$

with the further definition of $R''^{1}{}_{\sigma\mu}$ and $R''^{2}{}_{\sigma\mu}$:

$$R''^{1}{}_{\sigma\mu} = R'_{SU(3)}{}^{1}{}_{\sigma\mu} + R^{1}{}_{\sigma\mu}$$
$$R''^{2}{}_{\sigma\mu} = R'_{SU(3)}{}^{2}{}_{\sigma\mu} + R^{2}{}_{\sigma\mu} \tag{26.19}$$

Eq. 26.18 is the Ricci tensor. An additional Ricci-like tensor is

$$H_{\sigma\mu} = H^{\beta}{}_{\sigma\beta\mu} \tag{26.20}$$

The curvature scalar is

$$R' = g^{\sigma\mu}R'_{\sigma\mu} = + \partial^{\sigma}H^{1\beta}{}_{\sigma\beta} - \partial_{\beta}H^{1\beta}{}_{\sigma}{}^{\sigma} + H^{1\gamma}{}_{\beta\sigma}H^{1\beta}{}_{\gamma}{}^{\sigma} - H^{1\gamma}{}_{\mu\sigma}H^{1\beta}{}_{\gamma\beta} + \partial^{\sigma}H^{2\beta}{}_{\sigma\beta} - \partial_{\beta}H^{2\beta}{}_{\sigma}{}^{\sigma} +$$
$$+ H^{2\gamma}{}_{\beta\sigma}H^{2\beta}{}_{\gamma}{}^{\sigma} - H^{2\gamma\sigma}{}_{\sigma}H^{2\beta}{}_{\gamma\beta} + H^{1\gamma}{}_{\beta\sigma}H^{2\beta}{}_{\gamma}{}^{\sigma} - H^{1\gamma\sigma}{}_{\sigma}H^{2\beta}{}_{\gamma\beta} + H^{2\gamma}{}_{\beta\sigma}H^{1\beta}{}_{\gamma}{}^{\sigma} - H^{2\gamma\sigma}{}_{\sigma}H^{1\beta}{}_{\gamma\beta}$$

$$= g^{\sigma\mu}(R^{1\beta}{}_{\sigma\beta\mu} + R^{2\beta}{}_{\sigma\beta\mu}) \tag{26.21}$$

Additional curvature scalars are

$$H = g^{\sigma\mu}H_{\sigma\mu} \tag{26.22}$$
$$R'^{2} = g^{\sigma\mu}R'^{2}{}_{\sigma\mu} \tag{26.23}$$

27. Total Boson Lagrangian

In this chapter[233] we define a total vector boson and gravitational lagrangian as a generalization of the usual Einstein lagrangian with additional higher derivative terms added.

The major aspect of our extension is the introduction of higher derivative terms in such a manner that they can be handled by canonical lagrangian methods to obtain the dynamical equations of motion and the equal time commutation relations (for the 'free' field approximations.)

We separate the Ricci tensor into two parts in order to use Pseudoquantization field theory (with fields labeled '1' and '2') to implement canonical lagrangian methods, and to introduce the flat space metric $\eta^{\sigma\mu}$ by a Higgs Mechanism.[234] The constant flat space metric part $\eta^{\sigma\mu}$ of the weak field quantum gravity metric is usually an assumed quantity. But its close relation to the quantum field $g^{\sigma\mu}$ suggests that it could be generated by the same Higgs Mechanism that generates particle mass constants.[235]

We assume the lagrangian density:[236]

$$\mathfrak{L} = \mathrm{Tr}\,\sqrt{g}[MD_\nu R''^1{}_{\sigma\mu}D^\nu R''^{2\sigma\mu} + aR'^1{}_{\sigma\mu}R'^{2\sigma\mu} + bR' + cg^{\sigma\mu}g^2{}_{\sigma\mu} + c'g^{2\sigma\mu}g^2{}_{\sigma\mu} - dA_{SU(3)}{}^2{}_\mu A_{SU(3)}{}^{2\mu}] \tag{27.1}$$

where M, a, b, c, c', and d are constants to be determined later, and $R''^i{}_{\sigma\mu}$ for i = 1, 2 is determined by eqs. 26.12 and 26.13.[237]

---

[233] Most of the material in this chapter appeared in Blaha (2016h) and earlier books.

[234] In earlier books such as Blaha (2016f) we showed that the use of two fields for each particle type enables us to clearly separate the 'vacuum expectation value' from its associated second quantized 'Higgs' field. The application to the weak field approximation for gravitons is one example.

[235] See also Blaha (2016c).

[236] Since the lagrangian terms are matrices it is necessary to take the trace.

[237] One may ask why $R''^1{}_{\sigma\mu}$ and $R''^2{}_{\sigma\mu}$ appear in the first term of the lagrangian, and not other interaction terms. We believe the primary reason is: "The extended vierbein $l^{\mu ai}(x)$ can be viewed as located at a point in a 32-dimensional complex-valued space.

$$l^{\mu ai}(x) = (\partial\xi_X{}^{ai}(x)/\partial x_\mu)_{X=h(x)}$$

where $\xi_X{}^{ai}$ is a set of locally inertial coordinates located at a 32-dimensional point X, and x = h(x) is a 4-dimensional point in a tangent subspace of the 32-dimensional space:

$$X = h(x)$$

The relation between complex 4-dimensional coordinates x and the 32-dimensional coordinates X is an embedding of a 4-dimensional surface within a 32-dimensional complex space when account is taken of the range of possible x values. We have considered such embeddings in Blaha (2015a), and in earlier books, and developed a theory of a 16-

This higher derivative lagrangian maintains the locality of the theory but does entail a modest modification in the derivation of the Euler-Lagrange equations of motion. It also requires the use of principal value propagators rather than ordinary Feynman propagators for gluon and graviton interactions. Thus the Strong Interaction sector, and the Gravitation sector are Action-at-a-Distance theories that are similar in spirit to Wheeler-Feynman Electrodynamics. The two U(1) Electromagnetic sectors, the Generation group U(4) gauge field sector, the Layer group U(4) gauge field sector, the two SU2) Weak sectors, the U(4) A_S gauge field sector, the spinor connection sector, and the Ω-interaction sector may, or may not, be Action-at-a-Distance fields. They are not constrained to be Action-at-a-Distance by the present considerations.

Since we wish to apply our theory cosmologically, and within hadrons, where the gravitational spinor connections are negligible due to the smallness of the gravitational constant G and the 'smallness' of B spin on the cosmological scale, we set $B^1{}_{\nu\mu} = B^2{}_{\nu\mu} = 0$ and find[238]

$$\mathcal{L} = \text{Tr}\ \sqrt{g}[MD_\nu(R'^1{}_{SU(3)\sigma\mu} + R'_G{}^1{}_{\sigma\mu})D^\nu(R'_{SU(3)}{}^{2\sigma\mu} + R'_G{}^{2\sigma\mu}) + \\ + aR'^1{}_{\sigma\mu}R'^{2\sigma\mu} + bR' + cg^{\sigma\mu}g^2{}_{\sigma\mu} + c'g^{2\sigma\mu}g^2{}_{\sigma\mu} - dA_{SU(3)}{}^2{}_\mu A_{SU(3)}{}^{2\mu}] \qquad (27.2)$$

Since there are no strong interaction fields in 'empty' space and gravity is negligible within hadrons,[239] we can drop the interaction terms between the Strong interaction and the Gravity interaction. However, we cannot drop the interaction terms amongst Electromagnetism, the Weak interaction, the Strong Interaction, the Generation group U(4) interaction, the Layer group U(4) interaction, the U(4) Generaly Relativity Reality group interaction, and the SU(3)⊗U(64) Ω-interaction – within, and between, hadrons. The interaction terms between Electromagnetism and Gravitation are important cosmologically.

Eq. 27.2 can therefore be expressed as:[240]

$$\mathcal{L} = \mathcal{L}_E + \mathcal{L}_{SU(2)} + \mathcal{L}_{DE} + \mathcal{L}_{DSU(2)} + \mathcal{L}_{SU(3)} + \mathcal{L}_U + \mathcal{L}_V + \mathcal{L}_S + \mathcal{L}_\Omega + \mathcal{L}_G + \mathcal{L}_{int} \qquad (27.3)$$

---

dimensional complex-valued space (the *Megaverse*) that contains our universe and probably many other universes." Thus SU(3) and Gravitation have a special role in our particle dynamics based on geometry. The second reason is the common feature of color SU(3) and real-valued General Relativity is that they are the only interactions that do not participate in 'rotations of interactions' as described earlier and in chapter 31. The third, practical reason is the experimental reality that the Strong Interaction and Gravitation are known to have 'anomalous' features that will be seen to be remedied by these insertions while the other interactions are 'conventional.'

[238] The constants have the dimensions: M has the dimension of inverse mass squared, b has dimension mass squared, a is dimensionless, c and c' have dimension mass, and d has dimension mass squared.

[239] We show gravity weakens at very short distances using our Two-Tier Quantum Field Theory formalism. See Appendix A, and Blaha (2003) and (2005a) among other books by the author.

[240] We only consider the gauge field lagrangian terms in this chapter.

where taking traces of $\mathcal{L}$s terms is understood

$$\mathcal{L}_E = \mathrm{Tr}\,\sqrt{g}\{M\{[\partial_\nu + i(A_E{}^1{}_\nu + A_E{}^2{}_\nu)]F^1{}_{E\sigma\mu}[\partial^\nu + i(A_E{}^{1\nu} + A_E{}^{2\nu})]F^2{}_E{}^{\sigma\mu}\} + aF_E{}^1{}_{\sigma\mu}F_E{}^{2\sigma\mu}\} \quad (27.4)$$
$$\mathcal{L}_{SU(2)} = \mathrm{Tr}\,\sqrt{g}[aF_W{}^1{}_{\sigma\mu}F_W{}^{2\sigma\mu}]$$
$$\mathcal{L}_{DE} = \mathrm{Tr}\,\sqrt{g}\{M\{[\partial_\nu + i(A_{DE}{}^1{}_\nu + A_{DE}{}^2{}_\nu)]F^1{}_{DE\sigma\mu}[\partial^\nu + i(A_{DE}{}^{1\nu} + A_{DE}{}^{2\nu})]F_{DE}{}^{2\sigma\mu}\} + aF_{DE}{}^1{}_{\sigma\mu}F_{DE}{}^{2\sigma\mu}\}$$
$$\mathcal{L}_{DSU(2)} = \mathrm{Tr}\,\sqrt{g}[aF_W{}^1{}_{\sigma\mu}F_W{}^{2\sigma\mu}]$$
$$\mathcal{L}_{SU(3)} = \mathrm{Tr}\,\sqrt{g}\{M[\partial_\nu + i(A_{SU(3)}{}^1{}_\nu + A_{SU(3)}{}^2{}_\nu)]F_{SU(3)}{}^1{}_{\sigma\mu}[\partial^\nu + i(A_{SU(3)}{}^{1\nu} + A_{SU(3)}{}^{2\nu})]F_{SU(3)}{}^{2\sigma\mu} +$$
$$+ aF_{SU(3)}{}^1{}_{\sigma\mu}F_{SU(3)}{}^{2\sigma\mu} - dA_{SU(3)}{}^2{}_\mu A_{SU(3)}{}^{2\mu}\}$$
$$\mathcal{L}_U = \mathrm{Tr}\,\sqrt{g}[aF_U{}^1{}_{\sigma\mu}F_U{}^{2\sigma\mu}]$$
$$\mathcal{L}_V = \mathrm{Tr}\,\sqrt{g}[aF_V{}^1{}_{\sigma\mu}F_V{}^{2\sigma\mu}]$$
$$\mathcal{L}_S = \mathrm{Tr}\,\sqrt{g}[aF_S{}^1{}_{\sigma\mu}F_S{}^{2\sigma\mu}]$$
$$\mathcal{L}_\Omega = \mathrm{Tr}\,\sqrt{g}[aF_\Omega{}^1{}_{\sigma\mu}F_\Omega{}^{2\sigma\mu}]$$

$$\mathcal{L}_G = \mathrm{Tr}\,\sqrt{g}[MD_\nu R^1{}_{\sigma\mu}D^\nu R^{2\sigma\mu} + aR^1{}_{\sigma\mu}R^{2\sigma\mu} + bg^{\sigma\mu}(R^{1\beta}{}_{\sigma\beta\mu} + R^{2\beta}{}_{\sigma\beta\mu}) + cg^{\sigma\mu}g^2{}_{\sigma\mu} + c'g^{2\sigma\mu}g^2{}_{\sigma\mu}]$$
$$= \mathrm{Tr}\,\sqrt{g}[MD_\nu R^1{}_{\sigma\mu}D^\nu R^{2\sigma\mu} + aR^1{}_{\sigma\mu}R^{2\sigma\mu} + bH + cg^{\sigma\mu}g^2{}_{\sigma\mu} + c'g^{2\sigma\mu}g^2{}_{\sigma\mu}]$$

$$\mathcal{L}_{int} = \mathcal{L} - (\mathcal{L}_E + \mathcal{L}_{SU(2)} + \mathcal{L}_{DE} + \mathcal{L}_{DSU(2)} + \mathcal{L}_{SU(3)} + \mathcal{L}_U + \mathcal{L}_V + \mathcal{L}_S + \mathcal{L}_\Omega + \mathcal{L}_G)$$

Thus $\mathcal{L}_{SU(3)}$, $\mathcal{L}_{SU(2)}$, $\mathcal{L}_E$, $\mathcal{L}_{DE}$, $\mathcal{L}_{DSU(2)}$, $\mathcal{L}_U$,, $\mathcal{L}_V$, $\mathcal{L}_S$, $\mathcal{L}_\Omega$, and parts of $\mathcal{L}_{int}$ are the dominant interactions within hadrons, and $\mathcal{L}_G$, $\mathcal{L}_E$ and parts of $\mathcal{L}_{int}$ are the dominant interactions in space within the framework of this discussion.

The $D_\nu R^1{}_{\sigma\mu}$ and $D^\nu R^{2\sigma\mu}$ terms have the form:

$$D_\nu R^i{}_{\sigma\mu} = + \partial_\nu R^i{}_{\sigma\mu} - H^{1\beta}{}_{\sigma\nu}R^i{}_{\beta\mu} - H^{1\beta}{}_{\nu\mu}R^i{}_{\sigma\beta}$$

for i = 1, 2.

27.1 The Eleven Interaction Unified Theory

The unification implemented in an eleven interaction Riemann-Christoffel curvature tensor leads to new interaction terms beyond those in The Standard Model. These additional interactions in $\mathcal{L}_{int}$ imply new phenomena in this unified theory such as: 1) a possible relationship between the various coupling constants; 2) a possible relationship between parts of these interactions; 3) a possible explanation of the deviations of gravity at intra-galactic distances and inter-galactic distances; 4) a possible solution of the proton spin puzzle due to electromagnetic-gluon terms in $\mathcal{L}_{int}$ that have not been previously considered; and 5) a possible explanation of the proton radius puzzle found in experiment due to Generation interaction-Electromagnetic interaction terms in $\mathcal{L}_{int}$, as well as other results.

27.2 The Strong Interaction Sector

The Strong Interaction lagrangian density terms are:[241,242]

$$\mathcal{L}_{SU(3)} = \mathrm{Tr}\,\sqrt{g}[MD_{SU(3)\nu}R'^{1}{}_{SU(3)\sigma\mu}D_{SU(3)}{}^{\nu}R'^{2}{}_{SU(3)}{}^{\sigma\mu} + aR'^{1}{}_{SU(3)\sigma\mu}R'^{2}{}_{SU(3)}{}^{\sigma\mu} - dA_{SU(3)}{}^{2}{}_{\mu}A_{SU(3)}{}^{2\mu}] \tag{27.5}$$

with

$$D_{SU(3)\nu} = [\partial_\nu + if(A_{SU(3)}{}^{1}{}_{\nu} + A_{SU(3)}{}^{2}{}_{\nu})]$$

and the electromagnetic lagrangian density term is now

$$\mathcal{L}_E = \sqrt{g}\{aF_E{}^{1}{}_{\sigma\mu}F_E{}^{2\sigma\mu}\} \tag{27.6}$$

Corresponding changes take place in $\mathcal{L}_{SU(2)}$ and $\mathcal{L}_{int}$.

We now approximate the metric determinant as g = 1 within hadrons. Thus the Strong lagrangian part becomes

$$\mathcal{L}_{SU(3)} = \mathrm{Tr}\,\{MD_\nu R'^{1}{}_{SU(3)\sigma\mu}D^{\nu}R'^{2}{}_{SU(3)}{}^{\sigma\mu} + aR'^{1}{}_{SU(3)\sigma\mu}R'^{2}{}_{SU(3)}{}^{\sigma\mu} - dA_{SU(3)}{}^{2}{}_{\mu}A_{SU(3)}{}^{2\mu}\} \tag{27.7}$$

$$= \mathrm{Tr}\,\{MD_\nu R'^{1}{}_{SU(3)\sigma\mu}D^{\nu}R'^{2}{}_{SU(3)}{}^{\sigma\mu} + \zeta R'^{1}{}_{SU(3)\sigma\mu}R'^{2}{}_{SU(3)}{}^{\sigma\mu} - \varsigma A_{SU(3)}{}^{2}{}_{\mu}A_{SU(3)}{}^{2\mu}\} \tag{27.8}$$

where

$$D_\nu R'^{j}{}_{SU(3)\sigma\mu} = \partial_\nu R'^{j}{}_{SU(3)\sigma\mu} + g_{SU(3)}[A_{SU(3)}{}^{1}{}_{\nu}, R'^{j}{}_{SU(3)\sigma\mu}] \tag{27.9}$$

for j = 1, 2.

---

[241] The form is virtually identical to S. Blaha, Phys. Rev. D**11**, 2921 (1974), and Blaha's 1976 Gravity Research Foundation Essay, except for the initial derivative term. See Appendices A and D.

[242] We note the constant a, which appears in this chapter and elsewhere is NOT the Charmonium constant a.

28. Gravitational Potential on the Three Distance Scales

This chapter[243] and the next chapter describe some of the possible results of the unified eleven interaction theory described earlier. The discussions in this book, and earlier books, focus almost entirely on the boson sector of a Theory of Everything. The fermion sector is described in detail in earlier chapters of this book, and Blaha (2015a) and previous books.

In this chapter we pull together results on the gravitational potential found in Blaha (2016g), (2016) and (2017a). We note that a new experimental study of 33,000 galaxies indicated that the gravitational potential at inter-galactic distances deviates significantly from the Newtonian potential G/r. In Blaha (2017a) we showed that such a deviation occurs in our theory. This experimental result was not known at the time of its writing.[244]

28.1 Total Gravitational Potential

The gravitational potential generated from the real-valued affine connection term bR' in the lagrangian eq. 27.1 was shown in chapter 7 of Blaha (2007a), with a massless graviton term required, to be

$$V^{tot}_G(\mathbf{r}) = -G/r - a_1 G e^{-m_G r}/r + a_2 G\cos(m_G r)/r \qquad (28.1)$$

where

$$m_G^2 = 2b/a \cong 10^{55}\ ev^2 = 10^{27}\ GeV^2 \qquad (28.2)$$
$$a_1 = ½$$
$$a_2 = ½$$

The gravitational potential contribution due to the gravitational gauge field $A_S^{1\mu}$ was shown to be:

$$V_S(\mathbf{r}) \equiv V_{GA1}(\mathbf{r}) = -[1/(96\pi a)][1/r - e^{-m_A r}/r] \qquad (28.3)$$

where

$$m_A = (a/M)^{½} = m_{SI} \cong 10^{-71}\ GeV \qquad (28.4)$$

---

[243] Most of this chapter appears in Blaha (2007a) and earlier books by the author.

[244] The gravitational potential was found to be greater than G/r at inter-galactic distances in a survey of 33,000 galaxies by M. Brouwer and colleagues at the Leiden Observatory (The Netherlands) in an announcement on December 18, 2016.

Since a ≅ 1 the coupling constant

$$1/(96\pi a) \cong 0.0033 \tag{28.5}$$

In comparison the electromagnetic fine structure constant is

$$\alpha \cong 0.0073$$

Thus the A_S coupling constant is approximately ½ of the fine structure constant.

The total gravitational potential due to both sources of gravity is

$$V(\mathbf{r}) = -G/r - a_1Ge^{-m_G r}/r + a_1G\cos(m_G r)/r - [1/(96\pi a)][1/r - e^{-m_A r}/r] \tag{28.6}$$

We now examine $V(\mathbf{r})$ at short distances (within the solar system), distances of tens of thousands of light years (intra-galactic distances), and distances between galaxies (hundreds of thousands to millions of light years and beyond).

28.2 Intra-Solar System Distance Scale

Since m_G is very large and m_A is extremely small, the gravitational potential at distances of up to at least several light years is approximately

$$V(\mathbf{r}) \cong -G/r \tag{28.7}$$

to well within feasible experimental limits.

28.3 Galactic Distance Scale

At distances of several tens of thousands of light years up to perhaps 100,000's of light years we find eq. 28.6 becomes approximately

$$V(\mathbf{r}) \cong -[G + 1/(96\pi a)]/r + ½G\cos(m_G r)/r \tag{28.8}$$

Note that the approximate expansion of the terms in eq. 28.8 to third order yields

$$V(\mathbf{r}) \sim -[G + 1/(96\pi a)]/r + a_1Gm_G^{\,3}r^2 - a_1Gm_G^{\,2}r + \text{constants} \tag{28.9}$$

with the resultant force

$$\mathbf{F} = \nabla V^{tot}{}_G(\mathbf{r}) \sim [G + 1/(96\pi a)]\mathbf{r}/r^3 + 2a_1Gm_G^{\,3}\mathbf{r} - a_2Gm_G^{\,2}\mathbf{r}/r + \ldots \tag{28.10}$$

This result is to be compared to the MoND force of A. Balakin et al, Phys. Rev. **D70**, 064027 (2004):[245]

$$F = -\lambda Gm[M/r^2 - |\Pi_c|\, r/c^2] \qquad (28.11)$$

and the vector form suggested by H-S Zhao et al, Phys. Rev. **D82**, 103001 (2010):

$$\partial\Phi/\partial\mathbf{r} = Gm\mathbf{r}/r^3 + (Gm)^{1/2}\mathbf{r}/r^2 \qquad (28.12)$$

Recent studies of 153 galaxies confirm the MoND discrepancy from Newtonian gravitation.[246]

28.4 Intergalactic Distance Scale

We can estimate the gravitational potential of V(**r**) in eq. 28.1 for large distances of the order of many hundreds of thousands of light years, and beyond, we find

$$V(\mathbf{r}) \cong -[G + 1/(96\pi a)]/r \qquad (28.13)$$

to good approximation since $m_G^2 = 2b/a \cong 10^{55}$ $ev^2 = 10^{27}$ GeV^2 sets a distance scale of the order of tens of thousands of light years causing the oscillating term to 'wash out,' and the $a_1Ge^{-m_Gr}/r$ term to be negligible. The e^{-m_Ar}/r term whose distance scale is short range is also negligible.

Consequently we find a deeper potential and thus a larger attractive gravitational force at inter-galactic distances in agreement with the 33,000 galaxy survey of M. Brouwer and colleagues.

28.5 Theoretical Agreement With Gravitational Data at all Known Distances

Our unified theory agrees with known gravitational data at the three distance scales.

[245] The constant c in eq. 28.11 is the speed of light, and M is the mass (not the M used in our lagrangian equations.)
[246] S. S. McGaugh, F. Lelli, and J. M. Schombert, arXiv: 1609.0591 (2016).

29. The Strong Interaction Sector

In this chapter[247] we describe some of the implications of the Strong Interaction sector of our theory. We will find a linear quark potential and quark confinement follows.

29.1 Strong Interaction Lagragian Terms

The flat space-time Strong Interaction gauge field lagrangian terms (eq. 27.5) is

$$\mathcal{L}_{SU(3)} = \text{Tr}\,\sqrt{g}[MD_{SU(3)\nu}R'^{1}{}_{SU(3)\sigma\mu}D_{SU(3)}{}^{\nu}R'^{2}{}_{SU(3)}{}^{\sigma\mu} + aR'^{1}{}_{SU(3)\sigma\mu}R'^{2}{}_{SU(3)}{}^{\sigma\mu} - dA_{SU(3)}{}^{2}{}_{\mu}A_{SU(3)}{}^{2\mu}] \tag{27.5}$$

Dropping the subscript $_{SU(3)}$ for added clarity and adding color fermion terms we obtain[248]

$$\mathcal{L}_{SU(3)} = \text{Tr}\,\{MD_\nu F^{1}{}_{\sigma\mu}D^{\nu}F^{2\sigma\mu} + aF^{1}{}_{\sigma\mu}F^{2\sigma\mu} - dA^{2}{}_{\mu}A^{2\mu}\} + \bar{\psi}[i\nabla\!\!\!/ + f(A\!\!\!/^{1} + A\!\!\!/^{2}) - m]\psi \tag{29.1}$$

where (for j = 1, 2)

$$D_\nu F^{j}{}_{\sigma\mu} = \partial_\nu F^{j}{}_{\sigma\mu} + f[A^{1}{}_{\nu}, F^{j}{}_{\sigma\mu}] \tag{29.2}$$

The lagrangian is equal to eq. 17 of S. Blaha, Phys. Rev. D**11**, 2921 (1974) except for additional terms $MD_\nu F^{1}{}_{\sigma\mu}D^{\nu}F^{2\sigma\mu}$ and $[A^{2}{}_{\kappa}, A^{2}{}_{\mu}]$; and the following changes in parameters:

$$a = -\tfrac{1}{2} \qquad\qquad d = \tfrac{1}{2}\lambda^2$$

Since that paper essentially contains a complete description of our Strong Interaction theory (modulo the additional terms) we refer the reader to it and to its predecessor paper referenced therein. There are a few additional changes required to bring the 1974 papers into agreement with our current theory:

1. Eq. 30 of the above referenced paper must be modified to

$$F^{2}{}_{\kappa\mu} = \partial A^{2}{}_{\mu}/\partial x^{\kappa} - \partial A^{2}{}_{\kappa}/\partial x^{\mu} + \mathbf{if[A^{2}{}_{\kappa}, A^{2}{}_{\mu}]} + if[A^{1}{}_{\kappa}, A^{2}{}_{\mu}] + if[A^{2}{}_{\kappa}, A^{1}{}_{\mu}] \tag{29.3}$$

---

[247] Most of this chapter appears in Blaha (2006h) and earlier books by the author.

[248] We note the constant a, that appears in this chapter and chapter 4, is NOT the Charmonium constant a.

with the addition of the 'bolded' term $if[A^2_\kappa, A^2_\mu]$. There is also a trivial change of notation of coupling constant from 'g' to 'f'.

2. Eqs. 6 and 18 should have the interaction term expanded to

$$gA^1 \rightarrow g(A^1 + A^2) \quad \text{`` `} \tag{29.4}$$

and similarly in eq. 20. Eqs. 38 – 41 directly show that the additional interaction term leads to a gluon propagator[249] $<A^1 + A^2, A^1 + A^2> = 2<A^1, A^2> + <A^1, A^1>$, and introduces a 1/r term in the potential part of the gluon propagator.

As a result the effective gluon propagator in the theory, **if the $Mf^2D_\nu F^1{}_{\sigma\mu}D^\nu F^{2\sigma\mu}$ term is neglected**, combines eqs. 38 and 39 to give the *short-distance*[250] gluon propagator between quarks:

$$g_{\mu\nu}\delta_{ab}P[\lambda^2/k^4 - 1/k^2] \tag{29.5}$$

up to a constant factor.[251]

These changes *explicitly* leads to a Strong Interaction potential of the form

$$V(r) = -2f^2/r + f^2\lambda^2 r \tag{29.6}$$

Naturally one can expect perturbative corrections to eq. 29.6 in higher order in f. However, as will be seen later, the apparent relative smallness of f suggests eq. 29.6 is a good approximation to the *short-distance*, inter-quark interaction.

29.1.1 Canonical Equal Time Commutation Relations

The Euler-Lagrange equations of motion, eqs. 27 – 31 in the 1974 paper, are modified most significantly by the $Mf^2D_\nu F^1{}_{\sigma\mu}D^\nu F^{2\sigma\mu}$ lagrangian term in eq. 29.2. In order to follow the canonical method to obtain the contributions to the equations of motion of this higher derivative term we will use integration by parts and discard surface terms, as is usually done in quantum field theory. We should start with eq. 29.2. However, with a view towards perturbation theory which appears reasonable in view of the smallness of the strong interaction coupling constant $f^2/4\pi = 0.024$ seen below, we will abstract a quadratic expresssion in the fields from eq. 29.2

[249] Eqs. 40-41 in the above referenced paper.

[250] We anticipate that the $Mf^2D_\nu F^1{}_{\sigma\mu}D^\nu F^{2\sigma\mu}$ term will affect the short-distance behavior of the inter-quark interaction. The equivalent term in the gravitation sector influences the long-distance form of the gravity potential and leads to a MoND-like behavior. See chapter 28.

[251] This propagator is taken in Principal value to avoid potential unitarity problems. This topic is described in detail in earlier papers and books by the author.

and then proceed to develop gluon propagators and the strong interaction potential. The 'free' Strong Interaction lagrangian that we use is

$$\mathcal{L}_{SU(3)F} = \mathrm{Tr}\{MD_{F\nu}F_F{}^{1a}{}_{\sigma\mu}D_F{}^{\nu}F_F{}^{2a\sigma\mu} + aF_F{}^{1a}{}_{\sigma\mu}F_F{}^{2a\sigma\mu} - dA^{2a}{}_{\mu}A^{2a\mu}\} + \bar{\psi}[i\nabla\!\!\!/ + f(A\!\!\!/^1 + A\!\!\!/^2) - m]\psi \tag{29.7}$$

where

$$F^{1a}{}_{\mu\kappa} = \partial A^{1a}{}_{\mu}/\partial x^{\kappa} - \partial A^{1a}{}_{\kappa}/\partial x^{\mu} \tag{29.8}$$
$$F^{2a}{}_{\mu\kappa} = \partial A^{2a}{}_{\mu}/\partial x^{\kappa} - \partial A^{2a}{}_{\kappa}/\partial x^{\mu}$$

and

$$D_{F\nu} = \partial_{\nu} \tag{29.9}$$

The conjugate momenta to $A^{1a}{}_{\mu}$ and $A^{2a}{}_{\mu}$ are respectively

$$\pi^{1a}{}_{\mu} = \partial\mathcal{L}_{SU(3)F}/(\partial A^{1a}{}_{\mu}/\partial t) = aF_F{}^{2a\mu t} \tag{29.10}$$
$$\pi^{2a}{}_{\mu} = \partial\mathcal{L}_{SU(3)F}/(\partial A^{2a}{}_{\mu}/\partial t) = aF_F{}^{1a\mu t}$$

The non-zero, equal time commutation relations are

$$[\pi^{ia}{}_{\mu}(\mathbf{x}, t), A^{jb}{}_{\nu}(\mathbf{y}, t)] = i(1 - \delta^{ij})\delta^{ab}\delta^{G(\mu\nu)}(\mathbf{x} - \mathbf{y}) \tag{29.11}$$

where i and j label the fields, and G(μν) indicates the gauge[252] G and the associated index expressions, with

$$\delta^{G(\mu\nu)}(\mathbf{x} - \mathbf{y}) = \int d^4k \exp(-ik\cdot x)b^G{}_{\mu\nu}(k)/(2\pi)^4 \tag{29.12}$$

where $b^G{}_{\mu\nu}(k)$ is a polynomial in k with a δ-function factor restricting the integration over k.

29.1.2 Dynamical Equations

After performing partial integrations on the $MD_{F\nu}F_F{}^{1}{}_{\sigma\mu}D_F{}^{\nu}F_F{}^{2\sigma\mu}$ term (and discarding surface terms at 'infinity') the Euler-Lagrange dynamical equations (in the Landau gauge) due to independent variations with respect to $A^{1a}{}_{\mu}$ is

$$2M\Box^2 A^{2a}{}_{\mu} - 2a\Box A^{2a}{}_{\mu} = -f\bar{\psi}T^a\gamma_{\mu}\psi \tag{29.13}$$

and, with respect to $A^{2a}{}_{\mu}$, is

[252] Not the gravitational coupling constant.

$$2M\Box^2A^{1a}{}_\mu - 2a\Box A^{1a}{}_\mu - 2dA^{2a}{}_\mu = -f\,\bar{\psi}T^a\gamma_\mu\psi \qquad (29.14)$$

where T^a is an SU(3) generator. Subtracting the equations we find

$$2M\Box^2A^{1a}{}_\mu - 2a\Box A^{1a}{}_\mu - 2M\Box^2A^{2a}{}_\mu + 2a\Box A^{2a}{}_\mu - 2dA^{2a}{}_\mu = 0$$

or

$$A^{2a}{}_\mu = [2M\Box^2 - 2a\Box + 2d]^{-1}[2M\Box^2A^{1a}{}_\mu - 2a\Box A^{1a}{}_\mu] \qquad (29.15)$$

with the result

$$\{2M\Box^2 - 2a\Box - 2d[2M\Box^2 - 2a\Box + 2d]^{-1}[2M\Box^2 - 2a\Box]\}A^{1a}{}_\mu = -f\bar{\psi}T^a\gamma_\mu\psi$$

or

$$\{2M\Box^2 - 2a\Box - 2d[2M\Box^2 - 2a\Box + 2d]^{-1}[2M\Box^2 - 2a\Box]\}A^{1a}{}_\mu = -f\bar{\psi}T^a\gamma_\mu\psi$$

$$\{2M\Box^2 - 2a\Box - 2d + 4d^2[2M\Box^2 - 2a\Box + 2d]^{-1}\}A^{1a}{}_\mu = -f\bar{\psi}T^a\gamma_\mu\psi \qquad (29.16)$$

Eq. 29.16 leads to the Principal Value (Feynman) propagator:

$$\begin{aligned} D^{11}{}_{\mu\nu}(x - y) &= P\, -i\langle 0|T(A^1{}_\mu(x), A^1{}_\nu(y)|0\rangle \\ &= P\int d^4k\, \exp(-ik\cdot x)b_{\mu\nu}(k)D_1(k)/(2\pi)^4 \end{aligned} \qquad (29.17)$$

where $b_{\mu\nu}(k)$ is a Landau gauge polynomial in k, and

$$\begin{aligned} D_1(k) &= \{2Mk^4 - 2ak^2 - 2d + 4d^2[2Mk^4 - 2ak^2 + 2d]^{-1}\}^{-1} \\ &= [2Mk^4 - 2ak^2 + 2d](2Mk^4 - 2ak^2)^{-2} \end{aligned}$$

Thus

$$\begin{aligned} D^{11}{}_{\mu\nu}(x - y) &= P\int d^4k\, \exp(-ik\cdot x)b_{\mu\nu}(k)[2Mk^4 - 2ak^2 + 2d]/[(2\pi)^4(2Mk^4 - 2ak^2)^2] \\ &= P\int d^4k\, \exp(-ik\cdot x)b_{\mu\nu}(k)[2Mk^4 - 2ak^2 + 2d]/[(2\pi)^4k^4(2Mk^2 - 2a)^2] \end{aligned} \qquad (29.18)$$

indicating a linear potential r term as well as terms of lower powers in r and Yukawa-like terms with a mass of $(a/M)^{1/2}$. We will describe the resulting effective Strong Interaction potential in more detail later.

Eq. 29.11 leads to the other propagator:

$$D^{12}{}_{\mu\nu}(x - y) = P\, -i\langle 0|T(A^1{}_\mu(x), A^2{}_\nu(y)|0\rangle$$

$$= P \int d^4k \exp(-ik\cdot x) b_{\mu\nu}(k) D_2(k)/(2\pi)^4 \quad (29.19)$$

where

$$D_2(k) = [2Mk^4 - 2ak^2]^{-1} \quad (29.20)$$

Thus

$$D^{12}{}_{\mu\nu}(x-y) = P \int d^4k \exp(-ik\cdot x) b_{\mu\nu}(k)/[(2\pi)^4 k^2 (2Mk^2 - 2a)] \quad (29.21)$$

indicating a 1/r potential term plus a Yukawa term with a mass of $(a/M)^{1/2}$.

Due to the form of the interaction with quarks (eq. 29.7) the total effective gluon interaction between quarks is

$$D^{tot}{}_{\mu\nu}(x-y) = D^{11}{}_{\mu\nu}(x-y) + 2D^{12}{}_{\mu\nu}(x-y)$$
$$= P \int d^4k \exp(-ik\cdot x) b_{\mu\nu}(k)\{[3Mk^4 - 3ak^2 + 2d]/[(2\pi)^4 k^4 (2Mk^2 - 2a)^2]\} \quad (29.22)$$

29.2 Strong Interaction Potential

Eq. 29.22 leads to the form of the total Strong Interaction potential. We note that the $\mu = \nu = 0$ part of the Feynman propagator for transverse gluons has the form:

$$D^{tot}{}_{00}(x-y) = \ldots - \int d^4k\, V_{SI}(\mathbf{k}) \exp(-ik\cdot(x-y))/(2\pi)^4 = \ldots + V_{SI}(\mathbf{x}-\mathbf{y})\delta(x_0 - y_0) \quad (29.23)$$

where

$$V_{SI}(\mathbf{x}) = -\int d^3k \exp(i\mathbf{k}\cdot\mathbf{x}) V_{SI}(\mathbf{k})/(2\pi)^3$$

with

$$V_{SI}(\mathbf{k}) = [3M\mathbf{k}^4 + 3a\mathbf{k}^2 + 2d]/[\mathbf{k}^4(2M\mathbf{k}^2 + 2a)^2] \quad (29.24)$$
$$= (2M)^{-2}\{2d(M/a)^2/\mathbf{k}^4 + [3a(M/a)^2 - 4d(M/a)^3]/\mathbf{k}^2$$
$$+ 2d(M/a)^2/(\mathbf{k}^2 + a/M)^2 + [-3a(M/a)^2 + 4d(M/a)^3]/(\mathbf{k}^2 + a/M)\}$$

Letting

$$m_{SI} = (a/M)^{1/2} \quad (29.25)$$

we find

$$V_{SI}(\mathbf{k}) = (2a)^{-2}\{2d/\mathbf{k}^4 + [3a - 4dm_{SI}{}^{-2}]/\mathbf{k}^2 + 2d/(\mathbf{k}^2 + m_{SI}{}^2)^2 + [4dm_{SI}{}^{-2} - 3a]/(\mathbf{k}^2 + m_{SI}{}^2)\} \quad (29.26)$$

The constant, a, is dimensionless and of order 1. The constant M has the dimension of inverse mass squared. We anticipate M will be extremely large resulting in a very small gluon mass m_{SI}.

There are massless gluon terms that generate color confinement reducing the impact of the massive gluon terms to a negligible effect outside hadronic regions.

We also see that the value of the inverse of graviton masses is of the order of the average galactic radius (the average galactic radius is large) and thus generate a Modified Newtonian potential (MoND) as seen in chapter 28.

Substituting eq. 29.26 we obtain a sum of massless and Yukawa-like potentials. The Yukawa potential is

$$V_Y(\mathbf{r}) = \int d^3k \exp(i\mathbf{k}\cdot\mathbf{r})/[(2\pi)^3(\mathbf{k}^2 + m^2)] = \exp(-mr)/[4\pi r] \quad (29.27)$$

Thus we obtain

$$V_{SI}(\mathbf{r}) = -(2a)^{-2}\{2d(dV_Y(\mathbf{r})/dm^2)|_{m=0} + [3a - 4dm_{SI}^{-2}]/(4\pi r) - 2d(dV_Y(\mathbf{r})/dm^2|_{m=m_{SI}}) + [4dm_{SI}^{-2} - 3a]V_Y(\mathbf{r})|_{m=m_{SI}}\} \quad (29.28)$$

with the form

$$V_{SI}(\mathbf{r}) = \alpha_1 r + \alpha_2/r + \alpha_3 e^{-m_{SI}r}/(4\pi m_{SI}) + \alpha_4 e^{-m_{SI}r}/(4\pi r) \quad (29.29)$$

where the constants α_i are:

$$\alpha_1 = -(2a)^{-2}d/(8\pi) \quad \text{(up to an infrared divergent constant)} \quad (29.30)$$
$$\alpha_2 = -(2a)^{-2}[3a - 4dm_{SI}^{-2}]/(4\pi)$$
$$\alpha_3 = -(2a)^{-2}d$$
$$\alpha_4 = -(2a)^{-2}[4dm_{SI}^{-2} - 3a]$$

Thus we find the form of the potential of eq. 29.29 is linear plus Yukawa-like terms with small mass m_{SI} – perhaps near zero. As a result the form of the effective potential is

$$V(r) = -2g^2/r + g^2\lambda^2 r \quad (29.31)$$

29.3 Charmonium and the Strong Interaction

In 1974 a bound state of a charmed and an anti-charmed quark was discovered by two experimental groups. Since charmed quarks are quite massive theoretical attempts were made to understand the charmed quark bound states within the framework of non-relativistic quantum mechanics. The "Cornell group" developed a fairly satisfactory[253] charmed quark bound state spectrum in 1974-5 using a combination of a linear and inverse 1/r potential as the strong interaction. In a recent fit[254] they gave the potential energy:

---

[253] As did a Harvard group.

[254] E. J. Eichten, K. Lane, and C. Quigg, arXiv:hep-ph/ 0206018 (2002). See this paper for references to earlier work by the "Cornell group" and the "Harvard group" as well as papers by other researchers.

$$V(r) = -\kappa/r + r/a^2 \qquad (29.32)$$

where $\kappa = 0.61$, $a = 2.38\ \text{GeV}^{-1}$ and the charmed quark mass was 1.84 GeV.

29.4 The Origin of the Linear Potential

The linear potential appears to have originated in a suggestion of Feynman in the Spring of 1974. This author proposed[255] a non-Abelian gauge quantum field theory, which yielded a linear potential. These papers, which had 4th order dynamic equations for the gauge fields, showed how to avoid the problems previously associated with higher derivative theories by using principal-value gauge field propagators that were similar in concept to the action-at-a-distance propagators used by Feynman and Wheeler in the late 1940's to formulate action-at-a-distance Quantum Electrodynamics.

Thus a non-Abelian quantum field theory of the strong interaction yielding a linear potential was created. In parallel with this development, Professor Kenneth Wilson (later a Nobelist) was developing lattice gauge theory. Because lattice lines focus the field of gauge boson, lattice gauge theory exhibited a linear potential as well between quarks. Thus it offered an alternative to our gauge theory. However, the linearity of the lattice potential was "built-in" by the lattice theory formulation and thus was an artifact of the lattice formulation. This approach, and other proposed approaches, all share the problem that the linear potential that they produce cannot be proven to truly be a consequence. Rather the linear potential is the "likely result."

On the other hand our higher dimensional theory produces the linear potential if the standard rules of quantum field theory are followed with the proviso that gauge field propagators are principal-value propagators.

This author had several discussions with Professor Wilson in late 1974 in the author's office and while walking to lunch at the Cornell Faculty Club. Professor Wilson proposed possible flaws in the author's theory on almost a daily basis. The author was able to show these suggested flaws were not flaws but physically acceptable. The final discussion with Wilson ended with Wilson stating words to the effect, "Your theory may be a correct phenomenological approximation to my theory of the strong interaction and quark confinement. But my theory is the correct one. Your theory is only a phenomenology." In the forty plus years since this concluding discussion no one has proved that the conventional strong interaction theory truly has a linear potential and quark confinement although some approximations suggest it does.

In the absence of a demonstration of a linear potential in the standard strong interaction model we suggest our theory is a viable alternative. Since the linear potential appears to fairly successfully describe much of the charmonium spectrum we feel our theory with its explicit

[255] S. Blaha, Phys. Rev. D**10**, 4268 (July, 1974) and Phys. Rev. D**11**, 2921 (December, 1974). These papers appeared before the charmonium calculations of the Cornell and Harvard groups in 1975.

derivation of a linear potential is worthy of interest – especially because it is in agreement with experiment as far as we know. *An experimentally completely correct phenomenology is a theory.*

29.5 Comparison Between Our Theory and the Charmonium Analysis

Comparing our above results with the Charmonium calculation we find[256]

$$g^2\lambda^2 = -(2a)^{-2}d/(8\pi) \tag{29.33}$$
$$-2g^2 = -(2a)^{-2}[3a - 4dm_{SI}^{-2}]/(4\pi) \tag{29.34}$$

where $g = \sqrt{(\kappa/2)} = 0.552$ and $\lambda = 0.761$ GeV (the result of Charmonium analysis). Thus

$$\lambda^2 = d/[4dm_{SI}^{-2} - 3a] \tag{29.35}$$

Note the Strong 'fine structure' constant found by the Cornell group

$$g^2/(4\pi) = 0.024 \tag{29.36}$$

is small. The electromagnetic fine structure constant α is 0.0073, a factor of about 3 lower than the Strong 'fine structure' constant. Thus it seems perturbation theory in the Strong interactions is not necessarily unreasonable.

256 We note the constant, a, that appears in chapter 4 and this chapter is NOT the Charmonium constant, a, in eq. 2.1.

30. Other Effects of New Interactions Between Boson Interactions

30.1 Missing Nucleon Spin Puzzle

The estimates of nucleon spin that are obtained from parton analyses of deep inelastic electron – nucleon interactions are woefully short of the spin expected in quark models of nucleons. The missing spin has been attributed to a number of causes. However the Missing Spin Puzzle remains.

From eq. 27.3 - 27.4 it is clear that there are important new interaction terms between the Electromagnetic and Strong interaction fields. After taking traces we find

$$\mathcal{L}_{\text{intEM-S}} = -\text{Tr}\ iM\{(A_E{}^1{}_\nu + A_E{}^2{}_\nu)\ F_{SU(3)}{}^1{}_{\sigma\mu}D^\nu F_{SU(3)}{}^{2\sigma\mu} + iD_\nu F_{SU(3)}{}^1{}_{\sigma\mu}(A_E{}^{1\nu} + A_E{}^{2\nu})F_{SU(3)}{}^{2\sigma\mu}\} \qquad (30.1)$$

$\mathcal{L}_{\text{intEM-S}}$ generates a combined photon-gluon vertex insertion in gluon interactions between quarks within a hadron. Figs. 30.1 and 30.2 show two simple possible vertex insertions in a gluon line.

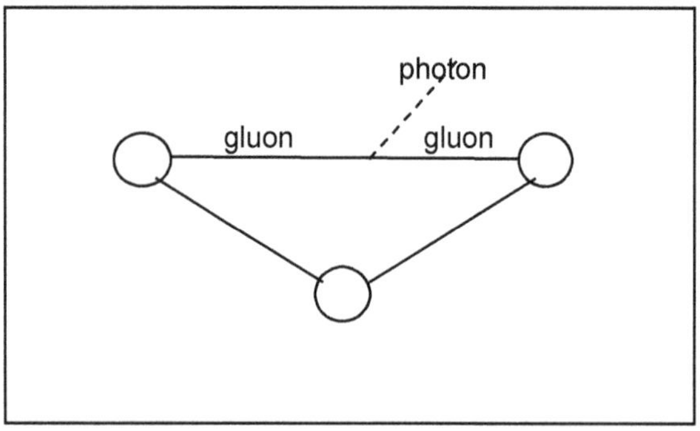

Figure 30.1. A single 'outgoing' photon vertex insertion in a gluon line. Only single gluon lines between the three quarks are displayed.

The gluon line, by itself, has $1/k^4$ and $1/k^2$ momentum space propagator terms. *The insertion of the vertex in Fig.30.1 generated by* $\mathcal{L}_{\text{intEM-S}}$ *yields a combined momentum factor of*

$k^3(k^4k^4)^{-1} = k^{-5}$ which would make it (summed over all gluon lines) a significant contribution to the proton spin determination in deep inelastic electron-nucleon scattering.[257] *The insertion of the vertex in Fig. 30.2 generated by $\mathcal{L}_{intEM\text{-}S}$ yields a combined momentum factor of $k^2(k^4k^4)^{-1} = k^{-6}$ which may have a less significant effect.*

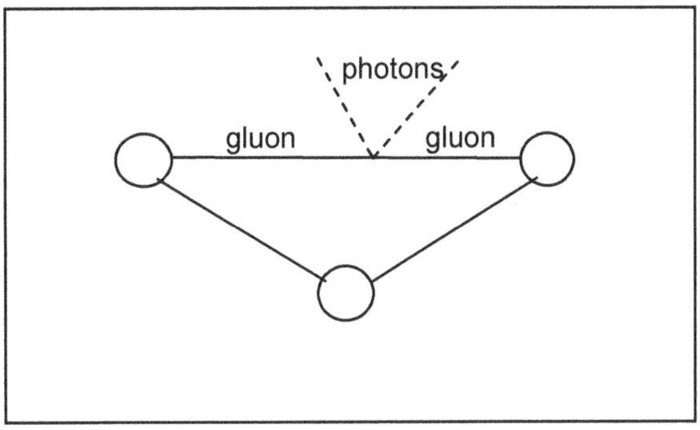

Figure 30.2. An 'outgoing' two photon vertex insertion in a gluon line. Only single gluon lines between the three quarks are displayed.

Thus our unified theory may solve the nucleon spin puzzle. The interactions in Figs. 30.1 and 30.2 introduce a new direct connection between photons and spin one gluons. Thus their contributions to the summations of proton spin interactions in parton models may account for the 'missing' two-thirds of proton spin. Our unified theory has a new gluon-photon interaction that is not found in the conventional Standard Model.

30.2 Discrepancies between Proton Radius Measurements

Recently experiments have confirmed that the radius of a proton in a muonic hydrogen atom is smaller than the proton radius measured in a conventional hydrogen atom composed of a proton encircled by an electron. The lagrangian in eq. 27.1 indicates that there are direct interactions between photons and Generation group gauge fields such as

$$A_E{}^1 A_E{}^{2v} U^{1i}{}_v U^{2iv}$$

[257] See C. A. Aidala, S. D. Bass, D. Hasch, and G. K. Mallot, arXiv: 1209.2803v2 (2013) and references therein for a review of the 'missing' nucleon spin puzzle.

These interactions result in Feynman diagrams that modify the electromagnetic field between a proton and a circling muon or electron. Fig. 30.3 shows the simplest forms of this interaction for a muon and a quark within a proton.

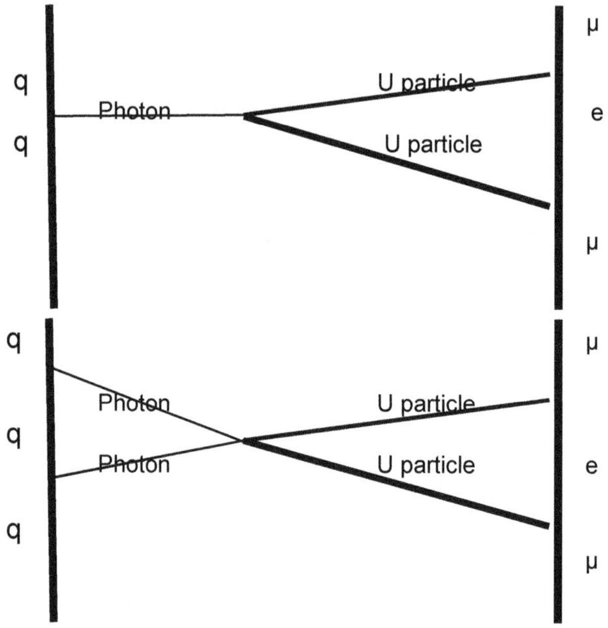

Figure 30.3. Photon-U gauge particle interactions between a muon and quark (within a proton).

The net effect of this additional interaction is to increase the force between the quark and muon beyond what one would expect using the electromagnetic potential, from the usual

$$V = -e^2/r$$

to

$$\begin{aligned} V' &= -(e+\delta)^2/r \\ &= -e^2/r' \end{aligned} \qquad (30.2)$$

where δ is small compared to e, and where

$$r' = r[e/(e+\delta)]^2 < r$$

giving an apparently smaller radial distance than the actual muon radial distance. This change would cause an energy shift to a lower value (more negative) in the muonic energy levels and, consequently, the proton radius would appear smaller for muonic atoms – as experiment confirms.[258]

Thus the new 'interactions between interactions' of our unified theory have significant effects that are already qualitatively verified by experiment.

[258] Fig. 30.3 is not of importance for 'normal' hydrogen since the electron is so much lighter than the muon.

31. The 'Interaction Rotation' Interaction A_Ω

31.1 The Eight Rotating Interactions

The plenitude of interactions that we have identified and summarized in chapter 25 leads us to consider the possibility of a unification principle based on the fourth principle of section 24.4 – the introduction of a form of a 'rotation of interactions' and a unification based on the Riemann-Christoffel tensor.[259]

We begin with the interactions 8-vectors of chapter 19:

$$\mathbf{A}_I^{1\mu}(x) = (g_1\mathbf{A}_{SU(3)}{}^{1\mu}(x),\ g_2\mathbf{W}^{1\mu}(x),\ g_3\mathbf{A}_E{}^{1\mu}(x),\ g_4\mathbf{W}_D{}^{1\mu}(x),\ g_5\mathbf{A}_{DE}{}^{1\mu}(x),\ g_6\mathbf{U}^{1\mu}(x),\ g_7\mathbf{V}^{1\mu}(x),\ g_8\mathbf{A}_S{}^{1\mu}(x))$$
$$\mathbf{A}_I^{2\mu}(x) = (g_1\mathbf{A}_{SU(3)}{}^{2\mu}(x),\ g_2\mathbf{W}^{2\mu}(x),\ g_3\mathbf{A}_E{}^{2\mu}(x),\ g_4\mathbf{W}_D{}^{2\mu}(x),\ g_5\mathbf{A}_{DE}{}^{2\mu}(x),\ g_6\mathbf{U}^{2\mu}(x),\ g_7\mathbf{V}^{2\mu}(x),\ g_8\mathbf{A}_S{}^{2\mu}(x))$$
$$(19.16)$$

where each element is a vector composed of the gauge fields in the group of the gauge field, and where the respective coupling constants were labeled $g_1, g_2, \ldots, g_8$.

The interactions' symmetry is

$$SU(3)\otimes SU(2)\otimes U(1)\otimes SU(2)\otimes U(1)\otimes U(4)\otimes U(4)\otimes U(4) \qquad (31.1)$$

The number of fields for each of the elements of the vectors is 8. 3, 1, 3, 1, 16, 16, and 16 respectively – totaling 64 fields.

Similarly we defined an 8-vector of 64 generators

$$\mathbf{T}_I = (\mathbf{T}_{SU(3)},\ \boldsymbol{\tau}_{SU(2)},\ \mathbf{I}_{U(1)},\ \boldsymbol{\tau}_{DSU(2)},\ \mathbf{I}_{DU(1)},\ \mathbf{G}_{U(4)},\ \mathrm{G}_{LU(4)},\ \mathbf{G}_S) \qquad (19.17)$$

Then the part of the gauge fields interactions within a covariant derivative corresponding to the eight interactions is

$$\mathbf{A}_I^{1\mu}(x)\cdot\mathbf{T}_I + \mathbf{A}_I^{2\mu}(x)\cdot\mathbf{T}_I = \mathbf{A}_I{}^{1\mu}{}_k(x)\mathbf{T}_{Ik} + \mathbf{A}_I{}^{2\mu}{}_k(x)\mathbf{T}_{Ik} \qquad (31.2)$$

summed over k for each gauge interaction.

---

[259] Much of this chapter first appeared in Blaha (2017a).

31.2 Form of the Eight Interactions

The eight interactions that appear in eq. 31.2 above, and in covariant derivatives (eqs. 26.2 and 26.3) are subject to 'Ω-rotations, which have the Pseudoquantum gauge fields[260], $A_\Omega^{1\mu}(x)$ and $A_\Omega^{2\mu}(x)$. Since the fields form a direct product (eq. 31.1), if we use the fundamental representations of each field, then we can consider them as having a Dirac field[261] with

$$3\times2\times2\times4\times4\times4 = 768$$

components counting spinor indices since the Species group gauge fields $\mathbf{A}_S^{1\mu}(x)$ act on spinor indices. Thus there are 768/4 = 192 distinct fermions—we identify the fermions as those of our Periodic Table of Fermions.

31.3 'Ω-Symmetry, 'Interaction Rotations' for Fermions

We define a column vector ψ containing the 4-spinors of all 192 normal and Dark fundamental fermions in our theory based on four generations in four layers as detailed earlier and in Blaha (2016a), (2016b) and (2016c). Then the Dirac equation has the form:[262]

$$\gamma_\mu D^\mu = \gamma_\mu\{\partial^\mu + i\,[g_\Omega A_\Omega^{1\mu}(x) + g_\Omega A_\Omega^{2\mu}(x) + \mathbf{A}_I^{1\mu}(x)\cdot\mathbf{T}_I + \mathbf{A}_I^{2\mu}(x)\cdot\mathbf{T}_I]\}\psi = 0 \quad (31.3)$$

with the 'Ω-Symmetry gauge field terms $A_\Omega^{1\mu}(x)$ and $A_\Omega^{2\mu}(x)$ inserted.

In general, the 'Ω-Symmetry group rotates the 64 = 8+3+2+1+2+1+4+4+4 field components of the eight interactions amongst each other. Thus the 'Ω-Symmetry group is U(64), to which we add an SU(3) factor to accommodate the 192 fermions in the combined fermion field ψ above, with the result the 'Ω-Symmetry group is SU(3)⊗U(64). Thus $\mathbf{A}_\Omega^{1\mu}(x)$ and $\mathbf{A}_\Omega^{2\mu}(x)$ gauge fields each separately have a 192×192 matrix representation of 4-component gauge fields labeled with the index μ. These rotations have 8×64^2 parameters.

The fields in $A_I^{1\mu}(x)$, and $A_I^{2\mu}(x)$ separately transform under an 'Ω-transformation consisting of a <u>1</u> representation of SU(3) and a <u>64</u> representation of U(64). They rotate the components of the eight interaction fields $\mathbf{A}_I^{1\mu}(x)$ and $\mathbf{A}_I^{2\mu}(x)$, each with 64 components, amongst each other. This rotation has 64^2 parameters.

In the case of the fermion field ψ we use the rotations of the <u>3</u> of the SU(3) and the <u>64</u> of U(64). factors to rotate a fermion vector ψ. These rotations have the same 8×64^2 parameters as the $\mathbf{A}_\Omega$ field.

Under an 'Ω-rotation we find the Dirac equation eq. 31.3 transforms to

---

[260] Those who prefer the usual formulation of Quantum Field Theory can use one field $A_\Omega^\mu(x)$.

[261] The discussion is analogous for other Dirac-like fields.

[262] We neglect the real-valued General Relativity interactions.

$$\gamma_\mu(x)\{\partial^\mu + i\,[g_\Omega A'_\Omega{}^{1\mu}(x) + g_\Omega A'_\Omega{}^{2\mu}(x) + \mathbf{A'}_I{}^{1\mu}(x)\cdot\mathbf{T'}_I + \mathbf{A'}_I{}^{2\mu}(x)\cdot\mathbf{T'}_I]\}\psi = 0 \tag{31.4}$$

where

$$\begin{aligned} A'_\Omega{}^{1\mu}(x) &= C_\Omega(x)A_\Omega{}^{1\mu}(x)C_\Omega{}^{-1}(x) - i\,C_\Omega(x)\partial^\mu C_\Omega{}^{-1}(x)/g_\Omega \\ A'_\Omega{}^{2\mu}(x) &= C_\Omega(x)A_\Omega{}^{2\mu}(x)C_\Omega{}^{-1}(x) \\ \mathbf{A'}_I{}^{1\mu}(x) &= \mathbf{A}_I{}^{1\mu}(x)C^{-1}{}_\Omega(x) \\ \mathbf{A'}_I{}^{2\mu}(x) &= \mathbf{A}_I{}^{2\mu}(x)C^{-1}{}_\Omega(x) \\ \mathbf{T'}_I &= C_\Omega(x)\mathbf{T}_I \end{aligned} \tag{31.5}$$

Note the Species group may cause the γ^μ Dirac matrix to become space-time dependent. (See section 23.1.2 for details.)

The effect of the 'Ω-transformation is to rotate the gauge fields components. It is accompanied by a rotation of the generator matrices components. Together they define an equivalent formulation of the original Dirac equation and thus the fermion sector. The next section provides an explicit simple example of 'Ω-transformations – an ElectroWeak theory.

Note that an 'Ω-transformation causes a change of gauge in the $A_\Omega{}^{1\mu}(x)$ field. The other gauge fields, $\mathbf{A}_I{}^{1\mu}(x)$ and $\mathbf{A}_I{}^{2\mu}(x)$, are 'rotated' but do not undergo a change of gauge. (Each of these other gauge fields do undergo their own particular changes of gauge for their transformation groups.)

The spinor connection field $B^{1\mu}$ (which Weinberg (1972) denotes as Γ^μ on p. 368) is not affected by 'Ω-transformations since it, as well as the General Coordinate affine connection, is in the gravitation sector, which all fermions (and bosons) experience uniformly.

31.4 Fermion Periodic Table and the 192 Dirac Spinor Vector ψ

The Periodic Table of 192 Fermions was constructed based on a number of reasonable (in the author's view) assumptions.[263] First we required that the fermion species (types) be constructed by complex Lorentz group transformations on a Dirac spinor at rest. This yielded four species of fermions that we identified with charged leptons, neutral leptons, up-type quarks, and down-type quarks. We then found that quarks occupied the 3 of color SU(3). Color SU(3) as well as the normal and Dark Weak interactions of group SU(2), and the normal and Dark electromagnetic interactions were 'derived' from the form of a Lorentz transformation.[264] Thus we found eight species of normal fermions, and four species of Dark fermions (since Dark fermions do not have color interactions – they are SU(3) singlets.) Thus in a generation there are eight normal fermions and four Dark fermions.

---

[263] The contents of this section and the figure appears in several of the author's earlier books in 2015 and 2016.
[264] These points all appear in Blaha (2015a).

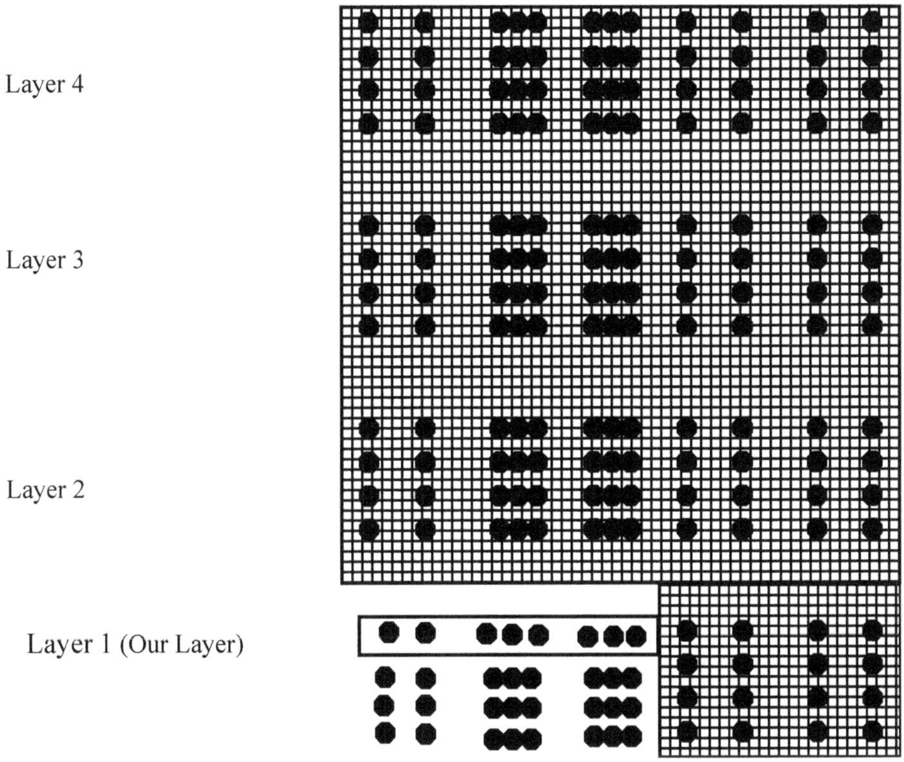

Figure 31.1. Dark parts of the periodic table are 'cross-hatched.' Light parts are the known fermions – with an additional, as yet not found, 4th generation of layer 1 is shown boxed. It is part of 'Dark matter' at present. When found experimentally it will be 'non-Dark.'

Then noting that there are four conserved particle number operators (baryon number,[265] lepton number, Dark baryon number and Dark lepton number) we were led to posit a U(4) Generation symmetry that generated four generations of fermions. Thus we found 32 normal fermions and 16 Dark fermions = 48 fermions in the four generations.

---

[265] The author has pointed out in previous books that disparities in measurements of the gravitational constant G could reflect the existence of a baryonic force that, like electromagnetism, leads to a conserved baryon number. Similar comments would apply to other conserved Generation group numbers and almost conserved Layer group numbers.

Next we noted that the number of particles in each of the four generations was (almost) conserved. This fact led to another group – the U(4) Layer group – that led to four layers of fermions. The total number of fermions then was found to be 4*48 = 192 fermions.

The Periodic Table that this construction implies appears in Fig. 31.1.

31.5 Structure of 192 Dirac Spinor Vector ψ

The Dirac spinor ψ for the 192 fermions is an eigenvector of SU3)$\otimes$U(64). We can express ψ in the form $\psi = \{\psi_{ij}\}$ where i = 1, 2, 3 is an SU(3) index and j = 1, 2, … , 64 is a U(64) index. Thus for each value of j there is a triplet of spinor fields. Since there are 32 normal quark triplets in the periodic table totaling 96 quarks it is natural to identify these 32 triplets as also SU(3) triplets of the Ω interaction.[266] The remaining 96 fermions can be structured as 32 Ω triplets as well. We define these triplets as follows:

- In each generation of each layer we form an UP-type triplet consisting of two normal charged leptons plus one up-type Dark quark.
- In each generation of each layer we form an DOWN-type triplet consisting of two normal neutral leptons plus one down-type Dark quark.

Thus we have a structuring of the 192 fermion field vector into 64 triplets.

It is possible to reorder the structure of ψ and $A_\Omega{}^{1\mu}(x)$ and $A_\Omega{}^{2\mu}(x)$ with U(192) transformations. These transformations will also change the ordering of the eight gauge interaction generator matrices $\mathbf{T}_I$. A U(192) reordering does not have appear to have physical consequences – it can be viewed as a bookkeeping change.

If, on the contrary, we take Ω-Symmetry to be U(192),[267] then it is broken at several levels to yield the eight interactions of which it is composed. Firstly, it is broken to SU3)$\otimes$U(64), and then it is broken to the eight interactions. We anticipate that the breakdown takes place in one step through some form of Higgs Mechanism for masses and, perhaps, for coupling constants as we described in Blaha(2015d).

31.6 Broken Ω-Symmetry

Ω-symmetry is broken by the complexon nature of color SU(3) gauge fields and quarks. The color SU(3) gauge fields $\mathbf{A}_{SU(3)}{}^{j\mu}(x)$ for j = 1,2 appearing in $\mathbf{A}_I{}^{j\mu}(x)$ are complexon fields in our formulation of The Extended Standard Model. Thus they are functions of complex spatial coordinates

---

266266 Although we identify the triplets we still regard the interactions as separate and distinct.

267 The choice of U(192) is dictated by the number of fundamental fermions, which, in turn, is ultimately dictated by space-time geometry.

$$x_c = (t, \mathbf{x}_r - i\mathbf{x}_i) \qquad (3.123a)$$

Therefore color SU(3) does not admit of 'Ω-transformations.Consequently 'Ω-Symmetry is broken to SU(3)⊗SU(3)$_{Color}$⊗U(56). This symmetry breaking reduces the number of 'Ω-Symmetry generators, and parameters, to $8 + 56^2$.

The symmetry breaking mechanism may be expected to lead to large A_Ω masses. Together with the likely smallness of the g_Ω coupling constant, this leads us to expet that the A_Ω interaction, which occurs between all 192 fermions, will not be detectable.

31.7 Example: ElectroWeak-like Theory

ElectroWeak model is an example of a global 'Ω-transformation which does not include the full gamut of features outlined above. Focussing on the Weak and Electromagnetic interactions we consider the covariant derivative

$$\{\partial^\mu + i\,[g\mathbf{W}^\mu\cdot\boldsymbol{\tau} + g'W_0{}^\mu\tau_0]\}\psi = 0 \qquad (31.6)$$

Rotating $W_3{}^\mu$ and $W_0{}^\mu$ with $C^{-1}{}_\Omega$

$$\begin{bmatrix} gW_3{}^\mu \\ g'W_0{}^\mu \end{bmatrix} \rightarrow \begin{bmatrix} g\cos\theta\ Z^\mu + g\sin\theta\ A^\mu \\ -g'\sin\theta\ Z^\mu + g'\cos\theta\ A^\mu \end{bmatrix}$$

and rotating the generator matrices by C_Ω

$$\begin{bmatrix} \tau_3 \\ \tau_0 \end{bmatrix} \rightarrow \begin{bmatrix} \cos\theta\ \tau_3\ - \sin\theta\ \tau_0 \\ \sin\theta\ \tau_3\ + \cos\theta\ \tau_0 \end{bmatrix} \qquad (31.7)$$

and making an appropriate choice of θ_W yields the electromagnetic field A and Z we find

$$g'W_0{}^\mu\ \tau_0 + gW_3{}^\mu\tau_3 \rightarrow Z^\mu[(g\cos^2\theta - g'\sin^2\theta)\tau_3 - \tfrac{1}{2}\sin(2\theta)(g + g')\tau_0 + A^\mu[\tfrac{1}{2}\sin(2\theta)(g + g')\tau_3 + + (g'\cos^2\theta - g\sin^2\theta)\tau_0] \qquad (31.8)$$

If we choose the coefficient of A^μ be e, then

$$e(\tau_3 + \tau_0) = [½\sin(2\theta)(g + g')\tau_3 + (g'\cos^2\theta - g\sin^2\theta)\tau_0] \tag{31.9}$$

and thus

$$½ \sin(2\theta)(g + g') = (g'\cos^2\theta - g\sin^2\theta) = e \tag{31.10}$$

Consequently

$$g' = -(½ \sin(2\theta) - \sin^2\theta)/(½ \sin(2\theta) - \cos^2\theta) \tag{31.11}$$

If we further choose g = g' (since equal coupling constants is an often stated goal of theorists) for simplicity we find the angle $\theta = \pi/8$ or 22.5° while the usual Weinberg angle θ_W is about 30°. Thus we find a close similarity to standard ElectroWeak theory.[268]

We conclude that ElectroWeak theory can be viewed as an example of an 'Ω-transformation.

31.8 'Ω-Symmetry and Higgs Fields

The Higgs boson fields η (or our Pseudoquantum alternative) also participate in 'Ω-symmetry rotations. Consider a composite boson field constructed from the concatenation of all Higgs fields. The term generating the gauge field masses has the form

$$(D^\mu\eta)^\dagger D^\mu\eta \tag{31.12}$$

as we saw repeatedly in previous chapters. Upon rotating gauge fields, mass terms will also correspondingly 'rotate' to maintain the covariance of the overall theory.

The A_Ω gauge fields will also all acquire masses since there is no evidence of long range A_Ω fields experimentally.

31.9 Path Integral Formulation, and the Faddeev-Popov Method

The path integral formulation of our theory with the complete set of eleven integrations is fairly straightforward with one exception. Since we use complex valued coordinates for the color SU(3) gauge theory a somewhat different approach must be followed for it. We detail this approach in Blaha (2015a).

In this section we will explicitly consider the Faddeev-Popov Method for the 'Ω gauge field $A_\Omega^{1\mu}(x)$ only – noting that the Yang-Mills gauge fields of the other interactions must also be subject to gauge fixing using the Faddeev-Popov Method.

The path integral we consider for the $A_\Omega^{1\mu}(x)$ gauge field is:[269]

---

[268] We can, of course, make the angles equal by adjusting the values of g ang g'.

[269] The $A_\Omega^{2\mu}(x)$ transforms homogeneously and thus does not have a Faddeev-Popov determinant $\Delta(A_\Omega^2)$.

$$Z(J^\mu) = N\int DA_\Omega^{1\mu}\,\Delta(A_\Omega^{1})\delta(F(A_\Omega))\exp\{i\int d^4y[\mathscr{L} + J^\mu(y)\,A_\Omega{}^1{}_\mu(y)]\} \tag{31.13}$$

where $\delta(F(A_\Omega))$ specifies the gauge, and $\Delta(A_\Omega{}^1)$ is its Faddeev-Popov determinant. The Faddeev-Popov determinant can be calculated in the standard way.[270]

First we consider the gauge fixing delta function. Note that it can be written as a delta function in the gauge times a determinant:

$$\delta(F(A_\Omega{}^\omega)) = \delta(\omega - \omega_0)|\det \delta F(A_\Omega{}^1{}_\mu{}^\omega(x))/\delta\omega(x)|^{-1}|_{F(A_\Omega)=0} \tag{31.14}$$

where ω_0 is a reference gauge, where

$$\begin{aligned} A_\Omega{}^{1a}{}_\mu{}^\omega(x) &= A_\Omega{}^{1a}{}_\mu(x) - g_\Omega{}^{-1}\partial_\mu\omega^a + f_\Omega{}^{abc}\,\omega^b(x)A_\Omega{}^{1c}{}_\mu(x) \\ &= A_\Omega{}^{1a}{}_\mu(x) + \delta A_\Omega{}^{1a}{}_\mu{}^\omega(x) \end{aligned} \tag{31.15}$$

and where

$$F_A{}^a{}_\mu{}^\omega = F_A{}^a{}_\mu + f_\Omega{}^{abc}\,\omega^b(x)F_A{}^c{}_\mu \tag{31.16}$$

under an infinitesimal gauge transformation where the $f_\Omega{}^{abc}$ are SU3)⊗U(64) structure constants, and a, b, and c label the 72 generators of SU3)⊗U(64).

Also

$$\Delta(A_\Omega{}^1) = |\det \delta F(A_\Omega{}^1{}_\mu{}^\omega(x))/\delta\omega(x)||_{F(A_\Omega) = 0,\ \omega = 0} \tag{31.17}$$

We will choose the Lorentz gauge to evaluate the Faddeev-Popov determinant:

$$F^a(A_\Omega) = \partial_\mu A_\Omega{}^{1a\mu}(x) = 0 \tag{31.18}$$

We find

$$\begin{aligned} F^a(A_{\Omega\mu}{}^\omega(x)) &= \partial^\mu[A_\Omega{}^{1a}{}_\mu(x) - g_\Omega{}^{-1}\partial_\mu\omega^a(x) + f_\Omega{}^{abc}\,\omega^b(x)A_\Omega{}^{1c}{}_\mu(x)] \\ &= -g_\Omega{}^{-1}\partial^\mu\partial_\mu\omega^a(x) + f_\Omega{}^{abc}A_\Omega{}^{1c}{}_\mu(x)\,\partial^\mu\omega^b(x) \end{aligned} \tag{31.19}$$

Thus

$$\delta F^a(A_{\Omega\mu}{}^\omega(x))/\delta\omega^b(x) = -g_\Omega{}^{-1}\delta^{ab}\partial^\mu\partial_\mu + f_\Omega{}^{abc}A_\Omega{}^{1c\mu}(x)\partial_\mu \tag{31.20}$$

and

$$\Delta(A_\Omega) = |\det(g_\Omega{}^{-1}\delta^{ab}\partial^\mu\partial_\mu - f_\Omega{}^{abc}A_\Omega{}^{1c\mu}(x)\partial_\mu)| \tag{31.21}$$

where | … | represents the absolute value.

We can rewrite the Faddeev-Popov determinant as a path integral over anti-commuting c-number fields χ^a with a ghost Lagrangian:

---

[270] See for example Huang (1992).

$$\Delta(A_\Omega) = \int D\chi^* D\chi \exp[\, i\int d^4x\, \mathscr{L}_\Omega^{ghost}(x)] \qquad (31.22)$$

where

$$\mathscr{L}_\Omega^{\ ghost}(x) = \chi^{a*}(x)[\delta^{ab}\partial^\mu\partial_\mu \ - g f_\Omega^{\ abc} A_\Omega^{\ 1c\mu}(x)\partial_\mu]\chi^b(x) \qquad (31.23)$$

with a, b and c ranging from 1 through 192.

The Ω gauge field lagrangian thus acquires a ghost lagrangian:

$$Z(J^\mu) = N\int DA_\Omega^{\ 1\mu}\,\delta(F(A_\Omega))\exp\{i\int d^4y[\mathscr{L}_\Omega^{\ ghost}(y) \ + J^\mu(y)\, A_\Omega{}^1{}_\mu(y)]\} \qquad 31.24)$$

in addition to other factors for the other gauge field interactions.

31.10 Interactions of $A_\Omega^{\ \mu}(x)$ and Ghost 'Fields'

In general the $A_\Omega^{\ \mu}(x)$ field with its 192 components embodies interactions between all 192 fermions in the Fermion Periodic Table. Similarly, the 192 ghost fields also yield interactions between all 192 fermions through their affect on $A_\Omega^{\ \mu}$ quanta propagated between frmions.

The $A_\Omega^{\ \mu}(x)$ field quanta has in and out states in perturbation theory. Ghost fields only exist as interactions between $A_\Omega^{\ \mu}$ quanta within Feynman diagrams and do not have in or out states. In this section we will overview the $A_\Omega^{\ \mu}(x)$ and ghost interactions.

31.10.1 $A_\Omega^{\ \mu}(x)$ Interactions

$A_\Omega^{\ \mu}$ quanta have self-interactions that are qualitatively similar to those of other Yang-Mills gauge fields. They are 'more numerous' because the $A_\Omega^{\ \mu}$ field is an SU3)⊗U(64) gauge field.

$A_\Omega^{\ \mu}$ quanta can be exchanged between any pair of fermions. As a result quarks and leptons interact via $A_\Omega^{\ \mu}$ quanta exchange (with possibly $A_\Omega^{\ \mu}$ self-interactions within the quanta exchanges) Since the only known interactions between quarks and leptons are Weak, and Electromagnetic interactions, the coupling constant g_Ω must be very small and/or the mass of the $A_\Omega^{\ \mu}$ quanta is very large. Since the Ω-symmetry is clearly broken, it is likely that $A_\Omega^{\ \mu}$ quanta acquire a mass via the Higgs Mechanism.

31.11 A New Form of Unification

Perhaps the most remarkable property of the SU3)⊗U(64) Ω-symmetry is its universality in coupling to all fermions. This unifying feature of Ω-symmetry is a new Unification Paradigm beyond the three possible forms of unification specified in I. It enables us to view the origin of the universe – the Big Bang – as unified about Ω-symmetry – not as the 'separate' appearance of 192 different fermions and 11 different interactions. The Ω-symmetry, the 11th interaction, ties the Big Bang universe together.

Indeed we can view the various fermions as being generated from a primordial sea of A_Ω^μ *quanta that 'decay' into pairs of different fermions of the set of 192 fermions. We note that all fermions (and bosons) may be expected to be massless at the Big Bang instant.*

31.12 Ghost Field Interactions

Ghost fields interact with A_Ω^μ quanta in a manner analogous to ghost field interactions in other gauge theories. They do not interact directly with fermions – but rather modify the A_Ω^μ interaction between fermions by changing the A_Ω^μ propagator.

Ghost fields do not have in-states or out-states in perturbation theory. Thus they are not part of states contributing to unitarity sums.

31.13 Speculation on the Big Bang

The origin of the Big Bang is a subject of much discussion. The Ω-symmetry and interaction raises an interesting point. It is the only interaction (excluding gravity) that acts between all fermions and bosons. It can create particle-antiparticle pairs.

For these reasons it is possible to consider the universe was originally a plasma of A_Ω quanta that spontaneously transformed to a particle-laden universe undergoing expansion. This process is consistent with our Big Bang theory (with built-in inflation) presented in Blaha (2015a) and earlier books.

It would be of interest to see if Ω-symmetry can support a mechanism for the predominance of matter over antimatter.

32. A Deeper Physics

We have developed a unified theory of elementary particles and gravity based ultimately on Complex General Relativity. It is clear that the development of the theory required a close attention to details and could not be painted with 'broad brush strokes' such as the lay reader would hope.

The question, to which our development brings us, is whether there is a yet deeper Physics that remains to be found. There are still clearly many open questions—the primary question is apparently the origin of four dimensions. So very much is directly and indirectly based on the dimensionality of space. Why not more dimensions?[271]

More generally, is there a deeper theory—possibly some form or variation of Superstring theory—that may be the ultimate theory? This topic requires more information and thought than we can seem to muster at present. So we finish this book with an enormous WHY?

[271] We hope to address the issue of more dimensions and more universes within a higher dimensional universe in a future book. Blaha (2015a) and other books by this author have considered this possibility.

Appendix A. Two-Tier Quantum Field Theory and Renormalization

This appendix[272] eliminates *all* infinities that would otherwise be present in unrenormalized perturbation theory calculations. Previously the divergences in ElectroWeak theory were handled by methods developed by t'Hooft and collaborators with contributions by other theoretical physicists. These methods are not sufficient for the Extended Standard Model and for Quantum Gravity.

The Two-Tier Quantum Field Theory that we presented initially in Blaha (2002) eliminates all divergences in perturbation theory calculations in the Extended Standard Model and Quantum Gravity, and in the extension of our theories to 16-dimensional space. In this appendix we will consider major features of Two-Tier Quantum Field Theories using material abstracted from Blaha (2005a)and earlier books leaving the interested reader to pursue additional details in that book.

Two-Tier Quantum Field Theory is based on our new paradigm in the Calculus of Variations that we present in Appendix B.

A.1 Perturbation Theory Divergences

Earlier we suggested that complex coordinates and the Complex Loreatz group are the true space-time of our universe (plus some small curvature due to General Relativistic effects.) Subsequently we showed that the form of The Standard Model, embedded in a larger Extended Standard Model, follows directly from the complexity of space-time upon the introduction of the Reality group.

We now address an issue in the Extended Standard Model – infinities (divergences) that appear when unrenormalized perturbation theory Feynman diagram calculations are undertaken. The elimination of these divergences in certain sectors of The Standard Model was a major achievement. But in the Extended Standard Model divergencs remain in the Standard Model secor as well as the extended sector. In addition the unification of the (Extended) Standard Model with quantized General Relativity is prevented by the divergences that appear in Quantum General Relativity despite attempts to get around them through novel approaches.

---

[272] This appendix is extracted from Blaha (2015a).

In the spirit of Ockham's Razor we proposed a new approach to eliminating all divergences[273] in all quantum field theories including Quantum Gravity. We called this approach Two-Tier Quantum Field Theory. Chapter 10 and appendix 10-A discussed applications of this approach.) In this appendix we present a detailed discussion, which is a slightly amended version of parts of Blaha (2005) and Blaha (2002).

The essence of Two-Tier Quantum Field Theory is to take the real coordinate systems that result after the application of Reality group transformations to the underlying complex coordinate system and to introduce an addition to the real coordinates of a q-number gauge field similar to Quantum Electrodynamics as an imaginary part of the now "slightly" complex q-number coordinates. With this addition, and the use of our 'new' paradigm (Appendix B) for the Calculus of Variations, we can develop finite quantum field theories for the Extended Standard Model and Quantum Gravity (using the Einstein or other lagrangian).

Turning to the larger question of a unified theory of the Extended Standard Model and Quantum Gravity we note that 1) our extended Standard Model is firmly grounded in space-time geometry just like Quantum Gravity, and 2) we eliminate divergences in both these through the Two-Tier formalism. Thus we fulfill the criteria of Ockham's Razor by providing the "simplest" solution to the divergence problems that plagued both theories by providing a minimal change in quantum field theory that gives a maximal benefit – finite, divergence-free theories in perturbation theory calculations with the low energy behavior of theories well-approximated by 'normal' perturbation theory.[274]

A.2 Quantization of Coordinate Systems

A.2.1 Non-commuting Coordinates

Field theories with non-commuting coordinates are an active field of study.[275] Investigators are studying gauge theories, and in particular Quantum Electrodynamics, with non-commuting coordinates. Non-commuting coordinates are usually implemented quantum mechanically by positing non-zero commutators for coordinates:

$$[x^i, x^j] = i\theta^{ij} \tag{A.2.1}$$

---

[273] The notorious divergence in the fermion triangle diagram is eliminated by our approach eliminating the Adler-Bell-Jackiw anomaly.

[274] Thus the magnificent higher order calculations of T. Kinoshita and others remain correct in Two-Tier QED.

[275] M. R. Douglas and N. A. Nekrasov, Rev. Mod. Phys. **73**, 977 (2002) and references therein; J. Harvey, hep-th/0102076; M. Hamanaka and K. Toda, hep-th/0211148; N. Seiberg and E. Witten, hep-th/9908142; R. J. Szabo, hep-th/0109162; G. Berrino, S. L. Cacciatori, A. Celi, L. Martucci, and A. Vicini, hep-th/0210171; S. Godfrey and A. Doncheski, DESY eprint 02-195; M. Caravati, A. Devoto, and W. W. Repko, hep-th/0211463; and references within these papers.

A.2.2 New Approach to Non-Commuting Coordinates

In this appendix we will consider an alternative approach that postulates a q-number coordinate system X^μ with which all particle fields are defined. This coordinate system is realized as a mapping from a more fundamental c-number coordinate system y^ν, which we will call the subspace for want of a better term. We will treat X^μ as a vector of quantum fields, thus realizing a new type of non-commutative coordinates at unequal subspace times.

This approach is radically different from the non-commutative coordinate realizations hitherto discussed in the literature. It has a number of beneficial results to recommend it – the main result is the finiteness of quantum field theories that are defined within its framework. We will explore some of these results in the following chapters.

The X^μ coordinate system, as we define it, has a c-number real part and a q-number imaginary part. Thus particle fields which are normally defined on four-dimensional real space-time will now be defined on a complex four-dimensional space-time where four imaginary dimensions will appear as *Quantum Dimensions* embodied in a vector quantum field $Y^\mu(y)$.

$$X_\mu(y) = y_\mu + i\, Y_\mu(y)/M_c^2$$

where M_c is an extremely large mass of the order of the Planck mass or perhaps much larger.

The $Y^\mu(y)$ field is a function of the subspace y coordinates. The real part of the space-time dimensions will be taken to be the subspace y coordinates.[276]

The imaginary part of space-time will simply be the quantum fluctuations of a massless vector quantum field that are suppressed further by a very large mass scale – perhaps of the order of the Planck mass – that reduces the imaginary Quantum Dimensions to the infinitesimal. The effects of Quantum Dimensions only become appreciable in quantum field theory at energies of the order of M_c. At these energies the exponential Gaussian factor in each particle (and ghost) propagator that is generated by the Quantum Dimensions serves to make perturbation theory calculations ultra-violet finite – including calculations in Quantum Gravity.

The formalism that we will describe introduces a new form of interaction that does not have the form of the simple polynomial interactions that have hitherto dominated quantum field theories. This form of interaction takes place via the composition of quantum fields and can be called a *Dimensional Interaction* or an *Interdimensional Interaction* since it affects particle behavior through Quantum Dimensions.

---

[276] In a deeper theory the real part might also be a quantum field that undergoes a condensation to generate c-number coordinates. We will not consider this possibility in this book.

A.2.3 Quantization Using a C-Number X^μ

We will begin by considering the case of a scalar quantum field theory. We assume a real underlying y subspace. Since X^μ is a set of coordinates, we choose to define a scalar field ϕ as a function of X^μ, which in turn is a function of the y^ν coordinates. We will provisionally second quantize ϕ treating X^μ as c-number coordinates using a conventional approach.[277]

We assume a Lagrangian, with the momentum conjugate to ϕ:

$$\pi_\phi = \partial L_F / \partial\phi' \equiv \partial L_F / \partial(\partial\phi/\partial X^0) \qquad \text{(A.2.2)}$$

Following the canonical quantization procedure, π and ϕ become hermitian operators with equal time ($X^0 = X^{0\prime}$) commutation rules:

$$[\phi(X), \phi(X')] = [\pi_\phi(X), \pi_\phi(X')] = 0 \qquad \text{(A.2.3)}$$

$$[\pi_\phi(X), \phi(X')] = -i\,\delta^3(\mathbf{X} - \mathbf{X'}) \qquad \text{(A.2.4)}$$

The hamiltonian is defined by eq. B.112. (Appendix B contains a detailed development of the formalism for the scalar particle case. It was placed there because there are many formal similarities to conventional quantum field and this approach allows us to proceed more quickly to the main points of difference between conventional quantum field theory and Two-Tier quantum field theory in the present appendix. Appendix B also describes a new type of method – the composition of extrema – for the Calculus of Variations. *Equations numbered B.xxx are in Appendix B.*) We assume a metric $\eta_{\mu\nu}$ where $\eta_{00} = +1$, $\eta_{0i} = 0$, and $\eta_{ij} = -1$ for i, j = 1,2,3.

The standard Fourier expansion of the solution to the Klein-Gordon equation (eq. B.34) is:

$$\phi(X) = \int d^3p\, N_m(p)\, [a(p)\, e^{-ip\cdot X} + a^\dagger(p)\, e^{ip\cdot X}] \qquad \text{(A.2.5)}$$

where

$$N_m(p) = [(2\pi)^3 2\omega_p]^{-1/2} \qquad \text{(A.2.6)}$$

and

$$\omega_p = (\mathbf{p}^2 + m^2)^{1/2} \qquad \text{(A.2.7)}$$

---

[277] Some texts are: Bogoliubov, N. N., Shirkov, D. V., *Introduction to the Theory of Quantized Fields* (Wiley-Interscience Publishers Inc., New York, 1959); Bjorken, J. D., Drell, S. D., *Relativistic Quantum Fields* (McGraw-Hill, New York, 1965); Huang, K., *Quarks, Leptons & Gauge Fields Second Edition* (World Scientific, River Edge, NJ, 1992); Kaku, M., *Quantum Field Theory* (Oxford University Press, New York, 1993); Weinberg, S., *The Quantum Theory of Fields* (Cambridge University Press, New York, 1995).

The commutation relations of the Fourier coefficient operators are:

$$[a(p), a^{\dagger}(p')] = \delta^3(\mathbf{p} - \mathbf{p'}) \qquad (A.2.8)$$

$$[a^{\dagger}(p), a^{\dagger}(p')] = [a(p), a(p')] = 0 \qquad (A.2.9)$$

The reader will recognize the quantization procedure is formally identical to the standard canonical quantization procedure of a free scalar quantum field.

In the case of spin ½, spin 1 and spin 2 fields the standard quantization procedure *in terms of the X coordinate system* can also be followed in a way similar to the procedure in standard texts. We will see these quantization procedures in the following chapters. In the next section we will quantize the transformation from the y coordinate system to the X coordinate system.

The procedures developed in this section and the following sections may disturb some readers since we are placing operators with Dirac delta functions and using other unusual operator expressions. These concerns should be put at rest when we show that a path integral formulation presented later gives precisely the same results as the present development.

A.2.4 Coordinate Quantization

In this section we will quantize the coordinates X^μ as a vector field defined on a fundamental c-number coordinate system y^ν of the same dimensionality. We will assume the y^ν space is a "normal" flat Minkowski space with three spatial and one time dimensions. Generalizations to spaces with more dimensions are straightforward but will not be considered here.

Thus we will assume X^μ has three spatial dimensions and one time dimension. For reasons primarily of simplicity (primarily to avoid multiple time coordinates) we will assume the X^μ fields are similar to the free electromagnetic vector potential A^μ with the Lagrangian:

$$\mathscr{L}_C = +\tfrac{1}{4} M_c^4 F^{\mu\nu} F_{\mu\nu} \qquad (A.2.10)$$

$$F_{\mu\nu} = \partial X_\mu/\partial y^\nu - \partial X_\nu/\partial y^\mu \qquad (A.2.11)$$

where M_c^4 is a mass scale to the fourth power that is required on dimensional grounds and serves to set the scale for new Physics as we will see later. *Note the sign in eq. A.2.10 is not negative – superficially contrary to the conventional electromagnetic Lagrangian. The reason for this difference is that the field part of X^μ is imaginary.* Thus L_C winds up having the correct sign after taking account of the factors of i in the field strength $F_{\mu\nu}$.

We assume X^μ is complex[278] with the form:

$$X_\mu(y) = y_\mu + i\, Y_\mu(y)/M_c^2 \qquad (A.2.12)$$

where $Y_\mu(y)$ is a quantum field, M_c is a mass scale, and the real part is the c-number 4-vector y_μ. If X^μ has this form, then

$$F_{\mu\nu} = i\, (\partial Y_\mu/\partial y^\nu - \partial Y_\nu/\partial y^\mu)/M_c^2 \qquad (A.2.13)$$

Defining

$$F_{Y\mu\nu} = (\partial Y_\mu/\partial y^\nu - \partial Y_\nu/\partial y^\mu) \qquad (A.2.14)$$

we see the Lagrangian assumes the form of the conventional electromagnetic Lagrangian:

$$L_C = -\tfrac{1}{4}\, F_Y^{\mu\nu} F_{Y\mu\nu} \qquad (A.2.15)$$

This Lagrangian can be used to develop field equations and a canonical quantization that is completely analogous to Quantum Electrodynamics.

A.2.5 Gauge Invariance

The gauge invariance of the Lagrangian allows us to choose a convenient gauge. The gauge invariance of the full Lagrangian

$$\mathscr{L}_s = L_F(\phi(X), \partial\phi/\partial X^\mu)\, J + \mathscr{L}_C(X^\mu(y), \partial X^\mu(y)/\partial y^\nu) \qquad (B.96)$$

is based on the standard gauge invariance of $\mathscr{L}_C$, and the gauge invariance of $J\mathscr{L}_F$ in the form of translational invariance

$$X^\mu(y) \to X^\mu(y) + \delta X^\mu(y) \qquad (B.97)$$

for the special case of a translation of X with the form of a gauge transformation:

$$\delta X^\mu(y) = \partial\Lambda(y)/\partial y_\mu$$

[278] Theories of quantum mechanics, and quantum fields, in complex and quaternion spaces have been considered by numerous authors. For example see C. M. Bender, D. C. Brody and H. F. Jones, "Complex Extension of Quantum Mechanics" Phys. Rev. Letters **89**, 270401-1 (2002) and references therein; S. L. Adler and A. C. Millard, "Generalized Quantum Dynamics as Pre-Quantum Mechanics", Princeton Univ. preprint arXiv:hep-th/9508076 (1995) and references therein. These theories are all very different from the theories presented herein.

In this case eq. B.106 implies

$$\int d^4y\ \Lambda(y)\ \partial\ [\ J\ \partial/\partial X^\mu\ \mathscr{T}_{F\mu\nu}\ ]/\partial y_\nu\ = 0$$

after a partial integration and so gives the differential conservation law:

$$\partial\ [\ J\ \partial\mathscr{T}_{F\mu\nu}/\partial X^\mu]/\partial y_\nu\ = 0 \qquad \text{(A.2.16)}$$

since Λ(y) is arbitrary. This conservation law is trivially obeyed since, by eq. B.108:

$$\partial/\partial X^\mu\ \mathscr{T}_{F\mu\nu}\ = 0 \qquad \text{(B.108)}$$

Thus translational invariance in the $\mathscr{L}_F$ sector together with standard gauge invariance in the $\mathscr{L}_C$ sector automatically guarantees Y field gauge invariance of the total Lagrangian. Basically we use the separate invariance of each term of

$$L = \int d^4y\ [\mathscr{L}_F J + \ \mathscr{L}_C\ ] = \int d^4X\ \mathscr{L}_F\ + \int d^4y\ \mathscr{L}_C = L_F + L_C$$

under a constant translation $X^\mu \to X^\mu + \delta X^\mu$ where δX^μ is constant to establish eq. B.108. Then we consider a position dependent translation/gauge transformation to derive eq. A.2.16, which taken together with eq. B.108, establishes the invariance under the position dependent translation/gauge transformation eq. B.97.

An alternate approach that leads to the same result is to start with the particle part of the Lagrangian $\mathscr{L}_F$ rewritten to be invariant under general coordinate transformations as it must when we generalize to include General Relativity. Since position dependent translations are a form of general coordinate transformation the full theory must be invariant under position dependent translations due to invariance under general coordinate transformations.

Having established invariance under gauge transformations we now choose to use the most convenient gauge – the Coulomb gauge[279]:

$$\partial Y^i/\partial y^i = 0 \qquad \text{(A.2.17a)}$$

which, in the absence of external sources, allows us to set

[279] It is also possible to quantize using an indefinite metric that preserves manifest Lorentz covariance as was done by Gupta and Bleuler for the electromagnetic field. We will use the Gupta-Bleuler approach later to establish covariance under special relativity later. Now we opt for manifest positivity and use the Coulomb gauge.

$$Y^0 = 0 \qquad \text{(A.2.17b)}$$

since Y^0 does not have a canonically conjugate momentum. A conventional treatment leads to the equal time commutation relations:

$$[Y^\mu(\mathbf{y}, y^0), Y^\nu(\mathbf{y}', y^0)] = [\pi^\mu(\mathbf{y}, y^0), \pi^\nu(\mathbf{y}', y^0)] = 0 \qquad \text{(A.2.18)}$$

$$[\pi^j(\mathbf{y}, y^0), Y_k(\mathbf{y}', y^0)] = -i\,\delta^{tr}{}_{jk}(\mathbf{y} - \mathbf{y}') \qquad \text{(A.2.19)}$$

(Note the locations of the j indexes in eq. A.2.19 introduce a minus sign.) where

$$\pi^k = \partial\mathcal{L}_C \,/\, \partial Y_k' \qquad \text{(A.2.20)}$$

$$\pi^0 = 0 \qquad \text{(A.2.21)}$$

$$\delta^{tr}{}_{jk}(\mathbf{y} - \mathbf{y}') = \int d^3k\; e^{i\,\mathbf{k}\cdot(\mathbf{y} - \mathbf{y}')}(\delta_{jk} - k_jk_k/\mathbf{k}^2)/(2\pi)^3 \qquad \text{(A.2.22)}$$

$$Y_k' = \partial Y_k/\partial y^0 \qquad \text{(A.2.23)}$$

The Coulomb gauge reveals the two degrees of freedom that are present in the vector potential. The Fourier expansion of the vector potential is:

$$Y^i(y) = \int d^3k\; N_0(k) \sum_{\lambda=1}^{2} \varepsilon^i(k, \lambda)[a(k,\lambda)\, e^{-ik\cdot y} + a^\dagger(k,\lambda)\, e^{ik\cdot y}] \qquad \text{(A.2.24)}$$

where

$$N_0(k) = [(2\pi)^3 2\omega_k]^{-\frac{1}{2}} \qquad \text{(A.2.25)}$$

and (since m = 0)

$$\omega_k = (\mathbf{k}^2)^{\frac{1}{2}} = k^0 \qquad \text{(A.2.26)}$$

with $\vec{\epsilon}(k, \lambda)$ being the polarization unit vectors for $\lambda = 1,2$ and $k^\mu k_\mu = 0$.

The commutation relations of the Fourier coefficient operators are:

$$[a(k,\lambda), a^\dagger(k',\lambda')] = \delta_{\lambda\lambda'}\,\delta^3(\mathbf{k} - \mathbf{k}') \qquad \text{(A.2.27)}$$

$$[a^\dagger(k,\lambda), a^\dagger(k',\lambda')] = [a(k,\lambda), a(k',\lambda')] = 0 \qquad \text{(A.2.28)}$$

and the polarization vectors satisfy

$$\sum_{\lambda=1}^{2} \varepsilon_i(k, \lambda)\varepsilon_j(k, \lambda) = (\delta_{ij} - k_i k_j/\mathbf{k}^2) \quad \text{(A.2.29)}$$

It will be convenient to divide the Y field into positive and negative frequency parts:

$$Y^+_i(y) = \int d^3k\, N_0(k) \sum_{\lambda=1}^{2} \varepsilon_i(k, \lambda)\, a(k,\lambda)\, e^{-ik\cdot y} \quad \text{(A.2.30)}$$

and

$$Y^-_i(y) = \int d^3k\, N_0(k) \sum_{\lambda=1}^{2} \varepsilon_i(k, \lambda)\, a^\dagger(k,\lambda)\, e^{ik\cdot y} \quad \text{(A.2.31)}$$

For later use we note the commutator between the positive and negative frequency parts is:

$$[\, Y^-_j(y_1), Y^+_k(y_2)] = -\int d^3k\, e^{ik\cdot(y_1 - y_2)}\, (\delta_{jk} - k_j k_k/\mathbf{k}^2)/[(2\pi)^3 2\omega_k] \quad \text{(A.2.32)}$$

A.2.6 Bare ϕ Particle States

We now turn to the ϕ particle states. The creation and annihilation operators can be used to define "bare" free particle states. Bare free particle states are states that are not dressed with coherent states of Y quanta. For example a bare one-particle state of momentum p is

$$|p> = a^\dagger(p)|0_\phi> \quad \text{(A.2.33)}$$

with corresponding bare bra state

$$<p| = <0_\phi|a(p) \quad \text{(A.2.34)}$$

where the vacuum is defined as usual:

$$a(p)|0_\phi> = 0 \quad \text{(A.2.35)}$$

$$<0_\phi|a^\dagger(p) = 0 \quad \text{(A.2.36)}$$

Multi-particle bare states can also be defined in the conventional way with products of creation and annihilation operators applied to the vacuum.

A.2.7 Y Fock Space Imaginary Coordinate States

States can also be defines for the quantized Y field. These states will be similar in form to electromagnetic photon states but play a different role in our approach since they are in fact coordinate excitation states for the imaginary part of X^μ. Thus the scalar field (and other particle fields) will exist in a real four-dimensional space with quantum excitations into imaginary Quantum Dimensions. These excitations become significant at high energies. At the low energies with which we are familiar, space-time appears real; at very high energies space-time becomes slightly complex.

There are two types of imaginary coordinate excitations: 1.) Quantum excitations into Fock states consisting of superpositions of states with a definite finite number of Y "particles" and 2.) Imaginary coordinate excitations into coherent Y states with an "infinite" number of particles. Coherent states can be viewed as representing "classical" fields.

In this section we will consider Y field states with a definite number of excitations ("particles"). The creation and annihilation operators of the Y field can be used to define free particle states. For example a one particle state can be defined by

$$|k, \lambda> = a^\dagger(k, \lambda)|0_Y> \tag{A.2.37}$$

with corresponding bra state

$$<k, \lambda| = <0_Y|a(k, \lambda) \tag{A.2.38}$$

where the "coordinate vacuum" is defined as usual:

$$a(k, \lambda)|0_Y> = 0 \tag{A.2.39}$$
$$<0_Y|a^\dagger(k, \lambda) = 0 \tag{A.2.40}$$

Multi-particle states can also be defined in the conventional way with products of the creation and annihilation operators applied to the vacuum. The set of all states containing a finite number of "particles" constitutes a Fock space.

A state with a finite number of Y "particles" represents a quantum fluctuation into imaginary Quantum Dimensions. Such states do not appear in Two-Tier quantum field theory since the Y field is a free field and has no source. Thus they appear only as part of normal particles. A normal particle, such as a ϕ particle, has a coherent state of Y quanta associated with it, which play a role in interactions. The Y coherent state part of a normal particle can be viewed as boring an infinitesimal "hole" into an extra pair of imaginary dimensions in a neighborhood of the particle of a radial extent set by the length M_c^{-1}.

A.2.8 Y Coherent Imaginary Coordinate States

Coherent Y states bring us closer what we might consider to be "classical" imaginary dimensions – dimensions that we can, in principle, experience as we do normal dimensions. Let us define the coherent state[280]

$$| \, y, p> = e^{-\mathbf{p}\cdot\mathbf{Y}^-(y)/M_c^2}|0_Y> \quad (A.2.41)$$

This state is an eigenstate of the coordinate operator $Y^+(y')$:

$$Y^+_j(y_1)\,|y_2, p> = -[Y^+_j(y_1), \mathbf{p}\cdot\mathbf{Y}^-(y_2)]/M_c^2\,|y, p> \quad (A.2.42)$$

$$= -\int d^3k\,[N_0(k)]^2\, e^{ik\cdot(y_2 - y_1)}\,(p_j - k_j\mathbf{p}\cdot\mathbf{k}/\mathbf{k}^2)/M_c^2\,|y, p> \quad (A.2.43)$$

$$= p^i\Delta_{Tij}(y_1 - y_2)/M_c^2\,|y, p> \quad (A.2.44)$$

where $p^i\Delta_{Tij}(y_1 - y_2)/M_c^2$ is the eigenvalue of $Y^+_j(y_1)$. The eigenvalue of Y^+ becomes large as $(y_1 - y_2)^2 \to 0$. Thus the imaginary Quantum Dimensions become significant at very short distances, and significantly modify the high-energy behavior of quantum field theories. In particular, Quantum Dimensions have a significant effect when

$$(y_1 - y_2)^2 \lessapprox (4\pi^2M_c^2)^{-1} \quad (A.2.45)$$

according to eq. A.3.13 in the next chapter. We are assuming the mass scale M_c is very large – perhaps of the order of the Planck mass (1.221×10^{19} GeV/c^2). Thus imaginary Quantum Dimensions are far from detectable in today's "low" energy experiments. Their effect are significant in the analysis of the first instants after the Big Bang.[281]

A.2.9 The Dynamical Generation of New Dimensions

Effectively, the imaginary dimensions that we have constructed raise the total number of real and Quantum Dimensions to 8 with 6 space dimensions and two time dimensions. As we will see later the requirement of gauge invariance for the quantized Y field reduces the number of time dimensions to one and constrains the six space dimensions to five degrees of freedom giving a 5+1 dimensional space. Since X is a function of y we can also view the four dimensional world that we live in as a four-dimensional surface in a 6-dimensional space-time.

---

[280] Coherent states are well known in the physics literature. See for example T. W. B. Kibble, J. Math. Phys. **9**, 315 (1968) and references therein; V. Chung, Phys. Rev. **140**, B1110 (1965); J. R. Klauder, J. McKenna, and E. J. Woods, J. Math. Phys. **7**, 822 (1966) and references therein.

[281] See Blaha (2004).

A.2.10 Generation of Quantum Dimensions by the ϕ (X) field

The $\phi(X)$ field generates Quantum Dimensions via coherent states from the vacuum. From eq. A.2.5 and A.2.12 we see

$$\phi(X) = \int d^3p\, N_m(p)\, [a(p)\, e^{-ip\cdot(y + iY/M_c^2)} + a^\dagger(p)\, e^{ip\cdot(y + iY/M_c^2)}] \qquad (A.2.46)$$

with the result

$$\phi(X)|0> = \int d^3p\, N_m(p)\, a^\dagger(p)\, e^{ip\cdot(y + iY/M_c^2)}|0> \qquad (A.2.47)$$

is a superposition of coherent Y states plus one scalar particle. The vacuum state |0> is the product of the ϕ and Y vacuum states $|0> = |0_Y>|0_\phi>$. We will use |0> in most of the following discussions.

We can also define coherent Y states with total momentum q using the expression:

$$|q\ Y> = \int d^4y\, e^{iq\cdot X(y)}|0> = \int d^4y\, e^{iq\cdot(y + iY/M_c^2)}|0> \qquad (A.2.48)$$

Expanding the Y part of the exponential in eq. A.2.48 gives

$$|q\ Y> = \sum_{n=0}^{\infty}(-1)^n(n!)^{-1}\prod_{j=1}^{n}\left(\int d^3k_j N_0(k_j)\right)\delta^4\left(q - \sum_{s=1}^{n} k_s\right)\prod_{r=1}^{n}\ \sum_{\lambda_r=1}^{2} \mathbf{q}\cdot \boldsymbol{\varepsilon}(k_r, \lambda_r)\, a^\dagger(k_r,\lambda_r)|0>$$

(A.2.49)

which indicates that the sum of the Y particle momenta for each term in the expansion is q.

A.2.11 Hamiltonian for Particle and Coordinate States

The hamiltonian for the separable (field hamiltonian term separate from the Y hamiltonian term – see Appendix B), coordinate quantized, scalar quantum field theory is:

$$\mathscr{H}_s = \int d^3y\, \mathscr{H}_s \qquad (B.79)$$

with

$$\mathscr{H}_s = J\mathscr{H}_F + \mathscr{H}_C \qquad (B.82)$$

$$\mathscr{H}_F(\phi(X), \pi_\phi, \partial\phi/\partial X^i) = \pi_\phi\, \phi' - \mathscr{L}_F \qquad (B.83)$$

$$\mathscr{H}_C(X^\mu(y), \pi_X^{\ \mu}, \partial X^\mu(y)/\partial y^j, y') = \pi_X^{\ \mu} X_\mu' - \mathscr{L}_C \qquad (B.84)$$

$$\mathcal{L}_F = ½ [(\partial\phi/\partial X^\nu)^2 - m^2\phi^2] \quad (B.33)$$

$$\mathcal{L}_C = -¼\, M_c^4 F_Y^{\mu\nu} F_{Y\mu\nu} \quad (A.2.15)$$

We note

$$\mathcal{H}_F = ½ [\pi_\phi^2 + (\partial\phi/\partial X^i)^2 + m^2\phi^2] \quad (A.2.50)$$

is the conventional scalar particle hamiltonian when viewed as a function of the X coordinates. $\mathcal{H}_C$ has the same form as the conventional electromagnetic hamiltonian when eq. A.2.12 is used to specify X in terms of the Y fields.

$$\mathcal{H}_C = ½ (E_Y^2 + B_Y^2) \quad (A.2.51)$$

where

$$E_Y^i = -\partial Y^i/\partial y^0 \quad (A.2.52)$$

$$B_Y^i = \varepsilon^{ijk}\, \partial Y_j/\partial y^k \quad (A.2.53)$$

Using the fourier expansions of ϕ and X^μ (eqs. A.2.5 and A.2.24) we obtain the following expression for the normal-ordered hamiltonian $\mathcal{H}_S$:

$$P_s^0 \equiv \mathcal{H}_s = \int :\mathcal{H}_s: d^3y \quad (A.2.54)$$

$$\mathcal{H}_s = \int d^3p\, (\mathbf{p}^2 + m^2)^{½} a^\dagger(p)a(p) + \int d^3k \sum_{\lambda=1}^{2} (\mathbf{k}^2)^{½}\, a^\dagger(k, \lambda)a(k, \lambda) \quad (A.2.55)$$

where : : indicates normal ordering and where we perform a functional integration over X (Note the Jacobian is present within H_s.) for the particle part of the hamiltonian H_F. The hamiltonian is manifestly positive definite.

The spatial momentum is specified by

$$P_s^j = -\int d^3X\, :\pi_\phi(X)\partial\phi(X)/\partial X_j: + \int d^3y\, :E_Y^i \partial Y^i/\partial y_j: \quad (A.2.56)$$

$$= \int d^3p\, p^j\, a^\dagger(p)a(p) + \int d^3k \sum_{\lambda=1}^{2} k^j\, a^\dagger(k, \lambda)a(k, \lambda) \quad (A.2.57)$$

where the first term in eq. A.2.57 follows because of $\int d^3X$ in eq. A.2.56. The momentum operator generates displacements in ϕ

$$[P_s^{\mu}, \phi(X)] = -\, i\partial\phi/\partial X_{\mu} \tag{A.2.58}$$

A.2.12 Second Quantized Coordinates

At this point we have developed a formalism for a scalar particle quantum field theory based on our non-commutative coordinates. In the following sections we will proceed to use this formalism to develop a unified quantum field theory of the known forces of nature.

A.3 Scalar Two-Tier Quantum Field Theory

A.3.1 Introduction

In this section we will examine a new formulation of quantum field theory that we call *Two-Tier quantum field theory* in more detail for the case of a free scalar particle. This type of quantum field theory incorporates a structure similar to a string-like substructure within a quantum field theoretic framework.

A.3.2 "Two-Tier" Space

In the preceding section we developed quantized coordinates X^{μ} defined on an underlying c-number coordinates y^{ν} with the equations:

$$X_{\mu}(y) = y_{\mu} + iY_{\mu}(y)/M_c^{\,2} \tag{A.2.12}$$

$$Y^i(y) = \int d^3k\, N_0(k) \sum_{\lambda=1}^{2} \varepsilon^i(k, \lambda)[a(k,\lambda)\, e^{-ik\cdot y} + a^{\dagger}(k,\lambda)\, e^{ik\cdot y}] \tag{A.2.24}$$

We also developed a free scalar quantum field theory with the Fourier expansion:

$$\phi(X) = \int d^3p\, N_m(p)\, [a(p)\, e^{-ip\cdot X} + a^{\dagger}(p)\, e^{ip\cdot X}] \tag{A.2.5}$$

We will now consider the implications of the separable Lagrangian:

$$\mathscr{L}_s = \mathscr{L}_F(\phi(X), \partial\phi/\partial X^{\mu})\, J + \mathscr{L}_C(X^{\mu}(y), \partial X^{\mu}(y)/\partial y^{\nu}) \tag{B.96}$$

where

$$\mathscr{L}_F = ½ [(\partial\phi/\partial X^\nu)^2 - m^2\phi^2] \quad (B.33)$$

and

$$\mathscr{L}_C = -¼ M_c^4 F_Y^{\mu\nu} F_{Y\mu\nu} \quad (A.2.10)$$

with

$$F_{Y\mu\nu} = \partial Y_\mu/\partial y^\nu - \partial Y_\nu/\partial y^\mu \quad (A.2.14)$$

M_c is the mass that sets the scale at which the imaginary part of X^μ becomes significant.

This quantum field theory behaves as a conventional quantum field theory until energies reach the magnitude of M_c. At energies of the order of M_c, and above, the imaginary part of X^μ becomes significant and alters the high-energy behavior of the theory in a major way. This modification leads to the elimination of divergences that normally appear in perturbation theory when interactions are introduced. Yet the low energy behavior of the theory remains the same remains the same as conventional scalar quantum field theory. Thus the precise calculations of QED that have been verified to an amazing degree of accuracy remain valid when a Two-Tier formulation of QED is created. And the "low energy" results found in other conventional quantum field theories such as Electroweak Theory and the Standard Model also are closely approximated by their corresponding Two-Tier versions.

The straightforward use of the above equations[282] (and the canonical quantization described in the preceding chapters) leads to a scalar quantum field with the Fourier expansion:

$$\phi(X) = \int d^3p\, N_m(p)\, [a(p)e^{-ip\cdot(y + iY/M_c^2)} + a^\dagger(p)e^{ip\cdot(y + iY/M_c^2)}] \quad (A.3.1)$$

using eq. A.2.5 above. We note the equal time commutation relations of ϕ and π_ϕ are the same as the conventional equal time commutation relations of a scalar field despite the fact that X^μ and Y^μ are themselves quantum fields since $[Y^\mu(\mathbf{y}, y^0), Y^\nu(\mathbf{y}', y^0)] = 0$ for $\mathbf{y} \neq \mathbf{y}'$. In addition, we note the ϕ and π_ϕ fields are not hermitean.

The Fourier expansion of ϕ does require one refinement – the exponential terms in X^μ must be *normal ordered* to avoid infinities in the unequal time commutation relations:

$$\phi(X) = \int d^3p\, N_m(p)\, [a(p) :e^{-ip\cdot(y + iY/M_c^2)}: + a^\dagger(p) :e^{ip\cdot(y + iY/M_c^2)}:] \quad (A.3.2)$$

Since the hamiltonian as well as other quantities are normal ordered in quantum field theory the additional requirement of normal ordering in the field operator is merely an extension

---

[282] The use of functionals in quantum field theory is, of course, far from new as one can see in texts such as Bogoliubov (1959) (see for example pp. 198-226).

of a standard procedure to a more complex situation and is not disturbing. The unequal time commutation relation of the normal ordered ϕ field is:

$$[\phi(X^\mu(y_1)), \phi(X^\mu(y_2))] = i\Delta(y_1 - y_2) + O(1/M_c^2) \quad \text{(A.3.3)}$$

where

$$\Delta(y_1 - y_2) = -i \int d^3k\, (e^{-ik\cdot(y_1 - y_2)} - e^{ik\cdot(y_1 - y_2)})/[(2\pi)^3 2\omega_k] \quad \text{(A.3.4)}$$

is a familiar c-number invariant singular function. The additional terms in eq. A.3.3 are q-number terms that become significant at very short distances of the order M_c^{-1}. Thus precise measurements of field strengths at larger distances are limited by standard quantum effects as indicated by the commutation relation.

The principle of *microscopic causality* is violated at extremely short distances of the order M_c^{-1} since the commutator (eq. A.3.3) is non-zero, in general, for space-like distances of the order of M_c^{-1} due to the q-number terms. This violation is not experimentally measurable now – and for the foreseeable future – and reflects a type of non-locality at extremely short distances.

We will see that the short distance behavior of Two-Tier quantum field theory leads to the elimination of divergences resulting in finite interacting quantum field theories.

A.3.3 Vacuum Fluctuations

While the expectation value of a *conventional* free scalar field $\phi_{conv}(X)$ is zero in a conventional quantum field theory:

$$<0|\phi_{conv}(X)|0> = 0 \quad \text{(A.3.5)}$$

the vacuum fluctuations of *conventional* scalar quantum field theory are quadratically divergent:

$$<0|\phi_{conv}(X)\phi_{conv}(X)|0> = \int d^3p/[(2\pi)^3 2\omega_p] \quad \text{(A.3.6)}$$

In "Two-Tier" quantum field theory we find the vacuum expectation value of a free field is zero (like eq. A.3.5) *and the expectation value of the square of the field is also zero:*

$$<0|\phi(X)\phi(X)|0> = \int d^3p\, e^{-p^ip^j\Delta_{Tij}(0)/Mc^4}/[(2\pi)^3 2\omega_p] = 0 \quad \text{(A.3.7)}$$

since the exponential factor in the integral is $-\infty$. The exponent contains

$$\Delta_{Tij}(z) = \int d^3k\, e^{-ik\cdot z}\, (\delta_{ij} - k_ik_j/\mathbf{k}^2)/[(2\pi)^3 2\omega_k] \quad \text{(A.3.8)}$$

where "T"is for "Two-Tier". Thus *vacuum fluctuations are zero in Two-Tier quantum field theory.* Correspondingly, we will see that renormalization constants are finite in the Two-Tier versions of QED, Electroweak Theory, the Standard Model and Quantum Gravity.

A.3.4 The Feynman Propagator

The Feynman propagator for a Two-Tier free scalar quantum field is:

$$i\Delta_F^{TT}(y_1 - y_2) = \langle 0|T(\phi(X(y_1)),\phi(X(y_2)))|0\rangle \quad \text{(A.3.9)}$$

$$\equiv \langle 0|\phi(X(y_1))\phi(X(y_2))|0\rangle\, \theta(y_1^0 - y_2^0) + \phi(X(y_2))\phi(X(y_1))|0\rangle\, \theta(y_2^0 - y_1^0) \quad \text{(A.3.10)}$$

Since $X^0 = y^0$ in the Coulomb gauge of the X^μ field there is no ambiguity in the choice of the relevant time variable. A straightforward calculation shows:

$$i\Delta_F^{TT}(y_1 - y_2) = i\int d^4p\, e^{-ip\cdot(y_1 - y_2)}\, R(\mathbf{p}, y_1 - y_2)/[(2\pi)^4(p^2 - m^2 + i\varepsilon)] \quad \text{(A.3.11)}$$

where

$$R(\mathbf{p}, y_1 - y_2) = \exp[-p^ip^j\Delta_{Tij}(y_1 - y_2)/M_c^4] \quad \text{(A.3.12)}$$

$$= \exp\{-p^2[A(v) + B(v)\cos^2\theta] / [4\pi^2M_c^4z^2]\} \quad \text{(A.3.13)}$$

with

$$z^\mu = y_1^\mu - y_2^\mu \quad \text{(A.3.14)}$$

$$z = |\mathbf{z}| = |\mathbf{y}_1 - \mathbf{y}_2| \quad \text{(A.3.15)}$$

$$p = |\mathbf{p}| \quad \text{(A.3.16)}$$

$$v = |z^0|/z \quad \text{(A.3.17)}$$

$$A(v) = (1 - v^2)^{-1} + .5v\ln[(v - 1)/(v + 1)] \quad \text{(A.3.18)}$$
$$B(v) = v^2(1 - v^2)^{-1} - 1.5v\ln[(v - 1)/(v + 1)] \quad \text{(A.3.19)}$$

$$\mathbf{p}\cdot\mathbf{z} = pz\cos\theta \quad \text{(A.3.20)}$$

and $|\mathbf{p}|$ denoting the length of a spatial vector $\mathbf{p}$ while $|z^0|$ is the absolute value of z^0.

As eq. A.3.11 indicates, the Gaussian damping factor R(p, z) for large momentum p is the same for both the positive and negative frequency parts of the Two-Tier Feynman propagator. It is also important to note that R(p, z) does not depend on p^0 (in the Y Coulomb gauge) and thus the integration over p^0 proceeds in the usual way to produce time-ordered positive and negative frequency parts.

A.3.5 Large Distance Behavior of Two-Tier Theories

The large distance behavior of the Two-Tier Feynman propagator approaches the behavior of the conventional Feynman propagator since

$$R(\mathbf{p}, y_1 - y_2) \rightarrow 1 \qquad \text{(A.3.21)}$$

when $(y_1 - y_2)^2$ becomes much larger than M_c^{-2} as eq. A.3.13 shows. Thus the behavior of a conventional quantum field theory naturally emerges at large distance. We will see that the conventional Standard Model is the large distance limit of the Two-Tier Standard Model thus *realizing a form of Correspondence Principle for Quantum Field Theory*. Some features of the conventional Standard Model that depend specifically on the existence of divergences, such as the axial anomaly, will be different in the Two-Tier Standard Model since it is a divergence-free theory.

A.3.6 Short Distance Behavior of Two-Tier Theories

At short distances the Gaussian factor dominates and radically changes the behavior of the Feynman propagator eliminating its short distance singular behavior, and thus paving the way to finite quantum field theories. Near the light cone, $M_c^{-2} \gg -(y_1 - y_2)^2 \rightarrow 0$, we can approximate eq. A.3.11 with

$$i\Delta_F^{TT}(y_1 - y_2) \approx \int d^3p \, [N(p)]^2 \, R(\mathbf{p}, y_1 - y_2) \qquad \text{(A.3.22)}$$

since $e^{-ip\cdot(y_1 - y_2)}$ is approximately unity for small $(y_1 - y_2)$. We assume the mass of the ϕ particle is zero or is negligible at high energies so we set m = 0 to study the high energy behavior of eq. A.3.22. Upon performing the integrations in eq. A.3.22 for space-like $(y_1 - y_2)^2$ (and analytically continuing to the time-like regions[283,284]) we find

[283] See S. Blaha, "Relativistic Bound State Models with Quasi-Free Constituent Motion", Phys. Rev. **D12**, 3921 (1975) and references therein.

[284] It should be noted that A and B have the same sign for $0 \leq v < 1.1243$ thus making for easy analytic continuation across the light cone (which corresponds to v = 1).

$$i\Delta_F{}^{TT}(y_1 - y_2) \approx [z^2 M_c{}^4/(4i\surd A\surd B)] \ln[(\surd A + i\surd B)/(\surd A - i\surd B)] \quad (A.3.23)$$

with A and B defined in eqs. A.3.18 and A.3.19. As $(y_1 - y_2)^2 \to 0$ from the space-like or time-like side of the light cone we find eq. A.3.23 becomes:

$$i\Delta_F{}^{TT}(y_1 - y_2) \to \pi M_c{}^4 |(y_1 - y_2)^\mu (y_1 - y_2)_\mu|/8 \quad (A.3.24)$$

Eq. A.3.24 has several noteworthy points:

1. The propagator is well behaved on the light cone and approaches zero smoothly from both space-like and time-like directions. In contrast, the conventional scalar Feynman propagator diverges as $[(y_1 - y_2)^\mu (y_1 - y_2)_\mu]^{-2}$. This good behavior near the light cone will be seen later for other particle propagators with the net result that the usual infinities found in conventional quantum field theory are absent in Two-Tier quantum field theories.

2. The quadratic form of the propagator in eq. A.3.24 is suggestive of attempts to formulate a relativistic harmonic oscillator model of elementary particles[285] and more recent attempts to achieve quark confinement. The fact that the absolute value of the quadratic term appears in eq. A.3.24 neatly avoids the common pitfall seen in fully relativistic harmonic oscillator attempts.

3. The quadratic behavior *in coordinate space* of the propagator at short distances is equivalent to a high-energy behavior of

$$p^{-6} \quad (A.3.25)$$

in momentum space. Thus we get the equivalent *of a higher derivative theory* in Two-Tier quantum field theory at high energies while retaining a positive definite energy spectrum. The problems of negative metric states that have plagued conventional higher derivative quantum field theories are avoided.[286]

---

[285] H. Yukawa, H., Phys. Rev. **91**, 416 (1953); Y. S. Kim and M. E. Noz, Phys. Rev. **D8**, 3521 (1973) and references therein.

[286] S. Blaha, Phys.Rev. **D10**, 4268 (1974); S. Blaha, Phys.Rev. **D11**, 2921 (1975); S. Blaha, Nuovo Cim. **A49**, :113 (1979); S. Blaha, "Generalization of Weyl's Unified Theory to Encompass a Non-Abelian Internal Symmetry Group" SLAC-PUB-1799, Aug 1976; S. Blaha, "Quantum Gravity and Quark Confinement" Lett. Nuovo Cim. **18**, 60 (1977); Nakanishi, N., Suppl. Prog. Theo. Phys. **51**, 1 (1972); and references therein.

A.3.7 String-like Substructure of the Theory

Imaginary Quantum Dimensions endow a particle with an extended structure that resembles to some extent the extended structure seen in bosonic string and Superstring theories. For example, Bailin (1994) use the operator[287]

$$V_\Lambda(k) = \int d^2\sigma \sqrt{-h}\, W_\Lambda(\tau, \sigma)\, e^{-ik\cdot X} \qquad (A.3.26)$$

where X^μ is a quantized fourier expansion of the string fields (see eq. 7.22 of Bailin (1994)).

We note our X^μ coordinate-field has two transverse degrees of freedom due to gauge invariance, which also invites comparison to the bosonic string. A point of difference is that we will create a well-defined quantum field theoretic formulation in conventional space-time that has the Standard Model as its "large distance" behavior thus introducing a note of reality that is not (yet?) very apparent in Superstring theories. We see that the interacting quantum field theories based on this approach also have good, finite, short distance behavior just as string theories.

The scalar, and other particles', Feynman propagators can be viewed as describing the propagation of a particle cloaked (accompanied) by a cloud of Y particles (which generates the $R(\mathbf{p}, y_1 - y_2)$ factor in the propagator of eq. A.3.11). If we examine the fourier transform of R(p, z) we see:

$$(2\pi)^4 R(\mathbf{p}, q) = \int d^4z\, e^{iq\cdot z} R(\mathbf{p}, z) = \int d^4z\, e^{iq\cdot z} \exp[-p^i p^j \Delta_{Tij}(z)/M_c^{\,4}] \qquad (A.3.27)$$

and we find

$$R(\mathbf{p},q) = \sum_{n=0}^{\infty} [i(2\pi M_c)^4]^{-n} (n!)^{-1} \prod_{j=1}^{n} \left[\int d^4k_j\, \theta(k_j^{\,0})(\mathbf{p}^2 - (\mathbf{p}\cdot\mathbf{k}_j)^2/\mathbf{k}_j^{\,2})/(k_j^{\,2} + i\varepsilon)\right] \delta^4(q - \Sigma\, k_r)$$

(A.3.28)

which can be interpreted as a "cloud" of Y particles dressing the "bare" particle propagator. (The manifest divergences in eq. A.3.28 for R(p, q) are an artifact of the expansion and the subsequent fourier transformation. They are not present in the $R(\mathbf{p}, y_1 - y_2)$ factor in the propagator of eq. A.3.11.) See Fig. A.3.1 for the Feynman diagram of the Two-Tier cloaked propagator as compared to the normal scalar particle Feynman propagator. The Two-Tier Feynman propagator is basically a conventional scalar propagator that is modified by coherent Y particle emission.[288]

---

[287] D. Bailin and A. Love, *Supersymmetric Gauge Field Theory and String Theory* (Institute of Physics Publishing, Philadelphia, PA, 1994) page 272.

[288] T. W. B. Kibble, Phys. Rev. **173**, 1527 (1968) and references therein. In particular see p. 1532 of Kibble's paper.

We note that R(p, q) satisfies the convolution theorem:

$$\int d^4k\, R(\mathbf{p}, k)\, R(\mathbf{p}, q - k) = [R(\mathbf{p}, q)]^2 \tag{A.3.29a}$$

or

$$(2\pi)^4 \int d^4z\, e^{iq\cdot z}\, R(\mathbf{p}, z)\, R(\mathbf{p}, z) = [\int d^4z\, e^{iq\cdot z}\, R(\mathbf{p}, z)\,]^2 \tag{A.3.29b}$$

The proof follows from eq. A.3.28 and the Binomial theorem.

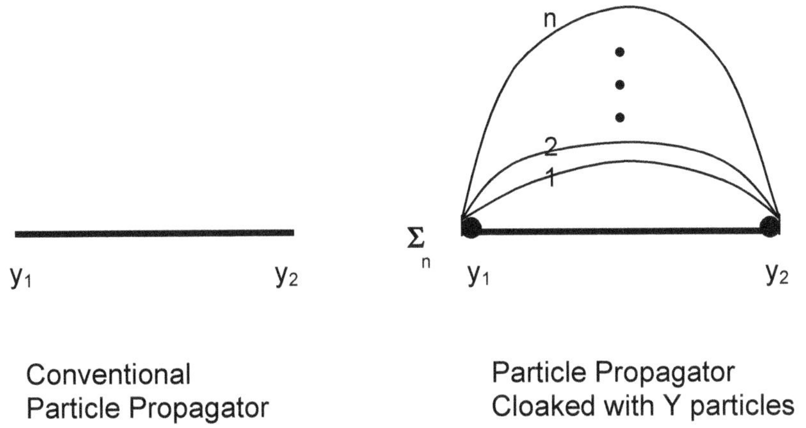

Figure A.3.1. Feynman diagram for conventional and cloaked Two-Tier propagators.

A.3.8 Parity

Parity can appear in two guises within the framework of Two-Tier quantum field theory. One can consider a parity operation where the space parts of X^μ are reversed while y^μ is unchanged. Or one can consider a second type of parity where the space parts of y^μ are reversed.

A.3.9 X Parity

Under this form of parity operation y^μ is unchanged while the arguments of ϕ *appear* to change by

$$X^i(y) \rightarrow - X^i(y) \tag{A.3.30}$$

$$X^0(y) \rightarrow X^0(y) \tag{A.3.31}$$

We will denote the parity operator of this type P_X. Under P_X the arguments of the scalar quantum field operator ϕ change according to eqs. A.3.30-1 so that ϕ transforms as

$$\mathscr{P}_X\phi(\mathbf{X}(y), X^0(y))\mathscr{P}_X^{-1} = \phi(-\mathbf{X}(y), X^0(y)) \tag{A.3.32}$$

We can implement eq. A.3.32 by requiring:

$$\mathscr{P}_X a(\mathbf{p}, p^0)\mathscr{P}_X^{-1} = a(-\mathbf{p}, p^0) \tag{A.3.33}$$

$$\mathscr{P}_X a^\dagger(\mathbf{p}, p^0)\mathscr{P}_X^{-1} = a^\dagger(-\mathbf{p}, p^0) \tag{A.3.34}$$

$$\mathscr{P}_X X^0(y)\mathscr{P}_X^{-1} = X^0(y) \tag{A.3.35}$$

$$\mathscr{P}_X X^i(y)\mathscr{P}_X^{-1} = X^i(y) \tag{A.3.36}$$

$$\mathscr{P}_X Y^i(y)\mathscr{P}_X^{-1} = Y^i(y) \tag{A.3.37}$$

where i = 1,2,3.

This parity transformation is analogous to the standard form of parity transformation in conventional quantum field theory. The separable Lagrangian in eq. B.96 (and listed at the beginning of this appendix) is invariant under this parity transformation.

A.3.10 y Parity

This form of parity transformation in which $y^i \rightarrow -y^i$ has significant differences from the normal parity transformation. We specify this parity transformation for a scalar quantum field by:

$$\mathscr{P}_y\phi(\mathbf{X}(\mathbf{y}, y^0), X^0(\mathbf{y}, y^0))\mathscr{P}_y^{-1} = \phi(\mathbf{X}(-\mathbf{y}, y^0), X^0(-\mathbf{y}, y^0)) \tag{A.3.38}$$

This transformation can be implemented through the following set of transformations:

$$\mathscr{P}_y a(\mathbf{p}, p^0)\mathscr{P}_y^{-1} = a(-\mathbf{p}, p^0) \tag{A.3.39}$$

$$\mathscr{P}_y a^\dagger(\mathbf{p}, p^0)\mathscr{P}_y^{-1} = a^\dagger(-\mathbf{p}, p^0) \tag{A.3.40}$$

$$\mathscr{P}_y X^0(\mathbf{y}, y^0)\mathscr{P}_y^{-1} = X^0(-\mathbf{y}, y^0) \tag{A.3.41}$$

$$\mathscr{P}_y Y^i(\mathbf{y}, y^0)\mathscr{P}_y^{-1} = -Y^i(-\mathbf{y}, y^0) \tag{A.3.42a}$$

$$\mathscr{P}_y a(\mathbf{k}, k^0,1)\mathscr{P}_y^{-1} = a(-\mathbf{k}, k^0,1) \quad \text{(A.3.42b)}$$

$$\mathscr{P}_y a(\mathbf{k}, k^0,2)\mathscr{P}_y^{-1} = -a(-\mathbf{k}, k^0,2) \quad \text{(A.3.42c)}$$

where i = 1,2,3 and assuming: $\boldsymbol{\varepsilon}(\mathbf{k},k^0,1)=-\boldsymbol{\varepsilon}(-\mathbf{k},k^0,1)$ and $\boldsymbol{\varepsilon}(\mathbf{k},k^0,2)=+\boldsymbol{\varepsilon}(-\mathbf{k},k^0,2)$.

A.3.11 Forms of the Parity Transformations

The parity transformations for a scalar particle are

$$\mathscr{P}_X = \exp\{-i\pi\int d^3p\, [a^\dagger(\mathbf{p}, p^0)a(\mathbf{p}, p^0) - a^\dagger(\mathbf{p}, p^0)a(-\mathbf{p}, p^0)]/2\} \quad \text{(A.3.43a)}$$

$$\mathscr{P}_y = \mathscr{P}_X \exp\{-i\pi \int d^3k\, [\sum_{\lambda=1}^{2} a^\dagger(\mathbf{k},k^0,\lambda)a(\mathbf{k},k^0,\lambda) - a^\dagger(\mathbf{k}, k^0, 1)a(-\mathbf{k},k^0,1) +$$

$$+ a^\dagger(\mathbf{k}, k^0, 2)a(-\mathbf{k},k^0,2)]/2\} \quad \text{(A.3.43b)}$$

The separable Lagrangian of eq. B.96 is invariant under these parity transformations.

A.3.12 Charge Conjugation

Charge conjugation is implemented in a way similar to that of conventional quantum field theory. In particular

$$\mathscr{C} X^\mu(\mathbf{y}, y^0)\mathscr{C}^{-1} = X^\mu(\mathbf{y}, y^0) \quad \text{(A.3.44)}$$

A.3.13 Time Reversal

Since $X^0 = y^0$ in the Y Coulomb gauge in Two-Tier quantum theory the only non-trivial form of time reversal transformation $\mathscr{T}$ is based on $y^0 = -y^0$. This time reversal transformation is similar in part to to the conventional time reversal transformation in conventional quantum field theory. Therefore we will define $\mathscr{T}$ as the product of the operation of taking the complex conjugate of all c-numbers times a unitary operator $\mathscr{U}_y$. Under $\mathscr{T}$ a scalar quantum field operator ϕ transforms as

$$\mathscr{T}\phi(\mathbf{X}(\mathbf{y}, y^0), X^0(\mathbf{y}, y^0))\mathscr{T}^{-1} = \phi(\mathbf{X}(\mathbf{y}, -y^0), X^0(\mathbf{y}, -y^0)) \quad \text{(A.3.45)}$$

From the form of in ϕ eq. A.3.2 we see that

$$\mathcal{T}a(\mathbf{p}, p^0)\mathcal{T}^{-1} = a(-\mathbf{p}, p^0) \quad (A.3.46)$$

$$\mathcal{T}a^{\dagger}(\mathbf{p}, p^0)\mathcal{T}^{-1} = a^{\dagger}(-\mathbf{p}, p^0) \quad (A.3.47)$$

$$\mathcal{T}X^i(\mathbf{y}, y^0)\mathcal{T}^{-1} = X^i(\mathbf{y}, -y^0) \quad (A.3.48)$$

$$\mathcal{T}Y^i(\mathbf{y}, y^0)\mathcal{T}^{-1} = -Y^i(\mathbf{y}, -y^0) \quad (A.3.49a)$$

$$\mathcal{T}a(\mathbf{k}, k^0,1)\mathcal{T}^{-1} = a(-\mathbf{k}, k^0,1) \quad (A.3.49b)$$

$$\mathcal{T}a(\mathbf{k}, k^0,2)\mathcal{T}^{-1} = -a(-\mathbf{k}, k^0,2) \quad (A.3.49c)$$

where i = 1,2,3 and assuming: $\boldsymbol{\varepsilon}(\mathbf{k},k^0,1)=-\boldsymbol{\varepsilon}(-\mathbf{k},k^0,1)$ and $\boldsymbol{\varepsilon}(\mathbf{k},k^0,2)=+\boldsymbol{\varepsilon}(-\mathbf{k},k^0,2)$.

The unitary operator U_y is given by

$$\mathcal{U}_X = \exp\{-i\pi\int d^3p\, [a^{\dagger}(\mathbf{p}, p^0)a(\mathbf{p}, p^0) - a^{\dagger}(\mathbf{p}, p^0)a(-\mathbf{p}, p^0)]/2\} \quad (A.3.50a)$$

and

$$\mathcal{U}_y = \mathcal{U}_X \exp\{-i\pi \int d^3k\, [\sum_{\lambda=1}^{2} a^{\dagger}(\mathbf{k},k^0,\lambda)a(\mathbf{k},k^0,\lambda) - a^{\dagger}(\mathbf{k}, k^0, 1)a(-\mathbf{k},k^0,1) + $$
$$+ a^{\dagger}(\mathbf{k}, k^0, 2)a(-\mathbf{k},k^0,2)]/2\} \quad (A.3.50b)$$

The separable Klein-Gordon Lagrangian (eq. 18-A.96) is invariant under our definition of time reversal.

We note

$$\mathcal{U}_y - \mathcal{P}_y \quad (A.3.50c)$$

Although the present theory is somewhat more complicated than conventional quantum field theory the overall nature of the $\mathcal{P}$, $\mathcal{C}$, and $\mathcal{T}$ transformations is the same.

A.4 Interacting Quantum Field Theory – Perturbation Theory

A.4.1 Introduction

The form of quantum field theory that we have developed in sections A.2 and A.3 can be used as the basis for new formulations of QED, Electroweak Theory, the Standard Model and a divergence-free, unified theory of all the known interactions. The development of these theories requires a number of topics be addressed. This section covers perturbation theory. As much as possible, we attempt to retain the features of the standard approach so that the reader will more readily follow the discussion and more readily accept this new formalism. In physics originality is secondary to reality. The perturbation theory that we will develop will be shown to be identical to the perturbation theory that we develop later using a path integral formalism.

A.4.2 An Auxiliary Asymptotic Field

The definition of the asymptotic "free" in and out states is an issue in Two-Tier quantum field theory because the "free particle field" of the theory $\phi(X(y))$ is a "dressed" particle, ab initio, since it is cloaked in a cloud of Y particles as discussed in the passage following eq. A.3.27.

While one could use $\phi(X(y))$ directly to define in and out asymptotic states it is more convenient initially to introduce a "fictitious" auxiliary asymptotic quantum field $\Phi(y)$ that will represent the equally fictitious "bare ϕ particle" in and out states.

We will consider the case of a scalar field. We define a free, scalar Klein-Gordon particle field with the physical mass m of the physical $\phi(X(y))$ particle.

$$\Phi(y) = \int d^3p\, N_m(p)\, [a(p)\, e^{-ip\cdot y} + a^\dagger(p)\, e^{ip\cdot y}] \qquad \text{(A.4.1)}$$

using the creation and annihilation operators of $\phi(X(y))$ (in eq. A.3.2). The set of particle states of $\Phi(y)$ has the familiar Fock space form

$$|\, p_1, p_2, \ldots p_n> = a^\dagger(p_1,)a^\dagger(p_2) \ldots a^\dagger(p_n)|0> \qquad \text{(A.4.2)}$$

with powers of creation operators allowed since Φ particles are bosons. The set of particle states constitutes a complete orthonormal set of states. The corresponding bra states are defined by hermitean conjugation:

$$<p_1, p_2, \ldots p_n| = (|\, p_1, p_2, \ldots p_n>)^\dagger \qquad \text{(A.4.3)}$$

We note that the energy spectrum of these states is positive definite with the hamiltonian

$$H_\Phi = P_\Phi^{\ 0} = \int d^3y\, \tfrac{1}{2}[\pi_\Phi^{\ 2} + (\partial\Phi/\partial X^i)^2 + m^2\Phi^2] \qquad \text{(A.4.4a)}$$

$$= \int d^3p\, (\mathbf{p}^2 + m^2)^{1/2} a^\dagger(p)a(p) \qquad \text{(A.4.4b)}$$

and momentum vector:

$$\mathbf{P}_\Phi = \int d^3p\, \mathbf{p}\, a^\dagger(p)a(p) \qquad \text{(A.4.5)}$$

We will use this set of energy-momentum eigenstates to define asymptotic "in" and "out" states in perturbation theory.

A.4.3 Transformation Between Φ(y) and ϕ(X(y))

For later use in the definition of the perturbation theory expansion, we will determine the transformations between the in and out Φ(y) fields, and the in and out ϕ(X(y)) fields. Let us define a transformation $W_a(y)$ that transforms in and out Φ(y) fields to in and out ϕ(X(y)) fields respectively:

$$\phi_a(X(y)) = :W_a(y)\Phi_a(y)W_a^{-1}(y): \qquad \text{(A.4.6)}$$

where the label a = "in" or a = "out", where : … : signifies normal ordering, and where

$$\Phi_{in}(y) = \int d^3p\, N_m(p)\, [a_{in}(p)\, e^{-ip\cdot y} + a_{in}^\dagger(p)\, e^{ip\cdot y}] \qquad \text{(A.4.7)}$$

$$\Phi_{out}(y) = \int d^3p\, N_m(p)\, [a_{out}(p)\, e^{-ip\cdot y} + a_{out}^\dagger(p)\, e^{ip\cdot y}] \qquad \text{(A.4.8)}$$

$$\phi_{in}(X) = \int d^3p\, N_m(p)\, [a_{in}(p)\, :e^{-ip\cdot(y + iY/M_c^2)}: + a_{in}^\dagger(p)\, :e^{ip\cdot(y + iY/M_c^2)}:] \qquad \text{(A.4.9)}$$

$$\phi_{out}(X) = \int d^3p\, N_m(p)\, [a_{out}(p)\, :e^{-ip\cdot(y + iY/M_c^2)}: + a_{out}^\dagger(p)\, :e^{ip\cdot(y + iY/M_c^2)}:] \qquad \text{(A.4.10)}$$

Note that the transformation eq. A.4.6 includes normal ordering. While this transformation may seem strange it is no stranger than the time reversal operator, in which the complex conjugate of all c-number terms is taken in addition to applying a unitary transformation.

In the Coulomb gauge of Y it is easy to show that

$$W_a(y) = \exp(-\mathbf{Y}(y)\cdot\mathbf{P}_{\Phi a}/M_c^2) \qquad \text{(A.4.11)}$$

and

$$W_a^{-1}(y) = \exp(\mathbf{Y}(y)\cdot\mathbf{P}_{\Phi a}/M_c^2) \qquad \text{(A.4.12)}$$

where the label a = "in" or a = "out", where the inner products in the exponentials are the usual spatial vector inner product, and where

$$\mathbf{P}_{\Phi a} = -\int d^3y\, \partial\Phi_a(y)/\partial y^0\, \nabla\Phi_a(y) = \int d^3p\, \mathbf{p}\, a_a^\dagger(p)a_a(p) \qquad \text{(A.4.12a)}$$

is a spatial vector (the Φ spatial momentum operator) that is written solely in terms of $\Phi_a(y)$'s creation and annihilation operators.

In addition to performing the transformation in eq. A.4.6, $W_a(y)$ also performs a "translation" in Y^μ:

$$W_a(y)Y^i(y')W_a^{-1}(y) = Y^i(y') + i\Delta^{trij}(y' - y)P_{\Phi a}{}^j/M_c^2 \qquad \text{(A.4.13a)}$$

where

$$i\Delta^{trij}(y' - y) = \int d^3k\, (e^{-ik\cdot(y' - y)} - e^{ik\cdot(y' - y)})(\delta_{jk} - k_jk_k/\mathbf{k}^2)/[(2\pi)^3 2\omega_k] \qquad \text{(A.4.13b)}$$

We note that $W_a(y)$ is not a unitary operator but it is pseudo-unitary:

$$W_a^{-1}(y) = V\, W_a^\dagger(y)\, V^{-1} = V\, W_a(y)\, V^{-1} \qquad \text{(A.4.14)}$$

where

$$V = \exp(-i\pi \sum_{\lambda=1}^{2} \int d^3k\, a^\dagger(k, \lambda)a(k, \lambda)) \qquad \text{(A.4.15)}$$

is a unitary operator with the property

$$V\, Y^j(y)\, V^{-1} = -Y^j(y) \qquad \text{(A.4.16)}$$

for j = 1,2,3. We note

$$V^\dagger = V^{-1} = V \qquad \text{(A.4.17)}$$

and thus

$$V^2 = I \qquad \text{(A.4.18)}$$

V will be shown to be a metric operator in the following discussion.[289] We note that the Y "particle" (hermitean) number operator appears in eq. A.4.9 in the expression for V:

$$N_Y = \sum_{\lambda=1}^{2} \int d^3k\, a^\dagger(k, \lambda)a(k, \lambda) \qquad \text{(A.4.19)}$$

Thus states with an even number of Y "particles" have a V eigenvalue of one, and states with an odd number of Y "particles" have a V eigenvalue of minus one.

A.4.4 Model Lagrangian with ϕ^4 Interaction

We will develop our perturbation theory using a scalar Lagrangian model with a ϕ^4 interaction term:

---

[289] P. A. M. Dirac, Proc. R. Soc. London A **180**, 1 (1942); T. D. Lee and G. C. Wick, Nucl. Phys. **B9**, 209 (1969); C. M. Bender, D. C. Brody and H. F. Jones, "Complex Extension of Quantum Mechanics" Phys. Rev. Letters **89**, 270401-1 (2002) and references therein.

$$\mathscr{L}_s = \text{JL}_F + \text{L}_C \qquad \text{(A.4.20)}$$

with

$$\mathscr{L}_F = ½ [(\partial\phi/\partial X^\nu)^2 - m^2\phi^2] + \mathscr{L}_{Fint} \qquad \text{(A.4.21)}$$

and

$$\mathscr{L}_C = -¼ F_Y^{\mu\nu}F_{Y\mu\nu} \qquad \text{(A.4.22)}$$

with

$$F_{Y\mu\nu} = \partial Y_\mu/\partial y^\nu - \partial Y_\nu/\partial y^\mu \qquad \text{(A.4.23)}$$

and

$$\mathscr{L}_{Fint} = 1/4!\ \chi_0\ \phi(X(y))^4 + ½ (m^2 - m_0^2)\phi^2 \qquad \text{(A.4.24)}$$

where J is the Jacobian (as in Appendix 18-A), χ_0 is the bare coupling constant, and m_0 is the bare mass.

The conserved momentum operator is:

$$P_{F\beta} = \int d^3X\ T_{F0\beta} \qquad \text{(A.4.25)}$$

where

$$\mathscr{T}_{F\mu\nu} = -\ g_{\mu\nu} L_F + \partial L_F / \partial(\partial\phi/\partial X_\mu)\ \partial\phi/\partial X^\nu \qquad \text{(A.4.26)}$$

is the ϕ field energy-momentum tensor with conservation law (eq. B.110):

$$\partial P_{F\beta}/\partial X^0 = 0 \qquad \text{(A.4.27)}$$

due to eq. B.108.

The hamiltonian density (eq. B.83) is

$$\mathscr{H}_F = \mathscr{T}_{F0\beta} = \mathscr{H}_{F0} + \mathscr{H}_{Fint} \qquad \text{(A.4.28)}$$

with

$$\mathscr{H}_{F0} = ½ [\pi_\phi^2 + (\partial\phi/\partial X^i)^2 + m^2\phi^2] \qquad \text{(A.4.29)}$$

$$\mathscr{H}_{Fint} = -\ 1/4!\ \chi_0\ \phi(X(y))^4 + ½ (m^2 - m_0^2)\phi(X(y))^2 \qquad \text{(A.4.30)}$$

A.4.5 In-states and Out-States

In this section we will develop properties of in-fields and out-fields. We will use a somewhat more complicated procedure to set up the perturbation theory for the S matrix due to the introduction of imaginary coordinates. The procedure can be schematized as:

$$\Phi_{in}(y) \Rightarrow \phi_{in}(X(y)) \Rightarrow \phi(X(y)) \Rightarrow \phi_{out}(X(y)) \Rightarrow \Phi_{out}(y) \quad \text{(A.4.31)}$$

In-states are constructed using the auxiliary field Φ_{in} which are then effectively transformed into $\phi_{in}(X(y))$ expressions in order to make contact with our Lagrangian formalism. Then $\phi_{in}(X(y))$ is related to the interacting field $\phi(X(y))$ as a limit ($y^0 \to -\infty$). Similarly out-states are constructed using the auxiliary field Φ_{out} which are then expressed in terms of $\phi_{out}(X(y))$. Then $\phi_{out}(X(y))$ is related to the interacting field $\phi(X(y))$ using the LSZ limiting process ($y^0 \to +\infty$).

Since much of the development differs only trivially from the standard treatment in textbooks we will simply "list" relevant equations and let the reader pursue them further in quantum field theory textbooks.

A.4.6 ϕ In-Field

In order to define a perturbation theory for particle scattering we will next specify features of the in-field $\phi_{in}(X(y))$ and in-field states – the field and states representing physical particles as $X^0 = y^0 \to -\infty$.

A. The in-field $\phi_{in}(X(y))$ satisfies the Klein-Gordon equation in the X variable:

$$(\Box_X + m^2)\, \phi_{in}(X) = 0 \quad \text{(A.4.32)}$$

where

$$\Box_X = (\partial/\partial X^\nu)(\partial/\partial X_\nu)$$

B. Under coordinate displacements and Lorentz transformations $\Phi_{in}(y)$, $\phi_{in}(X(y))$, and $\phi(X(y))$ transform in the same way:

$$[P^\mu, \Phi_{in}(y)] = -\, i\partial\Phi_{in}/\partial y_\mu \quad \text{(A.4.33a)}$$
$$[P^\mu, \phi_{in}(X)] = -\, i\partial\phi_{in}/\partial y_\mu \quad \text{(A.4.33b)}$$
$$[P^\mu, \phi(X)] = -\, i\partial\phi/\partial y_\mu \quad \text{(A.4.34)}$$

with the energy-momentum vector P^μ specified by eq. B.57.

C. We can relate the asymptotic in-field $\phi_{in}(X(y))$ to the interacting field $\phi(X(y))$ using the equation of motion of $\phi(X(y))$

$$(\Box_X + m^2)\,\phi(X) = j(X) \qquad \text{(A.4.35)}$$

where j(X) embodies the interaction. Using the physical mass m we find

$$(\Box_X + m^2)\,\phi(X) = j(X) + (m^2 - m_0^2)\phi(X) = j_{tot}(X) \qquad \text{(A.4.36)}$$

If the current is taken to be the source of the scattered waves we may write

$$\sqrt{Z}\,\phi_{in}(X(y)) = \phi(X(y)) - \int d^4X(y')\,\Delta_{ret}(y - y')\,j_{tot}(X(y')) \qquad \text{(A.4.37)}$$
$$= \phi(X(y)) - \int d^4y'\,J\,\Delta_{ret}(y - y')\,j_{tot}(X(y')) \qquad \text{(A.4.38)}$$

where Z is a wave function renormalization constant, J is the Jacobian, and Δ_{ret} is a retarded Green's function.

D. We can define Φ_{in} in-field states with expressions like

$$|\,p_1, p_2, \ldots p_n\ in\rangle = a_{in}^\dagger(p_1)a_{in}^\dagger(p_2)\ldots a_{in}^\dagger(p_n)|0\rangle \qquad \text{(A.4.39)}$$

with powers of creation operators allowed since Φ_{in} is a boson field. The set of all particle states constitutes a complete orthonormal set of states. The corresponding bra states are defined by hermitean conjugation:

$$\langle p_1, p_2, \ldots p_n\ in| = (|\,p_1, p_2, \ldots p_n\ in\rangle)^\dagger \qquad \text{(A.4.40)}$$

A.4.7 ϕ Out-Field

In order to define a perturbation theory for particle scattering we begin by listing aspects of the out-field $\phi_{out}(X(y))$ and out-field states – the field and states representing physical particles as $X^0 = y^0 \to -\infty$.

A. The out-field $\phi_{out}(X(y))$ satisfies the Klein-Gordon equation in the X variable:

$$(\Box_X + m^2)\,\phi_{out}(X) = 0 \qquad \text{(A.4.41)}$$

where

$$\Box_X = (\partial/\partial X^\nu)(\partial/\partial X_\nu)$$

B. Under coordinate displacements and Lorentz transformations $\Phi_{out}(y)$, $\phi_{out}(X(y))$, and $\phi(X(y))$ transform in the same way:

$$[P^\mu, \Phi_{out}(y)] = -i\partial\Phi_{out}/\partial y_\mu \tag{A.4.42a}$$
$$[P^\mu, \phi_{out}(X)] = -i\partial\phi_{out}/\partial y_\mu \tag{A.4.42b}$$
$$[P^\mu, \phi(X)] = -i\partial\phi/\partial y_\mu \tag{A.4.43}$$

with the energy-momentum vector P^μ specified by eq. B.57.

C. We can relate the asymptotic out-field $\phi_{out}(X(y))$ to the interacting field $\phi(X(y))$ using the equation of motion of $\phi(X(y))$ specified by eq. A.4.36:

$$\sqrt{Z}\,\phi_{out}(X(y)) = \phi(X(y)) - \int d^4X(y')\,\Delta_{adv}(y - y')\,j_{tot}(X(y')) \tag{A.4.44}$$

$$= \phi(X(y)) - \int d^4y'\,J\,\Delta_{adv}(y - y')\,j_{tot}(X(y')) \tag{A.4.45}$$

where Z is a wave function renormalization constant, J is the Jacobian, and Δ_{adv} is an advanced Green's function.

D. We can define Φ_{out} out-field states with expressions like

$$|\,p_1, p_2, \ldots p_n\ out\rangle = a_{out}^\dagger(p_1,)a\Phi_{out}^\dagger(p_2)\ \ldots\ a\Phi_{out}^\dagger(p_n)|0\rangle \tag{A.4.46}$$

with powers of creation operators allowed since Φ_{out} is a boson field. The set of all particle states constitutes a complete orthonormal set of states. The corresponding bra states are defined by hermitean conjugation:

$$\langle p_1, p_2, \ldots p_n\ out| = (|\,p_1, p_2, \ldots p_n\ out\rangle)^\dagger \tag{A.4.47}$$

A.4.8 The Y Field

The Y field in the present model Lagrangian (eq. A.4.20) is a free field and thus:

$$Y_{in}(y) = Y_{out}(y) = Y(y) \tag{A.4.48}$$

The states of the Y field have two general forms: 1) States in a Fock space consisting of particle states that are eigenstates of the Y particle number operator (eq. A.4.19); and 2) Coherent states in a non-Fock space of generalized coherent states in an infinite tensor product space.[290] The coherent ket states that arise in Two-Tier quantum field theory have the general form (eq. A.2.41):

$$|y,\ p> = e^{-p\cdot Y^{-}(y)/M_c^{\,2}}|0> \qquad \text{(A.2.41)}$$

as can be seen from an examination of $\phi_{in}(X(y))$. The corresponding bra state is:

$$<y,\ p| = (V|\ y,\ p>)^{\dagger} = <0|e^{+p\cdot Y^{+}(y)/M_c^{\,2}} \qquad \text{(A.4.49)}$$

with V, the metric operator, reversing the sign of Y in the exponential. The inner product of coherent states is:

$$<y_1\ ,\ p_1|\ y_2,\ p_2> = \exp[\,-p_1^{\ i}p_2^{\ j}\Delta_{Tij}(y_1 - y_2)/M_c^{\,4}] \qquad \text{(A.4.50)}$$

showing the set of coherent states is not orthonormal and, in fact, is overcomplete. Comparing eq. A.4.50 to eq. A.3.12 gives

$$<y_1\ ,\ p|\ y_2,\ p> = R(p,\ y_1 - y_2) \qquad \text{(A.4.50a)}$$

The completeness of the set of states for each time y^0 can be verified by examining the projection operator:

$$\mathscr{R}_Y(y^0) = \because \exp[-i \int d^3y\ Y^{-}_{\ i}(y)|0><0|\pi^{+j}(y)] \because \qquad \text{(A.4.51)}$$

where

$$\pi^{+j}(y) = -\partial Y^{+j}(y)/\partial y^0 \qquad \text{(A.4.52)}$$

and where $\because$ represents an extended normal ordering operator:

$$\because \ldots \because$$

which is defined as placing creation operators to the left, projection operators in the center, and annihilation operators to the right. Thus eq A.4.51 can be written

$$\mathscr{R}_Y = \sum_n (-i/n!)^n \int d^3y_1 \ldots \int d^3y_n Y^{-j_1}(y_1)Y^{-j_2}(y_2)\ldots Y^{-j_n}(y_n)|0><0|\pi^{+}_{\ j_1}(y_1)\pi^{+}_{\ j_2}(y_2)\ldots\pi^{+}_{\ j_n}(y_n) \qquad \text{(A.4.53)}$$

290 See Kibble and other references on coherent states.

where we have used the fact that |0><0| is a projection operator, and reduced |0><0| |0><0| … |0><0| to |0><0| in eq. A.4.53. The vacuum state is the product of the Y and ϕ vacuum states:

$$|0\rangle = |0_Y\rangle|0_\phi\rangle \tag{A.4.53a}$$

We note

$$\mathscr{R}_Y\,(y^0)|\mathbf{y},y^0\ p\rangle = |\mathbf{y},y^0\ p\rangle \tag{A.4.54}$$

using eq. A.2.22 and $\int d^3y_2\, p^i\, \Delta^{tr}{}_{ij}(y_1 - y_2)Y^{+j}(y_2) = \mathbf{p}\cdot\mathbf{Y}^+(y_1)$. Also

$$\mathscr{R}_Y(y^0)|n\rangle = |n\rangle \tag{A.4.55}$$

where |n> is any Y particle Fock state of finite particle number. In view of eqs. A.4.54 and A.4.55, we see that $\mathscr{R}_Y$ is the identity operator on the Fock space and the space of generalized coherent Y field states. Thus the set of Y coherent states forms an overcomplete set of states. We will define the S matrix for any combination of Φ Fock space states and coherent Y states. The R_Y operator can be generalized to include Φ Fock space states:

$$\mathscr{R}_{\Phi Y}(y^0) = \because \exp[-i \int d^3y\ Y^-_j(y)\mathscr{R}_\Phi \pi^{+j}(y)] \because \tag{A.4.56}$$

with

$$\mathscr{R}_\Phi = \sum_n |n\rangle\langle n| \tag{A.4.57}$$

is a sum over all Φ Fock space states with vacuum state given by eq. A.4.53a. Since R_Φ is a projection:

$$[\mathscr{R}_\Phi\,]^N = \mathscr{R}_\Phi$$

for any power N, we find:

$$\mathscr{R}_{\Phi Y}(y^0) = \sum_n (-i)^n \int d^3y_1 \ldots \int d^3y_n Y^{-j_1}(y_1)Y^{-j_2}(y_2)\ldots Y^{-j_n}(y_n)R_\Phi \pi^+{}_{j_1}(y_1)\pi^+{}_{j_2}(y_2)\ldots\pi^+{}_{j_n}(y_n) \tag{A.4.58}$$

As a result we have

$$\mathscr{R}_{\Phi Y}(y^0)|y,\ p;\ n_\Phi\rangle = |y,\ p;\ n_\Phi\rangle \tag{A.4.59}$$

for any combination of Y coherent states and Φ Fock space states n_Φ. Also

$$\mathscr{R}_{\Phi Y}(y^0)|n_\Phi> = |n_\Phi> \qquad (A.4.60)$$

Thus $\mathscr{R}_{\Phi Y}$ is the identity operator on this space – the (over) complete space of in and out states which we will use to define the S matrix of the scalar field theory specified by the Lagrangian eq. A.4.20.

A.4.9 S Matrix

Following the standard definition of the S matrix we have:

$$S_{\alpha\beta} = <\alpha \text{ out}|\beta \text{ in}> \qquad (A.4.61)$$
$$= <\alpha \text{ in}|S|\beta \text{ in}> \qquad (A.4.62)$$

$$|0> = |0 \text{ in}> = |0 \text{ out}> = S|0 \text{ in}> \qquad (A.4.63)$$

$$\Phi_{in}(y) = S\Phi_{out}(y)S^{-1} \qquad (A.4.64)$$

and the other standard properties of the S matrix with the sole exception being the form of the unitarity relation (discussed later).

A.4.10 LSZ Reduction for Scalar Fields

In this section we will determine the reduction formula for the S matrix for scalar ϕ fields. Consider the S matrix element corresponding to an in state of particles β plus one ϕ particle of momentum p, and an out state α:

$$S_{\alpha\beta p} = <\alpha \text{ out}|\beta p \text{ in}> \qquad (A.4.65)$$

After standard manipulations we have:

$$S_{\alpha\beta p} = <\alpha - p \text{ out}|\beta \text{ in}> - i<\alpha \text{ out}|\int d^3y\, f_p(y)\, \overleftrightarrow{\partial_0}\, [\Phi_{in}(y) - \Phi_{out}(y)]\, |\beta \text{ in}> \qquad (A.4.66)$$

where $<\alpha - p$ out| is an out state with a particle of momentum p removed (if present) and where

$$f(y^0)\, \overleftrightarrow{\partial_0}\, g(y^0) = f(y^0)\, \partial g(y^0)/\partial y^0 - \partial f(y^0)/\partial y^0\, g(y^0) \qquad (A.4.67)$$

and

$$f_p(y) = N_m(p)e^{-ip\cdot y} \qquad (A.4.68)$$

with $N_m(p)$ specified by eq. A.2.6.

We now express

$$S_{\alpha\beta p} = S_{\alpha-p\beta} - i\langle\alpha\ \text{out}|\int d^3y\ f_p(y)\ \overleftrightarrow{\partial}_0\ W^{-1}[\phi_{in}(X(y)) - \phi_{out}(X(y))]W|\beta\ \text{in}\rangle \qquad \text{(A.4.69)}$$

using $W(y) = W_{in}(y)$ with

$$\Phi_a(y) = W_a^{-1}(y)\phi_a(X(y))W_a(y) \qquad \text{(A.4.70)}$$

where the label a = "in" or a = "out", and where

$$W_a(y) = \exp(-\mathbf{Y}(y)\cdot\mathbf{P}_{\Phi a}/M_c^2) \qquad \text{(A.4.71)}$$

and

$$W_a^{-1}(y) = \exp(\mathbf{Y}(y)\cdot\mathbf{P}_{\Phi a}/M_c^2) \qquad \text{(A.4.72)}$$

in the Coulomb gauge of Y with $\mathbf{P}_{\Phi a}$ the momentum spatial vector defined by eq. A.4.12a.

We note that the interacting $\phi(X(y))$ approaches the in and out fields $\phi_{in}(X(y))$ and $\phi_{out}(X(y))$ in the limit that $y^0 \to -\infty$ and $y^0 \to +\infty$ respectively in the sense of Lehmann, Symanzik and Zimmermann[291] which we *symbolize* as:

$$\phi(X(y)) \to \sqrt{Z}\ \phi_{in}(X(y)) \quad \text{as} \quad y^0 \to -\infty \qquad \text{(A.4.73)}$$

$$\phi(X(y)) \to \sqrt{Z}\ \phi_{out}(X(y)) \quad \text{as} \quad y^0 \to +\infty \qquad \text{(A.4.74)}$$

with $\sqrt{Z}$ defined in eqs. A.4.37 and A.4.44. Thus we can rewrite eq. A.4.69 as

$$S_{\alpha\beta p} = S_{\alpha-p\beta} + iZ^{-1/2}\left(\lim_{y^0\to+\infty} - \lim_{y^0\to-\infty}\right)\langle\alpha\ \text{out}|\int d^3y\ f_p(y)\ \overleftrightarrow{\partial}_0\ W^{-1}\phi(X(y))W|\beta\ \text{in}\rangle \qquad \text{(A.4.75)}$$

which after standard manipulations becomes

$$S_{\alpha\beta p} = S_{\alpha-p\beta} + iZ^{-1/2}\int d^4y\ f_p(y)(\Box_y + m^2)\langle\alpha\ \text{out}|\ W(y)^{-1}\phi(X(y))W(y)|\beta\ \text{in}\rangle \qquad \text{(A.4.76)}$$

Eq. A.4.76 is similar to the usual LSZ reduction formula except for the appearance of the W(y) operator and its inverse. We note that $W(y) = W_{in}(y)$ still because $\mathbf{P}_{\Phi in}$ is independent of y^0.

---

[291] H. Lehmann, K. Symanzik and W. Zimmermann, Nuov. Cim., **1**, 1425 (1955); W. Zimmermann, Nuov. Cim., **10**, 567 (1958); O. W. Greenberg, Doctoral Dissertation, Princeton University 1956.

Similarly an out ϕ particle can be reduced from an S matrix part. For example,

$$<\alpha\ \text{out}|W^{-1}(y)\phi(X(y))W(y)|\beta\ \text{in}>=<\alpha-p'\ \text{out}|W^{-1}(y)\phi(X(y))W(y)|\beta-p'\ \text{in}> -$$
$$- i<\alpha-p'\ \text{out}|\int d^3y'\,[W^{-1}(y')\phi_{in}(X(y'))W(y')W^{-1}(y)\phi(X(y))W(y) -$$
$$- W^{-1}(y)\phi(X(y))W(y)W^{-1}(y')\phi_{out}(X(y'))W(y')]|\beta\ \text{in}>\overleftrightarrow{\partial_0} f_{p'}^{*}(y') \qquad (A.4.77)$$

which becomes

$$<\alpha\ \text{out}|W^{-1}(y)\phi(X(y))W(y)|\beta\ \text{in}> = <\alpha-p'\ \text{out}|\varphi(y)|\beta-p'\ \text{in}> +$$
$$+ iZ^{-1/2}\int d^4y'\,<\alpha-p'\ \text{out}|T(\varphi(y')\varphi(y))|\beta\ \text{in}>(\overleftarrow{\Box}_{y'} + m^2)\,f_{p'}^{*}(y') \qquad (A.4.78)$$

where the time ordered product is defined with respect to ordering with respect to y^0 and where

$$\varphi(y) = W^{-1}(y)\phi(X(y))W(y) \qquad (A.4.79)$$

These results directly generalize to multi-particle in and out states:

$$<p_1, p_2, \ldots p_n\ \text{out}|\ q_1, q_2, \ldots q_m\ \text{in}> = \ldots <0|T(\varphi(y'_1)\ \ldots\ \varphi(y'_n)\varphi(y_1)\ \ldots\ \varphi(y_m))|0> \ldots \qquad (A.4.80)$$

thus reducing the development of the perturbation theory of the S matrix to the evaluation of time ordered products such as

$$<0|T(\varphi(y_1)\ \ldots\ \varphi(y_n))|0> \qquad (A.4.81)$$

A.5 The U Matrix in Perturbation Theory

The U matrix for a Two-Tier theory is developed in a way similar to conventional field theory starting from the defining relations:

$$\phi(X(y)) = U^{-1}\phi_{in}(X(y))U \qquad (A.5.1)$$

$$\pi_\phi(X(y)) = U^{-1}\pi_{\phi in}(X(y))U \qquad (A.5.2)$$

From eq. A.4.29 we define the free field hamiltonian

$$H_{F0in}(\phi_{in}, \pi_{\phi in}) = \int d^3X\,\mathscr{H}_{F0}(\phi_{in}, \pi_{\phi in}) \qquad (A.5.3)$$

Noting $X^0 = y^0$ in the Y Coulomb gauge we find

$$\partial\phi_{in}/\partial y^0 = i[H_{F0in}, \phi_{in}(X)] \quad (A.5.4)$$

$$\partial\pi_{\phi in}/\partial y^0 = i[H_{F0in}, \pi_{\phi in}(X)] \quad (A.5.5)$$

For the entire hamiltonian (eq. A.4.28) we have

$$\partial\phi/\partial y^0 = i[H_F, \phi(X)] \quad (A.5.6)$$
$$\partial\pi_\phi/\partial y^0 = i[H_F, \pi_\phi(X)] \quad (A.5.7)$$

with

$$H_F(\phi, \pi_\phi) = :\int d^3X\, \mathscr{H}_F(\phi, \pi_\phi): \quad (A.5.8)$$

(Note the *entire* interaction term is normal ordered since d^3X is a q-number. Combining the above equations in the standard way yields a familiar differential equation for the U matrix:

$$i\partial U(y^0)/\partial y^0 = (H_{Fint} + E_0(t))U(y^0) \quad (A.5.9)$$

where $E_0(t)$ is a c-number function of y^0 that we can set equal to 0 (as it would be cancelled later in any case), and where

$$H_{Fint}(\phi_{in}, \pi_{\phi in}) = :\int d^3X\, \mathscr{H}_{Fint}(\phi_{in}, \pi_{\phi in}): \quad (A.5.10)$$

with $\mathscr{H}_{Fint}$ given by eq. A.4.30. Solving for U gives the familiar time ordered exponential:

$$U(y^0) = T(\exp[-i \int_{-\infty}^{t} dy^0\, H_{Fint}]) \quad (A.5.11a)$$

which is a symbolic notation for:

$$U(y^0) = 1 + \sum_{n=1}^{\infty} (-i)^{-n} (n!)^{-1} \int_{-\infty}^{y^0} dy_1{}^0 \ldots \int_{-\infty}^{y^0} dy_n{}^0\, T(H_{Fint}(y_1{}^0) \ldots H_{Fint}(y_n{}^0)) \quad (A.5.11b)$$

We note for later use that the hermiticity of H_{Fint} is not used in the derivation of eq. A.5.11. Thus eq. A.5.11 would still hold if H_{Fint} were not hermitean.

A.5.1 Reduction of Time Ordered φ Products

In the previous chapter we reduced the calculation of the S matrix to the evaluation of time ordered products of the form

$$\tau(y_1, \ldots, y_n) = \langle 0|T(\varphi(y_1) \ldots \varphi(y_n))|0\rangle \tag{A.5.12}$$

where φ(y) is specified by eq. A.4.79. Expanding the terms within eq. A.5.12 using eq. A.4.79 we find

$$\varphi(y_1) \ldots \varphi(y_n) = W^{-1}(y_1)\phi(X(y_1))W(y_1)W^{-1}(y_2)\phi(X(y_2))W(y_2) \ldots W^{-1}(y_n)\phi(X(y_n))W(y_n) \tag{A.5.13}$$

which can be re-expressed as

$$W^{-1}(y_1)U^{-1}(y_1^0)\phi_{in}(X(y_1))U(y_1^0)W(y_1)W^{-1}(y_2)U^{-1}(y_2^0)\phi_{in}(X(y_2))U(y_2^0)W(y_2) \ldots \tag{A.5.14}$$

using eq. A.5.1 and denoting $W_{in}(y)$ as W(y). Defining

$$\mathscr{U}(y_1, y_2) = U(y_1^0)W(y_1)W^{-1}(y_2)U^{-1}(y_2^0) \tag{A.5.15}$$

we see eq. A.5.14 can be rewritten as

$$W^{-1}(y_1)U^{-1}(y_1^0)\phi_{in}(X(y_1))U(y_1, y_2)\phi_{in}(X(y_2))\ U\ (y_2, y_3)\phi_{in}(X(y_3)) \ldots \phi_{in}(X(y_n))U(y_n^0)W(y_n) \tag{A.5.16}$$

From eqs. A.4.71 and A.4.72

$$\mathscr{U}(y_1, y_2) = U(y_1^0)\exp((\mathbf{Y}(y_2) - \mathbf{Y}(y_1))\cdot\mathbf{P}_{\Phi a}/M_c^2)U^{-1}(y_2^0) \tag{A.5.17}$$

Defining

$$W(y_1, y_2) = \exp((\mathbf{Y}(y_2) - \mathbf{Y}(y_1))\cdot\mathbf{P}_{\Phi a}/M_c^2) \tag{A.5.18}$$

and looking ahead to the Wick expansion of the time ordered product of eq. A.5.12 we note that the only time ordered products involving $W(y_1, y_2)$ that would appear in the expansion are

$$\langle 0|T(\phi_{in}(X(y))W(y_1, y_2))|0\rangle = 0 \tag{A.5.19a}$$
$$\langle 0|T(Y(y)W(y_1, y_2))|0\rangle = 0 \tag{A.5.19b}$$

$$<0|T(\partial Y(y)/\partial y^{\mu}\, W(y_1, y_2))|0> = 0 \quad \text{(A.5.19c)}$$
$$<0|T(\partial Y(y)/\partial y^{\mu}\, \phi_{in}(X(y)))|0> = 0 \quad \text{(A.5.19d)}$$
$$<0|T(W(y_1, y_2)W(y_3, y_4))|0> = 1 \quad \text{(A.5.19e)}$$

due to the factor of $\mathbf{P}_{\Phi a}$ that appears in $W(y_1, y_2)$. Also

$$<0|T(\phi_{in}(X(y))Y(y_1))|0> = 0 \quad \text{(A.5.20)}$$

due to the $a_{in}(p)$ and $a_{in}^{\dagger}(p)$ factors appearing in $\phi_{in}(X(y))$.

Thus the $W(y_1, y_2)$ factor in eq. A.5.17 may be set to the value one with the result

$$\mathscr{U}(y_1, y_2) \equiv U(y_1^{\,0})U^{-1}(y_2^{\,0}) = U(y_1^{\,0}, y_2^{\,0}) \quad \text{(A.5.21)}$$

where $U(y_1^{\,0}, y_2^{\,0})$ is the conventionally defined U matrix satisfying

$$i\partial\, U(y_1^{\,0}, y_2^{\,0})/\partial y_1^{\,0} = iH_{Fint}\, U(y_1^{\,0}, y_2^{\,0}) \quad \text{(A.5.22)}$$

with the boundary condition

$$U(y^0, y^0) = 1 \quad \text{(A.5.23)}$$

This result would still be true if the $W(y_1, y_2)$ exponentials were expanded in their “power series” form.

Then, paralleling the standard approach we find an expression for the U matrix:

$$U(y_1^{\,0}, y_2^{\,0}) = T(\exp[-i\int_{y_2^{\,0}}^{y_1^{\,0}} dy'^0{:}d^3X(y')\, \mathscr{H}_{Fint}(\phi_{in}(X(y')), \pi_{\phi in}(X(y'))){:}]) \quad \text{(A.5.24)}$$

The $U(y_1^{\,0}, y_2^{\,0})$ matrix satisfies the conventional multiplication rule:

$$U(y_1^{\,0}, y_3^{\,0}) = U(y_1^{\,0}, y_2^{\,0})U(y_2^{\,0}, y_3^{\,0}) \quad \text{(A.5.25)}$$

The inverse of $U(y_1, y_2)$ is

$$U^{-1}(y_1^{\,0}, y_2^{\,0}) = U(y_2^{\,0}, y_1^{\,0}) \quad \text{(A.5.26)}$$

We now return to eq. A.5.16, which can now be written in the form:

$$U^{-1}(y^0)U(y^0, y_1^{\,0})\phi_{in}(X(y_1))U(y_1^{\,0}, y_2^{\,0})\phi_{in}(X(y_2))U(y_2^{\,0}, y_3^{\,0}) \ldots \phi_{in}(X(y_n))U(y_n^{\,0}, -y^0)U(-y^0) \tag{A.5.27}$$

where y^0 is a reference time that is later than all other times, and $-y^0$ is earlier than all the other times, in the time-ordered product. As a result the time-ordered product in eq. A.4.80 can be expressed in a symbolic notation as:

$$<0|U^{-1}(y^0)T(\phi_{in}(X(y_1))\phi_{in}(X(y_2)) \ldots \phi_{in}(X(y_n))U(y^0, -y^0))U(-y^0)|0> \tag{A.5.28}$$

The analysis of eq. A.5.28 as $y^0 \to \infty$ follows the standard path, which begins by noting

$$U(-y)|0> = \lambda_-|0> \qquad \text{when } y^0 \to \infty \tag{A.5.29a}$$
$$U(y)|0> = \lambda_+|0> \qquad \text{when } y^0 \to \infty \tag{A.5.29b}$$

following a standard textbook proof, which, in turn, leads to:

$$\lambda_-\lambda_+^* = <0|T(\exp[\,+i \int_{-\infty}^{\infty} dy'^0{:}d^3X(y')H_{\,Fint}(\phi_{in}(X(y')), \pi_{\phi in}(X(y'))){:}])|0> \tag{A.5.30}$$

$$= [<0|T(\exp[\,-i \int_{-\infty}^{\infty} dy'^0 d^3X(y')\,\mathscr{H}_{Fint}(\phi_{in}(X(y')), \pi_{\phi in}(X(y')))])|0>]^{-1} \tag{A.5.31}$$

Thus the time ordered product of eq. A.5.12, which appears in the evaluation of the S matrix element in eq. A.4.80, can be symbolically written as:

$$\tau(y_1, \ldots, y_n) = \frac{<0|T(\phi_{in}(X(y_1)) \ldots \phi_{in}(X(y_n))U(\infty, -\infty))|0>}{<0|T(\exp[-i\int dy'^0{:}d^3X(y')\mathscr{H}_{Fint}(\phi_{in}(X(y')),\pi_{\phi in}(X(y'))){:}])|0>} \tag{A.5.32}$$

in the limit $y^0 \to \infty$.

A.5.2 The $\int d^3X$ Integration

The integration over the X space coordinates presents the difficulty of a functional integration of a q-number that needs to be properly defined. Since

$$X^\mu(y) = y^\mu + i\, Y^\mu(y)/M_c^{\,2} \tag{A.2.12}$$

by definition and since, in the Y Coulomb gauge we have $X^0(y) = y^0$ due to $Y^0 = 0$, the classical Jacobian for the transformation from y to X coordinates is the absolute value:

$$J = \left| \varepsilon^{ijk} \left(\delta^{1i} + \frac{i}{M_C^{\,2}} \frac{\partial Y^1}{\partial y^i}\right)\left(\delta^{2j} + \frac{i}{M_C^{\,2}} \frac{\partial Y^2}{\partial y^j}\right)\left(\delta^{3k} + \frac{i}{M_C^{\,2}} \frac{\partial Y^3}{\partial y^k}\right) \right| \tag{A.5.33}$$

The Jacobian appears in a change of integration variables:

$$\int d^3X = \int d^3y\; J \tag{A.5.34}$$

$$\int d^4X = \int d^4y\; J \tag{A.5.35}$$

in the Y Coulomb gauge.

A change of variables for c-number coordinate transformations is well known. The situation changes when one set of coordinates are in fact q-numbers. The second quantization of the Y field requires the definition of J to be clarified since the product of fields at the same position is normally undefined. The normal ordering of the interaction hamiltonian term in eqs. A.5.34 and A.5.32 resolves the issue. Therefore eq. A.5.33 must be considered as inserted within a normal ordered expression.

While normal ordering eliminates the infinities that would otherwise be present, J still presents a problem because it is still effectively part of the interaction term. This situation appears to be unsatisfactory in the present, scalar quantum field theory in which Y is not intended to play a direct dynamical role but rather a passive role as a coordinate. The normal ϕ field is supposed to be the only in, out, and interacting field.

The problem of J is resolved by eqs. A.5.19c and A.5.19d, which reduces the effect of the derivative terms in eq. A.5.33 to zero in the Wick expansion of the time ordered product in eq. A.5.32 if no Y quanta appear in or out S matrix states. Thus

$$J \equiv 1 \tag{A.5.36}$$

As a result the time ordered product (eq. A.5.32) becomes:

$$\tau(y_1,\ldots, y_n) = \frac{\langle 0|T(\phi_{in}(X(y_1))\ldots\phi_{in}(X(y_n))\exp[-i\int d^4y'\,\mathscr{H}_{Fint}(\phi_{in}(X(y')))])\,|0\rangle}{\langle 0|T(\exp\,[-i\int d^4y'\,\mathscr{H}_{Fint}(\phi_{in}(X(y')))])|0\rangle} \tag{A.5.37}$$

A.5.3 Y In and out states

The Y fields have no interactions and are thus free fields in the model Lagrangian under consideration and in the Two-Tier quantum field theories that we will construct later. Therefore "in" Y quanta are the same as "out" Y quanta.

Since the Lagrangians that we consider do not have interaction terms explicitly containing Y field factors, the S matrix is "block diagonal" in the sense that if an in-state does not contain Y quanta, (or Y coherent states) then out-states will not contain Y quanta (or coherent Y states). The proof is based on the expansion of S matrix elements using Wick's theorem in products of time ordered products of pairs of in field operators. Eqs. A.5.19, A.5.20 and A.5.36, and in particular,

$$<0|\, T(\phi_{in}(X(y_1))Y^j(y_2))|0> = 0 \qquad \text{(A.5.39)}$$

$$<0|T(\phi_{in}(X(y_1))e^{-q\cdot Y^-(y)/M_c^2})|0> = <0|T(\phi_{in}(X(y_1))e^{+q\cdot Y^+(y)/M_c^2})|0> = 0 \qquad \text{(A.5.40)}$$

prove S matrix elements with no incoming Y quanta or coherent states will have zero matrix elements to produce outgoing Y quanta or coherent states. In addition any non-zero S matrix element with n incoming Y quanta must have n outgoing Y quanta. For example an incoming state with 5 Y quanta and 2 ϕ particles can only become an outgoing state with 5 Y quanta and two or more ϕ particles. Therefore we have proved the general result:

Theorem A.5.I: ***Any non-zero S matrix element has the same number of incoming Y quanta and outgoing Y quanta.***

This theorem is true in any Two-Tier quantum field theory. In order to have a tractable theory we will require all in-states and out-states <u>*not*</u> *contain Y quanta or coherent states. All normal in-state and out-state particles will contain factors of* $:e^{\pm p\cdot Y/M_c^2}:$ *in the fourier expansions of their corresponding fields.*

A.5.4 Unitarity

For many years it has been evident that modified field theories[11, 17, 292] might offer some hope of avoiding the divergences of conventional quantum field theory. Usually these theories suffer from unitarity problems: negative norms and negative probabilities. In the absence of a

---

[292] S. Blaha, Phys.Rev. **D10**, 4268 (1974); S. Blaha, Phys.Rev. **D11**, 2921 (1975); S. Blaha, Nuovo Cim. **A49**, :113 (1979); S. Blaha, "Generalization of Weyl's Unified Theory to Encompass a Non-Abelian Internal Symmetry Group" SLAC-PUB-1799, Aug 1976; S. Blaha, "Quantum Gravity and Quark Confinement" Lett. Nuovo Cim. **18**, 60 (1977); S. Blaha, "The Local Definition of Asymptotic Particle States" Nuovo Cim. **A49**, 35 (1979) and references therein.

physically acceptable interpretation of negative probabilities, these theories have been thought to be unsatisfactory.

The Two-Tier type of quantum field theory *superficially* also appears to have a unitarity problem due to the non-hermitean nature of Two-Tier hamiltonians. The lack of hermiticity is due entirely to the appearance of iY^μ in the X^μ field coordinates. *In fact Two-Tier quantum field theories satisfy unitarity for physical states. Physical states are defined to consist of any number of normal Two-Tier particles and NO Y quanta.*

Two-tier interaction hamiltonians, such as the one in eq. A.5.37, are not hermitean. For example,

$$H_{Fint} = \int d^3y' \mathscr{H}_{Fint}(\phi_{in}(y' + iY(y')/M_c^2)) \qquad (A.5.41)$$

$$H_{Fint}^\dagger = \int d^3y' \mathscr{H}_{Fint}(\phi_{in}(y' - iY(y')/M_c^2)) \neq H_{Fint} \qquad (A.5.42)$$

The relation between H_{Fint} and its hermitean conjugate is

$$H_{Fint} = V\, H_{Fint}^\dagger\, V \qquad (A.5.43)$$

where $V^2 = I$ is the metric operator defined in eqs. A.4.15 – A.4.18. Thus the S matrix is not unitary; the S matrix is *pseudo-unitary*:

$$S^{-1} = V\, S^\dagger\, V \qquad (A.5.44)$$

$$VS^\dagger\, VS = I \qquad (A.5.45)$$

We will now show that the S matrix is *unitary between physical states.* To prove this point, consider eq. A.5.45 between physical states |i> and <f| – each consisting of a number of ϕ particles and no Y quanta.

$$\delta_{fi} = \langle f|I|i\rangle = \langle f|VS^\dagger VS|i\rangle$$

$$= \sum_{n,\, m,\, p} \langle f|V|p\rangle\langle p|S^\dagger|n\rangle\langle n|V|m\rangle\langle m|S|i\rangle$$

$$= \sum_{n,\, m,\, p} \langle f|S^\dagger|m\rangle\langle m|S|i\rangle \qquad (A.5.46)$$

since V has the eigenvalue 1 between states consisting of no Y quanta. Due to eqs. A.5.19a – A.5.19e and A.5.20 since there are no incoming Y quanta there are no outgoing Y quanta. The block diagonality of S (and the diagonality of V) limits the intermediate states |n> and |m> to states containing ϕ particles and no Y quanta – although normalization factors R(**p**, z) will appear (described later) due to the presence of $:e^{\pm p \cdot Y/M_c^2}:$ factors within quantum field fourier expansions that embody Y coherent state effects. Thus

$$S_{phys}^{\dagger} S_{phys} = I \tag{A.5.47}$$
$$S_{phys}^{\dagger} = S_{phys}^{-1} \tag{A.5.48}$$

proving unitarity between physical states – states consisting of ϕ particles and no Y quanta that are properly normalized.

A.5.5 Finite Renormalization of External Legs

In the previous section we showed the theory satisfies unitarity for states that are properly normalized. However the use of the non-unitary operator W(y) (eq. A.4.6) to transform $\Phi_{in}(y)$ fields into $\phi_{in}(X(y))$ fields in the LSZ procedure in eq. A.4.69, and related equations, does not preserve the norm of input and output ϕ particle legs. Thus a finite renormalization is needed for each external particle leg in order to have a unitary S-matrix.

We define this renormalization of external legs within the framework of a perturbation theory example in section A.5.9.

A.5.6 Perturbation Expansion

Perturbation theory in Two-Tier quantum field theory is very similar to conventional perturbation theory. The difference is in the form of the propagators, which have a high energy damping factor R(**p**, z) that eliminates infinities that normally appear at high energy in conventional quantum field theories.

In order to develop a feeling for Two-Tier perturbation theory we will calculate a few low order diagrams in the perturbation theory of the model scalar ϕ^4 theory that we have been using as an example in this appendix.

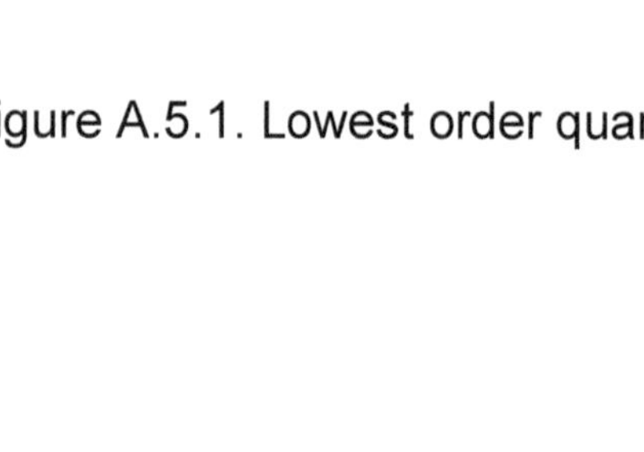

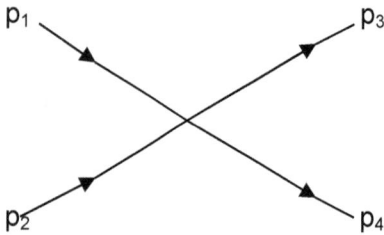

Figure A.5.1. Lowest order quartic interaction diagram.

Fig. A.5.1 contains the lowest order diagram for the scattering of two ϕ particles into a two ϕ particle out-state. The S matrix element for this diagram is

$$S_1 = i^4(\tfrac{1}{4!}\, i\varkappa_0) \prod_{j=1}^{4} \int d^4y_j\, d^4y\, f_{Zp_1}(y_1) f_{Zp_2}(y_2) f_{Zp_3}{}^*(y_3) f_{Zp_4}{}^*(y_4)(\Box_{y_1} + m^2)\cdot$$

$$\cdot(\Box_{y_2}+m^2)(\Box_{y_3}+m^2)(\Box_{y_4}+m^2)<0|T(\phi_{in}(X(y_1))\ldots\phi_{in}(X(y_4)):(\phi_{in}(X(y))^4:)|0> \qquad (A.5.49)$$

with $f_{Zp}(y)$ specified by

$$f_{Zp}(y) = [(2\pi)^3 2p^0 Z_p]^{-\frac{1}{2}}\, e^{-ip\cdot y} \qquad (A.5.49a)$$

where Z_p is a normalization factor that will be specified later.

Expanding the time ordered product and realizing there are 4! ways of combining the four field factors in the interaction hamiltonian leads to:

$$S_1 = i^4(i\varkappa_0) \prod_{j=1}^{4} \int d^4y_j\, d^4y\, f_{Zp_1}(y_1) f_{Zp_2}(y_2) f_{Zp_3}{}^*(y_3) f_{Zp_4}{}^*(y_4)(\Box_{y_1} + m^2)\cdot$$

$$\cdot(\Box_{y_2}+m^2)(\Box_{y_3}+m^2)(\Box_{y_4}+m^2)i\Delta_F{}^{TT}(y_1-y)i\Delta_F{}^{TT}(y_2-y)i\Delta_F{}^{TT}(y_3-y)i\Delta_F{}^{TT}(y_4-y) \quad (A.5.50)$$

where

$$i\Delta_F{}^{TT}(y_1 - y_2) = <0|T(\phi(X(y_1)),\phi(X(y_2)))|0> \qquad (A.5.51)$$

$$= i \int \frac{d^4p\, e^{-ip\cdot(y_1 - y_2)}\, R(\mathbf{p}, y_1 - y_2)}{(2\pi)^4\,(p^2 - m^2 + i\varepsilon)} \qquad (A.5.52)$$

with

$$R(\mathbf{p}, y_1 - y_2) = \exp[-p^ip^j\Delta_{Tij}(y_1 - y_2)/M_c{}^4] \qquad (A.5.53)$$

(summations are over space indices only in the Y Coulomb gauge) and

$$\Delta_{Tij}(z) = \int d^3k\, e^{-ik\cdot z}(\delta_{ij} - k_ik_j/\mathbf{k}^2)/[(2\pi)^3 2\omega_k] \qquad (A.5.54)$$

From section A.4 we have:

$$R(\mathbf{p}, y_1 - y_2) = \exp\{-p^2[A(v) + B(v)\cos^2\theta] / [4\pi^2 M_c^{\ 4} z^2]\} \qquad \text{(A.5.55)}$$

with

$$z^\mu = y_1^{\ \mu} - y_2^{\ \mu} \qquad \text{(A.5.56)}$$
$$z = |\mathbf{z}| = |\mathbf{y_1} - \mathbf{y_2}| \qquad \text{(A.5.57)}$$
$$p = |\mathbf{p}| \qquad \text{(A.5.58)}$$
$$v = |z^0| / z \qquad \text{(A.5.59)}$$
$$A(v) = (1 - v^2)^{-1} + .5v \ln[(v - 1)/(v + 1)] \qquad \text{(A.5.60)}$$
$$B(v) = v^2(1 - v^2)^{-1} - 1.5v \ln[(v - 1)/(v + 1)] \qquad \text{(A.5.61)}$$
$$\mathbf{p{\cdot}z} = pz\cos\theta \qquad \text{(A.5.62)}$$

and with $|\mathbf{p}|$ denoting the length of the spatial vector $\mathbf{p}$, while $|z^0|$ is the absolute value of z^0.

We note

$$R(\mathbf{p}, y) = R(\mathbf{p}, -y) \qquad \text{(A.5.62a)}$$

for later use.

Letting $y_i = w_i + y$ yields

$$S_1 = i^4(i\varkappa_0)(2\pi)^4\,\delta^4(p_3 + p_4 - p_1 - p_2)N^+(p_4)N^+(p_3)N(p_2)N(p_1) \qquad \text{(A.5.63)}$$

where

$$N(p) = iZ_p^{\ -\frac{1}{2}}\int d^4w\, f_p(w)(\Box + m^2)\Delta_F^{\ TT}(w) \qquad \text{(A.5.64)}$$
$$N^+(p) = iZ_p^{\ -\frac{1}{2}}\int d^4w\, f_p{}^*(w)(\Box + m^2)\Delta_F^{\ TT}(w) \qquad \text{(A.5.65)}$$

are "normalizations" of the "external legs" – the in and out states due to the Y field cloud around each particle with $Z^{-\frac{1}{2}}$ a renormalization factor to be determined later. In the limit of low momentum ($p \ll M_C$):

$$N(p) = N^+(p) \to -iZ_p^{\ -\frac{1}{2}}[(2\pi)^3\, 2p^0\,]^{-\frac{1}{2}} \qquad \text{(A.5.66)}$$

which the reader will note is the standard normalization factor for external scalar field legs in conventional quantum field theory modulo the $Z_p^{\ -\frac{1}{2}}$ factor. The factor $Z_p^{\ -\frac{1}{2}}$ performs the finite renormalization of external legs discussed in the preceding unitarity discussion.

A.5.7 Higher Order Diagram With a Loop

We will now consider the simplest one loop scattering diagrams in the scalar ϕ^4 theory.

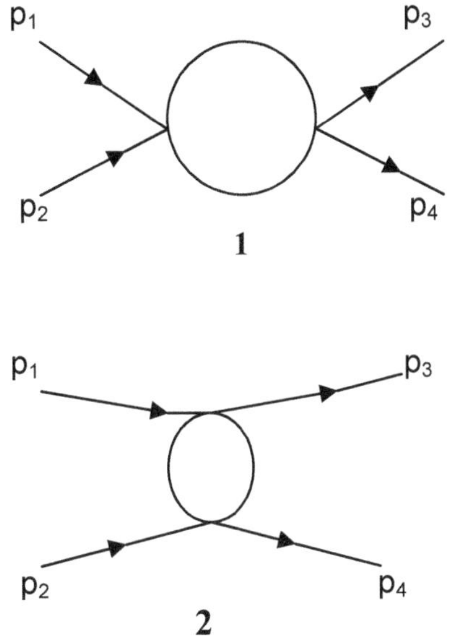

Figure A.5.2. Lowest order loop scattering diagrams.

The S matrix element for these diagrams (and some other disconnected diagrams) is contained in

$$S_2 = i^4(\tfrac{1}{4!}\, i\varkappa_0)^2 \prod_{j=1}^{4} \int d^4y_j\, d^4y'_1\, d^4y'_2\, f_{Zp_1}(y_1) f_{Zp_2}(y_2) f_{Zp_3}^{\,*}(y_3) f_{Zp_4}^{\,*}(y_4)(\Box_{y_1} + m^2)\cdot$$

$$\cdot(\Box_{y_2}+m^2)(\Box_{y_3}+m^2)(\Box_{y_4}+m^2)<0|T(\phi_{in}(X(y_1))\ldots\phi_{in}(X(y_4)):(\phi_{in}(X(y'_1))^4::(\phi_{in}(X(y'_2))^4:)|0>/2! \qquad \text{(A.5.67)}$$

together with some other disconnected diagrams.

Expanding the time ordered product and keeping only the terms corresponding to Fig. A.5.2 gives:

$$S_2 = i^4(i\varkappa_0)^2/2 \prod_{j=1}^{4} \int d^4y_j\, d^4y'_1\, d^4y'_2\, f_{Zp_1}(y_1) f_{Zp_2}(y_2) f_{Zp_3}^{\,*}(y_3) f_{Zp_4}^{\,*}(y_4)\cdot$$

$$\cdot(\Box_{y_1} + m^2)(\Box_{y_2}+m^2)(\Box_{y_3}+m^2)(\Box_{y_4}+m^2)\cdot\{i\Delta_F^{TT}(y_1-y'_1)i\Delta_F^{TT}(y_2-y'_1)i\Delta_F^{TT}(y_3-y'_2)i\Delta_F^{TT}(y_4-y'_2) +$$

$$+\, i\Delta_F^{TT}(y_1-y'_1)i\Delta_F^{TT}(y_2-y'_2)i\Delta_F^{TT}(y_3-y'_1)i\Delta_F^{TT}(y_4-y'_2)\}\, i\Delta_F^{TT}(y'_1-y'_2)i\Delta_F^{TT}(y'_1-y'_2) \qquad \text{(A.5.68)}$$

Following a similar procedure to the previous calculation yields

$$S_2 = i^4[(i\varkappa_0)^2/2](2\pi)^4\delta^4(p_3 + p_4 - p_1 - p_2)N^+(p_4)N^+(p_3)N(p_2)N(p_1) \int d^4z\, [e^{-i(p_1 + p_2)\cdot z} + e^{-i(p_1 - p_3)\cdot z}] \cdot$$
$$\cdot\, [i\Delta_F^{TT}(z)]^2 \qquad (A.5.69)$$

momentum conserving delta function as in eq. A.5.63. The loop integrals have the form:

$$I(q) = \int d^4z\, e^{-iq\cdot z}\, [i\Delta_F^{TT}(z)]^2 \qquad (A.5.70)$$

The behavior of the Two-Tier Feynman propagator $\Delta_F^{TT}(z)$ was studied at long and short distance in eqs. A.3.21-A.3.24. The large distance behavior of the Two-Tier Feynman propagator $\Delta_F^{TT}(z)$ approaches the behavior of the conventional Feynman propagator since

$$R(\mathbf{p}, z) \to 1 \qquad (A.5.71)$$

as $z^2 = z^\mu z_\mu$ becomes much larger than M_c^{-2} ($z^2 \gg M_c^{-2}$) (eq. A.5.55). Thus I(q) approaches the standard one loop expression of conventional field theory at large distance (or small momentum). Again we seee that Two-Tier *quantum field theory realizes a form of Correspondence Principle approaching conventional quantum field theory at large distance.*

At short distances the Gaussian factor $R(\mathbf{p}, z)$ dominates. The Two-Tier Feynman propagator $\Delta_F^{TT}(z)$ is radically different from the conventional Feynman propagator at very short distances (or very high momentum). The singular behavior of the conventional Feynman propagator is replaced with a well-behaved, high-energy (short distance) behavior. Near the light cone $M_c^{-2} \gg z^2 \to 0$ (or $p^2 \gg M_c^2$) we can approximate eq. A.5.52 with

$$i\Delta_F^{TT}(z) \approx \int d^3p\, [N(p)]^2\, R(\mathbf{p}, z) \qquad (A.5.72)$$

since $e^{-ip\cdot z}$ is approximately unity for small z. We assume the mass of the ϕ particle is negligible on this scale. Upon performing the integrations (see eq. A.3.23 for the exact result) we find eq. A.5.72 approaches:

$$i\Delta_F^{TT}(z) \to \pi\, M_c^4|z^2|/8$$

as $z^2 = z^\mu z_\mu \to 0$ from the space-like or time-like side of the light cone where | | represents the absolute value.

Therefore I(q) is finite and well-behaved. At high energy ($q^2 \gg M_c^2$)

$$I(q) \sim q^{-8}$$

since the fourier transform of $\Delta_F^{TT}(z)$ (momentum space) is

$$\Delta_F^{TT}(p) = \int d^4z\, e^{-ip\cdot z}\, \Delta_F^{TT}(z) \sim p^{-6}$$

for large p ($p^2 \gg M_c^2$). (Compare the preceding high energy behavior of I(q) with the conventional logarithmically divergent one loop result $I(q) \sim \ln(q^2/\Lambda^2)$ with Λ a cutoff.)

Thus Two-Tier quantum provides the benefits of a higher derivative theory without its drawbacks.

A.5.8 Finite Renormalization of External Particle Legs & Unitarity Example

The renormalization factor $Z_p^{-½}$ appearing in eqs. A.5.64 and A.5.65 that is due to the use of the non-unitary operator W(y) (eq. A.4.6) to transform $\Phi_{in}(y)$ fields into $\phi_{in}(X(y))$ fields in the LSZ procedure in eq. A.4.69, and related equations, does not preserve the norm of input and output ϕ particle legs. $Z_p^{-½}$ performs a finite renormalization for each external particle leg to compensate for the effects of W(y).

The required renormalization is nicely illustrated by considering the unitarity sum in the imaginary part of the preceding example.

The transition matrix T_{fi} is defined in terms of the S matrix by

$$S_{fi} = \delta_{fi} - i\,(2\pi)^4\, \delta^4(P_f - P_i)\, T^{(+)}{}_{fi}$$

The unitarity condition is

$$T^{(+)}{}_{fi} - T^{(-)}{}_{fi} = -i \sum_n (2\pi)^4\, \delta^4(P_n - P_i)\, T^{(-)}{}_{fn}\, T^{(+)}{}_{ni} \qquad \text{(A.5.73)}$$

Therefore we see that the first term on the right side of eq. A.5.69 gives a transition matrix term:

$$T^{(+)}{}_{2a} = -\, i[\chi_0^2/2]N^+(p_4)N^+(p_3)N(p_2)N(p_1)\int d^4z\, e^{-iP\cdot z}[i\Delta_F^{TT}(z)]^2 \qquad \text{(A.5.69a)}$$

where P = $p_1 + p_2$. Substituting for $i\Delta_F^{TT}$ (using eq. A.3.11) we find that the imaginary part of $T^{(+)}{}_{2a}$ is given by (Note R(**p**, z) is real.)

$$T^{(+)}{}_{2a} - T^{(-)}{}_{2a} = -i[\chi_0^2/2]N^+(p_4)N^+(p_3)N(p_2)N(p_1) \int d^4z\, e^{-iP\cdot z}\, [i \int d^4p\, \theta(p_0)\, \delta(p^2 - m^2)e^{-ip\cdot z} \cdot$$
$$\cdot\, R(\mathbf{p},z)/(2\pi)^3]^2$$

If we express the R(**p**, z) factors in terms of their fourier transforms (see eq. A.3.27):

$$R(\mathbf{p}, z) = \int d^4q\, e^{-iq\cdot z}\, R(\mathbf{p}, q)$$

Then we find

$$T^{(+)}{}_{2a} - T^{(-)}{}_{2a} = -i[\chi_0{}^2/2]N^+(p_4)N^+(p_3)N(p_2)N(p_1)\int d^4z\, e^{-iP\cdot z}\cdot$$
$$\cdot\left[i\int d^4k_1\, d^4q_1\theta(k_1{}^0)\,\delta(k_1{}^2 - m^2)e^{-ik1\cdot z}\, e^{-iq1\cdot z}\, R(\mathbf{k_1}, q_1)/(2\pi)^3\right]\cdot$$
$$\cdot\left[i\int d^4k_2\, d^4q_2\theta(k_2{}^0)\,\delta(k_2{}^2 - m^2)e^{-ik2\cdot z}\, e^{-iq2\cdot z}\, R(\mathbf{k_2}, q_2)/(2\pi)^3\right]$$

Performing the integral over z gives

$$T^{(+)}{}_{2a} - T^{(-)}{}_{2a} = +i[\chi_0{}^2/2]N^+(p_4)N^+(p_3)N(p_2)N(p_1)(2\pi)^4\cdot$$
$$\cdot\int d^4k_1d^4q_1d^4k_2d^4q_2\theta(k_1{}^0)\,\delta(k_1{}^2 - m^2)\,\theta(k_2{}^0)\,\delta(k_2{}^2 - m^2)\cdot$$
$$\cdot R(\mathbf{k_1}, q_1)R(\mathbf{k_2}, q_2)\,\delta^4(P + k_1 + q_1 + k_2 + q_2)/(2\pi)^6$$

Introducing delta functions enables us to re-express this equation as

$$T^{(+)}{}_{2a} - T^{(-)}{}_{2a} = +i[\chi_0{}^2/2]N^+(p_4)N^+(p_3)N(p_2)N(p_1)\int d^4r_1d^4r_2(2\pi)^4\,\delta^4(P - r_1 - r_2)\cdot$$
$$\cdot\int d^4k_1d^4q_1\,\delta^4(r_1 + k_1 + q_1)\theta(k_1{}^0)\,\delta(k_1{}^2 - m^2)\,R(\mathbf{k_1}, q_1)\cdot$$
$$\cdot\int d^4k_2d^4q_2\theta(k_2{}^0)\,\delta(k_2{}^2 - m^2)\,\delta^4(r_2 + k_2 + q_2)R(\mathbf{k_2}, q_2)/(2\pi)^6$$

which becomes

$$T^{(+)}{}_{2a} - T^{(-)}{}_{2a} = +i[\chi_0{}^2/2]N^+(p_4)N^+(p_3)N(p_2)N(p_1)\int d^4r_1d^4r_2(2\pi)^4\,\delta^4(P - r_1 - r_2)\cdot$$
$$\cdot\int d^4k_1\theta(k_1{}^0)\,\delta(k_1{}^2 - m^2)\,R(\mathbf{k_1}, -k_1 - r_1)\cdot$$
$$\cdot\int d^4k_2\theta(k_2{}^0)\,\delta(k_2{}^2 - m^2)R(\mathbf{k_2}, -k_2 - r_2)/(2\pi)^6$$

$R(\mathbf{k_2}, -k_2 - r_2)$ can be expressed in terms of its fourier transform $R(\mathbf{k_2}, z)$ using eq. A.3.27.

We can now rewrite the above expression in terms of intermediate states:

$$T^{(+)}{}_{2a} - T^{(-)}{}_{2a} = -i\int d^4r_1d^4r_2\,(2\pi)^4\delta^4(P - r_1 - r_2)\cdot$$
$$\cdot\, i\chi_0N^+(p_4)N^+(p_3)\int d^4k_1\theta(k_1{}^0)\,\delta(k_1{}^2 - m^2)\,[R(\mathbf{k_1}, -k_1 - r_1)/(2\pi)^3]\cdot$$

$$\cdot \int d^4k_2\theta(k_2^{\,0})\,\delta(k_2^{\,2} - m^2)[R(\mathbf{k_2}, -k_2 - r_2)/(2\pi)^3]i\,\varkappa_0 N(p_2)N(p_1)/2$$

which has the form:

$$T^{(+)}{}_{2a} - T^{(-)}{}_{2a} = -i\int d^4r_1 d^4r_2(2\pi)^4\delta^4(P - r_1 - r_2)[\int d^4k_1\theta(k_1^{\,0})\,\delta(k_1^{\,2} - m^2)\,R(\mathbf{k_1}, -k_1 - r_1)/(2\pi)^3]\cdot$$
$$\cdot[\int d^4k_2\theta(k_2^{\,0})\,\delta(k_2^{\,2} - m^2)R(\mathbf{k_2}, -k_2 - r_2)/(2\pi)^3]T^{(-)}{}_{fn}T^{(+)}{}_{ni}/2!$$

where

$$T^{(-)}{}_{fn} = \varkappa_0 N^+(p_4)N^+(p_3)N(r_2)N(r_1)$$

$$T^{(+)}{}_{ni} = \varkappa_0 N^+(r_2)N^+(r_1)N(p_2)N(p_1)$$

if

$$N^+(p) = N(p) = 1 \qquad \text{(A.5.74)}$$

Eq. A.5.74 implies the (finite) external leg renormalization must be

$$Z_p = -\left[\int d^4w\; f_p(w)(\Box + m^2)\Delta_F^{\,TT}(w)\right]^2 \qquad \text{(A.5.74a)}$$

by eqs. A.5.64 and A.5.65. Thus all external legs must be "lopped off."

The result is a theory that satisfies the unitarity condition (eq. A.5.73) as shown in the above detailed discussion.

If we define

$$N(r) = \int d^4k\;\theta(k^0)\delta(\,k^2 - m^2)R(\mathbf{k}, -k - r) \qquad \text{(A.5.75a)}$$

$$= (2\pi)^{-4}\int d^4k\; d^4z\;\theta(k^0)\delta(\,k^2 - m^2)\;e^{-i(k + r)\cdot z}\;R(\mathbf{k}, z) \qquad \text{(A.5.75b)}$$

then the Two-Tier completeness expression becomes:

$$S_{fi} = \sum_n (2\pi)^{-3n}(n!)^{-1}\int(\prod_{j=1}^{n} d^4r_jN(r_j))S_{fn}S_{ni}^{\dagger}\delta^4(P_n - \sum_{k=1}^{n} r_k) \qquad \text{(A.5.75c)}$$

This expression reflects the fact that ϕ particles are surrounded by a "cloud" of Y quanta. Thus we have achieved unitarity! For small momenta $r_j \ll M_c$, we find $N(r_j) \simeq \theta(r_j^{\,0})\delta(r_j^{\,2} - m^2)$ (eq. A.5.75b with $R(k, q) \simeq 1$.) $\theta(r_j^{\,0})\delta(r_j^{\,2} - m^2)$ is the form seen in conventional quantum field theory. For large momenta $N(r_j)$ is very different.

A.5.9 General Form of Propagators

In this section we have considered a scalar Two-Tier quantum field theory. We have seen that the Two-Tier Feynman propagator is well behaved near the light cone resulting in a finite ϕ^4 theory. This finite ϕ^4 theory approximates the results of conventional ϕ^4 theory at low energy thus implementing a correspondence principle: *At low energy results in Two-Tier quantum field theory approach the corresponding results of the corresponding conventional quantum field theory.*

The observations on Two-Tier field theory made in this appendix generally apply to Two-Tier versions of Quantum Electrodynamics, ElectroWeak Theory and the Standard Model as well as Two-Tier Quantum Gravity:

1. At low energy ($p^2 \ll M_c^2$ or large distances $z^2 \gg M_c^{-2}$) the Two-Tier quantum field theory is the same as the corresponding conventional quantum field theory to good approximation. (Correspondence Principle)

2. At high energy ($p^2 \gg M_c^2$ or short distances: $z^2 \ll M_c^{-2}$) Two-Tier quantum field theories (of physical interest) are well-behaved and finite.

3. Two-Tier quantum field theories (of physical interest) satisfy unitarity and Lorentz invariance (and in the case of quantum gravity their dynamical equations satisfy the requirements of general relativity).

The generality of these results is based on:

1. The expansion of the S matrix in time ordered products of field operators.
2. Wick's Theorem
3. The general form of all particle propagators in Two-Tier quantum field theories. All particle Feynman propagators have the form:

$$iG_F^{TT}{}_{\ldots}(y_1 - y_2) = \langle 0|T(\chi_{\ldots}(X(y_1)),\chi_{\ldots}(X(y_2)))|0\rangle \qquad (A.5.76)$$

$$= \int d^4p \; iG_{F\ldots}(p)e^{-ip\cdot(y_1 - y_2)} R(\mathbf{p}, y_1 - y_2) \qquad (A.5.77)$$

where $iG_{F\ldots}(p)$ is the conventional momentum space $\chi_{\ldots}$ particle propagator, and where … represents the relevant tensor and matrix indices. $R(\mathbf{p}, y_1 - y_2)$ introduces a damping factor in each particle propagator that eliminates divergences.

A.5.10 Scalar Particle Propagator

The Two-Tier propagator for the case of a free scalar particle is:

$$i\Delta_F^{TT}(y_1 - y_2) = <0|T(\phi(X(y_1)),\phi(X(y_2)))|0> \tag{A.5.51}$$

$$= i\int \frac{d^4p\, e^{-ip\cdot(y_1 - y_2)}\, R(\mathbf{p}, y_1 - y_2)}{(2\pi)^4\,(p^2 - m^2 + i\varepsilon)} \tag{A.5.52}$$

Since the mass m is not relevant at high energy we set m = 0. This enables us to obtain a more tractable expression for the propagator. After some manipulation the massless scalar propagator can be represented as:

$$i\Delta_F^{TT}(z) = -\beta[16\pi^3(AB)^{1/2}]^{-1}\int_{-\infty}^{\infty} dy_1 \int_{-\infty}^{\infty} dy_2\cdot$$
$$\cdot\{\theta(z_0)\exp[-\beta((y_1 - z_0)^2B + (y_2 + z)^2A)/(4AB)] +$$
$$+\,\theta(-z_0)\exp[-\beta((y_1 + z_0)^2B + (y_2 - z)^2A)/(4AB)]\}/(y_1^2 - y_2^2) \tag{A.5.78}$$

with $\beta = 4\pi^2M_c^4\mathbf{z}^2$. Using

$$(y_1^2 - y_2^2)^{-1} = -\,0.5\int_0^{\infty} dq_1 \int_{-\infty}^{\infty} dq_2\, \theta(q_1^2 - q_2^2)\exp[iq_1y_1 - iq_2y_2] \tag{A.5.79}$$

we obtain the representation

$$i\Delta_F^{TT}(z^\mu) = (8\pi^2)^{-1}\int_0^{\infty} dq_1 \int_{-\infty}^{\infty} dq_2\, \theta(q_1^2 - q_2^2)\exp\{iq_1|z_0| + iq_2z - [A'q_1^2 + B'q_2^2]/[\beta'(\mathbf{z}^2 - z_0^2)]\} \tag{A.5.80}$$

where $|z_0|$ is the absolute value of z_0, $z^2 - z_0^2 = -z^\mu z_\mu$ and

$$A = A'/(1 - v^2) \tag{A.5.81}$$
$$B = B'/(1 - v^2) \tag{A.5.82}$$
$$\beta = 4\pi^2M_c^4\mathbf{z}^2 = \beta' z^2 \tag{A.5.83}$$

with $\mathbf{z} = |\vec{\mathbf{z}}|$ – the magnitude of the spatial vector $\vec{\mathbf{z}}$, and A and B given by eqs. A.5.60 – A.5.61.

The representation of $i\Delta_F^{TT}$ in eq. A.5.80 is particularly useful in determining its low energy ($\ll M_c$), and its high energy ($\gg M_c$) behavior. The low energy behavior is governed by the linear terms in the exponential in eq. A.5.80 since $\beta'(\mathbf{z}^2 - z_0^2)$ is very large in this limit:

$$i\Delta_F{}^{TT}(z^\mu)_{low} \simeq (8\pi^2)^{-1} \int_0^\infty dq_1 \int_{-\infty}^\infty dq_2\ \theta(q_1{}^2 - q_2{}^2)\exp\{\ iq_1|z_0| + iq_2 z\} \qquad (A.5.84)$$

$$= [4\pi^2(\mathbf{z}^2 - z_0{}^2)]^{-1} \qquad (A.5.85)$$

$$= i\Delta_F(z^\mu)$$

equaling the exact massless, free, spin 0 Feynman propagator of conventional quantum field theory.

In the high energy limit when $\beta'(\mathbf{z}^2 - z_0{}^2)$ is small since $\mathbf{z}^2 \approx z_0{}^2$ (i.e. near the light cone), the quadratic terms in the exponential in eq. A.5.80 dominate and $A' \simeq B'$. We then find

$$i\Delta_F{}^{TT}(z^\mu)_{high} \simeq (8\pi^2)^{-1} \int_0^\infty dq_1 \int_{-\infty}^\infty dq_2 \theta(q_1{}^2 - q_2{}^2)\exp\{A'(q_1{}^2 + q_2{}^2)/[\ \beta'(\mathbf{z}^2 - z_0{}^2)]\} \qquad (A.5.86)$$

$$= \pi M_c{}^4|(\mathbf{z}^2 - z_0{}^2)|/8 \qquad (A.5.87)$$

as in eq. A.3.24. As pointed out earlier, eq. A.5.87 corresponds to k^{-6} behavior in momentum space:

$$i\Delta_F{}^{TT}(k)_{high} \backsim k^{-6} \qquad (A.5.87a)$$

A.5.11 Spin ½ Particle Propagator

For the case of a free, spin ½ particle the propagator is:

$$iS_F{}^{TT}(y_1 - y_2) = \langle 0|T(\bar\psi(X(y_1))\psi(X(y_2)))|0\rangle \qquad (A.5.88)$$

$$= i \int \frac{d^4p\ e^{-ip\cdot(y_1 - y_2)}\ (\not p + m)\ R(\mathbf{p}, y_1 - y_2)}{(2\pi)^4\ (p^2 - m^2 + i\varepsilon)}$$

Again setting m = 0 we find a convenient representation in the form:

$$S_F{}^{TT}(z^\mu) = i(8\pi^2)^{-1} \int_0^\infty dq_1 \int_{-\infty}^\infty dq_2\ \theta(q_1{}^2 - q_2{}^2)(\epsilon(z_0)q_1\gamma_0 - q_2\vec{\mathbf{z}}\cdot\vec{\gamma}/z\ )\cdot$$

$$\cdot \exp\{iq_1|z_0| + iq_2 z\ - [A'q_1{}^2 + B'q_2{}^2]/[\ \beta'(\mathbf{z}^2 - z_0{}^2)]\} \qquad (A.5.89)$$

using the same symbols and notation as eq. A.5.80, and with $\in(z_0) = +1$ if $z_0 \geq 0$ and -1 otherwise.

The representation of S_F^{TT} in eq. A.5.89 is useful in determining its low energy ($\ll M_c$), and high energy ($\gg M_c$) behavior. The low energy behavior is governed by the linear terms in the exponential in eq. A.5.89 since $\beta'(\mathbf{z}^2 - z_0^2)$ is large in this limit:

$$S_F^{TT}(z^\mu)_{low} \simeq (8\pi^2)^{-1} \int_0^\infty dq_1 \int_{-\infty}^\infty dq_2\, \theta(q_1^2 - q_2^2)(\in(z_0)q_1\gamma_0 - q_2\vec{\mathbf{z}}\cdot\vec{\gamma}/z) \exp\{iq_1|z_0| + iq_2z\} \quad (A.5.90)$$

$$= \not{z}[2\pi^2(\mathbf{z}^2 - z_0^2)^2]^{-1} \quad (A.5.91)$$

$$= S_F(z^\mu)$$

equaling the exact massless, spin ½ Feynman propagator of conventional quantum field theory. If we had not set m = 0 initially, we would have obtained the usual massive, spin ½ Feynman propagator.

In the high energy limit when $\beta'(\mathbf{z}^2 - z_0^2)$ is small since $\mathbf{z}^2 \approx z_0^2$ (i.e. near the light cone), the quadratic terms in the exponential in eq. A.5.89 dominate and $A' \simeq B'$. We then find

$$S_F^{TT}(z^\mu)_{high} \simeq (8\pi^2)^{-1} \int_0^\infty dq_1 \int_{-\infty}^\infty dq_2 \theta(q_1^2 - q_2^2)(\in(z_0)q_1\gamma_0 - q_2\vec{\mathbf{z}}\cdot\vec{\gamma}/z)\exp\{A'(q_1^2 + q_2^2)/[\beta'(\mathbf{z}^2 - z_0^2)]\} \quad (A.5.92)$$

$$= i(8\pi^2)^{-1}\{\mathbf{z}^{-1}(\mathbf{z}^2 - z_0^2)^{3/2}2^{3/2}\pi^{7/2}M_c^6 z_0\gamma_0 - 4i(\mathbf{z}^2 - z_0^2)^2\pi^5M_c^8\vec{\mathbf{z}}\cdot\vec{\gamma})\} \quad (A.5.93)$$

The leading momentum dependence of the fourier transform of $S_F^{TT}(z^\mu)_{high}$ is

$$S_F^{TT}(p)_{high} \backsim M_c^6 p^{-7}\gamma_0 \quad (A.5.94)$$

A.5.12 Massless Spin 1 Particle Propagator

The Two-Tier Feynman propagator for the case of a free, massless, spin 1, gauge field particle (coupled to a conserved current) such as a photon is:

$$iD_F^{TT}(z)_{\mu\nu} = -i \int \frac{d^4p\, e^{-ip\cdot z}\, g_{\mu\nu} R(\mathbf{p}, y_1 - y_2)}{(2\pi)^4 (p^2 + i\varepsilon)} \quad (A.5.95)$$

The form of eq. A.5.95 is the same as the scalar particle propagator multiplied by $-g_{\mu\nu}$. As a result we have the representation:

$$iD_F^{TT}(z)_{\mu\nu} = -(8\pi^2)^{-1} \int_0^\infty dq_1 \int_{-\infty}^\infty dq_2\ \theta(q_1^2 - q_2^2)\ g_{\mu\nu} \exp\{iq_1|z_0| + iq_2 z - [A'q_1^2 + B'q_2^2]/[\ \beta'(\mathbf{z}^2 - z_0^2)]\} \qquad (A.5.96)$$

As before in the scalar particle case, the low energy behavior is governed by the linear terms in the exponential in eq. A.5.96 since $\beta'(\mathbf{z}^2 - z_0^2)$ is very large in this limit:

$$iD_F^{TT}(z)_{\mu\nu\text{low}} \simeq -g_{\mu\nu}(8\pi^2)^{-1} \int_0^\infty dq_1 \int_{-\infty}^\infty dq_2\ \theta(q_1^2 - q_2^2)\exp\{iq_1|z_0| + iq_2 z\} \qquad (A.5.97)$$

$$= -g_{\mu\nu}[4\pi^2(\mathbf{z}^2 - z_0^2)]^{-1} \qquad (A.5.98)$$

$$= -ig_{\mu\nu}\Delta_F(z)$$

equaling the exact free, massless, spin 1 Feynman gauge field propagator of conventional quantum field theory.

In the high energy limit when $\beta'(\mathbf{z}^2 - z_0^2)$ is small since $\mathbf{z}^2 \approx z_0^2$ (i.e. near the light cone), the quadratic terms in the exponential in eq. A.5.96 dominate, and $A' \simeq B'$. We then find

$$iD_F^{TT}(z)_{\mu\nu\text{high}} \simeq -(8\pi^2)^{-1} \int_0^\infty dq_1 \int_{-\infty}^\infty dq_2 \theta(q_1^2 - q_2^2) g_{\mu\nu} \exp\{\ A'(q_1^2 + q_2^2)/[\ \beta'(\mathbf{z}^2 - z_0^2)]\} \qquad (A.5.99)$$

$$= -g_{\mu\nu}\pi\ M_c^4\ |(\mathbf{z}^2 - z_0^2)|/8 \qquad (A.5.100)$$

Eq. A.5.100 corresponds to k^{-6} behavior in momentum space:

$$iD_F^{TT}(k)_{\mu\nu\text{high}} \backsim g_{\mu\nu}\, M_c^4 k^{-6} \qquad (A.5.101)$$

A.5.13 Spin 2 Particle Propagator

The Two-Tier propagator for the case of a free, massless, spin 2 particle such as a graviton is:

$$i\Delta_{F2}^{TT}(z)_{\mu\nu\rho\sigma} = i \int \frac{d^4p\ e^{-ip\cdot z}\ b_{\mu\nu\rho\sigma}(p)R(\mathbf{p}, y_1 - y_2)}{(2\pi)^4\ (p^2 + i\varepsilon)} \qquad (A.5.102)$$

in an appropriate gauge where $b_{\mu\nu\rho\sigma}(p)$ is a tensor that is independent of the coordinates. We can express eq. A.5.102 in the form:

$$i\Delta_{F2}{}^{TT}(z)_{\mu\nu\rho\sigma} = (8\pi^2)^{-1} \int_0^\infty dq_1 \int_{-\infty}^\infty dq_2\, \theta(q_1{}^2 - q_2{}^2)\, \tilde{b}(z_0, z, q_1, q_2)_{\mu\nu\rho\sigma} \cdot$$

$$\cdot \exp\{ iq_1|z_0| + iq_2 z - [A'q_1{}^2 + B'q_2{}^2]/[\beta'(\mathbf{z}^2 - z_0{}^2)]\} \qquad (A.5.103)$$

where $\tilde{b}(z_0, z, q_1, q_2)_{\mu\nu\rho\sigma}$ is a tensor generated from the $b_{\mu\nu\rho\sigma}(p)$ tensor.

As before in the scalar particle case, the low energy behavior is governed by the linear terms in the exponential in eq. A.5.103 since $\beta'(\mathbf{z}^2 - z_0{}^2)$ is very large in this limit and we find that the covariant piece[293] behaves like:

$$i\Delta_{F2}{}^{TT}(z)_{\mu\nu\rho\sigma\text{lowCov}} \simeq \tilde{\tilde{b}}_{\mu\nu\rho\sigma}(8\pi^2)^{-1} \int_0^\infty dq_1 \int_{-\infty}^\infty dq_2\, \theta(q_1{}^2 - q_2{}^2)\exp\{ iq_1|z_0| + iq_2 z\} \qquad (A.5.104)$$

$$= \tilde{\tilde{b}}_{\mu\nu\rho\sigma}[4\pi^2(\mathbf{z}^2 - z_0{}^2)]^{-1} \qquad (A.5.105)$$

$$= i\Delta_F(z^\mu)\, \tilde{\tilde{b}}_{\mu\nu\rho\sigma}$$

where

$$\tilde{\tilde{b}}_{\mu\nu\rho\sigma} = \tfrac{1}{2}\, [\eta_{\mu\rho}\eta_{\nu\sigma} + \eta_{\mu\sigma}\eta_{\nu\rho} - \eta_{\mu\nu}\eta_{\rho\sigma}] \qquad (A.5.106)$$

so that the expression in eq. A.5.105 equals the corresponding covariant piece of the exact free, massless, spin 2 Feynman propagator of conventional quantum field theory.

In the high energy limit when $\beta'(\mathbf{z}^2 - z_0{}^2)$ is small since $\mathbf{z}^2 \approx z_0{}^2$ (i.e. near the light cone), the quadratic terms in the exponential in eq. A.5.103 dominate, and $A' \simeq B'$. We then find

$$i\Delta_{F2}{}^{TT}(z)_{\mu\nu\rho\sigma\text{high}} \simeq (8\pi^2)^{-1} \int_0^\infty dq_1 \int_{-\infty}^\infty dq_2 \theta(q_1{}^2 - q_2{}^2)\; \tilde{b}(z_0, z, q_1, q_2)_{\mu\nu\rho\sigma} \cdot$$

$$\cdot \exp\{A'(q_1{}^2 + q_2{}^2)/[\beta'(\mathbf{z}^2 - z_0{}^2)]\} \qquad (A.5.107)$$

and the covariant piece behaves like

$$i\Delta_{F2}{}^{TT}(z)_{\mu\nu\rho\sigma\text{highCov}} \simeq \tilde{\tilde{b}}_{\mu\nu\rho\sigma}\pi M_c{}^4|(\mathbf{z}^2 - z_0{}^2)|/8 \qquad (A.5.108)$$

The coordinate space behavior of eq. A.5.108 corresponds to k^{-6} behavior in momentum space:

---

[293] S. Weinberg, Phys. Rev. **135**, B1049 (1964); Phys. Rev. **138**, B988 (1965).

$$i\Delta_{F2}{}^{TT}(k)_{\mu\nu\rho\sigma\text{highCov}} \backsim \tilde{\tilde{b}}_{\mu\nu\rho\sigma}\, k^{-6} \qquad \text{(A.5.109)}$$

The high-energy behavior of the spin 2 propagator in momentum space results in a Two-Tier theory of quantum gravity that has no high-energy divergences and is thus finite.

A.6 Finite Quantum Gravity

A.6.1 Introduction - Two-Tier Quantum Gravity: Finite!

There are numerous excellent books and monographs on classical gravity and a large literature on quantum gravity.[294] Therefore our discussion will assume the reader is familiar with classical General Relativity and aware of attempts to create quantum theories of gravity.

We will begin by establishing the general form of Two-Tier classical General Relativity and then proceed to define a quantization procedure. We will work in Minkowski space with three space and one time dimension. The flat-space metric $\eta_{\alpha\beta}$ is defined as diagonal with $\eta_{00} = 1$ and $\eta_{ij} = -\delta_{ij}$ for i, j = 1, 2, 3.

A.6.2 Two-Tier General Relativity

In developing Quantum Gravity we will make the same ansatz that we have made throughout our development of Two-Tier quantum field theories: all field expressions are functions of the X coordinate field system, which in turn are functions of the "ordinary" y space-time coordinate system. Two-Tier Theory of Quantum Gravity is invariant under special relativistic transformations. The dynamical field equations, which are strictly functions of the X coordinates, are covariant under general relativistic transformations.

---

[294] H. Weyl, *Space, Time, Matter* (Dover, New York, 1950); L. D. Landau and E. M. Lifshitz, *The Classical Theory of Fields*, (Addison-Wesley, New York, 1962); S. Weinberg, *Gravitation and Cosmology*, (John Wiley & Sons, New York, 1972); C. W. Misner, K. S. Thorne and J. A. Wheeler, *Gravitation*, (W. H. Freeman, San Francisco, 1973); B. S. DeWitt, Phys. Rev. **162**, 1239 (1967), **162**, B1195 (1967); R. P. Feynman, Acta Physica Polonica **24**, 697 (1963); S. Deser and P. van Nieuwenhuizen, Phys. Rev. Letters **32**, 245 (1974); S. Deser, H.-S. Tsao and P. van Nieuwenhuizen, "One Loop Divergences of the Einstein-Yang-Mills System", Brandeis Univ. preprint (1974); S. Weinberg, Phys. Rev. **138**, B988 (1965); L. Smolin, *Three Roads to Quantum Gravity*, (Basic Books, New York, 2001); L. Smolin, "How Far are We From the Quantum Theory of Gravity", (Univ. Waterloo preprint (2003) and references therein; T. Thiemann, "Lectures on Loop Quantum Gravity", Preprint AEI-2002-087, Albert Einstein Insitute, Golm, Germany (2002) and references therein; A. pais and G. E. Uhlenbeck, Phys. Rev. **79**, 145 (1950); G. E. Uhlenbeck, "Lecture Notes on General Relativity", The Rockefeller University (1967), unpublished; S. Blaha, "Generalization of Weyl's Unified Theory to Encompass a Non-Abelian Internal Symmetry Group" SLAC-PUB-1799, Aug 1976; S. Blaha, "Quantum Gravity and Quark Confinement" Lett. Nuovo Cim. **18**, 60 (1977); R. Utiyama, Phys. Rev. **101**, 1597 (1956); T. W. B. Kibble, J. Math. Phys. **2**, 212 (1961); R. Arnowitt, S. Deser, and C. W. Misner, Phys. Rev. **117**, 1595 (1960); and references therein.

We define the proper time differential $d\tau$ as

$$d\tau^2 = g_{\mu\nu}(X(y))dX^\mu dX^\nu \tag{A.6.1}$$

where, as usual,

$$X^\mu(y) = y^\mu + i\, Y^\mu(y)/M_c^2$$

Thus eq. A.6.1 could be written:

$$d\tau^2 = g_{\mu\nu}(X(y))(\eta^\mu{}_\alpha + iM_c^{-2}\partial Y^\mu/\partial y^\alpha)(\eta^\nu{}_\beta + iM_c^{-2}\partial Y^\nu/\partial y^\beta)dy^\alpha dy^\beta \tag{A.6.2}$$

The inverse of $g_{\mu\nu}$, denoted $g^{\nu\lambda}$, satisfies

$$g_{\mu\nu}(X(y))g^{\nu\lambda}(X(y)) = \delta_\mu{}^\lambda \tag{A.6.3}$$

Since the algebraic manipulation of the tensor indices is the same as that of the conventional theory of gravitation the Two-Tier affine connection is:

$${}_X\Gamma^\sigma{}_{\lambda\mu} = ½\, g^{\nu\sigma}\{\partial g_{\mu\nu}/\partial X^\lambda + \partial g_{\lambda\nu}/\partial X^\mu - \partial g_{\lambda\mu}/\partial X^\nu\} \tag{A.6.4}$$

The Two-Tier Riemann-Christoffel curvature tensor is:

$${}_XR^\lambda{}_{\mu\nu\kappa} \equiv \partial_X\Gamma^\lambda{}_{\mu\nu}/\partial X^\kappa - \partial_X\Gamma^\lambda{}_{\mu\kappa}/\partial X^\nu + {}_X\Gamma^\alpha{}_{\mu\nu}\, {}_X\Gamma^\lambda{}_{\kappa\alpha} - {}_X\Gamma^\alpha{}_{\mu\kappa}\, {}_X\Gamma^\lambda{}_{\nu\alpha} \tag{A.6.5}$$

and the Two-Tier Ricci tensor is

$${}_XR_{\mu\nu} = {}_XR^\alpha{}_{\mu\alpha\nu} \tag{A.6.6}$$

The Two-Tier curvature scalar is

$${}_XR = g^{\mu\nu}{}_XR_{\mu\nu} \tag{A.6.7}$$

We also define

$${}_XR_{\lambda\mu\nu\kappa} = g_{\lambda\alpha}\, {}_XR^\alpha{}_{\mu\nu\kappa} \tag{A.6.8}$$

with the result

$${}_XR_{\lambda\mu\nu\kappa} = ½[\partial^2 g_{\lambda\nu}/\partial X^\kappa\partial X^\mu - \partial^2 g_{\mu\nu}/\partial X^\kappa\partial X^\lambda - \partial^2 g_{\lambda\kappa}/\partial X^\nu\partial X^\mu + \partial^2 g_{\mu\kappa}/\partial X^\nu\partial X^\lambda] + g_{\alpha\beta}[\, {}_X\Gamma^\alpha{}_{\nu\lambda}\, {}_X\Gamma^\beta{}_{\mu\kappa} - {}_X\Gamma^\alpha{}_{\kappa\lambda}\, {}_X\Gamma^\beta{}_{\mu\nu}] \tag{A.6.9}$$

We denote the fact that all quantities in eqs. A.6.4 – A.6.8 are only functions of X by placing a left subscript X on each quantity.

The algebraic properties, and the Bianchi identities, satisfied by ${}_X R_{\lambda\mu\nu\kappa}$ in the Two-Tier theory of gravitation are identical to those of the conventional theory with all derivatives being with respect to X.

The Two-Tier version of Einstein's field equations is:

$$ {}_X R_{\mu\nu} - \tfrac{1}{2} g_{\mu\nu}\, {}_X R = -8\pi G\, T_{\mu\nu} \tag{A.6.10}$$

where G is Newton's gravitational constant (6.674×10^{-11} $m^3 kg^{-1} s^{-2}$) and $T_{\mu\nu}$ is the energy-momentum tensor – also strictly a function of X. It is convenient to define the coupling constant

$$\kappa = \sqrt{4\pi G} \tag{A.6.11}$$

A.6.3 Lagrangian Formulation

We will now formulate a Two-Tier Quantum Gravity theory following the same ansatz that we have used throughout this book.

A.6.4 Unified Standard Model and Quantum Gravity Lagrangian

We define the Lagrangian, and action, for the unified quantum field theory of gravitation and the Standard Model as

$$L_{\text{Unified}} = \int d^4y\, \mathscr{L}_{\text{Unified}} \tag{A.6.12}$$

$$\mathscr{L}_{\text{Unified}} = J\sqrt{g(X)}\, (\mathscr{L}_F^{\text{Grav}}(X^\mu) + \mathscr{L}_F^{\text{SM}}(X^\mu)) + L_C \tag{A.6.13}$$

with

$$\mathscr{L}_F^{\text{Grav}}(X^\mu) = (2\kappa^2)^{-1}\, {}_X R \tag{A.6.14}$$

where $\mathscr{L}_F^{\text{SM}}$ is the complete "normal" Quantum Field theory Lagrangian for the Standard Model version under consideration written in a general covariant form, g(X) is the absolute value of the determinant of $g_{\mu\nu}$, and J is the Jacobian of eq. A.21. *All particle fields in $\mathscr{L}_F^{SM}$ are assumed to be functions of the X^μ coordinate only. The dependence of the particle fields on the "underlying" coordinates y^μ is assumed to be solely through X^μ.* The Lagrangian $\mathscr{L}_{\text{Unified}}$ is a separable Lagrangian of the type of eq. B.26 embodying the composition of extrema described in Appendix B.

As in all of cases that we have considered, we have specified the coordinate part of the Lagrangian $\mathscr{L}_C$ as

$$\mathscr{L}_C = -\tfrac{1}{4}\, F_Y{}^{\mu\nu} F_{Y\mu\nu} \qquad (A.2.15)$$

with

$$F_{Y\mu\nu} = \partial Y_\mu / \partial y^\nu - \partial Y_\nu / \partial y^\mu \qquad (A.2.14)$$

and

$$F_Y{}^{\mu\nu} = \eta^{\mu\alpha}\eta^{\nu\beta} F_{Y\alpha\beta} \qquad (A.6.15)$$

A.6.5 Why Are the Y Field Dynamics Independent of the Gravitational Field?

It is evident from eqs. A.6.12-3 and A.6.15 that the Y field is truly free *in a flat universe* and, in particular, does not depend on the gravitational field as represented by $\sqrt{g}$ and $g_{\mu\nu}$. For the moment our rationale is based on the following remarks. The Y field is a quantum field at each point in space-time including regions with ultra-strong gravitational fields such as the neighborhoods of black holes. If Y were to depend on the gravitational field then the Y field could be appreciable in such regions and might even be a "classical" field. In this case we would have new dimensions, allbeit imaginary, for which no evidence exists as yet.

Lastly, the non-invariance of the Y part of the action under general coordinate transformations effectively creates an "absolute" coordinate system – actually a class of "absolute coordinate systems" – namely the class of inertial reference frames that are related to each other by special relativistic transformations. This feature does not conflict with our knowledge of the universe. The universe appears to be almost flat. The large-scale distribution of masses is responsible for this flatness. The flatness, or flattened space if it is slightly curved, together with Mach's Principle (inertial forces are absent in the reference frame determined by the distribution of masses in the universe) selects a preferred class of local reference frames – local inertial reference frames. Since space is almost flat, or flat, these local reference frames occupy a large volume (if we exclude regions with intense gravitational fields.) We can define the Y field within this class of local inertial reference frames in each locale and establish a satisfactory quantum field theory. Thus we have a dynamics defined in the variable X, which we require to be covariant under general coordinate transformations, and a local "ground state" that "breaks" general coordinate invariance down to special relativistic invariance.

A.6.6 No "Space-time Foam"

The fact that our unified theory of the known forces of Nature *self-consistently* has a weak gravitational field at high energies (the graviton sector is finite to all orders in perturbation theory) supports the formulation of eq. A.6.12-3. Gravity becomes weaker at ultra-short distances. Therefore space-time is not quantum foam at ultra-short distances but rather smooth and flat a là special relativity – consistent with our formulation.

A.6.7 Quantum Gravity – Scalar Particle Model Lagrangian

While the application of the Two-Tier approach to the unified theory is a straightforward extension of the concepts and approaches described in the preceding chapters, it is useful to consider a simplified model that minimizes the tensorial verbiage so that the concepts and features might better stand out. The procedure differs only in detail from the case of gauge fields.

The introduction of spinor fields requires the use of a Two-Tier vierbein formalism, which is straightforward to develop. A Two-Tier vierbein field $e^{\mu}{}_a$ is a function of X, $e^{\mu}{}_a(X)$, with $g_{\mu\nu} = e_{\mu a}(X)e_{\nu}{}^a(X)$ where the index a is an index of a flat tangent space defined at each space-time point. The Two-Tier formulation of a vierbein theory is similar to the other Two-Tier formulations that we have considered and will not be developed here.

Thus we will consider the Lagrangian model for a scalar particle field interacting with the $g_{\mu\nu}$ gravitational field:

$$L_{GS} = \int d^4y \sqrt{g(X)}\ (\mathcal{L}_{GS} + \mathcal{L}_C) \tag{A.6.16}$$

$$\mathcal{L}_{GS} = J\ \mathcal{L}_F^{Grav}(X^{\mu}) + J\ \mathcal{L}_{F\phi}(X^{\mu}) \tag{A.6.17}$$

with covariant versions of eqs. 5.21 and 5.24:

$$\mathcal{L}_{F\phi} = ½\ [\ g^{\mu\nu}\partial\phi/\partial X^{\mu}\ \partial\phi/\partial X^{\nu} - m^2\phi^2] + \mathcal{L}_{F\phi int} \tag{A.6.18}$$

$$\mathcal{L}_{F\phi int} = 1/4!\ \chi_0\ \phi(X(y))^4 + ½\ (m^2 - m_0^2)\phi^2 \tag{A.6.19}$$

A.6.8 A Justifiable Weak Field Approximation for Quantum Gravity

Many discussions of quantizing conventional gravity make a weak field approximation for the gravity sector which, in view of divergences in the resulting quantum field theory, are impossible to justify:

$$g_{\mu\nu} = \eta_{\mu\nu} + \kappa h_{\mu\nu} \tag{A.6.20}$$

where $\eta_{\mu\nu}$ is the flat space-time metric and $h_{\mu\nu}$ is a "small" deviation ($<h_{\mu\nu}> \ll 1$) from the flat space-time metric.

The Two-Tier formulation of quantum gravity is finite and the effective field becomes increasingly weaker at short distances. Thus the weak field approximation becomes *more accurate* at short distances:

$$g_{\mu\nu}(X(y)) \simeq \eta_{\mu\nu} + \kappa h_{\mu\nu}(X(y)) \tag{A.6.21a}$$

At short distances space-time can be considered approximately flat (except possibly in the neighborhood of singularities) with quantum fluctuations embodied in $h_{\mu\nu}$. Thus eq. A.6.21a is reasonable within the context of Two-Tier Quantum Gravity.

To first order in $h_{\mu\nu}$ the square root of the absolute value of the determinant of the metric tensor is:

$$\sqrt{g(X)} \simeq 1 + ½\, \kappa h^{\sigma}{}_{\sigma}(X(y)) \qquad \text{(A.6.21b)}$$

A.6.9 Quantization of Quantum Gravity – Scalar Particle Model

We now proceed to quantize gravity based on the linearization of the gravitational field equations in the weak field approximation. Assuming eq. A.6.21a and keeping terms to first order in $h_{\mu\nu}$ gives the affine connection:

$$_X\Gamma^{\sigma}{}_{\mu\nu} = ½\, \kappa\eta^{\sigma\alpha}[\partial h_{\alpha\nu}/\partial X^{\mu} + \partial h_{\alpha\mu}/\partial X^{\nu} - \partial h_{\mu\nu}/\partial X^{\alpha}] + O\,(h^2) \qquad \text{(A.6.22)}$$

and the Ricci tensor:

$$_XR_{\mu\nu} = \partial_X\Gamma^{\lambda}{}_{\lambda\mu}/\partial X^{\nu} - \partial_X\Gamma^{\lambda}{}_{\mu\nu}/\partial X^{\lambda} + O\,(h^2) \qquad \text{(A.6.23)}$$

Thus the linearized gravitation lagarangian terms are

$$L^{Grav} = \int d^4y\, \sqrt{g(X)}\; J\mathscr{L}_F^{Grav}(X^{\mu}) \rightarrow L^{Grav}{}_{linear} = \int d^4y\; J\mathscr{L}^{Grav}{}_{linear}(X^{\mu}) \qquad \text{(A.6.24)}$$

The scalar particle Lagrangian terms become

$$L^{\phi} = \int d^4y\sqrt{g(X)}J\mathscr{L}_{F\phi} \rightarrow \int d^4yJ\{[½(\eta^{\mu\nu}\partial_{\mu}\phi\partial_{\nu}\phi - m^2\phi^2) + \mathscr{L}_{F\phi int}] +$$
$$+ ½\kappa h^{\mu\nu}\partial_{\mu}\phi\partial_{\nu}\phi + ¼\, \kappa h(\eta^{\mu\nu}\partial_{\mu}\phi\partial_{\nu}\phi - m^2\phi^2) + ½\, \kappa h\mathscr{L}_{F\phi int}\} \qquad \text{(A.6.25)}$$

with the notation $h = h^{\sigma}{}_{\sigma}$ and using

$$\partial_{\mu} \equiv \partial/\partial X^{\mu}$$

$\eta^{\mu\nu}$ and $\eta_{\mu\nu}$ are used to raise and lower indices in keeping with the linearized, weak field approximation.

The Y terms in the Lagrangian are (as previously):

$$L^{Y} = \int d^4y\mathscr{L}_C = -¼\int d^4y\eta^{\mu\nu}\eta^{\alpha\beta}F_{Y\mu\alpha}F_{Y\nu\beta} \qquad \text{(A.6.27)}$$

We will lump the higher order terms (in h) in the gravity part of the Lagrangian, and the scalar particle part of the Lagrangian, into

$$L_{Higher} = \int d^4y\, J \mathscr{L}_{Higher}(h, \phi) \tag{A.6.28}$$

Thus the complete lagragian for a scalar particle interacting with gravitons is

$$L_{GS} = L^{Grav}{}_{linear} + L^{\phi}{}_{linear} + L^{Y} + L_{Higher} \tag{A.6.29}$$

The linearized gravitational Lagrangian term $L^{Grav}{}_{linear}$ generates the field equations:

$$\Box_X\, h_{\mu\nu} + \partial_\nu\partial_\mu h - \partial_\alpha\partial_\nu h^\alpha{}_\mu - \partial_\alpha\partial_\mu h^\alpha{}_\nu = \kappa S_{\mu\nu} \tag{A.6.30}$$

where

$$\partial_\mu S^\mu{}_\nu = \tfrac{1}{2}\, \partial_\nu S^\sigma{}_\sigma \tag{A.6.31}$$

to 0th order in h and where

$$\Box_X = (\partial/\partial X^\nu)(\partial/\partial X_\nu) \tag{A.6.32}$$

The most general coordinate transformation that maintains the weakness of the gravitational field has the form:

$$y^a \rightarrow\ y'^a = y^a + \chi^a(X(y)) \tag{A.6.33}$$

This transformation induces a gauge transformation in $h_{\mu\nu}$ to:

$$h'_{\mu\nu} = h_{\mu\nu} - \partial_\mu \chi_\nu - \partial_\nu \chi_\mu \tag{A.6.34}$$

It is easy to verify that eq. A.6.30 is satisfied by $h'_{\mu\nu}$ if it is satisfied by $h_{\mu\nu}$.

Let us assume that we perform a gauge transformation making $h_{\mu\nu}$ traceless:

$$h^\sigma{}_\sigma = 0 \tag{A.6.35}$$

and choose the gauge

$$\partial^\mu h_{\mu\nu} = 0 \tag{A.6.36}$$

then eq. A.6.30 becomes the wave equation:

$$\Box_X\, h_{\mu\nu} = \kappa S_{\mu\nu} \tag{A.6.37}$$

Another gauge transformation of the free field $h_{\mu\nu}$ (if $S_{\mu\nu} = 0$) makes

$$h_{\mu 0} = h_{0\mu} = 0 \quad \text{(A.6.38)}$$

while retaining

$$h_{\mu\nu} = h_{\nu\mu} \quad \text{(A.6.39)}$$

The general solution[295] for the free field $h_{\mu\nu}$ (with $S_{\mu\nu} = 0$ in eq. A.6.37) can be expressed as a fourier expansion:

$$h_{\mu\nu}(X(y)) = \int d^3k\, N_0(k) \sum_{\lambda=1}^{2} \varepsilon_{\mu\nu}(k, \lambda)[a(k,\lambda)\, e^{-ik\cdot X} + a^\dagger(k,\lambda)\, e^{ik\cdot X}] \quad \text{(A.6.40)}$$

where $\lambda = 1,2$ labels the ± 2 helicity states, and where $N_0(k)$ is specified by eq. A.2.25. The equal time ($y'^0 = y^0$) commutation relations are:

$$[h_{\mu\nu}(X(y)), h_{\alpha\beta}(X(y'))] = [\pi_{\mu\nu}(X(y)), \pi_{\alpha\beta}(X(y'))] = 0 \quad \text{(A.6.41)}$$

$$[h_{\alpha\beta}(X(y')), \pi_{\mu\nu}(X(y))] = i\, D_{\alpha\beta,\mu\nu}(\mathbf{X}(y) - \mathbf{X}(y')) \quad \text{(A.6.42)}$$

for μ, $\nu = 1, 2, 3$ and where

$$\pi_{\mu\nu}(X(y)) = \partial h_{\mu\nu}(X(y))/\partial y^0 \quad \text{(A.6.43)}$$

in the Y Coulomb gauge where $X^0 = y^0$. $\mathscr{D}_{\alpha\beta,\mu\nu}$ is specified by:

$$\mathscr{D}_{\alpha\beta,\mu\nu}(X(y) - X(y')) = \int d^3k\, e^{i\,\mathbf{k}\cdot(\mathbf{X}(y) - \mathbf{X}(y'))}\, \Pi_{\alpha\beta\mu\nu}(\mathbf{k})/(2\pi)^3 \quad \text{(A.6.44)}$$

$$\Pi_{\alpha\beta\mu\nu}(\mathbf{k}) = \tfrac{1}{2}\, [(\delta_{\alpha\mu} - k_\alpha k_\mu/\mathbf{k}^2)(\delta_{\beta\nu} - k_\beta k_\nu/\mathbf{k}^2) + (\delta_{\alpha\nu} - k_\alpha k_\nu/\mathbf{k}^2)(\delta_{\beta\mu} - k_\beta k_\mu/\mathbf{k}^2) -$$

$$- (\delta_{\alpha\beta} - k_\alpha k_\beta/\mathbf{k}^2)(\delta_{\mu\nu} - k_\mu k_\nu/\mathbf{k}^2)] \quad \text{(A.6.45)}$$

where α, β, μ, $\nu = 1, 2, 3$.

The "transverse" graviton propagator can be represented as a time-ordered product of field operators:

$$i\Delta_{F2}{}^{TT}(y_1 - y_2)_{\lambda\tau\rho\sigma} = \langle 0|T(h_{\lambda\tau}(X(y_1)), h_{\rho\sigma}(X(y_2)))|0\rangle \quad \text{(A.6.46)}$$

295 S. Weinberg, Phys. Rev. **135**, B1049 (1964); Phys. Rev. **138**, B988 (1965)

$$= -i \int \frac{d^4k\, e^{-ik\cdot(y_1 - y_2)}\, b_{\lambda\tau\rho\sigma}(k) R(\mathbf{k}, y_1 - y_2)}{(2\pi)^4\,(k^2 + i\varepsilon)}$$

where $R(\mathbf{k}, y_1 - y_2)$ is the gaussian factor appearing in propagators throughout Two-Tier theories, $\mathbf{k}$ is a spatial 3-vector, and where $b_{\mu\nu\rho\sigma}(k)$ is a function of k only:

$$b_{\alpha\beta\mu\nu}(k) = \tfrac{1}{2}[(\eta_{\alpha\mu} - k_\alpha k_\mu/\mathbf{k}^2)(\eta_{\beta\nu} - k_\beta k_\nu/\mathbf{k}^2) + (\eta_{\alpha\nu} - k_\alpha k_\nu/\mathbf{k}^2)(\eta_{\beta\mu} - k_\beta k_\mu/\mathbf{k}^2) - (\eta_{\alpha\beta} - k_\alpha k_\beta/\mathbf{k}^2)(\eta_{\mu\nu} - k_\mu k_\nu/\mathbf{k}^2)] \tag{A.6.47}$$

where $\alpha, \beta, \mu, \nu = 0, 1, 2, 3$.

The quantum gravitational interaction also has an "instantaneous" part (similar to the instantaneous Coulomb interaction of QED) in addition to the transverse interaction embodied in eq. A.6.46. This "instantaneous" interaction contains the Newtonian potential (described later) as its large distance limit. The sum of the instantaneous interaction and the transverse interaction gives the total gravitational interaction.

The above graviton propagator has the form given earlier. The caculation of the leading behavior is the same as that of the Two-Tier scalar boson propagator except for the presence of factors such as $\eta_{\rho\sigma}$. The leading momentum dependence of the graviton propagator in momentum space is

$$i\Delta_{F2}{}^{TT}(p)_{\lambda\tau\rho\sigma} \backsim p^{-6} \tag{A.6.48}$$

The graviton vertices in Two-Tier Quantum Gravity will be described within the framework of the path integral formulation.

A.6.10 Quantum Gravity–Scalar Particle Model Path Integral

A path integral formalism can be developed for Two-Tier Quantum Gravity interacting with matter fields. In this section we will consider the case of a matter field consisting of massive scalar bosons with a quartic interaction. The path integral formalism that we develop is similar to that of Yang-Mills theories in the previous chapter.

The Two-Tier path integral for a Quantum Gravity–Scalar Particle Theory can be written as:

$$Z(J, J^{\mu\nu}) = N \int D\phi Dh DY \Delta_{FPG}(h)\delta(F(h)) \exp\{i\mathscr{J}\int d^4y[\mathscr{J}(\mathscr{L}^{Grav}{}_{linear}(X^\mu) +$$

$$+ \mathscr{L}^{\phi}{}_{\text{linear}}(X^{\mu}) + \mathscr{L}_{\text{Higher}}(h, \phi)) + \mathscr{L}_{C}(X, y) +$$

$$+ j_{\mu}(y)Y^{\mu}(y) + J(y)\phi(X) + \mathscr{J}J^{\mu\nu}(y)h_{\mu\nu}(X)]\}|_{j_{\mu}=0} \qquad \text{(A.6.49)}$$

where $\delta(F(h))$ specifies the gauge as a functional delta function, and $\Delta_{FPG}(h)$ is the corresponding Fadeev-Popov determinant. $\mathscr{J}$ is the Jacobian for the transformation from y coordinates to X coordinates. The Fadeev-Popov determinant $\Delta_{FPG}(h)$ can be calculated in the standard way. First we note

$$\delta(F(h^{\chi})) = \delta(\chi - \chi_0)\ |\det \delta F(h_{\mu\nu}{}^{\chi}(X))/\delta\chi(X)|^{-1}|_{F(h)=0} \qquad \text{(A.6.50)}$$

where

$$h_{\mu\nu}{}^{\chi} = h_{\mu\nu} - \partial_{\mu}\chi_{\nu} - \partial_{\nu}\chi_{\mu} \qquad \text{(A.6.34)}$$

Then

$$\Delta_{FPG}(h) = \ |\det \delta F(h^{\chi}(X))/\delta\chi(X)||_{F(h)=0} \qquad \text{(A.6.51)}$$

We will choose the gauge of eq. A.6.36 to evaluate the Fadeev-Popov determinant. Under an infinitesimal gauge transformation of the form:

$$h_{\mu\nu}{}^{\chi}(X) = h_{\mu\nu}(X) - \partial_{\mu}\chi_{\nu} - \partial_{\nu}\chi_{\mu} \qquad \text{(A.6.52)}$$

which preserves the weak field nature of $h_{\mu\nu}$, we find

$$\begin{aligned} F_{\nu}(h^{\chi}) &= \partial^{\mu}(h_{\mu\nu}(X) - \partial_{\mu}\chi_{\nu} - \partial_{\nu}\chi_{\mu}) \\ &= -\Box_X\,\chi_{\nu}(X) - \partial_{\nu}\partial^{\mu}\chi_{\mu} \end{aligned} \qquad \text{(A.6.53)}$$

Thus

$$\delta F_{\mu}(h^{\chi}(X))/\delta\chi^{\nu}(X) = -\,\eta_{\mu\nu}\Box_X - \partial_{\mu}\partial_{\nu} \qquad \text{(A.6.54)}$$

and

$$\Delta_{FP}(A) = |\det(-\,\eta_{\mu\nu}\Box_X - \partial_{\mu}\partial_{\nu})||_{F(h)=0} \qquad \text{(A.6.55)}$$

We note the Two-Tier Fadeev-Popov determinant is solely a function of the X coordinates. The determinant only introduces an overall multiplicative constant that can be absorbed into the normalization constant N. This fact becomes evident if we follow the standard procedure and rewrite the determinant as a path integral over anti-commuting c-number fields with a ghost

Lagrangian. Then we see that the ghost does not interact with the other fields and thus only generates an overall multiplicative constant that can be absorbed in N:

$$\Delta_{FPG}(h) = \int Dc^* Dc \exp[\, i\int d^4X\, \mathscr{L}^{ghost}(X^\mu)] \qquad \text{(A.6.56)}$$

where

$$\mathscr{L}^{ghost}(X^\mu) = c^{\mu *}(X)[\eta_{\mu\nu}\Box_X + \partial_\mu\partial_\nu]c^\nu(X) \qquad \text{(A.6.57)}$$

We now go through the same analysis as we did in the ϕ^4 theory path integral example and the Yang-Mills path integral example (with some superficial differences). First we integrate the linear part of the Y field Lagrangian as we did previously. Then we integrate the linear part of the ϕ field Lagrangian as done previously. Lastly we integrate the linear part of the gravitation Lagrangian to obtain the path integral for the perturbative expansion with the result:

$$Z(J, J^{\mu\nu}) = N\,\{\exp[i\int d^4y\, \mathscr{L}_{Higher}(\partial/\partial y^\nu, -i\delta/\delta J^{\mu\nu}(y), -i\delta/\delta J(y))]\cdot$$
$$\cdot \exp[-\tfrac{1}{2}\, i\int d^4y_1 d^4y_2\, J^{\mu\nu}(y_1)\Delta_{F2}{}^{TT}(y_1 - y_2, z)_{\mu\nu\rho\sigma} J^{\rho\sigma}(y_2)]\cdot$$
$$\cdot \exp[-\tfrac{1}{2}\, i\int d^4y_1 d^4y_2\, J(y_1)\Delta_F{}^{TT}(y_1 - y_2, z)J(y_2)]\}\Big|_{z=y_1-y_2} \qquad \text{(A.6.58)}$$

There are two issues that arise in the development of eq. A.6.58:

1.) The integral over y in $\int d^4y \mathscr{L}_{Higher}$ which began as the integral $\int d^4y\, \mathscr{J}\mathscr{L}_{Higher} = \int d^4X \mathscr{L}_{Higher}$ in eq. A.6.49; and
2.) The handling of derivatives with respect to X in $\mathscr{L}_{Higher}$.

These are resolved by the following respective observations:

1.) See the discussion following eqs. A.6.34 that apply here as well without change.

2.) We note that the derivative with respect to X of the graviton propagator (eq. A.6.46-A.6.47) is specified by the following:

$$\partial i\Delta_{F2}{}^{TT}(y_1 - y_2)_{\lambda\tau\rho\sigma}/\partial X^\mu(y_1) = \partial[i\Delta_{F2}{}^{TT}(y_1 - y_2, z)_{\lambda\tau\rho\sigma}]/\partial y_1{}^\mu\,\Big|_{z=y_1-y_2} \qquad \text{(A.6.59)}$$

where

$$i\Delta_{F2}{}^{TT}(y_1 - y_2, z)_{\lambda\tau\rho\sigma} = -i \int \frac{d^4k\, e^{-ik\cdot(y_1 - y_2)}\, b_{\lambda\tau\rho\sigma}(k)R(\mathbf{k}, z)}{(2\pi)^4\,(k^2 + i\varepsilon)} \quad \text{(A.6.60)}$$

Thus

$$\frac{\partial\, i\Delta_{F2}{}^{TT}(y_1 - y_2)_{\lambda\tau\rho\sigma}}{\partial X^{\mu}(y_1)} = -i \int \frac{d^4k\, e^{-ik\cdot(y_1 - y_2)}\,(-ik_{\mu})b_{\lambda\tau\rho\sigma}(k)R(\mathbf{k}, y_1 - y_2)}{(2\pi)^4\,(k^2 + i\varepsilon)}$$

(A.6.61)

Therefore derivatives with respect to X in the interaction Lagrangian terms can be replaced by derivatives with respect to y if the graviton propagator is generalized to eq. A.6.60. After taking all derivatives with respect to y, we set z equal to the respective $y_1 - y_2$ (actually the difference of the appropriate variables) in each propagator with results similar to eq. A.6.61.

$$Z(J, J^{\mu\nu}) = N\,\{\exp[i\int d^4y\,\mathscr{L}_{Higher}(\partial/\partial y^{\nu}, -i\delta/\delta J^{\mu\nu}(y), -i\delta/\delta J(y))]\cdot$$
$$\cdot \exp[-\tfrac{1}{2}\, i\int d^4y_1 d^4y_2\, J^{\mu\nu}(y_1)\Delta_{F2}{}^{TT}(y_1 - y_2, z)_{\mu\nu\rho\sigma} J^{\rho\sigma}(y_2)]\cdot$$
$$\cdot \exp[-\tfrac{1}{2}\, i\int d^4y_1 d^4y_2\, J(y_1)\Delta_F{}^{TT}(y_1 - y_2, z)J(y_2)]\}\big|_{z=y_1-y_2} \quad \text{(A.6.58a)}$$

To be precise eq. A.6.58a is interpreted as executing the following steps:

1. For a given process take appropriate functional derivatives of Z(J) with respect to J and $J^{\mu\nu}$.

2. Then expand the exponential factors in a perturbation series applying any derivatives with respect to y in L_{Higher}. Do not perform any of the $\int d^4y_1 d^4y_2$ integrals.

3. Then set $z = y_1 - y_2$ in each $\Delta_{Fk}{}^{TT}(y_1 - y_2, z)$ and $\Delta_{F2}{}^{TT}(y_1 - y_2, z)_{\mu\nu\rho\sigma}$ propagator.

4. Lastly perform all $\int d^4y_1 d^4y_2$ integrals.

Thus we achieve a path integral formulation that is very similar to the corresponding expression in conventional field theory – the only difference is in the form of the free field propagators, which each now contain a Gaussian factor. The net consequence is that graviton vertices result in exactly the same polynomials in momenta as the conventional theory.

Thus Two-Tier gravity generates a perturbative expansion identical to conventional quantum gravity except that each graviton propagator has a gaussian damping factor $R(\mathbf{k},y_1-y_2)$. At low energies the tree diagrams of conventional gravity theory emerge to good approximation in Two-Tier gravity. All diagrams with loops converge. Thus Two-Tier gravity is finite.

A.6.11 Finiteness of Quantum Gravity–Scalar Particle Model

Two-Tier Quantum Gravity perturbation theory is finite. Calculations are highly convergent at large momentum ($\gtrsim M_c$). At low momentum the Two-Tier theory is similar to conventional gravity – particularly for tree diagrams and other convergent diagrams in conventional quantum gravity.

For pure *conventional* Quantum Gravity DeWitt[296] finds the superficial degree of divergence of a diagram to be:

$$D = -2L_i + 2\sum_n V_n + 4K \qquad (A.6.62)$$

where L_i is the number of internal lines, V_n is the number of n-pronged vertices, and K is the number of independent momentum integrations. DeWitt further points out

$$K = L_i - \sum_n V_n + 1 \qquad (A.6.63)$$

Thus the superficial degree of divergence of a conventional Quantum Gravity diagram is:

$$D = 2(K + 1) \qquad (A.6.64)$$

for $K \geq 1$, displaying an ever increasing degree of divergence as the order of the diagram increases.

In the case of *Two-Tier Quantum Gravity* the superficial degree of divergence of a diagram is:

$$D_{TT} = -6L_i + 2\sum_n V_n + 4K \qquad (A.6.65)$$

---

[296] B. S. DeWitt, Phys. Rev. **162**, 1239 (1967).

(from eq. A.6.48) with the result (taking account of eq. A.6.63):

$$D_{TT} = -2L_i - 2\sum_n V_n + 2 \quad \text{(A.6.66)}$$

Since any diagram with a loop has $L_i \geq 1$ and $\sum_n V_n \geq 1$ we see that $D \leq -2$. Thus *all* diagrams are convergent and *the Two-Tier formulation of Quantum Gravity theory is finite. The addition of arbitrary species of other Two-Tier fields – matter and gauge fields – does not introduce divergences in the combined Two-Tier theory.*

A.6.12 Unitarity of Quantum Gravity–Scalar Particle Model

The Two-Tier Quantum Gravity – Scalar Particle Model *superficially* appears to have a unitarity problem due to the non-hermitean nature of its hamiltonian. The lack of hermiticity is due entirely to the appearance of iY^μ in the X^μ field coordinates.

Thus interaction Lagrangian is not hermitean:

$$L_{Higher} = \int d^3y' \mathscr{L}_{Higher}(y' + iY(y')/M_c^2) \quad \text{(A.6.67)}$$

and

$$L_{Higher}^\dagger = \int d^3y' \mathscr{L}_{Higher}(y' - iY(y')/M_c^2) \neq L_{Higher} \quad \text{(A.6.68)}$$

The relation between L_{Higher} and its hermitean conjugate is

$$L_{Higher} = V\, L_{Higher}^\dagger\, V \quad \text{(A.6.69)}$$

where $V^2 = I$ is a metric operator. As a result Two-Tier S matrix is not unitary – it is pseudo-unitary:

$$S^{-1} = V\, S^\dagger\, V \quad \text{(A.6.70)}$$

Therefore

$$S^\dagger\, VS = V \quad \text{(A.6.71)}$$

The S matrix satisfies the unitarity condition between physical asymptotic states – states consisting of only scalar ϕ particles and gravitons. The proof is identical in form to that given earlier. The S matrix of the unified theory of the Standard Model and Quantum Gravity can be similarly shown to satisfy the unitarity condition.

A.6.13 The Mass Scale M_c

The mass scale of Two-Tier theories is set by M_c. This mass scale cannot be ascertained with any degree of certainty at current, experimentally accessible, accelerator energies. Cosmic ray data also does not seem to give any clues as to the value of M_c. It appears that M_c is probably above 10^3 GeV/c^2 and may be of the order of (or equal to) the Planck mass:

$$M_{planck} = \sqrt{\hbar c/G} = 1.22\times10^{19}\ \text{GeV}/c^2 \qquad \text{(A.6.75)}$$

If M_c is of the 1,000 GeV/c^2 or larger the differences between our theory's predictions at current accelerator energies, and the predictions of conventional renormalized perturbation theory will be negligible. Actually a much lower value of M_c would still be consistent with the current stringent QED theoretical predictions as well as other predictions of conventional renormalized perturbation theory.

A.6.14 Planck Scale Physics

A finite theory of Quantum Gravity can provide information on the issues that have been of concern for many years – including the short distance behavior of the gravitational metric and ultra-small black holes.

A.6.15 Quantum Foam

Some theorists have conjectured that the classical view of smooth, almost flat space-time does not hold in the quantum regime at energies of the order of the Planck mass. Suggestions that space-time dissolves into quantum foam have appeared.

The finite Two-Tier formulation of Quantum Gravity is well-behaved at short distances and suggests that the quantum behavior of gravity and space-time in the short distance limit does not have limitless quantum fluctuations that result in a foam-like space-time picture.

A.6.16 Measurement of the Quantum Gravity Field

A number of conceptual problems have been raised about the effects of quantized General Relativity. Two-tier Quantum Gravity seems to resolve these issues.

A.6.17 Measurement of Time Intervals

Wigner[297] has studied the measurement of time intervals in General Relativity and sees a problem in the measurement of extremely short intervals. According to Wigner: the measurement of a time inteval in a region of space requires the measurement of the length of

[297] E. P. Wigner, Rev. Mod. Phys. **29**, 255 (1957); J. Math. Phys. **2**, 207 (1961).

time required for an event to happen. The measurement requires an accurate clock. But the accuracy of the clock is limited by the energy-time uncertainty relation:

$$\Delta E \Delta t \geq \hbar \qquad \text{(A.6.76)}$$

Thus the uncertainty in the clock's time measurement is related to the uncertainty in the clock's energy which is, in turn, related to the uncertainty in the clock's mass:

$$\Delta E = (\Delta m)c^2 \qquad \text{(A.6.77)}$$

To obtain "infinite" accuracy the uncertainty (fluctuations) in the clock's mass must be infinite and thus the clock's mass must be infinite. Infinite fluctuations in the clock's mass will produce corresponding infinite fluctuations in the gravitational field.

$$\Delta h \propto \Delta E \qquad \text{(in conventional General Relativity)} \qquad \text{(A.6.78)}$$

As a result the notion of space-time and time intervals (which depend on the geometry through General Relativity) become uncertain. Thus, according to Wigner, and others, the concept of time intervals and space-time points becomes questionable.

The Two-Tier version of Quantum Gravity offers a potential way out of this dilemma. The gravitational force becomes stronger as one goes to shorter distances (higher energies) down to a distance (up to an energy) whose scale is set by M_c. At shorter distances (higher energies) the gravitational force becomes weaker and declines to zero at zero distance. Thus at very high energy the gravitational field fluctuations (Δh) are at worst inversely proportional to the energy (and probably decline by a higher power of inverse energy.) (The same considerations would apply if one chooses to consider fluctuations in the Riemann-Christoffel symbols.)

$$\Delta h < c_1/E < c_1/(\Delta E) \qquad \text{(in Two-Tier Quantum Gravity)} \qquad \text{(A.6.79)}$$

Thus Wigner's conclusion does not hold in the Two-Tier version of Quantum Gravity as gravitational fluctuations actually become smaller at energies above a critical energy whose scale is set by M_c.

In fact, combining eqs. A.6.79 and A.6.76 we see

$$c_1\Delta t/\Delta h \geq \hbar \qquad \text{(A.6.80)}$$

at sufficiently high energy. Therefore the time uncertainty Δt, and the gravitational field fluctuations Δh, can both decrease while maintaining the energy-time uncertainty relation. *Thus the notion of a space-time point "is saved" in Two-Tier quantum gravity.*

A.6.18 Vacuum Fluctuations in the Gravitation Fields

While the expectation value of the free graviton field $h_{\mu\nu conv}(X)$ is zero in a conventional quantum field theoric approach:

$$<0|h_{\mu\nu conv}(X)|0> = 0 \qquad (A.6.81)$$

the vacuum fluctuations of the *conventional* quantum graviton field is quadratically divergent since

$$<0|h_{\mu\nu conv}(X)h_{\alpha\beta conv}(X)|0> = \int d^3p\, b'_{\mu\nu\alpha\beta}(p)/[(2\pi)^3\, 2\omega_p] = \infty \qquad (A.6.82)$$

where $b'_{\mu\nu\alpha\beta}(p)$ is a rational function of the momentum p.

In "Two-Tier" quantum field theory we find

$$<0|h_{\mu\nu}(X)h_{\alpha\beta}(X)|0> = \int d^3p\, b'_{\mu\nu\alpha\beta}(p)\, e^{-p^ip^j\Delta_{Tij}(0)}/[(2\pi)^3 2\omega_p] = 0 \qquad (A.6.83)$$

since the exponential factor in the integrand is $-\infty$. The exponent contains

$$\Delta_{Tij}(z) = \int d^3k\, e^{-ik\cdot z}(\delta_{ij} - k_ik_j/\mathbf{k}^2)/[(2\pi)^3 2\omega_k]$$

Thus the vacuum fluctuations of $h_{\mu\nu}$ are zero in "Two-Tier" quantum field theory.

A.6.19 The Two-Tier Gravitational Potential vs. Newton's Gravitational Potential

The familiar gravitational potential of Newton is:

$$V_{Newton} = - G/|\mathbf{r}| \qquad (A.6.84)$$

The Two-Tier gravitational potential is:

$$V_{Two\text{-}Tier} = - G\Phi(M_c^2\pi|\mathbf{r}|^2)/|\mathbf{r}| \qquad (A.6.85)$$

where $\Phi(y)$ is the error function.[298] It can be calculated in Two-Tier Quantum Gravity from Two-Tier Quantum Gravity propagator terms similar to corresponding terms in the Two-Tier photon propagator that led to the Two-Tier Coulomb potential. At small distances ($\pi r^2 \ll M_c^{-2}$)

$$V_{\text{Two-Tier}} \to -\,G2\surd\pi\, M_c^2|\mathbf{r}| \qquad (A.6.86)$$

a linear potential, and at large distances ($\pi r^2 \gg M_c^{-2}$)

$$V_{\text{Two-Tier}} \to V_{\text{Newton}} = -\,G/|\mathbf{r}| \qquad (A.6.87)$$

the Newtonian potential.

The Two-Tier gravitational potential has a minimum at

$$M_c^2 \pi r_{\text{MIN}}^2 = 1 \qquad (A.6.88)$$

At the minimum $V_{\text{Two-Tier}}$ has the value:

$$V_{\text{Two-TierMIN}} = -\,.8427G\surd\pi\, M_c \qquad (A.6.89)$$

Figs.A.6.1 – A.6.2 display plots of $V_{\text{Two-Tier}}$ for $M_c = 1$ TeV/c^2, and $M_c = 1.22\ 10^{19}$ GeV/c^2 $= G^{-\frac{1}{2}}$ – the Planck mass.

[298] W. Magnus and F. Oberhettinger, *Formulas and Theorems for the Special Functions of Mathematical Physics* (Chelsea Publishing Co., New York, 1949) page 96.

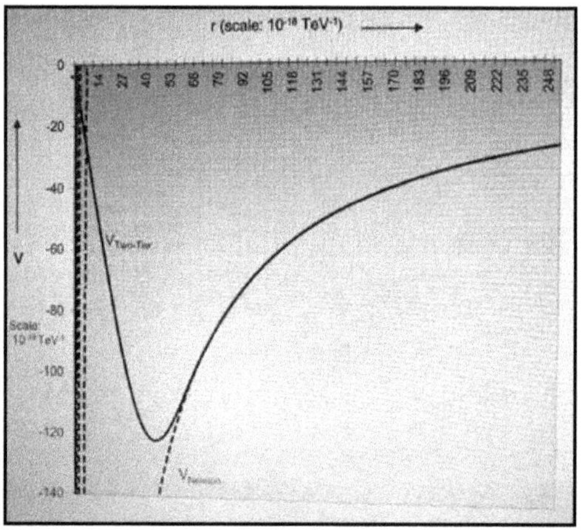

Figure A.6.1. Plot of Two-Tier gravitational potential for M_c = 1 TeV/c^2 and Newton's gravitational potential. The potentials are measured in units of 10^{-36} GeV^{-1}. The radial distance is measured in units of 10^{-5} GeV^{-1}. The plot of the Two-Tier potential shows the force of gravity is repulsive for small r < 5.7 $\times$ 10^{-4} GeV^{-1}.

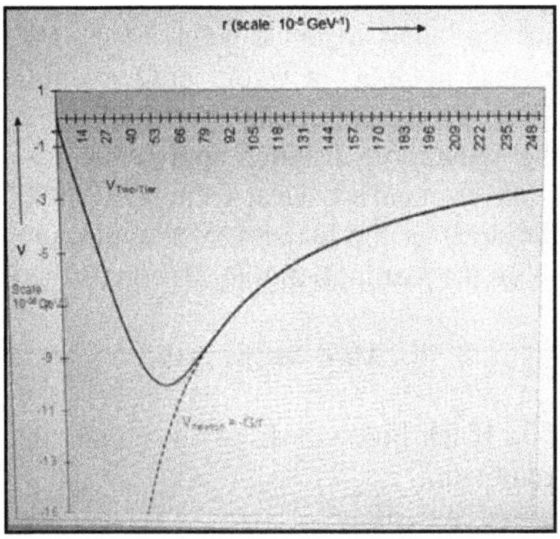

Figure A.6.2. Plot of Two-Tier gravitational potential for $M_C = 1.22 \times 10^{19}$ GeV/c^2 (the Planck mass) and Newton's gravitational potential. Potentials are measured in units of 10^{-18} TeV^{-1}. The radial distance is measured in units of 10^{-18} TeV^{-1}.

A.6.20 Black Holes

The existence of microscopic black holes has been the subject of much speculation. It appears that arbitrarily small black holes can exist in classical General Relativity. The divergences associated with the short distance behavior of its conventional quantization raise the possibility of additional singular behavior at short distances as well.

On the other hand, in Two-Tier Quantum Gravity, at short distances, when the distance scale becomes less than M_c^{-1} (and thus the energy scale becomes greater than M_c), the Two-Tier gravitational force grows smaller and become zero in the limit of zero distance or infinite energy. The preceding figures (Figs. A.6.1 – A.6.2) show the Two-Tier gravitational potential linearly approaches zero at short distances unlike the Newtonian gravitational potential which approaches $-\infty$ as r approches zero. (The transverse gravitational propagator also approaches zero at short distances.) Thus the short distance behavior of Two-Tier gravity suggests that black holes of ultra-small size may not exist in Two-Tier Quanum Gravity.

If we examine the Two-Tier gravitational potential we note that it is similar to the Newtonian potential until the separation distance approaches the minimum of the potential. Thus we might expect that conventional classical General Relativity would be approximately valid down to distances of the order of the location of the minimum of the Two-Tier potential. Based on this assumption and on the assumption that M_c equals the Planck mass:

Assumption: $$M_c = M_{Planck} = G^{-\frac{1}{2}} \qquad \text{(A.6.90)}$$

we can calculate the mass of a black hole whose radius equals the minimum of the Two-Tier potential. From eq. A.6.88 we obtain

$$r_{MIN} = (G/\pi)^{\frac{1}{2}} = r_{BlackHole} = 2GM_{BlackHoleMIN} \qquad \text{(A.6.91)}$$

with the result

$$M_{BlackHoleMIN} = (4\pi G)^{-\frac{1}{2}} = \kappa^{-1} \qquad \text{(A.6.92)}$$

by eq. A.6.11 and

$$M_{BlackHoleMIN} \cong 0.282\ M_{Planck} \qquad \text{(A.6.93)}$$

or 6.15×10^{-6} grams. This lower limit on black hole mass is substantially greater than the collision energy than can be achieved in any current particle accelerator. Thus the production of ultra-small black holes in particle accelerators is highly unlikely.

Since corrections to conventional quantum gravity are at most of the order of M_c^{-2} it appears that the value of $M_{BlackHoleMIN}$ is consistent with the approximate validity of classical expression for a black hole radius. We note

$$(M_{BlackHoleMIN}/M_c)^2 \cong .0795 \qquad (A.6.94)$$

and so corrections to eq. A.6.93 would be very small.

A.7 Curved Space-time Generalization of Two-tier Quantum Gravity

The preceding sections developed a divergence-free theory of scalar particles and quantum gravity in a flat space-time. In this section we show that a curved space-time version of Two-Tier quantum field theories including quantum gravity can be developed along the lines pioneered by DeWitt and collaborators. Two-tier curved space-time quantum field theory is based on a mapping from a flat space-time parametrized by y coordinates to a curved space-time parametrized by X coordinates.

The physical picture of the mapping can be visualised using the simple example of a sphere of radius one in three-dimensional space with a coordinate system on the sphere and two planes – one above the sphere and one below it – each with its own flat space coordinate system. Both planes are assumed to be parallel to the disk defined by the crossection of the sphere bounded by the equator of the sphere. A minimum of two coordinate patches are needed to cover a sphere in three dimensions since it necessarily has coordinate singularities.

Let us place a rectangular coordinate system on the top plane. Points on this plane can be mapped onto its northern hemisphere of the sphere in a simple one-to-one fashion. Similarly a rectangular coordinate system can be placed on the bottom plane which can be mapped in a one to one fashion onto the southern hemisphere of the sphere. The top and bottom planes each have a two-dimensional coordinate system that we can choose to be a Cartesian coordinate system in both cases. We will label the coordinates on the top plane x_t^1 and x_t^2, and the points on the bottom plane as x_b^1 and x_b^2. Each plane has a flat space metric $g_{tij} = g_{bij} = \delta_{ij}$ for i, j = 1,2 with δ_{ij} the Kronecker delta.

In addition, just for concreteness, we will place the origin of the top plane coordinate system vertically above the north pole of the sphere, and the origin of the bottom plane coordinate system vertically below the south pole of the sphere.

If we place the sphere at the center of a three dimensional, coordinate system then the points on the sphere (x,y,z) all satisfy:

$$x^2 + y^2 + z^2 = 1 \qquad (A.7.3)$$

We can defined coordinates u^1 and u^2 for each hemisphere on the surface of the sphere with equations of the form:

$$x_n = f_{1n}(u_n^1, u_n^2) \tag{A.7.4}$$
$$y_n = f_{2n}(u_n^1, u_n^2) \tag{A.7.5}$$
$$z_n = f_{3n}(u_n^1, u_n^2) \tag{A.7.6}$$

for the northern hemisphere, and

$$x_s = f_{1s}(u_s^1, u_s^2) \tag{A.7.7}$$
$$y_s = f_{2s}(u_s^1, u_s^2) \tag{A.7.8}$$
$$z_s = f_{3s}(u_s^1, u_s^2) \tag{A.7.9}$$

for the southern hemisphere.

In addition, we choose $u_n^1 = u_n^2 = 0$ at the north pole and $u_s^1 = u_s^2 = 0$ at the south pole. The surface of the sphere is curved and each (u^1, u^2) coordinate system has a metric, g_{nij} and g_{sij} for i, j = 1,2 respectively, and a non-zero curvature tensor R_{nijkl} and R_{sijkl}.

Now we are allowed to define a simple map of points on the northern hemisphere of the sphere to points on the top plane such as:

$$x_t^1 = u_n^1 \tag{A.7.10}$$
$$x_t^2 = u_n^2 \tag{A.7.11}$$

and of points on the southern hemisphere of the sphere to points on the bottom plane:

$$x_b^1 = u_s^1 \tag{A.7.12}$$
$$x_b^2 = u_s^2 \tag{A.7.13}$$

Thus we can specify the location of events on the sphere on our planes. Note that eqs. A.7.4 – A.7.9 are *not* a coordinate transformation of the (u^1, u^2) coordinate systems on the sphere and thus the plane can have a different (flat) metric from the sphere.

The preceding example can be simplified by using a cylinder enclosing the sphere instead of two planes. The cylinder, which is a flat surface technically, is aligned so that its axis is parallel to, and cenetered on, the north-south axis of the sphere. Then a map can be made from points on the sphere to points on the cylinder that is similar to a Mercator projection, or from points on the sphere to the cylinder that maps the poles to the ends of the cylinder at + and – infinity.

The preceding discussion shows a clear analogy to our map from the y Minkowski space-time to the curved X space-time using

$$X^\mu = y^\mu + i\, Y^\mu(y)/M_c^2 \tag{A.7.14}$$

modulo the imaginary term. The y Minkowski space-time has a flat space-time in which we are allowed to choose the Minkowski metric $\eta_{\mu\nu}$. The curved X space-time has an appropriate metric $g_{\mu\nu}(X)$ that can only be transformed to locally inertial coordinates with perhaps a Minkowski metric in the neighborhood of a point. The additional imaginary term does not alter this picture except that the curved X space-time is now a slightly complex manifold in complex space-time.

Therefore we conclude that our Two-Tier quantum field theoretic formalism that is erected on eq. A.7.14, where the real part of the X space-time was flat, can be extended to curved space-time while maintaining eq. A.7.14 if the y space-time consists of coordinate patches analogous to the two planes (or the cylinder) in the example of the sphere. The difference is that we now use a curved space-time background metric $g_{\mu\nu}(X)$ instead of $\eta_{\mu\nu}$ throughout the lagrangian with the exception of L^Y (eq. A.6.27).

In L^Y we continue to use $\eta_{\mu\nu}$ as the metric. As a result L^Y breaks the invariance of the complete lagrangian under general coordinate transformations. Thus an implicit absolute space-time is implied – as it is implicitly in classical General Relativity and in cosmological experiments. This consequence is not disturbing and is physically acceptable for the following reasons:

1. As Bergmann and Synge point out classical general relativity implicitly embodies an absolute space-time.
2. Experiment shows that space in the large (of the order of the Hubble length) is nearly flat although space does appear to be closed. CBR, and other, experimental data suggests that an absolute reference frame exists.

Thus our universe does appear to be in a state of broken general coordinate transformation invariance. Two-tier quantum field theory in curved space-time is not in contradiction with our previous classical general relativistic theories or with our experimental knowledge of the universe. *The full lagrangian theory L is invariant under special relativity. $L - L^Y$ is formally invariant under general coordinate transformations in the X coordinates.*

A.8 Why Are the Y Field Dynamics Independent of the Gravitational Field?

It is evident that the Y^a field is a truly free field in our formulation. In particular, it does not depend on, or interact directly with, the gravitational field as represented by $\sqrt{g}$ and $g_{\mu\nu}$ factors. On the other hand, these quantities depend on the Y^a field through their dependence on the variable X^μ.

Thus the role of Y^a is strictly that of a coordinate, and of a field that is parametrized by a set of inertial frame coordinates y^μ. The arguments of Mach supplemented by the arguments of

Bergmann and Synge show that a de facto absolute reference frame exists (actually it is the set of inertial reference frames). Therefore we can chose to formulate our theory in an inertial reference frame and require that the theory only be invariant under Lorentz transformations to other inertial reference frames.

In this context it is allowed to have one or more fields like Y^a whose dynamics are not invariant under general coordinate transformations. It it is reasonable to require the particle and gravitational dynamical equations be covariant under general coordinate transformations in X. *Thus a part of the dynamics is invariant under Lorentz transformations – the Y^a sector – but this part of the dynamics is not directly observable; and a part of the dynamics – the observable part – is invariant under general coordinate transformations.*

Some reasons for having a free Y^a field are:

1. It is required to avoid divergences that would appear in perturbation theory if the Y^a were allowed to interact with gravitons. For example an hhYY interaction term causes a divergence to appear by generating a Y particle loop in graviton-graviton scattering.
2. If the Y^a particle interacted with gravity then measurable, classical Y^a fields could be generated in regions with ultra-strong gravitational fields such as the neighborhoods of black holes. In this case we would have new dimensions, allbeit imaginary, for which no experimental evidence currently exists.
3. The Principle of Equivalence has only been shown to apply on the classsical level for real coordinates. Any quantization that uses Minkowskian coordinates, or quasi-Minkowskian coordinates, causes general coordinate transformation invariance to be abandoned ab initio in the quantum regime.

A.9 Renormalization of the Standard Model

The Standard Model that we have developed has the same renormalizability as the conventional Standard Model except that we have assigned complex 3-momenta to quarks. This was an arbitrary choice. However it made a point of difference between quarks and leptons, and we see that they differ very much in experiments. In addition it made the introduction of the SU(3) color interaction seem natural.

However, the extra three imaginary spatial dimensions would make our theory non-renormalizable in the quark sector <u>if</u> we followed the standard renormalization approach.

Therefore, we suggest that a different form of quantum field theory[299] be used in which our Standard Model, and a unified theory of the Standard Model and Quantum Gravity, are both fully renormalizable. In fact all diagrams calculated in perturbation theory are finite in this new

[299] Blaha (2005a). This book is the second book in this volume and the second edition of Blaha (2003). Blaha (2005a) discusses this type of quantum field theory in detail including issues such as anomalies and unitarity.

approach. The reader is directed to Blaha (2005) for a detailed account of this new form of quantum field theory.

Thus our form of the Standard Model is fully renormalizable in the new Two-Tier form of quantum field theory.

A.10 Adler-Bell-Jackiw Anomalies

The axial anomaly (Adler-Bell-Jackiw anomaly) follows from the linear divergence of a fermion triangle graph (Fig. A.10.1) in the conventional Standard Model. All higher order terms are divergence-free. These terms do not contribute to the axial anomaly. Thus the axial anomaly can properly be regarded as an artifact of the regularization of the divergence of the fermion triangle diagram.

In Two-Tier theory the axial anomaly does not appear to be present. Fermion triangle diagrams in Two-Tier quantum field theories are finite. Thus the source of the anomaly in conventional theories is absent in Two-Tier theories.

A massless Dirac field theory is formally invariant under a chiral transformation implying a conserved axial-vector current. The Two-Tier axial-vector current is

$$j_5^{\mu}(X(y)) = \bar{\psi}(X(y))\gamma^{\mu}\gamma_5\psi(X(y)) \tag{A.10.1}$$

with formal conservation law:

$$\partial j_5^{\mu}(X(y))/\partial X^{\mu} = 2m\, j_5(X(y)) = 2m\, \bar{\psi}(X(y))\gamma_5\psi(X(y)) \tag{A.10.2}$$

Eq. A.10.2 implies

$$\partial j_5^{\mu}(X(y))/\partial X^{\mu} = 0 \tag{A.10.3}$$

in the limit $m \rightarrow 0$. The question we now address is whether eq. A.10.3 holds in Two-Tier perturbation theory – perhaps in the same form as the conventional axial anomaly:

$$\partial j_5^{\mu}(X(y))/\partial X^{\mu} = 2m\, j_5(X(y)) + \alpha_0(4\pi)^{-1}\varepsilon^{\mu\nu\alpha\beta}F_{\alpha\beta}F_{\mu\nu} \quad ? \tag{A.10.4}$$

where α_0 is the unrenormalized fine structure constant.

The simplest manifestation of the axial anomaly in conventional field theory is the fermion triangle diagram, which we will now examine in Two-Tier quantum field theory. As stated earlier, the Two-Tier triangle diagram is finite and zero unlike the conventional quantum field theory result. *Thus the axial anomaly does not appear to exist in Two-Tier quantum field theory. The axial anomaly is a result of the divergence of the triangle diagram in conventional quantum field theory.*

The absence of the anomaly reflects the absence of divergences in Two-Tier quantum field theory, which preserves chiral invariance. Unlike Pauli-Villars regularization, for example, the finiteness of Two-Tier theory follows from the Gaussian factors. Unlike the dimensional regularization approach (where there is no equivalent to γ_5), Two-Tier theory can use the normal γ_5 matrix.

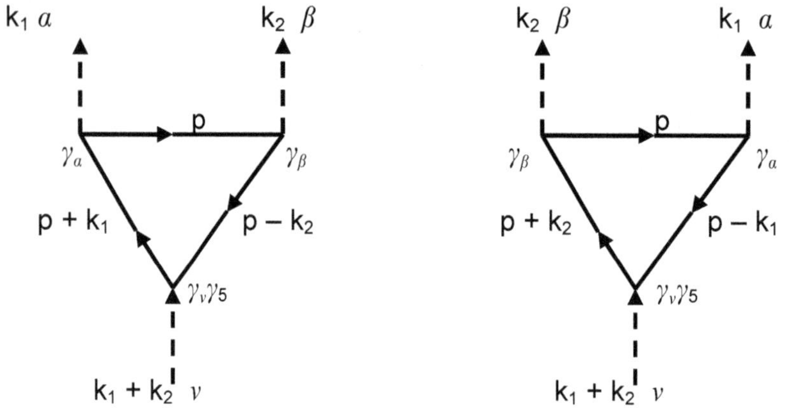

Figure A.10.1. The V-V-A triangle diagrams.

The expression for the Two-Tier triangle diagrams is:

$$T_{\alpha\beta\nu}(k_1, k_2) = S_{\alpha\beta\nu}(k_1, k_2) + S_{\beta\alpha\nu}(k_2, k_1) \tag{A.10.5}$$

where

$$S_{\alpha\beta\nu}(k_1, k_2)\delta^4(k_1 + k_2 - q) = -iN\int d^4y_1 d^4y_2 d^4y_3\, e^{ik_1\cdot y_1 + ik_2\cdot y_2 - iq\cdot y_3}\,\cdot$$
$$\cdot\, Tr\{S_F^{TT}(y_1 - y_3)\gamma_\alpha S_F^{TT}(y_2 - y_1)\gamma_\beta S_F^{TT}(y_3 - y_2)\gamma_\nu\gamma_5\}/(2\pi)^4 \tag{A.10.6}$$

where N is a constant, and where $S_F^{TT}(z)$ is specified previously. We now define the fourier transform:

$$S_F^{TT}(z) = -i\int d^4p\, e^{-ip\cdot z}\, \mathscr{S}^{TT}(p)/(2\pi)^4 \tag{A.10.7}$$

where $S^{TT}(p)$ defined previously. We then substitute the fourier transform in eq. A.10.6 and perform the coordinate integrations to obtain:

$$S_{\alpha\beta\nu}(k_1, k_2) = N\int d^4p\ Tr\{\mathscr{S}^{TT}(p + k_1)\gamma_\alpha \mathscr{S}^{TT}(p)\gamma_\beta \mathscr{S}^{TT}(p - k_2)\gamma_\nu\gamma_5\}/(2\pi)^4$$

We note that

$$k_1{}^\alpha T_{\alpha\beta\nu}(k_1, k_2) \neq 0 \tag{A.10.8}$$
$$k_2{}^\beta T_{\alpha\beta\nu}(k_1, k_2) \neq 0 \tag{A.10.9}$$
$$(k_1 + k_2)^\nu T_{\alpha\beta\nu}(k_1, k_2) \neq 0 \tag{A.10.10}$$

in Two-Tier quantum field theory because the conservation laws are expressed with respect to the X coordinates – not the y coordinates. Thus since $k_1{}^\alpha$ corresponds $\partial/\partial y_\alpha$, and not $\partial/\partial X_\alpha$ there is no reason for eqs. A.10.8-A.10.10 to be zero. However at "large distances" relative to M_c^{-1} we see

$$k_1{}^\alpha T_{\alpha\beta\nu}(k_1, k_2) \cong 0 \tag{A.10.11}$$
$$k_2{}^\beta T_{\alpha\beta\nu}(k_1, k_2) \cong 0 \tag{A.10.12}$$
$$(k_1 + k_2)^\nu T_{\alpha\beta\nu}(k_1, k_2) \cong 0 \tag{A.10.13}$$

to very good approximation since the gaussian damping factor in the fermion propagators is approximately unity and thus the Two-Tier expression becomes essentially the same as the conventional field theory expression.

On the other hand at very short distances the anomaly appears to be absent since Two-Tier theory is very well behaved at high energy with

$$\mathscr{S}^{TT}(p) \backsim \gamma^0 M_c{}^6\, p^{-7} + O(p^{-9}) \tag{A.10.14}$$

As a result we see

$$k_1{}^\alpha T_{\alpha\beta\nu}(k_1, k_2) \sim p^{4-21} \sim p^{-17} \tag{A.10.15}$$

as $p \to \infty$ is highly convergent. Thus there is no high energy divergence unlike conventional field theory where the integral is linearly divergent. And so no anomaly is generated.

A.11 Gravity and Y^μ (y)

Our discussion until this point has been based on the assumption of a flat space-time. The introduction of gravity and the requirement of covariance under coordinate transformations requires the introduction of a $\sqrt{g}$ factor in the Y^μ (y) lagrangian kinetic terms:

$$L_C = \int d^4y \sqrt{g(X^\mu(y))}\, \mathscr{L}_C(X^\mu(y), \partial X^\mu(y)/\partial y^\nu) \qquad \text{(A.11.1)}$$

Then Y^μ (y) interacts with the gravitational field and it is no longer a free field. However the weakness of gravitation and the almost flat nature of space-time makes the effect of this interaction negligible with respect to the Extended Standard Model.

Appendix B. A New Paradigm Supporting Two-Tier QED: Composition of Extrema in the Calculus of Variations

This Appendix appeared originally in Blaha (2005a) and earlier books. In it we describe a new paradigm in the Calculus of Variations that supports the formulation of Two-Tier Quantum Field Theory.

B.1. A New Paradigm in the Calculus of Variations

The Calculus of Variations has a long and venerable history in Physics and Mathematics. Many problems in Physics and Mathematics have been treated with approaches based on techniques in the Calculus of Variations.[300] In this book (Blaha (2003)) we have developed a unified quantum field theory of the known forces of nature based on a new type of problem, or paradigm, in the Calculus of Variations. One way of viewing the spectrum of problems in the Calculus of Variations is the following progression.

B.1.1. A Classification of Variational Problems

1. Variational problems in a Euclidean, or Minkowski, flat space such as the minimal distance between two points or the extrema of a field theory Lagrangian.

2. Variational problems seeking extrema on a curved surface such as the shortest distance between points on the surface of a sphere.

The development in this book suggests a third and fourth, possibility, that to the author's knowledge, has not been addressed in the literature:

3. Variational problems where the extrema are determined on a surface that is itself defined as an extremum. The discussions in this book exemplify this pardigm.

4. Variational problems where the extrema are determined on a surface that is itself defined as an extremum that depends on the extrema on the surface. More simply put, the

---

[300] See Akhiezer (1962), Blaha (2003), Gelfand (2000), Giaquinta (1996) and (1998), Jost (1998), and Sagan (1993),

extrema, and the surface upon which they are defined, are jointly determined and are interrelated. Fortunately, our unified theory does not use this paradigm. A future theory might.

In the unified theory that we will develop all particle fields including the graviton field are defined as a mapping of a Minkowski space-time y to a "particle" space-time X with the mapping determined as an extremum of a variation of a fundamental field (a type 3 variational problem in the above classification). Our theory could be generalized to include a back-reaction of the particle fields on the fundamental field (a type 4 variational problem in the above classification). We will not discuss this possibility in this book.

B.2. Simple Physical Example – Strings On Springs

In this section we describe a simple physical example that illustrates a variational problem of type 3 in the Calculus of Variations. We view it as a composition of extrema. (This problem can be addressed using other calculus of Variations techniques.) The approach used in the solution of this problem is similar to the approach used in Two-Tier quantum field theory.

B.2.1 A Strings on Springs Mechanics Problem

Consider a long string or bar that can oscillate (undulate) in a direction perpendicular to its length. Further assume that one end of this bar or string is attached to a spring that cause the entire bar or string to oscillate back and forth in a direction parallel to its long side.

Let x denote the distance to a point on the string when the spring is at equilibrium. If 2π times the frequency of the spring is ω_1, then the location of this point when the spring is oscillating is

$$X(t) = x + A \sin(\omega_1 t + \phi_1) \qquad (B.1)$$

where ϕ_1 is a phase, and A is the amplitude of the spring oscillation. Then the vertical displacement of a traveling wave on the *string* can take the form

$$\psi(t) = B \sin(\omega_2 t - k_2(x + A \sin(\omega_1 t + \phi_1)) + \phi_2) \qquad (B.2)$$

where B is the amplitude of the string wave, and k_2, ω_2 and ϕ_2 are the parameters of the string wave. These simple mechanical formulae are well known. But they lead to an interesting new application of the ideas of the Calculus of Variations.

Suppose we treat X as an independent variable with X given by eq. (B.1), and with eq. (B.2) written as:

$$\psi(t) = B \sin(\omega_2 t - k_2 X + \phi_2) \qquad (B.3)$$

Defining

$$\psi = \psi(X(t), t) \quad \text{(B.4)}$$

we can specify the dynamics of the above motion by finding the extrema of

$$I = \int \mathscr{L}_\psi \, dX(t) + \int \mathscr{L}_X \, dt \quad \text{(B.5)}$$

where the Lagrangian terms are

$$\mathscr{L}_\psi = \tfrac{1}{2} \{ \mu \, (\partial\psi/\partial t)^2 - Y \, (\partial\psi/\partial X)^2 \} \quad \text{(B.6)}$$

with μ and Y being constants, and

$$\mathscr{L}_X = \tfrac{1}{2} \{ m(\partial X/\partial t)^2 - k(X - x)^2 \} \quad \text{(B.7)}$$

where m and k are constants, and where x is a parameter. Applying Hamilton's Principle, and performing independent variations of X and ψ yields the Lagrangian equations:

$$\frac{\partial \mathscr{L}_\psi}{\partial \psi} - \frac{\partial}{\partial X} \frac{\partial \mathscr{L}_\psi}{\partial(\partial\psi/\partial X)} - \frac{\partial}{\partial t} \frac{\partial \mathscr{L}_\psi}{\partial(\partial\psi/\partial t)} = 0 \quad \text{(B.8)}$$

and

$$\frac{\partial \mathscr{L}_X}{\partial X} - \frac{\partial}{\partial t} \frac{\partial \mathscr{L}_X}{\partial(\partial X/\partial t)} = 0 \quad \text{(B.9)}$$

The resulting equations of motion are:

$$\mu \, \partial^2\psi/\partial t^2 - Y \, \partial^2\psi/\partial X^2 = 0 \quad \text{(B.10)}$$

and

$$m \, \partial^2 X/\partial t^2 + k(X - x) = 0 \quad \text{(B.11)}$$

with the solutions given in eqs. B.1 and B.2.

The procedure that we use to obtain these results may look a bit strange but they illustrate a type 3 problem in the Calculus of Variations involving the composition of extrema—the composition of an extremum that specifies a manifold in a space (possibly including all of space in a $R^n \rightarrow R^n$ mapping) with an extremum determining a function on that manifold. The procedure is described in detail in the next section.

B.3. The Composition of Extrema – A Lagrangian Formulation

In this section we will explore the general case of the composition of extrema for fields. We will discuss the case of a scalar field ϕ that is a function of a vector field X^μ in a D-dimensional space with coordinate variables that we will denote as y^μ. (The discussion for other types of fields is a straightforward extension of this discussion.) Thus let

$$\phi = \phi(X) \tag{B.12}$$

and

$$X^\mu = X^\mu(y) \tag{B.13}$$

We assume that the dynamics can be described by a Lagrangian formulation using an extension of Hamilton's principle:

$$I = \int \mathscr{L} d^4y \tag{B.14}$$

with

$$\mathscr{L} = \mathscr{L}(\phi(X), \partial\phi/\partial X^\nu, X^\mu(y), \partial X^\mu(y)/\partial y^\nu, y) \tag{B.15}$$

If we perform a standard variation[301] in ϕ for fixed y (and thus fixed X) we find

$$\delta I = \int [\delta\phi\, \partial\mathscr{L}/\partial\phi + \delta(\partial\phi/\partial X^\nu)\, \partial\mathscr{L}/\partial(\partial\phi/\partial X^\nu)]\, d^4y \tag{B.16}$$

We can rewrite the variation in the derivative of ϕ as

$$\delta(\partial\phi/\partial X^\nu) = \partial(\delta\phi)/\partial X^\nu \tag{B.17}$$

$$= \partial y^\mu/\partial X^\nu\, \partial(\delta\phi)/\partial y^\mu \tag{B.18}$$

[301] Bogoliubov, N. N., & Shirkov, D. V., Volkoff, G. M. (tr), *Introduction to the Theory of Quantized Fields* (Wiley-Interscience, New York, 1959); Goldstein H., *Classical Mechanics* (Addison-Wesley, Reading, MA 1965).

with an implied summation over repeated indices. After substituting eq. B.18 in eq. B.16, and performing an integration by parts (and discarding the surface term which is assumed to yield zero in the standard fashion) we obtain:

$$\delta I = \int \delta\phi \, \{\partial\mathscr{L}/\partial\phi - \partial/\partial y^\mu \, [\partial\mathscr{L}/\partial(\partial\phi/\partial X^\nu) \, \partial y^\mu/\partial X^\nu) \,]\} \, d^4y$$

Since the variation of $\delta\phi$ is arbitrary we conclude

$$\partial\mathscr{L}/\partial\phi - \partial/\partial y^\mu \, [\partial\mathscr{L}/\partial(\partial\phi/\partial X^\nu) \, \partial y^\mu/\partial X^\nu)] = 0 \qquad \text{(B.19a)}$$

The second term in eq. B.19a shows the effect of the dependence of ϕ on the field X, $\phi = \phi(X)$, rather than directly on the coordinate system y.

Similarly we can perform a variation in X^μ and obtain

$$\partial\mathscr{L}/\partial X^\mu - \partial/\partial y^\nu \, [\partial\mathscr{L}/\partial(\partial X^\mu/\partial y^\nu)] = 0 \qquad \text{(B.19b)}$$

The X field defines a "manifold" or, more properly, specifies a transformation from $R^n \to R^n$. If we make standard assumptions about the mapping: that it is continuous and piece-wise invertible, then we can establish the following lemmas:

Lemma 1: *If the transformation $X^\mu = X^\mu(y)$ is a transformation from $R^n \to R^n$ that is of class C' and piece-wise invertible, then*

$$\frac{\partial}{\partial y^\nu}\frac{\partial y^\nu}{\partial X^\mu} = -\frac{\partial \ln J}{\partial X^\mu} \qquad \text{(B.20)}$$

where

$$J = |\partial(X)/\partial(y)| \qquad \text{(B.21)}$$

is the absolute value of the Jacobian of the transformation.

Proof:

Consider two equivalent forms of an integral:

$$I = \int \mathscr{L} J \, d^4y = \int \mathscr{L} \, d^4X$$

where $\mathscr{L}$ is specified as in eq. B.15. Then the first expression for I leads to eq. B.19a which can be written in the form

$$\partial\mathscr{L}/\partial\phi - \partial/\partial X^\mu\,[\partial\mathscr{L}/\partial(\partial\phi/\partial X^\mu)] - \partial\mathscr{L}/\partial(\partial\phi/\partial X^\mu)\{\partial[J\partial y^\nu/\partial X^\mu]/\partial y^\nu\} = 0$$

Using the second expression for I above we obtain the following equation by variation in ϕ:

$$\partial\mathscr{L}/\partial\phi - \partial/\partial X^\mu\,[\partial\mathscr{L}/\partial(\partial\phi/\partial X^\mu)] = 0$$

Comparing these two expressions and realizing that $\partial[J\partial y^\nu/\partial X^\mu]/\partial y^\nu$ is totally independent of ϕ and its derivatives leads us to conclude

$$\partial\,[J\partial y^\nu/\partial X^\mu]/\partial y^\nu = 0 \tag{B.22}$$

It is a general relationship for a transformation between X and y based on continuity and piece-wise invertibility. After a few elementary manipulations eq. B.22 can be rewritten in the form of eq. B.20. ■

Lemma 2: *If the transformation* $X^\mu = X^\mu(y)$ *is a transformation from* $R^n \rightarrow R^n$ *that is of class* C' *and piece-wise invertible and* $\mathscr{L} = \mathscr{L}(\phi(X), \partial\phi/\partial X^\nu, X^\mu(y), \partial X^\mu(y)/\partial y^\nu, y)$, *then*

$$\partial\mathscr{L}/\partial(\partial\phi/\partial X^\nu)\,\partial y^\mu/\partial X^\nu = \partial\mathscr{L}/\partial(\partial\phi/\partial y^\mu) \tag{B.23}$$

Proof:

Let us express $\mathscr{L}$ as a power series in derivatives of ϕ:

$$\mathscr{L} = \sum_{n=0} a_{n\mu_1\mu_2\ldots\mu_n}(\phi(X), X^\mu(y), \partial X^\mu(y)/\partial y^\nu, y)\prod_{j=1}^{n}\partial\phi/\partial X^{\mu_j}$$

which can rewritten using piece-wise invertibility as

$$\mathscr{L} = \sum_{n=0} a_{n\mu_1\mu_2\ldots\mu_n}(\phi(X), X^\mu(y), \partial X^\mu(y)/\partial y^\nu, y)\prod_{j=1}^{n}\partial\phi/\partial y^{\nu_j}\,\partial y^{\nu_j}/\partial X^{\mu_j}$$

Taking the derivative of this equation with respect to $\partial\phi/\partial y^\mu$ immediately yields the result. ■

Eq. B.23 enables us to rewrite eq. B.19a as:

$$\partial\mathscr{L}/\partial\phi - \partial/\partial y^\mu\,[\partial\mathscr{L}/\partial(\partial\phi/\partial y^\mu)] = 0 \qquad \text{(B.24)}$$

which is as one would expect.

In order to get a feeling for the effect of eq. B.19a we will look at a simple example where we specify the relation of the X and y variables directly. Then we will look at the composition of extrema where the transformation between X and y is itself determined as an extremum solution.

B.3.1. Example: a hyperplane

We assume eq. B.19b yields the transformation:

$$X^i = ay^i \qquad \text{for } i = 1,2,3$$

$$X^0 = 0$$

Then eq. B.19a becomes

$$\partial\mathscr{L}/\partial\phi - \partial/\partial y^i\,[\partial\mathscr{L}/\partial(\partial\phi/\partial y^i)] = 0 \qquad \text{(B.25)}$$

with the time derivative disappearing. Effectively the variation of ϕ on the hyperplane $X^0 = 0$ is determined by the differential equation generated by B.25. On this hyperplane the transformation between the X and y variables is invertible.

B.3.2 Coordinate Transformation Determined as an Extremum Solution

We now develop a formalism that determines a mapping from space onto itself as the solution of an extremum problem and also determines the dynamics of one or more fields as a function of this mapping. To this author's knowledge, this area in the Calculus of Variations – the determination of an extremum on a manifold where the manifold itself is determined by an extremum – has not been previously explored. We will also develop a hamiltonian formulation. Then we will proceed to quantize the theory.

B.3.3. Separable Lagrangian Case

Although there are many forms that the composition of extrema could take, one fairly general form that is directly useful in quantum field theory applications is based on a Lagrangian that can be split into two parts which we will call a *separable Lagrangian*:

$$\mathscr{L} = \mathscr{L}_F J + \mathscr{L}_C(X^\mu(y), \partial X^\mu(y)/\partial y^\nu, y) \tag{B.26}$$

where J is defined in eq. B.21, where $\mathscr{L}_F$ contains all the dynamics of the fields and their interactions, and where $\mathscr{L}_C$ defines the coordinate mapping as an extremum solution. The procedure to determine the differential equations that specify the mapping, and the field equations that specify field interactions and evolution, is to vary in the coordinates X^μ and in the fields independently, using Hamilton's Principle. The extrema are to be determined for

$$I = \int \mathscr{L} \, d^4y \tag{B.27}$$

We will begin by considering the case of one scalar field:

$$\mathscr{L}_F = \mathscr{L}_F(\phi(X), \partial\phi/\partial X^\nu) \tag{B.28}$$

and

$$\mathscr{L}_C = \mathscr{L}_C(X^\mu(y), \partial X^\mu(y)/\partial y^\nu, y) \tag{B.29}$$

Eq. B.27 can be written in the form:

$$I = \int \mathscr{L}_F(\phi(X), \partial\phi/\partial X^\nu) \, dX + \int \mathscr{L}_C(X^\mu(y), \partial X^\mu(y)/\partial y^\nu, y) \, d^4y \tag{B.30}$$

using the Jacobian to transform to an integral over dX in the first term. A standard variation of ϕ and the application of Hamilton's Principle yields

$$\partial\mathscr{L}_F/\partial\phi - \partial/\partial X^\mu \, [\partial\mathscr{L}_F/\partial(\partial\phi/\partial X^\mu)] = 0 \tag{B.31}$$

reflecting the fact that ϕ is a function of X^μ only, with X^μ a function of the y coordinates.

Next we perform a variation of X^μ determining the mapping from $y \to X$ as an extremum of the integral in eq. B.27. We note the piece-wise invertibility of the coordinate mapping $X^\mu(y)$ allows us to write the Jacobian J as a function of y^μ only. A standard variation of X^μ and the application of Hamilton's Principle yields

$$\partial\mathscr{L}_C/\partial X^\mu - \partial/\partial y^\nu\,[\partial\mathscr{L}_C/\partial(\partial X^\mu/\partial y^\nu)] = 0 \qquad \text{(B.32)}$$

B.3.4 Klein-Gordon Example

The Klein-Gordon scalar field theory furnishes us with a simple example of the application of the preceding development. The Lagrangian is

$$\mathscr{L}_F = ½\,[\,(\partial\phi/\partial X^\nu)^2 - m^2\phi^2\,] \qquad \text{(B.33)}$$

From eq. B.31 we obtain the field equation:

$$(\Box + m^2)\,\phi(X) = 0 \qquad \text{(B.34)}$$

where

$$\Box = \partial/\partial X^\nu\,\partial/\partial X_\nu \qquad \text{(B.34a)}$$

A fourier representation of the solution of eq. B.34 is:

$$\phi(X) = \int dp\,\delta(p^2 - m^2)\theta(p^0)\,[A(p)\,e^{-ip\cdot X} + A(p)^*\,e^{ip\cdot X}] \qquad \text{(B.35)}$$

where A(k) is a function of k and * indicates complex conjugation.

The determination of $X^\mu(y)$ depends on the Lagrangian $\mathscr{L}_C$ and the solutions of eq. B.3A. If we chose

$$\mathscr{L}_C = -½\,(\partial X^\mu/\partial y^\nu)^2 \qquad \text{(B.36)}$$

Then we obtain the equation

$$\Box\,X^\mu = 0 \qquad \text{(B.37)}$$

with the solution

$$X^\mu = \int dk\,\delta(k^2)\theta(k^0)\,[a^\mu(k)\,e^{-ik\cdot y} + a^\mu(k)^*\,e^{ik\cdot y}] \qquad \text{(B.38)}$$

where $a^{\mu}(k)$ are complex vector functions of k in general. (We ignore positivity issues for the moment.) Substitution of eq. B.38 in eq. B.35 yields an expression with a form reminiscent of bosonic string expressions.[302] We will take up this point later in subsequent chapters.

B.4. The Composition of Extrema – Hamiltonian Formulation

The previous section established a Lagrangian formulation of dynamics based on the composition of extrema. In this section we will develop an equivalent hamiltonian formulation. We will assume a Minkowskian space-time with X^0 and y^0 playing the role of the time coordinates in the respective coordinate systems.

Initially, we will assume a scalar field ϕ with a Lagrangian of the form in eq. B.15 and define canonical momenta with

$$\Pi_{\phi} = \partial\mathscr{L}/\partial\dot{\phi} \equiv \partial\mathscr{L}/\partial(\partial\phi/\partial X^{\mu})\,\partial y^0/\partial X^{\mu} \qquad (B.39)$$

$$\Pi_X{}^{\mu} = \partial\mathscr{L}/\partial\dot{X}_{\mu} \qquad (B.40)$$

where

$$\dot{\phi} = \partial\phi/\partial y^0 \equiv \partial\phi/\partial X^{\mu}\,\partial X^{\mu}/\partial y^0 \qquad (B.41)$$

$$\dot{X}^{\mu} = \partial X^{\mu}/\partial y^0 \qquad (B.42)$$

Then we define the hamiltonian density as

$$\mathscr{H} = \Pi_{\phi}\dot{\phi} + \Pi_X{}^{\mu}\dot{X}_{\mu} - \mathscr{L}(\phi(X), \partial\phi/\partial X^{\nu}, X^{\mu}(y), \partial X^{\mu}(y)/\partial y^{\nu}, y) \qquad (B.43)$$

and the hamiltonian

$$H = \int \mathscr{H}\, d^3y \qquad (B.44)$$

The hamiltonian density has the general form

[302] See for example Polchinski (1998) and Bailin (1994).

$$\mathscr{H} = \mathscr{H}(\phi(X), \partial\phi/\partial X^i, \Pi_\phi, X^\mu(y), \partial X^\mu(y)/\partial y^j, \Pi_X{}^\mu, y^\nu) \qquad (B.45)$$

for the case of one scalar field where the indices i and j represent space coordinates; time coordinates are assigned index value 0.

If we calculate the differential change in H using eq. B.45 we obtain

$$dH = \int \{\partial\mathscr{H}/\partial\phi\, d\phi + \partial\mathscr{H}/\partial\Pi_\phi\, d\Pi_\phi - \partial/\partial y^\nu[\partial\mathscr{H}/\partial(\partial\phi/\partial X^j)\partial y^\nu/\partial X^j]d\phi + $$
$$+ \partial\mathscr{H}/\partial X^\mu\, dX^\mu + \partial\mathscr{H}/\partial\Pi_X{}^\mu\, d\Pi_X{}^\mu - \partial/\partial y^j\, [\partial\mathscr{H}/\partial(\partial X^\mu/\partial y^j)]\, dX^\mu\}\, d^3y \qquad (B.46)$$

after some partial integrations. (Repeated indices indicate summations. Indices labeled i and j indicate space coordinates. Greek indices include all space-time components of a variable.)

Expressing the differential in H using eq. B.43 we obtain

$$dH = \int dy\, \{\Pi_\phi\, d\dot{\phi} + \dot{\phi}\, d\Pi_\phi - \partial\mathscr{L}/\partial\phi\, d\phi - \partial\mathscr{L}/\partial(\partial\phi/\partial X^\mu)d(\partial\phi/\partial X^\mu) + $$
$$+ \Pi_X{}^\mu\, d\dot{X}^\mu + \dot{X}^\mu\, d\Pi_X{}^\mu - \partial\mathscr{L}/\partial X^\mu\, dX^\mu - \partial\mathscr{L}/\partial(\partial X^\mu/\partial y^j)d(\partial X^\mu/\partial y^j)\} \qquad (B.47a)$$

After some manipulations we find

$$dH = \int\{\, \dot{\phi}\, d\Pi_\phi + \dot{X}_\mu d\Pi_X{}^\mu \; - \partial/\partial y^0\, \Pi_\phi\, d\phi - \partial/\partial y^0\, \Pi_X{}^\mu\, dX_\mu\}\, dy \qquad (B.47b)$$

using the equations of motion eqs. B.19a and B.19b.

Comparing eqs B.46 and B.47 we obtain Hamilton's equations in the case of the composition of extrema:

$$\dot{\phi} = \partial\mathscr{H}/\partial\Pi_\phi \qquad (B.48a)$$

$$\dot{\Pi}_\phi = -\partial\mathscr{H}/\partial\phi + \partial/\partial y^\nu\, [\partial\mathscr{H}/\partial(\partial\phi/\partial X^i)\, \partial y^\nu/\partial X^i] \qquad (B.48b)$$

$$\dot{X}_\mu = \partial\mathscr{H}/\partial\Pi_X{}^\mu \qquad (B.48c)$$

where

$$\dot{\Pi}_X{}^{\mu} = -\partial\mathscr{H}/\partial X^{\mu} + \partial/\partial y^j\,[\partial\mathscr{H}/\partial(\partial X^{\mu}/\partial y^j)] \tag{B.49a}$$

$$\dot{\Pi}_{\phi} = \partial\,\Pi_{\phi}/\partial y^0 \tag{B.49b}$$

$$\dot{\Pi}_X{}^{\mu} = \partial\Pi_X{}^{\mu}/\partial y^0$$

B.5. Translational Invariance

If the Lagrangian of a field theory has no explicit dependence on the coordinates then one expects translational invariance accompanied by a conservation law for an energy-momentum stress tensor. We will show this is the case for Lagrangians implementing the composition of extrema. We assume a Lagrangian without an explicit dependence on the coordinates y^{ν}:

$$\mathscr{L} = \mathscr{L}(\phi(X), \partial\phi/\partial X^{\nu}, X^{\mu}(y), \partial X^{\mu}(y)/\partial y^{\nu}) \tag{B.50}$$

Under an infinitesimal displacement,

$$y'^{\nu} = y^{\nu} + \epsilon^{\nu} \tag{B.51a}$$

$$\delta\phi = \phi(X(y + \epsilon)) - \phi(X(y))$$

$$= \epsilon^{a}\,\partial\phi/\partial y^{a} \tag{B.51b}$$

$$\delta X^{\mu} = \epsilon^{a}\,\partial X^{\mu}/\partial y^{a} \tag{B.51c}$$

$$\delta(\partial\phi/\partial X^{\mu}) = \epsilon^{a}\,\partial(\partial\phi/\partial y^{a})/\partial X^{\mu} \tag{B.51d}$$

$$\delta(\partial X^{\mu}/\partial y^{\nu}) = \epsilon^{a}\,\partial(\partial X^{\mu}/\partial y^{a})/\partial y^{\nu} \tag{B.51e}$$

and the Lagrangian changes by

$$\delta\mathscr{L} = \in^a \partial\mathscr{L}/\partial y^a \tag{B.52}$$

The change can also be expressed in terms of the changes in the fields, their derivatives and the mapping X^μ:

$$\delta\mathscr{L} = \partial\mathscr{L}/\partial\phi\ \delta\phi + \partial\mathscr{L}/\partial(\partial\phi/\partial X^\mu)\ \delta(\partial\phi/\partial X^\mu) + \partial\mathscr{L}/\partial X^\mu\ \delta X^\mu + $$
$$+ \partial\mathscr{L}/\partial(\partial X^\mu/\partial y^\nu)\ \delta(\partial X^\mu/\partial y^\nu) \tag{B.53}$$

Combining eqs. B.51, B.52 and B.53 we obtain (after some manipulations):

$$\in^\nu \partial/\partial y_\mu\ \mathscr{T}_{\mu\nu} = 0 \tag{B.54}$$

where

$$\mathscr{T}_{\mu\nu} = -g_{\mu\nu}\mathscr{L} + \partial\mathscr{L}/\partial(\partial\phi/\partial X^\delta)\ \partial y_\mu/\partial X^\delta\ \partial\phi/\partial y^\nu + \partial\mathscr{L}/\partial(\partial X^\delta/\partial y_\mu)\partial X^\delta/\partial y^\nu \tag{B.55a}$$

or, alternately using Lemma 2,

$$\mathscr{T}_{\mu\nu} = -\ g_{\mu\nu}\mathscr{L} + \partial\mathscr{L}/\partial(\partial\phi/\partial y_\mu)\ \partial\phi/\partial y^\nu + \partial\mathscr{L}/\partial(\partial X^\delta/\partial y_\mu)\ \partial X^\delta/\partial y^\nu \tag{B.55b}$$

Since $\in^a$ is an arbitrary displacement we obtain the conservation law:

$$\partial/\partial y_\mu\ \mathscr{T}_{\mu\nu} = 0 \tag{B.56}$$

Eq. B.56 implies the energy-momentum vector

$$P_\beta = \int d^3y\ \mathscr{T}_{0\beta} \tag{B.57}$$

is conserved. We note

$$\partial/\partial y^0\ P_\beta = 0 \tag{B.58}$$

since eq. B.56 and B.57 can be used to obtain the integral of a divergence, which results in zero.

The hamiltonian (eqs. B.43-44) is

$$H = P_0 \tag{B.59}$$

We note for later use that the total energy, H, which is conserved, contains a term that represents the energy in the X^μ mapping. Thus energy can be exchanged in principle between the ϕ field sector and the X^μ sector.

B.6. Lorentz Invariance and Angular Momentum Conservation

We can also verify Lorentz invariance and obtain the form of the conserved angular momentum by considering the effect of an infinitesimal Lorentz transformation. We will consider the case of a scalar field ϕ.

Under an infinitesimal Lorentz transformation ($\epsilon_{\mu\nu} = -\epsilon_{\nu\mu}$):

$$y'_\mu = y_\mu + \delta y_\mu = y_\mu + \epsilon_{\mu\nu} y^\nu \tag{B.60a}$$

$$\delta\phi = \phi(X(y')) - \phi(X(y))$$

$$= \epsilon^{\mu\nu} y_\nu \, \partial\phi/\partial X^a \, \partial X^a/\partial y^\mu \tag{B.60b}$$

$$\delta X^\mu = S^\mu{}_a X^a(y') - X^\mu(y) \tag{B.60c}$$

$$= \epsilon^\mu{}_a X^a(y) + \partial X^\mu/\partial y^\beta \, \delta y^\beta \tag{B.60d}$$

where $S^\mu{}_a$ is the matrix for the Lorentz transformation of a vector. (If X^μ were a gauge field then an additional operator gauge term would have to be added to eq. B.60d.)

The Lagrangian changes by

$$\delta\mathscr{L} = \epsilon^{\mu\nu} y_\nu \partial\mathscr{L}/\partial y^\mu \tag{B.61}$$

under the infinitesimal Lorentz transformation. The change in the Lagrangian can also be expressed as:

$$\delta\mathscr{L} = \partial\mathscr{L}/\partial\phi \, \delta\phi + \partial\mathscr{L}/\partial(\partial\phi/\partial X^\mu) \, \delta(\partial\phi/\partial X^\mu) + \partial\mathscr{L}/\partial X^\mu \, \delta X^\mu +$$

$$+ \partial\mathscr{L}/\partial(\partial X^\mu/\partial y^\nu)\ \delta(\partial X^\mu/\partial y^\nu) \tag{B.62}$$

Combining eqs. B.61 and B.62, and substituting and simplifying terms leads to:

$$\epsilon_{\mu\nu}\ \partial/\partial y^\sigma \mathscr{M}^{\sigma\mu\nu} = 0 \tag{B.63}$$

where

$$\mathscr{M}^{\sigma\mu\nu} = (g^{\mu\sigma}y^\nu - g^{\nu\sigma}y^\mu)\mathscr{L} + \partial\mathscr{L}/\partial(\partial\phi/\partial X^\alpha)\ \partial y^\sigma/\partial X^\alpha\ (y^\mu\partial\phi/\partial y_\nu - y^\nu\partial\phi/\partial y_\mu) +$$
$$+ \partial\mathscr{L}/\partial(\partial X^\delta/\partial y^\sigma)\ (g^{\delta\nu}X^\mu - g^{\delta\mu}X^\nu + y^\mu\ \partial X^\delta/\partial y_\nu - y^\nu\ \partial X^\delta/\partial y_\mu) \tag{B.64}$$

The conserved angular momentum is:

$$M^{\mu\nu} = \int d^3y\ \mathscr{M}^{0\mu\nu} \tag{B.65}$$

with

$$\partial M^{\mu\nu}/\partial y^0 = 0 \tag{B.66}$$

The angular momentum density can be written in the familiar form:

$$\mathscr{M}^{\sigma\mu\nu} = y^\mu\ \mathscr{T}^{\sigma\nu} - y^\nu\ \mathscr{T}^{\sigma\mu} + \partial\mathscr{L}/\partial(\partial X^\delta/\partial y^\sigma)\ (g^{\delta\nu}X^\mu - g^{\delta\mu}X^\nu) \tag{B.67}$$

taking account of the vector nature of X^μ. The spatial part of $M^{\mu\nu}$ is the angular momentum.

B.7. Internal Symmetries

We will now consider the case of a set of scalar fields ϕ_r in a Lagrangian with an internal symmetry. Under a local transformation

$$\phi_r(X) \rightarrow \phi_r(X) - i\epsilon\lambda_{rs}\ \phi_s(X) \tag{B.68}$$

If the Lagrangian is invariant under this transformation, then

$$\delta\mathscr{L} = 0 = \partial\mathscr{L}/\partial\phi_r\delta\phi_r + \partial\mathscr{L}/\partial(\partial\phi_r/\partial X^\alpha)\ \delta(\partial\phi_r/\partial X^\alpha) \tag{B.69}$$

Using the equation of motion eq. B.19a satisfied by all the components ϕ_r we obtain a conserved current:

$$\mathscr{J}^{\nu} = -i\, \partial\mathscr{L} / \partial(\partial\phi_r/\partial X^{\delta})\, \partial y^{\nu}/\partial X^{\delta}\, \lambda_{rs}\, \phi_s \tag{B.70}$$

which satisfies

$$\partial\mathscr{J}^{\nu}/\partial y^{\nu} = 0 \tag{B.71}$$

The conserved charge is

$$Q = \int d^3y\, \mathscr{J}^0 \tag{B.72}$$

$$\partial Q/\partial y^0 = 0 \tag{B.73}$$

B.8. Separable Lagrangians

We now consider the case of a separable Lagrangian such as in eq. B.26. Adopting the definitions:

$$\phi' = \partial\phi/\partial X^0 \tag{B.74}$$

$$X_{\mu}' = \partial X_{\mu}/\partial y^0 \tag{B.75}$$

we define canonical momenta as

$$\pi_{\phi} = \partial\mathscr{L} / \partial\phi' \equiv \partial\mathscr{L} / \partial(\partial\phi/\partial X^0) \tag{B.76}$$

$$\pi_X{}^{\mu} = \partial\mathscr{L} / \partial X_{\mu}' \equiv \partial\mathscr{L} / \partial(\partial X_{\mu}/\partial y^0) \tag{B.77}$$

We now define the separable hamiltonian density as

$$\mathscr{H}_s = J\pi_{\phi}\, \phi' + \pi_X{}^{\mu}\, X_{\mu}' - \mathscr{L}_s \tag{B.78}$$

where J is the Jacobian (eq. B.21) and

$$H_s = \int \mathscr{H}_s\, d^3y \tag{B.79}$$

The separable Lagrangian (from eq. B.26) is:

$$\mathscr{L}_s = \mathscr{L}_F(\phi(X), \partial\phi/\partial X^\mu)\, J + \mathscr{L}_C(X^\mu(y), \partial X^\mu(y)/\partial y^\nu, y) \qquad (B.80)$$

In the case of one scalar field the separable hamiltonian density has the general form

$$\mathscr{H}_s = \mathscr{H}_S(\phi(X), \pi_\phi, \partial\phi/\partial X^i, X^\mu(y), \pi_X{}^\mu, \partial X^\mu(y)/\partial y^j, y^\nu) \qquad (B.81)$$

where the indices i and j indicate spatial components. In particular, the terms in the separable hamiltonian are:

$$\mathscr{H}_s = \mathscr{H}_F\, J + \mathscr{H}_C \qquad (B.82)$$

with

$$\mathscr{H}_F(\phi(X), \pi_\phi, \partial\phi/\partial X^i) = \pi_\phi\, \phi' - \mathscr{L}_F \qquad (B.83)$$

$$\mathscr{H}_C(X^\mu(y), \pi_X{}^\mu, \partial X^\mu(y)/\partial y^j, y^\nu) = \pi_X{}^\mu\, X_\mu{}' - \mathscr{L}_C \qquad (B.84)$$

where J is the absolute value of the Jacobian defined in B.21.

We now define the time integral of H as we did in eq. B.14 when considering the Lagrangian formulation:

$$G = \int dy^0\, H_s \qquad (B.85)$$

Thus G is an integral over all space-time coordinates. Using G we can develop a hamiltonian formulation. First we calculate the differential change in G. Using eqs. B.81-2 and B.85 we obtain

$$\begin{aligned} dG = \int \{ & J\, \partial\mathscr{H}_F/\partial\phi\, d\phi + J\, \partial\mathscr{H}_F/\partial\pi_\phi\, d\pi_\phi + \\ & + J\, \partial\mathscr{H}_F/\partial(\partial\phi/\partial X^i)\, d(\partial\phi/\partial X^i) + \partial\mathscr{H}_C/\partial X^\mu\, dX^\mu + \\ & + \partial\mathscr{H}_C/\partial\pi_X{}^\mu\, d\pi_X{}^\mu + \partial\mathscr{H}_C/\partial(\partial X^\mu/\partial y^j)\, d(\partial X^\mu/\partial y^j) \}\, d^4y \end{aligned} \qquad (B.86)$$

with summations implied by repeated indices. (Index labels i and j label spatial coordinates only; Greek indices label space-time coordinates.) Rewriting dG as two integrals and performing partial integrations yields:

$$dG = \int d^4X \{ \partial\mathscr{H}_F/\partial\phi\, d\phi + \partial\mathscr{H}_F/\partial\pi_\phi\, d\pi_\phi - \partial/\partial X^j [\partial\mathscr{H}_F/\partial(\partial\phi/\partial X^j)]\, d\phi \} +$$

$$+ \int d^4y \{\partial\mathscr{H}_C/\partial X^\mu\, dX^\mu + \partial\mathscr{H}_C/\partial\pi_X{}^\mu\, d\pi_X{}^\mu - \partial/\partial y^j[\partial\mathscr{H}_C/\partial(\partial X^\mu/\partial y^j)]\, dX^\mu\} \tag{B.87}$$

Alternately, expressing the differential in G using eqs. B.82-4 we obtain

$$dG = \int d^4X \{\pi_\phi\, d\phi' + \phi' d\pi_\phi - \partial\mathscr{L}_F/\partial\phi\, d\phi - \partial\mathscr{L}_F/\partial(\partial\phi/\partial X^\mu) d(\partial\phi/\partial X^\mu)\} +$$
$$+ \int d^4y \{\pi_{X\mu}\, dX^{\mu\prime} + X^{\mu\prime}\, d\pi_{X\mu} - \partial\mathscr{L}_C/\partial X^\mu\, dX^\mu - \partial\mathscr{L}_C/\partial(\partial X^\mu/\partial y^j) d(\partial X^\mu/\partial y^j)\} \tag{B.88}$$

which becomes

$$dG = \int d^4X \{-\pi_\phi'\, d\phi + \phi'\, d\pi_\phi\} + \int d^4y \{-\pi_{X\mu}'\, dX^\mu + X^{\mu\prime}\, d\pi_{X\mu}\} \tag{B.89}$$

using the equations of motion eqs. B.31-2.

Comparing eqs B.87 and B.89 we obtain Hamilton's equations for the case of the composition of extrema for a separable Lagrangian:

$$\phi' = \partial\mathscr{H}_F/\partial\pi_\phi \tag{B.90}$$

$$\pi_\phi' = -\partial\mathscr{H}_F/\partial\phi + \partial/\partial X^j\, [\partial\mathscr{H}_F/\partial(\partial\phi/\partial X^j)] \tag{B.91}$$

$$X_\mu' = \partial\mathscr{H}_C/\partial\pi_X{}^\mu \tag{B.92}$$

$$\pi_{X\mu}' = -\partial\mathscr{H}_C/\partial X^\mu + \partial/\partial y^j\, [\partial\mathscr{H}_C/\partial(\partial X^\mu/\partial y^j)] \tag{B.93}$$

where

$$\pi_\phi' = \partial\, \pi_\phi/\partial X^0 \tag{B.94}$$

$$\pi_{X\mu}' = \partial\, \pi_{X\mu}/\partial X^0 \tag{B.95}$$

Notice that $\mathscr{L}_F$, $\mathscr{H}_F$ and π_ϕ have precisely the same form, as a function of X^μ, as one sees in a conventional field theory formalism. Yet X^μ is a mapping/function of the coordinates y. In reality, it can be viewed as a field as we shall see.

B.9. Separable Lagrangians and Translational Invariance

The general rule for conventional Lagrangians is: if a Lagrangian has no explicit dependence on the coordinates then translational invariance follows accompanied by a conservation law for an energy-momentum tensor. We will show that this rule needs modification for separable Lagrangians that implement the composition of extrema.

Consider the Lagrangian:

$$\mathscr{L}_s = J\ \mathscr{L}_F(\phi(X), \partial\phi/\partial X^\mu) + \mathscr{L}_C(X^\mu(y), \partial X^\mu(y)/\partial y^\nu) \qquad \text{(B.96)}$$

in which the X^μ play a dual role as both fields and coordinates. Let us consider a variation in X^μ:

$$X^\mu(y) \rightarrow X^\mu(y) + \delta X^\mu(y) \qquad \text{(B.97)}$$

where $\delta X^\mu(y)$ is an arbitrary function of y that vanishes at the endpoints of the integration region of the integral. The action is:

$$I = \int \mathscr{L}_s d^4y \qquad \text{(B.98)}$$

We will show that a variation in $X^\mu(y)$ leads to a conserved energy-momentum tensor. But we will use integrals of the Lagrangian density since it provides a simpler derivation of the result. Under the variation of eq. B.97 we find

$$\delta\phi = \phi(X(y) + \delta X^\mu(y)) - \phi(X(y))$$

$$= \delta X^\mu\ \partial\phi/\partial X^\mu \qquad \text{(B.99a)}$$

$$\delta(\partial\phi/\partial X^\nu) = \delta X^\mu\ \partial(\partial\phi/\partial X^\mu)/\partial X^\nu \qquad \text{(B.99b)}$$

$$\delta(\partial X^\mu/\partial y^\nu) = \partial(\delta X^\mu)/\partial y^\nu \qquad \text{(B.99c)}$$

The integral in eq. B.98 changes by

$$\delta I - \int d^4y\ \delta\mathscr{L}_s = \int d^4y\ [\delta(J\mathscr{L}_F) + \delta\mathscr{L}_C] \qquad \text{(B.100a)}$$

which becomes:

$$\delta I = \int d^4y\ [\delta X^\mu\, \partial(J\mathscr{L}_F)/\partial X^\mu + \partial(\delta X^\mu \partial\mathscr{L}_C\ /\ \partial(\partial X^\mu/\partial y^\nu))/\partial y^\nu] \qquad \text{(B.100b)}$$

due to the equations of motion of X^μ (eq. B.19b) in X^μs role. Since the second term is a total divergence its contribution to δI is zero. Thus we can express eq. B.100b as:

$$\delta I = \int d^4y\ [\, J\, \delta\mathscr{L}_F + \mathscr{L}_F\ \delta J\,] \qquad \text{(B.101)}$$

realizing that the Jacobian J depends on y and thus X:

$$\delta J = \delta X^\mu\, \partial J/\partial X^\mu \qquad \text{(B.102)}$$

A partial integration gives

$$\mathscr{L}_F\, \delta J = \delta X^\mu\, \partial(\, J\mathscr{L}_F)/\partial X^\mu - \delta X^\mu J\, \partial\mathscr{L}_F/\partial X^\mu \qquad \text{(B.103)}$$

Evaluating $\delta\mathscr{L}_F$ we find:

$$\delta\mathscr{L}_F = \partial\mathscr{L}_F/\,\partial\phi\ \delta\phi + \partial\mathscr{L}_F\ /\,\partial(\partial\phi/\partial X^\mu)\ \delta(\partial\phi/\partial X^\mu) \qquad \text{(B.104)}$$

which gives

$$\delta\mathscr{L}_F = \delta X^\nu\ \partial/\partial X^\mu\ [\partial\mathscr{L}_F/\,\partial(\partial\phi/\partial X^\mu)\ \partial\phi/\partial X^\nu\,] \qquad \text{(B.105)}$$

using the equations of motion eq. B.31, and using eq. B.99b. Combining eqs. B.100, B.101, B.103 and B.105 we obtain:

$$\int d^4y\, J\, \delta X^\nu\ \partial/\partial X_\mu\, \mathscr{T}_{F\mu\nu} = \int d^4X\, \delta X^\nu\ \partial/\partial X_\mu\, \mathscr{T}_{F\mu\nu} = 0 \qquad \text{(B.106)}$$

where

$$\mathscr{T}_{F\mu\nu} = -\, g_{\mu\nu}\, \mathscr{L}_F + \partial\mathscr{L}_F/\,\partial(\partial\phi/\partial X_\mu)\ \partial\phi/\partial X^\nu \qquad \text{(B.107)}$$

after some manipulations. Since δX^ν is an arbitrary function of y the differential conservation law follows:

$$\partial/\partial X_\mu \mathscr{T}_{F\mu\nu} = 0 \tag{B.108}$$

Eq. B.108 implies the energy-momentum vector

$$P_{F\beta} = \int d^3X \, \mathscr{T}_{F0\beta} \tag{B.109}$$

is conserved:

$$\partial/\partial X^0 P_{F\beta} = 0 \tag{B.110}$$

The hamiltonian density (eq. B.83) is

$$\mathscr{H}_F = \mathscr{T}_{F0\beta} \tag{B.111}$$

Thus the field energy

$$H_F = P_{F0} = \int d^3X \, \mathscr{T}_{F00} \tag{B.112}$$

is conserved with respect to the "time" X^0. Later we will see that H_F is trivially conserved in the Coulomb gauge of X_μ. (We will also establish an electromagnetic-like quantum field theory for X_μ with gauge invariance.) In other gauges the conservation of H_F is not trivial.

B.10. Separable Lagrangians and Angular Momentum Conservation

We can also verify Lorentz invariance and obtain the form of the conserved angular momentum for a separable Lagrangian by considering the effect of an infinitesimal Lorentz transformation. We will consider the case of a scalar field ϕ.

Under an infinitesimal Lorentz transformation as specified by eqs. B.60a – B.60d the separable Lagrangian changes by

$$\delta\mathscr{L}_s = \epsilon^{\mu\nu} y_\nu \partial\mathscr{L}_s / \partial y^\mu \tag{B.113}$$

which can also be expressed as

$$\delta\mathscr{L}_s = \partial\mathscr{L}_s / \partial\phi\, \delta\phi + \partial\mathscr{L}_s / \partial(\partial\phi/\partial X^\mu)\, \delta(\partial\phi/\partial X^\mu) + \partial\mathscr{L}_s / \partial X^\mu\, \delta X^\mu + \\ + [\partial\mathscr{L}_s / \partial(\partial X^\mu/\partial y^\nu)]\, \delta(\partial X^\mu/\partial y^\nu) \qquad \text{(B.114)}$$

Combining eqs. B.113 and B.114 leads to:

$$\epsilon_{\mu\nu}\, \partial/\partial y^\sigma \mathscr{M}_s^{\ \sigma\mu\nu} = 0 \qquad \text{(B.115)}$$

where

$$\mathscr{M}_s^{\ \sigma\mu\nu} = J\, \mathscr{M}_F^{\ \sigma\mu\nu} + \mathscr{M}_C^{\ \sigma\mu\nu} + \mathscr{M}_M^{\ \sigma\mu\nu} \qquad \text{(B.116)}$$

$$\mathscr{M}_F^{\ \sigma\mu\nu} = (g^{\mu\sigma}y^\nu - g^{\nu\sigma}y^\mu)\mathscr{L}_F + \partial\mathscr{L}_F / \partial(\partial\phi/\partial y_\sigma)\, (y^\mu\partial\phi/\partial y_\nu - y^\nu\partial\phi/\partial y_\mu) \qquad \text{(B.117)}$$

$$\mathscr{M}_C^{\ \sigma\mu\nu} = (g^{\mu\sigma}y^\nu - g^{\nu\sigma}y^\mu)\mathscr{L}_C + \\ + \partial\mathscr{L}_C / \partial(\partial X^\delta/\partial y^\sigma)(g^{\delta\nu}X^\mu - g^{\delta\mu}X^\nu + y^\mu\, \partial X^\delta/\partial y_\nu - y^\nu\, \partial X^\delta/\partial y_\mu) \qquad \text{(B.118)}$$

$$\mathscr{M}_M^{\ \sigma\mu\nu} = \mathscr{L}_F\, \partial J / \partial(\partial X^\delta/\partial y^\sigma)(g^{\delta\nu}X^\mu - g^{\delta\mu}X^\nu + y^\mu\, \partial X^\delta/\partial y_\nu - y^\nu\, \partial X^\delta/\partial y_\mu) \qquad \text{(B.119)}$$

where the third term originates in the dependence of J on derivatives of X^μ. Eq. B.117 was obtained in part by using the identity:

$$\partial\mathscr{L} / \partial(\partial\phi/\partial y^\sigma) = \partial\mathscr{L} / \partial(\partial\phi/\partial X^\alpha)\, \partial y^\sigma/\partial X^\alpha \qquad \text{(B.120)}$$

where $\mathscr{L}$ and ϕ have the form specified in eq. B.15.

The conserved angular momentum is:

$$M_s^{\mu\nu} = \int dy\, \mathscr{M}_s^{0\mu\nu} \qquad \text{(B.121)}$$

with

$$\partial M_s^{\mu\nu}/\partial y^0 = 0 \qquad \text{(B.122)}$$

B.10.1. Angular Momentum and $\mathscr{L}_F$

An alternate conserved angular momentum can be obtained by considering the "field" part of the Lagrangian $\mathscr{L}_F$ under an infinitesimal Lorentz transformation ($\epsilon_{\mu\nu} = -\epsilon_{\nu\mu}$):

$$X'_\mu = X_\mu + \delta X_\mu \tag{B.123a}$$

$$\begin{aligned}\delta\phi &= \phi(X'(y)) - \phi(X(y)) \\ &= \delta X^\mu\, \partial\phi/\partial X^\mu\end{aligned} \tag{B.123b}$$

$$\delta X^\mu = S^\mu{}_a X^a(y) - X^\mu(y) \tag{B.123c}$$

$$= \epsilon^\mu{}_a X^a(y) \tag{B.123d}$$

where $S^\mu{}_a$ is the Lorentz transformation matrix for a vector. (If X^μ is a gauge field then an additional operator gauge term would have to be added to eq. B.123d.)

The Lagrangian changes by

$$\delta\mathscr{L}_F = \epsilon^{\mu\nu} X_\nu\, \partial\mathscr{L}_F/\partial X^\mu \tag{B.124}$$

under an infinitesimal Lorentz transformation. The change can also be expressed as:

$$\delta\mathscr{L}_F = \partial\mathscr{L}_F/\partial\phi\, \delta\phi + \partial\mathscr{L}_F/\partial(\partial\phi/\partial X^\mu)\, \delta(\partial\phi/\partial X^\mu) \tag{B.125}$$

Combining eqs. B.124 and B.125 leads to:

$$\epsilon_{\mu\nu}\partial/\partial X^\sigma\, \mathscr{M}_{FX}{}^{\sigma\mu\nu} = 0 \tag{B.126}$$

Where

$$\mathscr{M}_{FX}{}^{\sigma\mu\nu} = (g^{\mu\sigma}X^\nu - g^{\nu\sigma}X^\mu)\mathscr{L}_F + \partial\mathscr{L}_F/\partial(\partial\phi/\partial X^\sigma)\, (X^\mu\partial\phi/\partial X_\nu - X^\nu\partial\phi/\partial X_\mu) \tag{B.127}$$

The conserved angular momentum associated with the X coordinates is:

$$M_{FX}^{\mu\nu} = \int d^3X\, \mathscr{M}_{FX}^{\;0\mu\nu} \tag{B.128}$$

with

$$\partial M_{FX}^{\mu\nu}/\partial X^0 = 0 \tag{B.129}$$

The angular momentum density can be written in the familiar form:

$$\mathscr{M}_{FX}^{\sigma\mu\nu} = X^\mu\, \mathscr{T}_F^{\sigma\nu} - X^\nu\, \mathscr{T}_F^{\sigma\mu} \tag{B.130}$$

using eq. B.107.

B.11. Separable Lagrangians and Internal Symmetries

We will now consider the case of a set of scalar fields ϕ_r in a separable Lagrangian with an internal symmetry under a local transformation

$$\phi_r(X) \rightarrow \phi_r(X) - i\epsilon\lambda_{rs}\, \phi_s(X) \tag{B.131}$$

If the Lagrangian is invariant under this transformation, then

$$\delta\mathscr{L}_S \equiv \delta\mathscr{L}_F = 0 = \partial\mathscr{L}_F/\partial\phi_r\, \delta\phi_r + \partial\mathscr{L}_F/\partial(\partial\phi_r/\partial X^\alpha)\, \delta(\partial\phi_r/\partial X^\alpha) \tag{B.132}$$

Using the equation of motion eq. B.31, which is satisfied by all components ϕ_r, we obtain a conserved current:

$$\mathscr{J}^\nu = -i\, \partial\mathscr{L}_F/\partial(\partial\phi_r/\partial X^\nu)\, \lambda_{rs}\, \phi_s \tag{B.133}$$

satisfying

$$\partial\mathscr{J}^\nu/\partial X^\nu = 0 \tag{B.134}$$

The conserved charge is

$$Q = \int d^3X\, \mathscr{J}^0 \tag{B.135}$$

$$\partial Q/\partial X^0 = 0 \tag{B.136}$$

We note eq. 18-A.71 provides a corresponding conservation law for the y coordinate system.

REFERENCES

Akhiezer, N. I., Frink, A. H. (tr), 1962, *The Calculus of Variations* (Blaisdell Publishing, New York, 1962).

Bailin, D. & Love, A., 1994, *Supersymmetric Gauge Field Theory and String Theory* (Institute of Physics Publishing, Philadelphia, PA, 1994).

Bjorken, J. D., Drell, S. D., 1964, *Relativistic Quantum Mechanics* (McGraw-Hill, New York, 1965).

Bjorken, J. D., Drell, S. D., 1965, *Relativistic Quantum Fields* (McGraw-Hill, New York, 1965).

Blaha, S., 1998, *Cosmos and Consciousness* (Pingree-Hill Publishing, Auburn, NH, 1998).

_______, 2002, *A Finite Unified Quantum Field Theory of the Elementary Particle Standard Model and Quantum Gravity Based on New Quantum Dimensions™ & a New Paradigm in the Calculus of Variations* (Pingree-Hill Publishing, Auburn, NH, 2002).

_______, 2003, *A Finite Unified Quantum Field Theory of the Elementary Particle Standard Model and Quantum Gravity Based on New Quantum Dimensions™ and a New Paradigm in the Calculus of Variations* (Pingree-Hill Publishing, Auburn, NH, 2003).

_______, 2004, *Quantum Big Bang Cosmology: Complex Space-time General Relativity, Quantum Coordinates™ Dodecahedral Universe, Inflation, and New Spin 0, ½, 1 & 2 Tachyons & Imagyons* (Pingree-Hill Publishing, Auburn, NH, 2004).

______, *2005a, Quantum Theory of the Third Kind: A New Type of Divergence-free Quantum Field Theory Supporting a Unified Standard Model of Elementary Particles and Quantum Gravity based on a New Method in the Calculus of Variations* (Pingree-Hill Publishing, Auburn, NH, 2005).

______, 2005b, *The Metatheory of Physics Theories, and the Theory of Everything as a Quantum Computer Language* (Pingree-Hill Publishing, Auburn, NH, 2005).

______, 2005c, *The Equivalence of Elementary Particle Theories and Computer Languages: Quantum Computers, Turing Machines, Standard Model, Superstring Theory, and a Proof that Gödel's Theorem Implies Nature Must Be Quantum* (Pingree-Hill Publishing, Auburn, NH, 2005).

______, 2006a, *The Foundation of the Forces of Nature* (Pingree-Hill Publishing, Auburn, NH, 2006).

______, 2006b, *A Derivation of ElectroWeak Theory based on an Extension of Special Relativity; Black Hole Tachyons; & Tachyons of Any Spin.* (Pingree-Hill Publishing, Auburn, NH, 2006).

______, 2007a, *Physics Beyond the Light Barrier: The Source of Parity Violation, Tachyons, and A Derivation of Standard Model Features* (Pingree-Hill Publishing, Auburn, NH, 2007).

______, 2007b, *The Origin of the Standard Model: The Genesis of Four Quark and Lepton Species, Parity Violation, the ElectroWeak Sector, Color SU(3), Three Visible Generations of Fermions, and One Generation of Dark Matter with Dark Energy* (Pingree-Hill Publishing, Auburn, NH, 2007).

______, *2008a, A Direct Derivation of the Form of the Standard Model From GL(16) (Pingree-Hill Publishing, Auburn, NH, 2008).*

______, 2008b, *A Complete Derivation of the Form of the Standard Model With a New Method to Generate Particle Masses Second Edition* (Pingree-Hill Publishing, Auburn, NH, 2008)

______, 2009, *The Algebra of Thought & Reality: The Mathematical Basis for Plato's Theory of Ideas, and Reality Extended to Include A Priori Observers and Space-Time Second Edition* (Pingree-Hill Publishing, Auburn, NH, 2009).

______, 2010a, *Operator Metaphysics: A New Metaphysics Based on a New Operator Logic and a New Quantum Operator Logic that Lead to a Mathematical Basis for Plato's Theory of Ideas and Reality* (Pingree-Hill Publishing, Auburn, NH, 2010).

______, 2010b, *The Standard Model's Form Derived from Operator Logic, Superluminal Transformations and GL(16)* (Pingree-Hill Publishing, Auburn, NH, 2010).

______, 2011a, *21st Century Natural Philosophy Of Ultimate Physical Reality* (McMann-Fisher Publishing, Auburn, NH, 2011).

______, 2011b, *All the Universe! Faster Than Light Tachyon Quark Starships & Particle Accelerators with the LHC as a Prototype Starship Drive Scientific Edition* (Pingree-Hill Publishing, Auburn, NH, 2011).

______, 2011c, *From Asynchronous Logic to The Standard Model to Superflight to the Stars* (Blaha Research, Auburn, NH, 2011).

______, 2012a, *From Asynchronous Logic to The Standard Model to Superflight to the Stars volume 2: Superluminal CP and CPT, U(4) Complex General Relativity and The Standard Model, Complex Vierbein General Relativity, Kinetic Theory, Thermodynamics* (Blaha Research, Auburn, NH, 2012).

______, 2012b, *Standard Model Symmetries, And Four And Sixteen Dimension Complex Relativity; The Origin Of Higgs Mass Terms* (Blaha Reasearch, Auburn, NH, 2012).

______, 2013a, *Multi-Stage Space Guns, Micro-Pulse Nuclear Rockets, and Faster-Than-Light Quark-Gluon Ion Drive Starships* (Blaha Research, Auburn, NH, 2013).

______, 2013b, *The Bridge to Dark Matter; A New Sister Universe; Dark Energy; Inflatons; Quantum Big Bang; Superluminal Physics; An Extended Standard Model Based on Geometry* (Blaha Reasearch, Auburn, NH, 2013).

______, 2014a, *Universes and Multiverses: From a New Standard Model to a Physical Multiverse; The Big Bang; Our Sister Universe's Wormhole; Origin of the Cosmological Constant, Spatial Asymmetry of the Universe, and its Web of Galaxies; A Baryonic Field between Universes and Particles; Flatverse Extended Wheeler-DeWitt Equation* (Blaha Reasearch, Auburn, NH, 2014).

______, 2014b, *All the Multiverse! Starships Exploring the Endless Universes of the Cosmos Using the Baryonic Force* (Blaha Research, Auburn, NH, 2014).

______, 2014c, *All the Multiverse! II Between Multiverse Universes: Quantum Entanglement Explained by the Multiverse Coherent Baryonic Radiation Devices – PHASERs Neutron Star Multiverse Slingshot Dynamics Spiritual and UFO Events, and the Multiverse Microscopic Entry into the Multiverse* (Blaha Research, Auburn, NH, 2014).

______, 2015a, *PHYSICS IS LOGIC PAINTED ON THE VOID: Origin of Bare Masses and The Standard Model in Logic, U(4) Origin of the Generations, Normal and Dark Baryonic Forces,*

Dark Matter, Dark Energy, The Big Bang, Complex General Relativity, A Megaverse of Universe Particles (Blaha Research, Auburn, NH, 2015).

______, 2015b, *PHYSICS IS LOGIC Part II: The Theory of Everything, The Megaverse Theory of Everything, U(4)⊗U(4) Grand Unified Theory (GUT), Inertial Mass = Gravitational Mass, Unified Extended Standard Model and a New Complex General Relativity with Higgs Particles, Generation Group Higgs Particles* (Blaha Research, Auburn, NH, 2015).

______, 2015c, *The Origin of Higgs ("God") Particles and the Higgs Mechanism: Physics is Logic III, Beyond Higgs – A Revamped Theory With a Local Arrow of Time, The Theory of Everything Enhanced, Why Inertial Frames are Special, Universes of the Mind* (Blaha Research, Auburn, NH, 2015).

______, 2015d, *The Origin of the Eight Coupling Constants of The Theory of Everything: U(8) Grand Unified Theory of Everything (GUTE), S^8 Coupling Constant Symmetry, Space-Time Dependent Coupling Constants, Big Bang Vacuum Coupling Constants, Physics is Logic IV* (Blaha Research, Auburn, NH, 2015).

______, 2016a, *New Types of Dark Matter, Big Bang Equipartition, and A New U(4) Symmetry in the Theory of Everything: Equipartition Principle for Fermions, Matter is 83.33% Dark, Penetrating the Veil of the Big Bang, Explicit QFT Quark Confinement and Charmonium, Physics is Logic V* (Blaha Research, Auburn, NH, 2016).

______, 2016b, *The Periodic Table of the 192 Quarks and Leptons in The Theory of Everything: The U(4) Layer Group, Physics is Logic VI* (Blaha Research, Auburn, NH, 2016).

______, 2016c, *New Boson Quantum Field Theory, Dark Matter Dynamics, Dark Matter Fermion Layer Mixing, Genesis of Higgs Particles, New Layer Higgs Masses, Higgs Coupling Constants, Non-Abelian Higgs Gauge Fields, Physics is Logic VII* (Blaha Research, Auburn, NH, 2016).

______, 2016d, *Unification of the Strong Interactions and Gravitation: Quark Confinement Linked to Modified Short-Distance Gravity; Physics is Logic VIII* (Blaha Research, Auburn, NH, 2016).

______, 2016e, *MoND: Unification of the Strong Interactions and Gravitation II, Quark Confinement Linked to Large-Scale Gravity, Physics is Logic IX* (Blaha Research, Auburn, NH, 2016).

______, 2016f, *CQMechanics: A Unification of Quantum & Classical Mechanics, Quantum/Semi-Classical Entanglement, Quantum/Classical Path Integrals, Quantum/Classical Chaos* (Blaha Research, Auburn, NH, 2016).

______, 2016g, *GEMS: Unified Gravity, ElectroMagnetic and Strong Interactions: Manifest Quark Confinement, A Solution for the Proton Spin Puzzle, Modified Gravity on the Galactic Scale* (Pingree Hill Publishing, Auburn, NH, 2016).

______, 2016h, *Unification of the Seven Boson Interactions based on the Riemann-Christoffel Curvature Tensor* (Pingree Hill Publishing, Auburn, NH, 2016).

______, 2017a, *Unification of the Eleven Boson Interactions based on 'Rotations of Interactions'* (Pingree Hill Publishing, Auburn, NH, 2017).

Chrystal, G., 1961, *Textbook of Algebra Part One* (Dover Publications, Inc., New York, 1961).

Eddington, A. S., 1952, *The Mathematical Theory of Relativity* (Cambridge University Press, Cambridge, U.K., 1952).

Fant, Karl M., 2005, *Logically Determined Design: Clockless System Design With NULL Convention Logic* (John Wiley and Sons, Hoboken, NJ, 2005).

Gelfand, I. M., Fomin, S. V., Silverman, R. A. (tr), 2000, *Calculus of Variations* (Dover Publications, Mineola, NY, 2000).

Giaquinta, M., Modica, G., Souchek, J., 1998, *Cartesian Coordinates in the Calculus of Variations* Volumes I and II (Springer-Verlag, New York, 1998).

Giaquinta, M., Hildebrandt, S., 1996, *Calculus of Variations* Volumes I and II (Springer-Verlag, New York, 1996).

Heitler, W., 1954, *The Quantum Theory of Radiation* (Claendon Press, Oxford, UK, 1954).

Huang, Kerson, 1992, *Quarks, Leptons & Gauge Fields 2nd Edition* (World Scientific Publishing Company, Singapore, 1992).

Jost, J., Li-Jost, X., 1998, *Calculus of Variations* (Cambridge University Press, New York, 1998).

Kaku, M., 1993, *Quantum Field Theory* (Oxford University Press, Oxford, UK, 1993).

Misner, C. W., Thorne, K. S., and Wheeler, J. A., 1973, *Gravitation* (W. H. Freeman, New York, 1973).

Sagan, H., 1993, *Introduction to the Calculus of Variations* (Dover Publications, Mineola, NY, 1993).

Sakurai, J. J., 1964, *Invariance Principles and Elementary Particles* (Princeton University Press, Princeton, NJ, 1964).

Streater, R. F. and Wightman, A. S., 2000, *PCT, Spin, Statistics, and All That* (Princeton University Press, Princeton, NJ 2000).

Weinberg, S., 1972, *Gravitation and Cosmology* (John Wiley and Sons, New York, 1972).

Weinberg, S., 1995, *The Quantum Theory of Fields Volume I* (Cambridge University Press, New York, 1995).

Weyl, H., 1950, *Space, Time, Matter* (Dover, New York, 1950).

Weyl, H., (Tr. S. Pollard et al), 1987, *The Continuum* (Dover Publications, New York, 1987).

INDEX

About the Author

Stephen Blaha is a well known Physicist and Man of Letters with interests in Science, Society and civilization, the Arts, and Technology. He had an Alfred P. Sloan Foundation scholarship in college. He received his Ph.D. in Physics from Rockefeller University. He has served on the faculties of several major universities. He was also a Member of the Technical Staff at Bell Laboratories, a manager at the Boston Globe Newspaper, a Director at Wang Laboratories, and President of Blaha Software Inc and of Janus Associates Inc. (NH).

Among other achievements he was a co-discoverer of the "r potential" for heavy quark binding developing the first (and still the only demonstrable) non-abelian gauge theory with an "r" potential; first suggested the existence of topological structures in superfluid He-3; first proposed Yang-Mills theories would appear in condensed matter phenomena with non-scalar order parameters; first developed a grammar-based formalism for quantum computers and applied it to elementary particle theories; first developed a new form of quantum field theory without divergences (thus solving a major 60 year old problem that enabled a unified theory of the Standard Model and Quantum Gravity without divergences to be developed); first developed a formulation of complex General Relativity based on analytic continuation from real space-time; first developed a generalized non-homogeneous Robertson-Walker metric that enabled a quantum theory of the Big Bang to be developed without singularities at $t = 0$; first generalized Cauchy's theorem and Gauss' theorem to complex, curved multi-dimensional spaces; received Honorable Mention in the Gravity Research Foundation Essay Competition in 1978; first developed a physically acceptable theory of faster-than-light particles; first derived a composition of extrema method in the Calculus of Variations; first quantitatively suggested that inflationary periods in the history of the universe were not needed; first proved Gödel's Theorem implies Nature must be quantum; provided a new alternative to the Higgs Mechanism, and Higgs particles, to generate masses; first showed how to resolve logical paradoxes including Gödel's Undecidability Theorem by developing Operator Logic and Quantum Operator Logic; first developed a quantitative harmonic oscillator-like model of the life cycle, and interactions, of civilizations; first showed how equations describing superorganisms also apply to civilizations. A recent book shows his theory applies successfully to the past 14 years of history and to *new* archaeological data on Andean and Mayan civilizations as well as Early Anatolian and Egyptian civilizations.

He first developed an axiomatic derivation of the forms of The Standard Model from geometry – space-time properties – The Extended Standard Model. It has a Dark Matter sector that approximates the ElectroWeak sector with Dark doublets and Dark gauge interactions. It also uses quantum coordinates to remove infinities that crop up in most interacting quantum field theories and additionally to remove the infinities that appear in the Big Bang and generate an inflationary growth of the universe. The Extended Standard Model has an ultra-high energy GUT (Grand Unified Theory) limit with a $U(4)\otimes U(4)$ symmetry; and can be united with gravitation to form a Theory of Everything. (See *Physics is Logic Part II.*)

Blaha has had a major impact on a succession of elementary particle theories: his Ph.D. thesis (1970), and papers, showed that quantum field theory calculations to all orders in ladder approximations could not give scaling deep inelastic electron-nucleon scattering. He later showed the eigenvalue equation for the fine structure constant α in Johnson-Baker-Willey QED had a zero at $\alpha = 1$ not 1/137 by solving the Schwinger-Dyson equations to all orders in an approximation that agreed with exact results to 4th order in α thus ending interest in this theory. In 1979 at Prof. Ken Johnson's (MIT) suggestion he calculated the proton-neutron mass difference in the MIT bag model and found the result had the wrong sign reducing interest in the bag model. These results all appear in Physical Review papers. In the 2000's he repeatedly pointed out the shortcomings of SuperString theory and showed that The Standard Model's form could be derived from space-time geometry by an extension of Lorentz transformations to faster than light transformations. This deeper space-time basis greatly increases the possibility that it is part of THE fundamental theory.Recently, Blaha showed that the Weak interactions differed significantly from the Strong, electromagnetic and gravitation interactions in important respects while these interactions had similar features, and suggested that ElectroWeak theory, which is essentially a glued union of the Weak interactions and Electromagnetism, possibly modulo unknown Higgs particle features, be replaced by a unified theory of the other interactions combined with a stand-alone Weak interaction theory. Blaha also showed that, if Charmonium calculations are taken seriously, the Strong interaction coupling constant is only a factor of five larger than the electromagnetic coupling constant, and thus Strong interaction perturbation theory would make sense and yield physically meaningful results.

In graduate school (1965-71) he wrote substantial papers in elementary particles and group theory: The Inelastic E- P Structure Functions in a Gluon Model. Phys. Lett. B40:501-502,1972; Deep-Inelastic E-P Structure Functions In A Ladder Model With Spin 1/2 Nucleons, Phys.Rev. D3:510-523,1971; Continuum Contributions To The Pion Radius, Phys. Rev. 178:2167-2169,1969; Character Analysis of U(N) and SU(N), J. Math. Phys. 10, 2156 (1969); and The Calculation of the Irreducible Characters of the Symmetric Group in Terms of the Compound Characters, (Published as Blaha's Lemma in D. E. Knuth's book: *The Art of Computer Programming Vols. 1 – 4*).

In the early 1980's Blaha was also a pioneer in the development of UNIX for financial, scientific and Internet applications: benchmarked UNIX versions showing that block size was critical for UNIX performance, developing financial modeling software, starting database benchmarking comparison studies, developing Internet-like UNIX networking (1982) and developing a hybrid shell programming technique (1982) that was a precursor to the PERL programming language. He was also the manager of the AT&T ten-year future products development database. His work helped lead to commercial UNIX on computers such as Sun Micros, IBM AIX minis, and Apple computers.

In the 1980's he pioneered the development of PC Desktop Publishing on laser printers. and was nominated for three "Awards for Technical Excellence" in 1987 by PC Magazine for PC software products that he designed and developed.

Recently he has developed a theory of Megaverses – actual universes of which our universe is one – with quantum particle-like properties based on the Wheeler-DeWitt equation of Quantum Gravity. He has developed a theory of a baryonic force, which had been conjectured many years ago, and estimated the strength of the force based on discrepancies in measurements of the gravitational constant G. This force, operative in 15-dimensinal space, can be used to escape from our universe in "uniships" which are the equivalent of the faster-than-light starships proposed in the author's earlier books. Thus travel to other universes, as well as to other stars is possible.

Blaha also considered the complexified Wheeler-DeWitt equation and showed that its limitation to real-valued coordinates and metrics generated a Cosmological Constant in the Einstein equations.

The author has also recently written a series of books on the serious problems of the United States and their solution as well as a book on the decline of Mankind that will follow from current social and genetic trends in Mankind.

In the past twelve years Dr. Blaha has written over 40 books on a wide range of topics. Some recent major works are: *From Asynchronous Logic to The Standard Model to Superflight to the Stars*, *All the Universe!*, *SuperCivilizations: Civilizations as Superorganisms*, *America's Future: an Islamic Surge, ISIS, al Qaeda, World Epidemics, Ukraine, Russia-China Pact, US Leadership Crisis*, *The Rises and Falls of Man – Destiny – 3000 AD: New Support for a Superorganism MACRO-THEORY of CIVILIZATIONS From CURRENT WORLD TRENDS and NEW Peruvian, Pre-Mayan, Mayan, Anatolian, and Early Egyptian Data, with a Projection to 3000 AD*, and *Mankind in Decline: Genetic Disasters, Human-Animal Hybrids, Overpopulation, Pollution, Global Warming, Food and Water Shortages, Desertification, Poverty, Rising Violence, Genocide, Epidemics, Wars, Leadership Failure.*

He has taught approximately 4,000 students in undergraduate, graduate, and postgraduate corporate education courses primarily in major universities, and large companies and government agencies.

The above paragraphs summarize much of his work over the past fifty years. This work is fully documented. He continues to engage in research and writing at Blaha Research.